Lecture Notes in Computer Science   1689

Edited by G. Goos, J. Hartmanis and J. van Leeuwen

Springer

Berlin
Heidelberg
New York
Barcelona
Hong Kong
London
Milan
Paris
Singapore
Tokyo

Franc Solina Aleš Leonardis (Eds.)

Computer Analysis of Images and Patterns

8th International Conference, CAIP'99
Ljubljana, Slovenia, September 1-3, 1999
Proceedings

Springer

Series Editors

Gerhard Goos, Karlsruhe University, Germany
Juris Hartmanis, Cornell University, NY, USA
Jan van Leeuwen, Utrecht University, The Netherlands

Volume Editors

Franc Solina
Aleš Leonardis
Faculty of Computer and Information Science
Tržaška cesta 25, 1001 Ljubljana, Slovenia
E-mail: {franc/alesl}@fri.uni-lj.si

Cataloging-in-Publication data applied for

Die Deutsche Bibliothek - CIP-Einheitsaufnahme

Computer analysis of images and patterns : 8th international
conference ; proceedings / CAIP '99, Ljubliana, Slovenia, September
1 - 3, 1999. Franc Solina ; Aleš Leonardis (ed.). - Berlin ; Heidelberg
; New York ; Barcelona ; Hong Kong ; London ; Milan ; Paris ;
Singapore ; Tokyo : Springer, 1999
  (Lecture notes in computer science ; Vol. 1689)
  ISBN 3-540-66366-5

CR Subject Classification (1998): I.4, I.5, I.3.3, I.3.7, J.2

ISSN 0302-9743
ISBN 3-540-66366-5 Springer-Verlag Berlin Heidelberg New York

© Springer-Verlag Berlin Heidelberg 1999
Printed in Germany

Typesetting: Camera-ready by author
SPIN: 10704452   06/3142 – 5 4 3 2 1 0   Printed on acid-free paper

Preface

This volume presents the articles accepted for the 8th International Conference on Computer Analysis of Images and Patterns (CAIP'99), held in Ljubljana, Slovenia, 1–3 September 1999. The CAIP series of conferences started 14 years ago in Berlin. The series served initially as a forum for meetings between scientists from Western and Eastern-bloc countries. Political circumstances have changed dramatically since the inception of the conference and such contacts are fortunately no longer subject to obstacle. While CAIP conferences are still rooted in Central Europe, they now attract participants from all over the world.

We received 120 submissions, which went through a thorough double blind review process by the program committee members who, had the option of assigning additional reviewers. The final program consists of 47 oral and 27 poster presentations, with authors from 25 different countries. The proceedings also include 2 of the 5 invited lectures given at the conference.

In the name of the steering committee we would like to thank the program committee members and the additional reviewers for their time and efforts. Our thanks also go to the authors for their cooperation and meeting of all deadlines. Most of the communication with the authors and reviewers was done through the World Wide Web. This was made possible by the dedicated work of our conference web master, Bor Prihavec. We thank the International Association for Pattern Recognition for taking CAIP'99 under its wing, the Faculty of Computer and Information Science at University of Ljubljana for hosting the conference, and the industrial and institutional sponsors for giving financial support. Our thanks also go to Springer-Verlag Heidelberg for publishing the CAIP'99 proceedings in the LNCS series. Finally, we thank the local organizers for their commitment, which made the conference possible.

Ljubljana, June 1999

Franc Solina and Aleš Leonardis
Program Co-Chairs
CAIP'99

Organization

CAIP'99 is organized by the Faculty of Computer and Information Science, University of Ljubljana, Slovenia.

Conference Co-Chairs

Franc Solina, University of Ljubljana, Slovenia
Aleš Leonardis, University of Ljubljana, Slovenia

Steering Committee

Chair: G. Sommer (Christian-Albrechts-University of Kiel, Germany)
D. Chetverikov (Computer and Automation Research Institute, Budapest, Hungary)
V. Hlaváč (Czech Technical University, Prague, Czech Republic)
R. Klette (University of Auckland, New Zealand)
F. Solina (University of Ljubljana, Slovenia)

Program Committee

S. Ablameyko, Belarus
J. Arnspang, Denmark
R. Bajcsy, USA
I. Bajla, Slovakia
A. M. Bruckstein, Israel
V. M. Chernov, Russia
D. Chetverikov, Hungary
A. Del Bimbo, Italy
J.-O. Eklundh, Sweden
V. Hlaváč, Czech Republic
J. Kittler, United Kingdom

R. Klette, New Zealand
W. G. Kropatsch, Austria
A. Leonardis, Slovenia (Co-Chair)
R. Mohr, France
M. Schlesinger, Ukraine
W. Skarbek, Poland
F. Solina, Slovenia (Co-Chair)
G. Sommer, Germany
L. Van Gool, Belgium
M. A. Viergever, Netherlands
S. W. Zucker, USA

Reviewers

Sergey Ablameyko
Jens Arnspang
Ruzena Bajcsy
Ivan Bajla
Alberto Del Bimbo
Horst Bischof

Alfred M. Bruckstein
Jörg Bruske
Thomas Bülow
Vladimir M. Chernov
Dmitry Chetverikov
Marina A. Chichyeva

Carlo Colombo
Kostas Daniilidis
Jan-Olof Eklundh
Roman Englert
James Gee
Roland Glantz

Local Organization

Conference Web Master:	Bor Prihavec
Computing Facilities:	Aleš Jaklič
Local Arrangements:	Bojan Kverh
	Igor Lesjak
	Jasna Maver
	Peter Peer
	Danijel Skočaj

Sponsoring Organizations and Institutions

International Association for Pattern Recognition (IAPR)
Slovenian Pattern Recognition Society
IEEE Slovenia Section
Ministry of Science and Technology, Republic of Slovenia
Hermes SoftLab

Table of Contents

Color

Image processing

Image databases

Image compression and watermarking

Invited lecture

Poster session I

Object and face recognition

Classification and fitting

3D reconstruction and shape representation

Motion

Range image registration

Applications

Invited lecture

Poster session II

Shape from shading, texture and stereo

Real-time tracking

Panoramic images

Grouping

Image rendering

Alignment and matching

Author Index

Spectro-Spatial Gradients for Color-Based Object Recognition and Indexing

Daniel Berwick and Sang Wook Lee

University of Michigan, Ann Arbor MI 48109-2110, USA,
{dberwick, swlee}@eecs.umich.edu

Abstract. This paper presents illumination pose- and illumination color-invariant color feature descriptors for object recognition and indexing which are derived from spectral (color) and spatial derivatives of logarithmic image irradiance. While the use of spatial gradients and spatial ratios of image irradiance have been suggested for limited viewing-pose invariance and illumination-color invariance, respectively, gradients in the spectral direction and combination of spectral and spatial gradients have not been fully investigated. We present a unified framework for analyzing spatial and spectral gradients of logarithmic image irradiance, and suggest that spectro-spatial gradients have rich potential for developing local and global descriptors of object color. Experimental results are presented to demonstrate the efficacy of the proposed descriptors.

1 Introduction

Recognition/indexing approaches have recently emerged that do not explicitly use object geometry and shape, but instead rely on object appearances or collections of photometric features such as color, texture and local image variation [6][14]. While the importance of shape and geometry is not diminishing, the feature-collection methods are gaining popularity because of their efficiency especially for content-based image retrieval.

Color is a powerful cue when objects are distinguishable by their distribution of color reflectances. The earliest approach for recognizing objects based on their color distributions was developed by Swain and Ballard [18], and most of the content-based image retrieval systems use color distribution as a primary cue [6]. Color information is highly susceptible to illumination conditions. To effectively employ color information as a reliable descriptor of an object, color reflectance intrinsic to the object surface must be extracted. Variation of color appearance due to object shape, viewing and illumination pose, illumination color, and specularity needs to be discounted. The knowledge gained through color constancy research [4][7][11][13] is reflected in recent recognition/indexing methods that explore illumination color-invariant descriptors [8][9][15][17].

Most of the previous work focuses on finding feature descriptors invariant to illumination color. These approaches assume the color features used are located on 2-D planar surfaces, or the illumination pose is fixed for 3-D objects. Only recently has the issue of illumination-pose invariance been explicitly addressed, and the use of normalized color, i.e., chromaticity suggested [1][2][3][5][12]. These spectral approaches use the ratios of color bands of a pixel in contrast to previous methods which employ the ratios of spatially different pixels. Spatial method rely on the local ratios of reflectances, and thus are insensitive to smoothly varying illumination across a scene. However, spatial ratios are sensitive to illumination pose and viewing pose. While the spectral methods are insensitive to illumination pose, color, and viewing pose, on the other hand, they use global signatures of object color that can be eroded when illumination color varies across a scene.

In this paper, we describe our development of color feature descriptors insensitive to: (1) illumination pose, (2) global and local changes of illumination color, and (3) viewing pose. We investigate the utility of spectral and spatial gradients of logarithmic image irradiance to achieve this end. We present a unified framework for deriving spectral and spatial differential invariants, and discuss the approximation of spectral gradients from a limited number of spectral bands.

2   Spectral and Spatial Gradients

Object recognition and indexing should be based on the features that are derived only from object reflectance. Illumination conditions should be discounted. A photometric feature is constructed from image irradiance represented as:

$$\mathcal{I}(x, \lambda) = g(x)e(x, \lambda)s(x, \lambda) \tag{1}$$

where $g(x)$ is the geometric factor, $e(x, \lambda)$ is the illumination light, $s(x, \lambda)$ is the diffuse surface reflectance of the object, all projected to $x = (x, y)$ in the image plane, and λ represents wavelength direction of visible light spectrum. It is the object reflectance $s(x, \lambda)$ from which useful object features are obtained. For diffuse reflections, the geometric factor for Lambertian shading is given as $g(x) = n_s \cdot n(x)$, where n_s is the illumination direction and $n(x)$ is the normal of the surface projected to x.

The image irradiance $\mathcal{I}(x, \lambda)$ is influenced by the object pose with respect to illumination ($g(x)$) and illumination intensity and color ($e(x, \lambda)$). The distribution of $\mathcal{I}(x, \lambda)$ in an image is determined by object pose with respect to a sensor. The model database contains only some of the possible appearance information since diverse viewing conditions make it implausible for the database to include all possible appearances of an object. Consequently, some measures that are invariant to $g(x)$, $e(x, \lambda)$ and x need to be employed.

The image irradiance given in Equation 1 includes confounded effects of geometry, illumination and surface reflectance, we take the logarithm of image irradiance to separate the multiplicative terms into additive terms. The logarithmic irradiance is given as:

$$\mathcal{L}(x, \lambda) = ln\mathcal{I}(x, \lambda) = ln\, g(x) + ln\, e(x, \lambda) + ln\, s(x, \lambda) = \mathcal{G}(x) + \mathcal{E}(x, \lambda) + \mathcal{S}(x, \lambda) \tag{2}$$

We first consider $\mathcal{L}$ in a continuous space $(x, \lambda) \in \mathbf{R}^2 \times \mathbf{R}$ and develop invariant image features using gradients in the λ and x directions. We then discuss the effects of discrete sampling in the λ direction.

2.1 Spectral Gradients

The key to our approach is to investigate the use the gradients of $\mathcal{L}$ in the λ direction as illumination pose- and color-invariant signatures. Since $g(\boldsymbol{x})$ is independent of λ, the illumination-pose variation will be eliminated in $\mathcal{L}_\lambda$ as:

$$\mathcal{L}_\lambda(\boldsymbol{x}, \lambda) = \frac{\partial \mathcal{L}(\boldsymbol{x}, \lambda)}{\partial \lambda} = \frac{e_\lambda(\boldsymbol{x}, \lambda)}{e(\boldsymbol{x}, \lambda)} + \frac{s_\lambda(\boldsymbol{x}, \lambda)}{s(\boldsymbol{x}, \lambda)} = \mathcal{E}_\lambda + \mathcal{S}_\lambda \ . \tag{3}$$

As the result of λ-differentiation, $\mathcal{L}_\lambda$ consists only of normalized e_λ and s_λ, i.e., *chromaticity* gradients of illumination and reflectance, respectively. This means that since $g(\boldsymbol{x})$ is removed $\mathcal{L}_\lambda$ is independent of shading change due to illumination pose differences. The modification of $\mathcal{L}_\lambda$ by illumination e_λ/e is only additive. The collection of λ-gradients at spectral locations $\lambda_k, k = 1, 2, ..., L$ forms a L-dimensional feature vector:

$$\boldsymbol{\xi}^{\mathcal{L}} = (\mathcal{L}_\lambda|_{\lambda_1}, \ \mathcal{L}_\lambda|_{\lambda_2}, \ ..., \ \mathcal{L}_\lambda|_{\lambda_L}) = \boldsymbol{\xi}^{\mathcal{E}} + \boldsymbol{\xi}^{\mathcal{S}}. \tag{4}$$

When the illumination color does not vary over the region where $\boldsymbol{\xi}^{\mathcal{L}}$ is extracted, $\boldsymbol{\xi}^{\mathcal{L}}$ is invariant to illumination color up to the the bias generated by the normalized illumination gradient $\boldsymbol{\xi}^{\mathcal{E}}$. Therefore, the feature vector $\boldsymbol{\xi}^{\mathcal{L}}$ can be used for forming a descriptor of the object in a feature space $\boldsymbol{\xi}$. The most notable disadvantage of using $\boldsymbol{\xi}^{\mathcal{L}}$ as an object signature is that it will be distorted when illumination color varies over the scene. If $\boldsymbol{\xi}^{\mathcal{L}}$ is to be used for successful local descriptors, there must be some means of discounting $\boldsymbol{\xi}^{\mathcal{E}}$ in each local region.

2.2 Spatial Gradients

Combination of spatial derivatives have been suggested for constructing local shape descriptor invariant to viewing pose. However, illumination color and pose invariance have not been paid notable attention until recently. The spatial derivative of $\mathcal{L}$ in the x direction is given as:

$$\mathcal{L}_x(\boldsymbol{x}, \lambda) = \frac{\partial \mathcal{L}(\boldsymbol{x}, \lambda)}{\partial x} = \mathcal{G}_x + \mathcal{E}_x + \mathcal{S}_x = \frac{g_x(\boldsymbol{x})}{g(\boldsymbol{x})} + \frac{e_x(\boldsymbol{x}, \lambda)}{e(\boldsymbol{x}, \lambda)} + \frac{s_x(\boldsymbol{x}, \lambda)}{s(\boldsymbol{x}, \lambda)} \tag{5}$$

When illumination color and intensity vary gradually over the small region where the spatial gradient is computed, $\mathcal{E}_x$ will be small and thus can be ignored. If $\mathcal{L}_x$ is invariant to illumination pose, we may construct a feature space based on the x- and y-derivatives of $\mathcal{L}$ (such as $\| \nabla_{\bullet}\mathcal{L} \|$ or $\nabla^2 \mathcal{L}$) at spectral locations λ_k's. However, $\mathcal{G}_x$ is only conditionally invariant to illumination pose, and $\mathcal{S}_x$ is not invariant to general 3-D rotation and scale.

The use of spatial gradients requires a full investigation of geometric characteristics of differential invariants. The CCCI approach by Funt and Finlayson utilizes the distribution of simple ratios from spatially neighboring pixels, but they are sensitive to illumination pose and their invariance to viewing pose is limited [8]. The $\mathcal{S}_x$ term in Equation 5 is not invariant to viewing pose. Although $\| \nabla_{\bullet}\mathcal{S} \|$ is insensitive to a 2-D rotation about the optical axis, it is not invariant to general 3-D rotation and scale. Differential gradients have been suggested for identifying local features and recently tested for real images [10][16]. While most of these combinations of derivatives computed from a 2-D image are not theoretically invariant to 3-D rotations of features and such differential invariants to features are still under investigation, experimental studies showed that they can achieve some degree of invariance over mild change of viewing pose[16]. However, the effect of illumination pose has received little attention. We suggest to use $\mathcal{L}$ instead of $\mathcal{I}$ for deriving differential invariants for recognition and indexing to reduce the influence of $g(\boldsymbol{x})$ on the differential descriptors.

2.3 Spectro-Spatial Gradients

The most notable contrast between the spectral and spatial gradients is that while $\mathcal{L}_\lambda$ is invariant to illumination pose, $\mathcal{L}_x$ can be insensitive to illumination color. A natural reaction to this observation would be to consider taking derivatives both in the λ and x directions. The basic component of spectro-spatial gradients we suggest to use is:

$$\mathcal{L}_{x\lambda}(\boldsymbol{x},\lambda) = \frac{\partial^2 \mathcal{L}(\boldsymbol{x},\lambda)}{\partial x \partial \lambda} = \frac{e_{x\lambda}(\boldsymbol{x},\lambda)}{e(\boldsymbol{x},\lambda)} - \frac{e_x(\boldsymbol{x},\lambda)e_\lambda(\boldsymbol{x},\lambda)}{e^2(\boldsymbol{x},\lambda)} + \frac{s_{x\lambda}(\boldsymbol{x},\lambda)}{s(\boldsymbol{x},\lambda)} - \frac{s_x(\boldsymbol{x},\lambda)s_\lambda(\boldsymbol{x},\lambda)}{s^2(\boldsymbol{x},\lambda)} \quad (6)$$

A feature vector that consists of various spectral, spatial and spectro-spatial can provide rich information about objects and illumination conditions. Although various combinations of appropriate spatial gradients may be used, we have chosen to adapt those presented in [16]. We have modified the feature vector developed there to include a spectral gradient. This creates a high order spectro-spatial gradient of logarithmic image irradiance shown as:

$$
\begin{aligned}
f_1 =\ & \mathcal{L}_{x\lambda}\mathcal{L}_{x\lambda} + \mathcal{L}_{y\lambda}\mathcal{L}_{y\lambda} \\
f_2 =\ & \mathcal{L}_{xx\lambda}\mathcal{L}_{x\lambda}\mathcal{L}_{x\lambda} + 2\mathcal{L}_{xy\lambda}\mathcal{L}_{x\lambda}\mathcal{L}_{y\lambda} + \mathcal{L}_{yy\lambda}\mathcal{L}_{y\lambda}\mathcal{L}_{y\lambda} \\
f_3 =\ & \mathcal{L}_{xx\lambda} + \mathcal{L}_{yy\lambda} \\
f_4 =\ & \mathcal{L}_{xx\lambda}\mathcal{L}_{xx\lambda} + 2\mathcal{L}_{xy\lambda}\mathcal{L}_{yx\lambda} + \mathcal{L}_{yy\lambda}\mathcal{L}_{yy\lambda} \\
f_5 =\ & \mathcal{L}_{xxx\lambda}\mathcal{L}_{y\lambda}\mathcal{L}_{y\lambda}\mathcal{L}_{y\lambda} + 3\mathcal{L}_{xyy\lambda}\mathcal{L}_{x\lambda}\mathcal{L}_{x\lambda}\mathcal{L}_{y\lambda} - 3\mathcal{L}_{xxy\lambda}\mathcal{L}_{x\lambda}\mathcal{L}_{y\lambda}\mathcal{L}_{y\lambda} - \mathcal{L}_{yyy\lambda}\mathcal{L}_{x\lambda}\mathcal{L}_{x\lambda}\mathcal{L}_{x\lambda} \\
f_6 =\ & \mathcal{L}_{xxx\lambda}\mathcal{L}_{x\lambda}\mathcal{L}_{y\lambda}\mathcal{L}_{y\lambda} - \mathcal{L}_{xxy\lambda}(2\mathcal{L}_{x\lambda}\mathcal{L}_{x\lambda}\mathcal{L}_{y\lambda} - \mathcal{L}_{y\lambda}\mathcal{L}_{y\lambda}\mathcal{L}_{y\lambda}) \\
& - \mathcal{L}_{xyy\lambda}(2\mathcal{L}_{x\lambda}\mathcal{L}_{y\lambda}\mathcal{L}_{y\lambda} - \mathcal{L}_{x\lambda}\mathcal{L}_{x\lambda}\mathcal{L}_{x\lambda}) + \mathcal{L}_{yyy\lambda}\mathcal{L}_{x\lambda}\mathcal{L}_{x\lambda}\mathcal{L}_{y\lambda} \\
f_7 =\ & \mathcal{L}_{xxx\lambda}\mathcal{L}_{x\lambda}\mathcal{L}_{x\lambda}\mathcal{L}_{y\lambda} - \mathcal{L}_{xxy\lambda}(\mathcal{L}_{x\lambda}\mathcal{L}_{x\lambda}\mathcal{L}_{x\lambda} - 2\mathcal{L}_{x\lambda}\mathcal{L}_{y\lambda}\mathcal{L}_{y\lambda}) \\
& - \mathcal{L}_{xyy\lambda}(2\mathcal{L}_{x\lambda}\mathcal{L}_{x\lambda}\mathcal{L}_{y\lambda} - \mathcal{L}_{y\lambda}\mathcal{L}_{y\lambda}\mathcal{L}_{y\lambda}) - \mathcal{L}_{yyy\lambda}\mathcal{L}_{x\lambda}\mathcal{L}_{y\lambda}\mathcal{L}_{y\lambda} \\
f_8 =\ & \mathcal{L}_{xxx\lambda}\mathcal{L}_{x\lambda}\mathcal{L}_{x\lambda}\mathcal{L}_{x\lambda} + 3\mathcal{L}_{xxy\lambda}\mathcal{L}_{x\lambda}\mathcal{L}_{x\lambda}\mathcal{L}_{y\lambda} - 3\mathcal{L}_{xyy\lambda}\mathcal{L}_{x\lambda}\mathcal{L}_{y\lambda}\mathcal{L}_{y\lambda} - \mathcal{L}_{yyy\lambda}\mathcal{L}_{y\lambda}\mathcal{L}_{y\lambda}\mathcal{L}_{y\lambda} \quad (7)
\end{aligned}
$$

2.4 Spectral Sampling

So far we have considered spectral derivatives of $\mathcal{L}$ in the continuous domain of $(\boldsymbol{x},\lambda) \in \mathbf{R}^2 \times \mathbf{R}$. For computing accurate spectral gradients in multiple spectral points, narrow-band filters or specially tuned photodiodes can be used for multispectral sampling. However, most ordinary CCD-based color cameras use a small number of film-type (gelatin) filters each with the bandwidth of approximately

100 [nm]. In this section, we discuss the approximation of λ derivatives using a small number of coarse spectral samples from broadband color filters.

The spectral sampling of $\mathcal{I}$ can be expressed as:

$$\mathcal{I}(\boldsymbol{x}_{ij}, \lambda_k) = \int_\lambda g(\boldsymbol{x}_{ij})e(\boldsymbol{x}_{ij}, \lambda)s(\boldsymbol{x}_{ij}, \lambda)Q_k(\lambda)d\lambda, \tag{8}$$

where $Q_k(\lambda)$'s are the spectral sampling functions, i.e., color filter functions and $\boldsymbol{x}_{ij} = (x_i, y_j)$ is the sampled image points. Since $Q_k(\lambda)$'s are broad in λ, a restriction on illumination is needed for the approximation of spectral gradients. If we assume that $e(\boldsymbol{x}, \lambda)$ is constant throughout the passband of each $Q_k(\lambda)$, the sampled values become:

$$\mathcal{I}(\boldsymbol{x}_{ij}, \lambda_k) \cong g(\boldsymbol{x}_{ij})e(\boldsymbol{x}_{ij}, \lambda_k)\rho(\boldsymbol{x}_{ij}, \lambda_k), \tag{9}$$

where $\rho(\boldsymbol{x}_{ij}, \lambda_k) = \int_\lambda s(\boldsymbol{x}_{ij}, \lambda)Q_k(\lambda)d\lambda$, and the logarithmic image irradiance is approximated as:

$$\mathcal{L}(\boldsymbol{x}_{ij}, \lambda_k) \cong ln\left[g(\boldsymbol{x}_{ij})e(\boldsymbol{x}_{ij}, \lambda_k)\rho(\boldsymbol{x}_{ij}, \lambda_k)\right] \tag{10}$$

Although, the spectrally constant illumination spectrum over $Q_k(\lambda)$ is a strong requirement on the illuminant, this approximation has long been used in many color constancy algorithms [4][7][8][15][11].

Based on the approximation in Equation 10, the spectral gradient in the discrete domain of $(\boldsymbol{x}_{ij}, \lambda_k)$ is given by the finite difference:

$$\mathcal{L}_{\lambda_k}(\boldsymbol{x}_{ij}, \lambda_k) \cong \mathcal{L}(\boldsymbol{x}_{ij}, \lambda_k) - \mathcal{L}(\boldsymbol{x}_{ij}, \lambda_{k-1}). \tag{11}$$

For the three bands of RGB colors, where $\mathcal{I}(\boldsymbol{x}_{ij}, \lambda_1) = B(\boldsymbol{x}_{ij})$, $\mathcal{I}(\boldsymbol{x}_{ij}, \lambda_2) = G(\boldsymbol{x}_{ij})$ and $\mathcal{I}(\boldsymbol{x}_{ij}, \lambda_3) = R(\boldsymbol{x}_{ij})$, the spectral gradients are limited to two logarithmic chromaticity values:

$$\boldsymbol{\xi} = (\xi_1, \xi_2) = (ln\frac{B}{G}, ln\frac{G}{R}), \tag{12}$$

The spectro-spatial gradient in the x_i and λ directions are are computed from spatial and spectral finite differences as:

$$\mathcal{L}_{x_i, \lambda_k}(\boldsymbol{x}_{ij}, \lambda_k) \cong \mathcal{L}(\boldsymbol{x}_{ij}, \lambda_k) - \mathcal{L}(\boldsymbol{x}_{ij}, \lambda_{k-1}) - \mathcal{L}(\boldsymbol{x}_{(i-1)j}, \lambda_k) + \mathcal{L}(\boldsymbol{x}_{(i-1)j}, \lambda_{k-1}), \tag{13}$$

and higher-order terms in the vector in Equation 7 can be obtained similarly from higher-order finite differences. For the stable implementation for the computation of high-order derivatives, some form of smoothing should precede differencing. A popular choice of smoothing function is the Gaussian function.

3 Experimental Result and Discussion

This section presents preliminary experimental results to provide support for the framework presented in the previous section. The proposed algorithm was tested using a database of five objects. Color images were generated using a Sony XC-77RR and Kodak Wratten filters numbers 29, 47, and 61. Illumination color

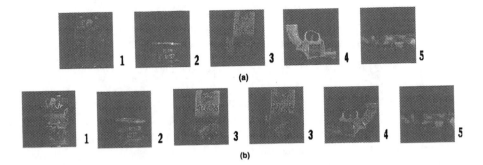

Fig. 1. (a) Database object images. (b) Test images under the changes in viewing pose, illumination pose and global/local illumination color. Two test images under different illumination conditions are shown for Object 3.

was altered by placing gel filters in front of the lamps. The database images were taken with blue filtered illumination. Test images were created by varying the object pose, illumination pose, and color. We present the results from the five objects shown in Figure 1.

Descriptors of each image were generated in the following way. The logarithm of each channel of the image was taken, and smoothed with a Gaussian filter. The two chromaticity images $\xi_1(x)$ and $\xi_2(x)$ were computed using Equation 12, and spatial derivatives of $\xi_1(x)$ and $\xi_2(x)$ were also computed to construct the feature vector in Equation 7. The resulting feature vector consists of sixteen spectro-spatial components. The f_1 components for $\xi_1(x)$ and $\xi_2(x)$ were used to discard feature vectors below a pre-determined threshold.

Binary projections of some of the components of the feature vector for some images are shown in Figure 2. They show the distinctiveness of the feature vectors as object descriptors. These feature vector distributions could be matched a variety of similarity-measuring methods. We choose to use a voting method similar to the one presented in [16] which uses the Mahalanobis distance and a spatial proximity constraint.

The results of the recognition algorithm are presented in Table 1. The table shows that the uniqueness of database objects is reasonably good and that the test objects find correct database objects under varying viewing and illumination conditions. The set of features detected in a test image may not be identical to that in the corresponding database image. Some of the features in the database images may not appear in the corresponding test images since different illumination pose and color make their intensities too low. Different illumination may also generate new features in a test image that are not detected in its database image. This explains why the number of matches is smaller that the number of detected features in the test images. There is similarity between Object 1 and Object 2, but they are distinctive enough to be correctly identified. Table 2 shows the results using a feature vector constructed from spatial gradients in R, G and B channels without any spectral processing. The matching of these features is significantly reduced for objects with shape variation.

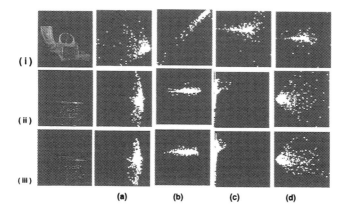

Fig. 2. 2-D projections of feature vector distributions for (i) database Object 4 (ii) database Object 2 and (iii) test Object 2 under different illumination. The vertical and horizontal axes of the displays are: (a) $f_1^{\xi_2}$ and $f_1^{\xi_1}$ (b) $f_2^{\xi_2}$ and $f_2^{\xi_1}$ (c) $f_3^{\xi_2}$ and $f_3^{\xi_1}$ (d) $f_4^{\xi_2}$ and $f_4^{\xi_1}$, respectively.

The experimental results provide good support for our models for spectral and spatial gradients. There are a number of sources of errors that may elude better results presented in this paper. They include: electronic sensor noise, quantization error and the approximation of spectral gradients from coarse spectral sampling. A more effective feature matching method may marginally improve the result. The spectral gradients can be better approximated by using filters of narrower bandwidths than those used in our experiments.

4 Conclusion

The framework outlined in this paper provides color-based object recognition and indexing methods with invariance to changing illumination conditions. We show that the use of spectro-spatial gradients reduce the effect of shading change due to illumination pose change and that of globally and locally varying illumination color. A feature vector that is composed of various spectro-spatial gradients provides a descriptor of object color distribution and limited invariance to viewing pose. The experimental results show the validity of our analyses under the framework and the effectiveness of the object color descriptors proposed.

References

1. D. Berwick and S. W. Lee. Specualrity-, illumination color- and pose-invariant chromaticity space for 3-d object recognition. In *ICCV*, pages 165–170, Bombay, India, January, 1998.
2. M.S. Drew, J. Wei, and Z. Li. Illumination invariant color object recognition via compressed chromaticity histograms of color-channel-normalize images. In *ICCV*, pages 533–540, Bombay, India, January, 1998.
3. G. D. Finlayson. Color in perspective. *PAMI*, 18(10):1034–1038, 1996.
4. G. D. Finlayson, M. S. Drew, and B. V. Funt. Color constancy: Generalized diagonal transforms suffice. *JOSA*, 11(11):3011–3019, 1994.
5. G.D. Finlayson, B. Schiele, and J.L. Crowley. Comprehensive colour image normalization. In *ECCV*, Freiburg, Germany, June, 1998.

Table 1. Feature matches using spectro-spatial gradients. The first and second columns represent the database images and the number of feature vectors detected in the database images, respectively. The feature matching is performed to the database objects in the first five columns and to the test objects in the next six columns. The notation (') is used for test objects. The first and second rows are the database and test images and the number of feature vectors detected in the test and database images, respectively. The remaining cells indicate how many votes each model received from the test image.

		Obj 1	Obj 2	Obj 3	Obj 4	Obj 5	Obj 1'	Obj 2'	Obj 3'	Obj 3"	Obj 4'	Obj 5'
		259	366	363	296	280	291	362	405	396	287	644
Obj 1	259	259	57	79	0	41	195	99	45	54	77	0
Obj 2	366	160	366	10	20	27	169	246	14	0	0	0
Obj 3	363	60	14	363	74	27	29	31	295	298	32	0
Obj 4	296	0	0	0	296	226	0	0	0	0	254	403
Obj 5	280	0	0	0	28	280	0	0	0	0	53	494

Table 2. Feature matches using spatial gradients in RGB

		Obj 1	Obj 2	Obj 3	Obj 4	Obj 5	Obj 1'	Obj 2'	Obj 3'	Obj 3"	Obj 4'	Obj 5'
		839	468	1016	417	509	599	539	1000	889	469	474
Obj 1	839	839	113	190	737	107	93	54	33	77	271	407
Obj 2	468	745	468	687	672	90	217	329	520	607	324	212
Obj 3	1016	554	214	1016	234	33	105	55	900	650	252	224
Obj 4	417	488	102	123	417	495	149	61	89	294	385	470
Obj 5	509	0	0	0	0	509	0	0	0	0	0	35

6. M. Flickner et al. Query by image and video content: The qbic system. *IEEE Computer*, 28:23–32, 1991.
7. D. A. Forsyth. A novel approach to colour constancy. *IJCV*, 5(1):5–36, 1990.
8. B. V. Funt and G. D. Finlayson. Color constant color indexing. *PAMI*, 17(5):522–529, 1995.
9. G.H. Healey and D. Slater. Global color constancy: Recognition of objects by use of illumination-invariant properties of color distributions. *JOSA*, 11(11), 1994.
10. J.J. Koenderink and A.J. van Doorn. Representation of local geometry in the visual system. *Biological Cybernetics*, 55:367–375, 1987.
11. E.H. Land and J. J. McCann. Lightness and retinex theory. *JOSA*, 61:1–11, 1971.
12. S. Lin and S.W. Lee. Using chromaticity distributions and eigenspace for pose-, illumination-, and specularity-invariant 3d object recognition. In *CVPR*, pages 426–431, Puerto Rico, 1997.
13. L. T. Maloney and B. A. Wandell. A computational model of color constancy. *JOSA*, 1:29–33, 1986.
14. H. Murase and S.K. Nayar. Visual learning and recognition of 3-d objects from appearance. *IJCV*, 14, 1995.
15. S.K. Nayar and R.M. Bolle. Reflectance based object recognition. *IJCV*, 17, 1996.
16. C. Schmid and R. Mohr. Local greyvalue invariants for image retrieval. *PAMI*, 19(5):830–835, 1997.
17. D. Slater and G.H. Healey. The illumination-invariant recognition of 3d objects using local color invariants. *PAMI*, 18(2), 1996.
18. M.J. Swain and D.H. Ballard. Color indexing. *IJCV*, 7(1):11–32, 1991.

MCEBC – A Blob Coloring Algorithm for Content-Based Image Retrieval System

Sue J. Cho[1] and Suk I. Yoo[2]

Department of Computer Science, Seoul National University,
Shillim-dong, Kwanak-gu, Seoul, Korea
Phone: +82-2-880-6577
Fax: +82-2-874-2884
[1]sue@cs.snu.ac.kr
[2]siyoo@hera.snu.ac.kr

Abstract. In most content-based image retrieval systems, colors are used as very significant feature for indexing and retrieval purposes. Many of them use the quantized colors for various reasons. In some systems, especially the systems accepting user-drawn queries, color quantization improves the overall performance as well as the execution time. In this paper, a new color quantization method, which uses some heuristics to minimize the color matching errors, is proposed. The proposed MCEBC(Main Color Emphasized Blob Coloring) algorithm quantizes the colors into the predefined color classes, using the heuristic information that humans tend to emphasize the dominant color component when they perceive and memorize colors. The experimental results indicate that the proposed method approximates the user's classification better than the other methods that use the mathematical color difference formulas. The proposed MCEBC algorithm is implemented in a content-based image retrieval system, called QBM system. The retrieval results of the system using MCEBC algorithm were quite satisfactory and showed higher success rates than those using other methods.

Keywords: color quantization, content-based retrieval, image database

1 Introduction

In content-based image retrieval system that accepts queries drawn by human users, color quantization is performed for three major reasons. First, the color images can be segmented through color quantization. Since pixels can be classified into some fixed set of representative colors, image can be segmented easily. Second, it helps the user select colors while drawing a query image. By providing a global palette of some fixed number of colors, the user has only to select a color that he thinks the nearest to the intended color. Since the colors of the images in the database are quantized, queries can be drawn in such a way. Third, it simplifies the similarity measure used in image retrieval. With color quantization, the complicated calculation of color differences is not needed.

Color quantization is the process of reducing the number of colors in a digital image and mapping them to the nearest representative colors[7]. Since the submitted query image is compared with every image in the database, the image dependent palette cannot be used in image retrieval system. Therefore, the color quantization problem is narrowed to the mapping of image pixels to the palette colors. The simplest approach to this problem is to map each pixel to its nearest neighbor in the palette[6]. The definition of the "nearest color" is varied according to the difference function used by the color quantization algorithm. Generally, colors are represented based on Young's three-color theory that any color can be reproduced by mixing an appropriate set of three primary colors[5]. There are several color coordinate systems, which have come into existence for a variety of reasons[2,3,4,5,6]. Although several color-difference formulas have been proposed to yield perceptually uniform spacing of colors, the existing color spaces do not provide the difference function that is identical to the human perception. In color quantization process, the palette color that a human thinks nearest to a certain color does not always have the smallest difference function value. It is observed that humans tend to emphasize the dominant color component when they perceive colors. In this paper, a heuristic color quantization method that uses heuristic information to approximate the process of human's color quantization is proposed. The proposed color quantization algorithm is implemented in a content-based image retrieval system, QBM(Query by Blob Map) system[1]. QBM accepts a user's query as rough map of colored blobs. It considers an image as a set of colored blobs. The pixels in a blob have the same quantized color although they may have different colors in the original image. Some experiments show that the proposed simple algorithm approximates the human's color mapping better than other methods using mathematical color differences. The retrieval results of the system using the proposed algorithm were quite satisfactory and showed higher success rates than those using other methods.

2 QBM System Overview

QBM (Query by Blob Map) is a content-based image retrieval system, which provides the convenient and powerful query-specification scheme. The user of the QBM system submits a query by laying out colored blobs roughly on the screen so that it looks similar to the target image wholly or partially. The position of each individual blob may not be identical to that in the target image because position information is not included in the feature set used in QBM system. In QBM, the following three features are used in image indexing and retrieval; (a) the color of each object in the image, (b) the relative size of each object, and (c) the topological structure of the objects. Therefore, QBM system is robust to imperfect queries such as shifted queries and partial queries.

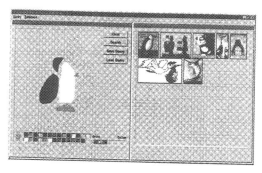

Fig. 1. A screen dump of the running application

The system consists of indexing module, querying module, and retrieving module. In indexing module, an image is segmented into a collection of objects. The term "object" does not refer the thing that has semantic meaning in real life. An "object" is defined as a set of pixels that are connected and have the same quantized color. Each object is represented as a set of attributes such as color, size, and position. Whenever an image is inserted into the database, it is converted into a collection of objects and these attributes are automatically extracted. Querying module accepts the user-drawn queries. Since a user draws a query in object basis, the query image can be recognized as a collection of objects without any segmentation process. This query image is converted into a graph called *prime edge graph*, which represents the topological structure of objects in the image. In a prime edge graph, each node represents an object and an edge between two nodes represents the positional relationship of corresponding objects. Retrieving module measures the similarity score of each image in the database through two-step matching – object matching and prime edge matching. Finally, the accepted images are displayed in the order of similarity. A screen dump of the running application is shown in figure 1. The user draws a query image in the rectangular area on the left side of the application window. The system converts the query image to a prime edge graph and tests it against all the images in the database.

3 MCEBC Algorithm

When a human memorizes a color of certain object, he quantizes it into several classes such as apples being red, cucumbers being green, etc. The same is true when a human reproduces the image in his memory on a computer monitor. He does not require thousands of colors as he does in creating an artistic picture. Actually, he generally uses only about 20 colors in reproducing an image. In indexing an image database, the process of labeling colors to the stored images is similar to the process of memorizing a color. Since the query image is generated by a human, color quantization of database image must approximate the process of human quantization.

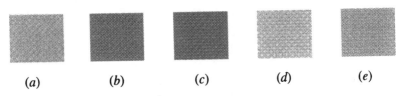

Fig. 2. Color blocks of RGB components (*a*) (0.746,0.598,0.313), (*b*) (0.500, 0.500, 0.500), (*c*) (0.500, 0.500, 0.000), (*d*) (1.000, 0.500, 0.500), and (*e*) (1.000, 0.500, 0.000)

Color quantization is the process of reducing the numbers used in an image and mapping each pixel to the nearest representative color class. The first problem is how many colors and what colors to be used as representative colors. Second, the meaning of the "nearest" class must be defined. In this paper, the word "nearest" does not simply mean "which has the shortest mathematical distance". It means that "which a human thinks as most similar". In most cases, a human considers the mathematically closest color as the nearest color. However, near gray, he becomes more sensitive to one of the components which constitute the color. For example, although the color of the block in figure 2(*a*) is closest to the color in figure 2(*b*) in both *RGB* and *L*a*b** spaces, a human tends to select (*c*) or (*e*) as the nearest color. In fact, 18 out of 20 subjects selected (*c*) rather than (*b*). Two subjects selected (*e*).

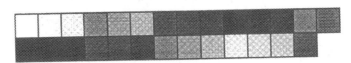

Fig. 3. The global palette used in QBM system.

Generally, the color quantization is performed in two steps; the design of palette and the mapping of image pixels. Since the proposed color quantization algorithm is designed for an image retrieval system, in which submitted query image is compared with every image in the database, the image dependent palette cannot be used in this case. Therefore, it constructs a simple global palette. The global palette consists of 27 colors which are on the RGB color space and each R, G, and B component has one of the three values {0, 0.5, 1} respectively. The constructed global palette is shown in figure 3. Next, it classifies each pixel into 27 color classes defined in the global palette. Basically, each Red, Green, and Blue component of a pixel is quantized independently into three bands {0, 0.5, 1} according to one-dimensional distance. For example, if RGB coordinate of a pixel is (0.9, 0.1, 0.1), it is quantized into (1, 0, 0). However, if all components have values between 0.25 and 0.75, they are not always quantized into value 0.5 although 0.5 is closer than 0 or 1. In this case, the value of one component may be emphasized or suppressed according to its dominance over other components. If a color component c has the greatest value and is dominant over other components, it is emphasized and replaced with 1. On the other hand, if c has the least value and is dominant over other components, it is suppressed and replaced with 0. A component is said to be dominant, if its degree of dominance is greater than that of any other component by some threshold t. The degree of dominance (DoD) of each component of the color (r, g, b) is calculated as follows:

$$DoD(R) = |r - g| + |r - b|$$

$$DoD(G) = |g - r| + |g - b| \tag{1}$$

$$DoD(B) = |b - r| + |b - g|.$$

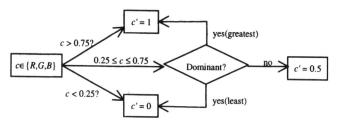

Fig. 4. the block diagram of MCEBC algorithm

The dominant component is what has much larger or smaller value than other components. A threshold t is used in deciding whether a component is dominant or not. Given a color $C = (r, g, b)$, one of the components, say, R is said to be greatest (least) dominant if (a) all r, g, and b are between 0.25 and 0.75, (b) $r > g$ and $r > b$ ($r < g$ and $r < b$), and (c) $DoD(R) > DoD(G) + t$ and $DoD(R) > DoD(B) + t$. If a component of certain color is greatest dominant or least dominant, it is said to be dominant. If t takes a large value, the color quantization becomes less sensitive, i.e., most of value between 0.25 and 0.75 is replaced with 0.5. On the other hand, if t takes a small value, many gray-tone colors are shifted toward their dominant colors. Experimental results showed that the appropriate value for t, which best approximated the human perception, was 0.4.

4 Experimental Results

4.1 Evaluation measures

The objective of the MCEBC algorithm is to map each color pixel to the same color class as a human user does. In order to set evaluation criteria for the proposed color quantization algorithm, two types of human perception data in the form of pairs of colors were collected from 20 subjects. The first set, called "similarity data" was obtained by asking each subject to select the nearest colors of presented color set out of the global palette displayed in common screen. An evaluation function F_s for this criterion is defined as

$$F_s = \frac{1}{N} \sum_{i=1}^{N} H_s(\text{Result}(D_i)) \tag{2}$$

where N is the number of gathered data, $\text{Result}(D_i)$ is the resulted color by quantization algorithm for D_i, and H_s is the percentage of the subjects who answered that the nearest color of D_i is $\text{Result}(D_i)$. The larger the value of F_s, the better the quantization algorithm approximates the similarity measure that a human uses in comparing colors.

The second set, called "memory quantization data" was obtained in similar manner except that each subject was asked to observe a color in one screen and select the nearest color out of the palette displayed in another screen. An evaluation function F_m for this criterion is defined as

$$F_m = \frac{1}{M} \sum_{i=1}^{M} H_m(\text{Result}(D_i)) \tag{3}$$

where M is the number of gathered data, $\text{Result}(D_i)$ is the resulted color by quantization algorithm for D_i, and H_m is the percentage of the subjects who answered that the nearest color of D_i is $\text{Result}(D_i)$. The larger the value of F_m, the better the quantization algorithm approximates the quantization process that occurs when human memorizes a color as well as the similarity measure that a human uses.

4.2 Determination of the dominant component

The selection of threshold t used in determining whether a component is dominant influences the performance of the algorithm. We found appropriate value for t using two types of gathered data as a training set. The result is shown in figure 5. If t takes the value 0, every greatest or least component is determined to be dominant.

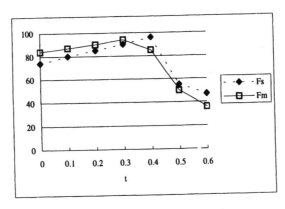

Fig. 5. Selecting the threshold used in determination of the dominant component

4.3 Performance of the heuristic approach

In order to compare MCEBC algorithm with other algorithms which use the mathematical distance as a similarity metric, we implemented another two color quantization algorithms. One uses RGB distance and the other uses $L*a*b*$ distance as color similarity metrics. Table 1 compares our heuristic approach employed in MCEBC algorithm with $t = 0.3$ and 0.4 to RGB distance and $L*a*b*$ distance using the evaluation functions F_s and F_m.

Table 1. Comparison of the color similarity metrics with human's.

Metric	Data Set	
	F_s	F_m
MCEBC ($t = 0.3$)	90	94
MCEBC ($t = 0.4$)	96	85
RGB Distance	47	36
$L*a*b*$ Distance	50	38

4.4 Performance on image retrieval

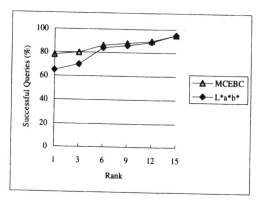

Fig. 6. Comparison of success rates.

To evaluate the retrieval performance on image queries, two image databases were constructed; one with MCEBC algorithm and the other with $L*a*b*$ distance algorithm. Both image databases are of 1000 images. The success rates over 100 user-drawn queries are plotted in the graph of figure 6. For most of queries, the similarity score of the intended image was higher in MCEBC database than in $L*a*b*$ database. However, success rate of raking the target image in top 15 was the same because color is not the only factor that determines the similarity score. The database indexing speed is much faster with MCEBC algorithm because the color-conversion and distance-calculation steps are not required.

5 Conclusion

The proposed color quantization algorithm approximates the human's quantization process better than other algorithms that use mathematical color difference. In addition, the quantization speed is much faster because it does not perform any color conversion or distance calculation. It manifests these advantages when used in the image retrieval system that accepts user-drawn queries. Our future work includes adopting additional heuristics that reflects the effect of neighboring colors to the human-perceived colors.

References

1. Cho, S.J., Yoo, S.I.: Image Retrieval Using Topological Structure of User Sketch. Proc. IEEE SMC98 (1998)
2. Foley, J.D., van Dam, A., Feiner, S.K., Hughes, J.F.: Computer Graphics Principles and Practice 2nd Edition. Addison-Wesley (1990).
3. Gong, Y., Proietti, G., Faloutsos, C.: Image Indexing and Retrieval Based on Human Perceptual Color clustering. Proc. CVPR98 (1998) 578-583
4. Gonzalez, R.C., Woods, R.E.: Digital Image Processing. Addison-Wesley (1992).
5. Jain, A.K.: Fundamentals of Digital Image Processing. Prentice Hall, Englewood Cliffs (1989)
6. Sharma, G., Trussell, H.J.: Digital Color Imaging. IEEE Trans. on Image Processing, Vol. 6, No. 7. (1997)
7. Uysal, M., Yarman-Vural, F.T.: A Fast Color Quantization Algorithm Using a Set of One Dimensional Color Intervals. Proc. ICIP98. (1998) 191-195

EigenHistograms: Using Low Dimensional Models of Color Distribution for Real Time Object Recognition

Jordi Vitriá, Petia Radeva, and Xavier Binefa

Centre de Visió per Computador, Dept. Informàtica, Universitat Autònoma de Barcelona, 08193 Bellaterra (Barcelona), Spain,
{ jordi, petia}@cvc.uab.es

Abstract. Distribution of object colors has been used in computer vision for recognition and indexing. Most of the recent approaches to this problem have been focused on defining optimal spaces for representing pixel values that are related to physical models and that present some invariance. We propose a new approach to identify individual object color distributions by using statistical learning techniques and to allow their compact representation in low dimensional spaces. This approach outperforms generic "optimal" spaces when color illumination is constant, allowing changes in object pose and illumination direction. This approach has been tested for real time industrial inspection of multicolored objects.

1 Introduction

Object appearance in an image is caused by many factors, including object pose, illumination directions and illumination color. Traditional recognition methods [7],[6] have mainly used geometric cues as shape models and feature matching for detecting and identifying objects in images, but they have not demonstrated the level of performance that allow them to be systematically used in real time for a large number of applications (i.e. face recognition). Recently, some new approaches based on the direct representation of object appearance have been developed for object recognition and pose estimation. Turk and Pentland [3] used principal component analysis to describe face patterns in a low dimensional appearance space. Murase and Nayar in [4] have shown real time recognition of complex 3D objects based on Principal Component Analysis (PCA) of geometrical shape of the objects. The PCA approach is very appropriate for real time applications because of the low cost of the recognition algorithms, however it is limited to the analysis of the geometric shape and depends on the object's pose.

Color distributions can be efficiently used as signatures for object recognition in the appearance-based framework. The earliest approach [2] showed the usefulness of color histograms for indexing large object databases independently of object's pose. Most of the recent approaches focus on illumination color invariance [1],[5] known as color constancy, but although these methods perform better than histogram indexing when color illumination changes, they use color information only where surface color varies and are very sensitive to noise.

The above mentioned approaches do not consider color distribution changes due to illumination pose, which can induce significant changes in color histograms. In this work we consider the problem of non constant illumination and introduce statistical techniques to construct a low dimensional space for representing this effect. To achieve pose invariant object's recognition we consider the image (or a Region Of Interest (ROI)) histogram and perform the recognition process in the color distribution of the object applying the principal component analysis technique. In section 2 and 3 we discuss the color representation and the PCA approach as a statistical technique to recognize objects. In section 4 we define the EigenHistogram approach as a PCA applied to the object histograms in order to recognize multicolor objects. Finally, we discuss the experimental results obtained in an industrial (pharmaceutical) application, compare the eigenhistogram technique to other color inspection techniques and finish the article with conclusions.

2 Color Indexing

We describe the histogram intersection algorithm as developed by Swain and Ballard [2]. Colors in an image are mapped into a discrete color space containing n colors. A color histogram is a vector in an n-dimensional space where each element represents the number of pixels of color j in the image. Each object in the image database is represented by its histogram M which will be used to be compared to the histogram I of another image presented to the system.

To measure the distance d between a model M and an image I we can use the following expression:

$$l = \frac{1}{d} = \frac{\sum_{i=1}^{n} min(I_i, M_i)}{\sum_{i=1}^{n} M_i} \tag{1}$$

It is easy to see that when the image coincides with the model, the distance measure (1) is 1 and when their colors differ the measure tends to 0. Two histograms are called α-similar if l is greater than or equal to α. For a fixed threshold t, a model is going to be retrieved if its histogram is t-similar to the histogram of I.

Histogram representation is computationally simple and presents some interesting invariances (object's rotation and translation) but difficulties arise when there are variations in the object pose in space, illumination color and illumination intensity.

3 Principal Component Analysis as a Classification Technique

Principal Component Analysis is a dimension reduction method which its first goal is to minimize the dimension of n-dimensional vectors to m-dimensional vectors (where $m < n$). PCA can be seen as a linear transformation that extracts a lower dimensional space that preserves the major linear correlations in the data and discards the minor ones. Vector projections can be used as representatives of original vectors for recognition purposes. In computer vision, the eigenspace approach has led to a powerful alternative to standard techniques such as correlation or template matching for appearance-based object recognition [4],[3]. The reconstruction error of the eigenspace decomposition is an effective indicator of similarity. In [8] it has been shown that reconstruction error can be interpreted as an estimate of a marginal component of the probability density of the object.

Having a data vector x ($x \in R^n$), PCA projects it onto the m ($m < n$) dimensional linear subspace spanned by the leading eigenvectors of the data covariance matrix: $\Sigma = E[(x - \mu)(x - \mu)^T]$, where $\mu = E[x]$ and E denotes an expectation with respect to x. The leading m eigenvectors $\{e_i | i_1, \ldots, i_m\}$ of a positive semidefinite matrix are the m eigenvectors which correspond to the m largest eigenvalues. The indices are assigned such that the corresponding eigenvalues in decreasing order are given by $\lambda_1 \geq \lambda_2 \geq \ldots \geq \lambda_m$.

Let be $f : R^n \to R^m$ an encoding function from a vector $x \in R^n$ to a vector $z = f(x) \in R^m$, where $m < n$. Let be $g : R^m \to R^n$ a decoding function from z to $x' = g(f(x)) \in R^n$, a reconstruction of x. PCA encodes x as: $z = f(x) = V(x - \mu) = (e_1^T(x - \mu), \ldots, e_n^T(x - \mu))$ where V is an $m \times n$ matrix whose rows, e_i are the leading m orthonormal eigenvectors of Σ and z is the m dimensional encoding. The components of z are called the principal components. PCA reconstructs/decodes x' from z as follows: $x' = g(z) = V^T z + \mu$

The mean squared error in reconstructing the original data is: $\epsilon = E[||x - g(f(x))||^2]$

Using PCA we can obtain the least error in terms of reconstructing the original vector. PCA builds a global linear model of the data: an m dimensional hyperplane spanned by the leading eigenvectors of the data covariance matrix.

Using the retroprojection technique and having a vector to recognize, it is only necessary to project this vector into the m-dimensional space, retroproject to the original space and calculate the error with respect to a set of model vectors in order to decide which is the corresponding one.

4 Color Object Recognition by Principal Component Analysis of Color Histograms

Given p images of the same object k corresponding to different views taken under different illumination conditions, we compute the set of histograms that represents it as follows: $O_k = \{H_{1k}, \ldots, H_{pk}\}$

This set of histograms represents all possible variations for the object color distribution. Considering that color change is derived from different illumination directions, we assume that these variations span a low dimensional linear subspace of R^n (being n the number of colors represented in the histogram) which can be computed using PCA. Principal Component Analysis of a histogram set defines an encoding function that projects each histogram from the set on a low dimensional subspace defined by the m principal eigenvectors e_i of the histogram covariance matrix. Given that e_i form an orthonormal base for object histograms, we call each e_i an *EigenHistogram*.

Given a set of objects we represent each one by its *EigenHistogram* set e_{ik}. When the system is presented with a new image, the recognition process performs as follows:

1. The color histogram I of the new image is computed.
2. For each object r in the database, compute $\epsilon_r = |I - g(f(I))|^2$ using the EigenHistogram set e_{ir}.
3. Classify I as O_{r*}, where $r* = argmin\{\epsilon_r\}$.

This technique has shown a great capacity to represent complex color distributions under illumination changes in pose and intensity.

5 Experimental Results

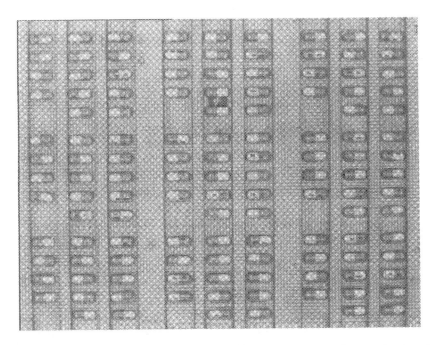

Fig. 1. Example of a blister used to learn the color distribution of a certain capsule

In order to test our approach, the method has been used in a large object database composed of thousands (20 520) of pharmaceutical tablets and capsules. Different blisters with up to 135 capsules are learned and inspected in real time (see Fig. 1 and Fig. 2). Capsules belong to 20 different types where each type can be of 1 or two colors. Some of the capsules have black letters that can be

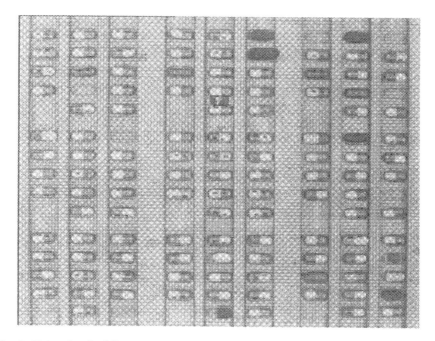

Fig. 2. Example of a blister used to inspect the capsules from Fig.1

completely or partly visible. The illumination spatially differs as well as due to the plastic material of the capsules, shines can appear. Therefore, the colors of each product should be represented by a color distribution instead of a constant color.

For each product we manually define the rois of the capsules and use their histogram to determine the color distribution by the PCA. We determine the EigenHistogram extracted from the covariance matrix of the rois histogram of each type of capsules and use them to project and retroproject the rois during the inspection process. Other blister faults can occur in case of broken capsules or open capsules. Rois are determined using the blister alveolus position. capsules are limited to the alveolus location, yet have freedom to move inside the alveolus.

Our work has focused on comparing EigenHistograms in RGB sensor color space to the classical color indexing in different color spaces. In particular, we have used the following color spaces:

1. $(\mathbf{R}, \mathbf{G}, \mathbf{B})$: sensor color space.
2. $(\mathbf{RG}, \mathbf{BY}, \mathbf{WB})$: three opponent color axes, where $RG = R - G$, $BY = 2B - R - G$, $WB = R + G + B$.
3. $(\mathbf{l_1}, \mathbf{l_2}, \mathbf{l_3})$: photometric invariant color features for both matte and shiny surfaces, as described in [5].
4. $(\mathbf{r}, \mathbf{g}, \mathbf{b})$: normalized colors $r(R, G, B) = R/(R + G + B)$, $g(R, G, B) = G/(R + G + B)$, $b(R, G, B) = B/(R + G + B)$.
5. $(\mathbf{c_1}, \mathbf{c_2}, \mathbf{c_3})$: photometric invariant color features for matte, dull surfaces, as described in [5].

Table 1. Error Probability (in %) as a function of the training capsules

Method	1 Capsule	5 Capsules	128 Capsules
(R,G,B) Indexing	0.022	0.006	0.004
(RG,BY,WB) Indexing	0.066	0.030	0.030
(l_1, l_2, l_3) Indexing	0.082	0.006	0.060
(r,g,b) Indexing	0.028	0.020	0.045
(c_1, c_2, c_3) Indexing	0.015	0.012	0.022
EigenHistograms	0.083	0.033	0.002

Tables 1 and 2 exhibit the performance of our method on the test objects. Table 1 shows the error rate as a function of the number of training capsules. In case of small number of training capsules it seems that the best approaches of color indexing use the (R, G, B) and (c_1, c_2, c_3) methods. When the number of training capsules is increased to improve the color object recognition, the best results are obtained using the EigenHistogram approach.

Table 2. Discrimination capacity of different color indexing techniques

Method	Time	m_1	s_1	m_2	s_2
(R,G,B) Indexing	16ms	0.243	0.006	0.796	0.002
(RG,BY,WB) Indexing	25ms	0.309	0.006	0.829	0.003
(l_1, l_2, l_3) Indexing	662ms	0.329	0.009	0.866	0.001
(r,g,b) Indexing	161ms	0.401	0.031	0.927	0.002
(c_1, c_2, c_3) Indexing	313ms	0.460	0.030	0.892	0.004
EigenHistograms	60ms	0.093	0.007	0.822	0.008

Table 2 shows the results obtained for each method of color object inspection considering 21 blisters of different rois where each blister contains correct and wrong objects. The purpose of this test is to observe the global probabilistic distributions of wrong and correct color objects. In the table 2 the third and fourth column show the mean and variance of the distance measure l for the classification of the wrong objects, the fifth and sixth columns show the mean and variance of the probability error in the classification of the correct products. One can note that the smallest mean of the probability error[1] for wrong products is achieved in case of EigenHistograms that shows the discrimination power of our method. The small variance of the probability error for the class of wrong

---

[1] Remember that according to (1) the distance measure of similar objects tends to 1 and for different objects tends to 0.

capsules obtained from the EigenHistogram approach shows that the wrong objects are identified even when the inspection rois do not exactly coincide with the training rois. The fifth column shows that the eigenhistogram tecnique is still good in recognizing the correct objects with the high distance rate of 0.822.

Figure 3 shows both mean distributions of the inspection values l obtained in different experiments with different rois obtained by small translations and scaling. The x axis denotes the means obtained for both classes where the distance mean for each experiment of the wrong objects was between 0 and 0.196 and of the correct objects was between 0.705 and 0.936. The y axis of Fig. 3 denotes the number of tests where the means corresponding to the different intervals have been obtained. From the graphics it can be appreciated that both distributions are well separated illustrating the discrimination "power" of the EigenHistogram classification technique. As a result, we can summarize from the tables that the EigenHistograms have the better recognition results and the better discrimination capacity.

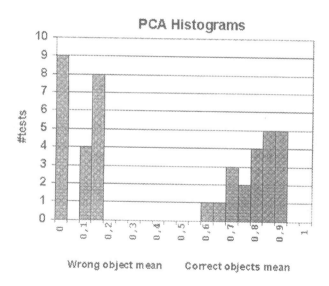

Fig. 3. Inspection values obtained on a blister with 200 correct and wrong objects

From the table it can be seen that the fastest inspection techniques are the (R,G,B), (RG, BY, WB) and the EigenHistogram methods where the time of inspection of up to 135 capsules is in less than 0.065s using an INTEL PENTIUM 200MMX with 32Mb of RAM. The fact that the EigenHistogram method consists of a projection and a retroprojection of a histogram to measure its distance to the original histogram determines the O(nm) complexity of the algorithm. In practice, we obtained that a number of EigenHistograms $m = 8$ and working with color histograms of $8 \times 8 \times 8$ $n = 512$ is sufficient to represent and recognize the color distribution of the pharmaceutical objects.

We should note that different color indexing methods are useful for color object recognition but can not cope with the problems of small defaults in the geometric shape (e.g. capsules of different shape and size like broken capsules or open capsules). It is due to the fact that these approaches analyze the color of the image. Another issue is that we model the color distribution of the roi (not only the product) using the fact that the background in the trained and inspected rois belongs to the same distribution. In many industrial applications the environment of inspection is constant, hence such an assumption is appliable.

6 Conclusions

This paper presents a new approach to color-based indexing and recognition that describes object colors using probabilistic models of their histograms. Histogram distributions of multicolored objects are represented and generated using eigen-basis vectors, that we call EigenHistograms. Object recognition is achieved by examining its proximity to database subspaces. As a result object multicolored recognition is achieved invariantly to object pose and changes in illumination directions. This approach has been tested and validated on a large object database consisting of multicolored pharmaceutical products and is applied for real-time inspection of capsules and other pharmaceutical objects. Our future plans involve extending the EigenHistogram approach to the problem of color object recognition in unknown environments as well as to the tracking problem.

Acknowledgments

This work is supported by CICYT and EU grants TAP98-0631, TIC98-1100 and 2FD97-0220.

References

1. Funt B., Finlayson G.: Color Constant Color Indexing. IEEE Trans. PAMI, **17**, (1995) 522-528
2. Swain M., Ballard D.: Color Indexing. Intern. J. of Computer Vision, **7**, (1991) 11-32
3. Turk M.A., Pentland A.: Eigenfaces for recognition. Journal of Cognitive Neuroscience, **3** (1), (1991) 71–86
4. Murase H., Nayar S.K.: Learning and recognition of 3D objects from appearance. Proc. IEEE Qualitative Vision Workshop, NY, (1993) 39–49
5. Gevers T., Smeulders A.: Color Based Object Recognition. In Image Analysis and Processing, Alberto del Bimbo (Ed), LNCS 1310 (1997).
6. Ullman S.: High-Level Vision, MIT Press (1996).
7. Grimson W.: Object Recognition by Computer, MIT Press (1990).
8. Moghaddam B. and Pentland A. (1996) Probabilistic Visual Learning for Object Recognition, in Nayar S. and Poggio T. (eds.) "Early Visual Learning", Oxford University Press.

A Novel Approach to the 2D Analytic Signal[*]

Thomas Bülow and Gerald Sommer

Christian–Albrechts–Universität zu Kiel
Institute of Computer Science, Cognitive Systems
Preußerstraße 1–9, 24105 Kiel
Tel:+49 431 560433, Fax: +49 431 560481 {tbl,gs}@informatik.uni-kiel.de

Abstract. The analytic signal of a real signal is a standard concept in 1D signal processing. However, the definition of an analytic signal for real 2D signals is not possible by a straightforward extension of the 1D definition. There rather occur several different approaches in the literature. We review the main approaches and propose a new definition which is based on the recently introduced quaternionic Fourier transform. The approach most closely related to ours is the one by Hahn [8], which defines the analytic signal, which he calls complex signal, to have a single quadrant spectrum. This approach suffers form the fact that the original real signal is not reconstructible from its complex signal. We show that this drawback is cured by replacing the complex frequency domain by the quaternionic frequency domain defined by the quaternionic Fourier transform. It is shown how the new definition comprises all the older ones. Experimental results demonstrate that the new definition of the analytic signal in 2D is superior to the older approaches.

1 Introduction

The notion of the analytic signal of a real one-dimensional signal was introduced in 1946 by Gabor [6]. It can be written as $f_A(x) = f(x) + i f_{\mathcal{H}i}(x)$, where f is the original signal and $f_{\mathcal{H}i}(x)$ is its Hilbert transform. Thus, the analytic signal is the generalization of the complex notation of harmonic signals given by Eulers equation $\exp(i\omega x) = \cos(\omega x) + i\sin(\omega x)$. The construction of the analytic signal can also be understood as suppressing the negative frequency components of f.

The analytic signal plays an important role in one-dimensional signal processing. One of the main reasons for this fact is, that the instantaneous amplitude and the instantaneous phase of a real signal f at a certain position x can be defined as the magnitude and the angular argument of the complex-valued analytic signal f_A at the position x. The analytic signal is a global concept, i.e. the analytic signal at a position x depends on the whole original signal and not only on values at positions near x.

Often local concepts are more reasonable in signal processing: They are of lower computational complexity than global concepts. Furthermore, it is reasonable that the local signal structure, like local phase and local amplitude should

---

[*] This work was supported by the Studienstiftung des deutschen Volkes (Th.B.) and by the DFG (So-320/2-1) (G.S.).

only depend on local neighborhoods. The "local version" of the analytic signal was also introduced by Gabor in [6]. This is derived from the original signal by applying Gabor filters which are bandpass filters with an approximately one-sided transfer function.

Complex Gabor filters are widely used in 1D signal-processing as well as in image-processing. However, their theoretical basis – the analytic signal – is only uniquely defined in 1D. There have been many attempts to generalize the notion of the analytic signal to higher dimensions. However, there is no unique, straightforward generalization but rather different ones with different advantages and disadvantages. We will propose a new definition of the analytic signal in 2D which is based on the recently introduced quaternionic Fourier transform (QFT) [2,3]. From this definition there follows a new kind of 2D Gabor filters (so-called quaternionic Gabor filters (QGF)) which already have found applications in texture segmentation and disparity estimation [1]. In the present article we restrict ourselves to the motivation, definition and analysis of the analytic signal.

The structure of this article is as follows: In Sect. 2 and Sect. 3 we give a short introduction to the one-dimensional analytic signal and to the main approaches towards a two-dimensional analytic signal, respectively. A short review of the QFT will be given in Sect. 4 followed by the new definition of the analytic signal in Sect. 5. In Sect. 6 we compare the different approaches to a two-dimensional analytic signal and present experimental results. Finally conclusions are drawn.

2 The 1D Analytic Signal

The analytic signal f_A can be derived from a real 1D signal f by taking the Fourier transform F of f, suppressing the negative frequencies and multiplying the positive frequencies by two. Applying this procedure, we do not lose any information about f because of the Hermite symmetry of the spectrum of a real function. The formal definition of the analytic signal is as follows:

Definition 1. *Let f be a real 1D signal. Its analytic signal is then given by*

$$f_A(x) = f(x) + if_{\mathcal{H}i}(x) = f(x) * \left(\delta(x) + \frac{i}{\pi x} \right) , \qquad (1)$$

where the Hilbert transform of f is defined as $f_{\mathcal{H}i} = f * (1/(\pi x))$ and $*$ denotes the convolution operation.

In the frequency domain this definition reads:

$$F_A(u) = F(u)(1 + \text{sign}(u)) \quad \text{with} \quad \text{sign}(u) = \begin{cases} 1 & \text{if } u > 0 \\ 0 & \text{if } u = 0 \\ -1 & \text{if } u < 0 \end{cases}$$

$$= F(u) + iF_{\mathcal{H}i}(u) . \qquad (2)$$

As an example we give the analytic signal of $f(x) = \cos(\omega x)$ which is $\cos_A(x) = \cos(x) + i \sin(x) = \exp(i\omega x)$. Thus, *cos* and *sin* constitute a *Hilbert pair*, i.e. one is the Hilbert transform of the other. The effect of the Hilbert transform is, that it shifts each frequency component of frequency $u = 1/\lambda$ by $\lambda/4$ to the right. We state three properties of the 1D analytic signal.

1. The spectrum of an analytic signal is causal $(F_A(u) = 0$ for $u < 0)$.
2. The original signal is reconstructible from its analytic signal, particularly, the real part of the analytic signal is equal to the original signal.
3. The envelope of a real signal is given by the magnitude of its analytic signal which is called the *instantaneous amplitude* of f.

While the first property is a construction rule which has not necessarily to be extended to 2D, we expect an extension of the analytic signal to fulfill the last two properties: The second one guarantees that two different signals can never have the same analytic signal, while the third one is the property of the analytic signal which is mainly used in applications.

3 Approaches to an Analytic Signal in 2D

In this section we will mention some of the extensions of the analytic signal to two-dimensional signals which have occurred in the literature . All of these have a straightforward extension to n-dimensional signals. We will use the notation $\boldsymbol{x} = (x, y)$ and $\boldsymbol{u} = (u, v)$.

The first definition is based on the 2D Hilbert transform [9] which is given by

$$f_{\mathcal{H}i}(\boldsymbol{x}) = f(\boldsymbol{x}) * * \left(\frac{1}{\pi^2 xy}\right) \; , \tag{3}$$

where $* *$ denotes the 2D convolution. In analogy to 1D an extension of the analytic signal can be defined as follows:

Definition 2. *The analytic signal of a real 2D signal f is defined as*

$$f_A(\boldsymbol{x}) = f(\boldsymbol{x}) + i f_{\mathcal{H}i}(\boldsymbol{x}) \; ,$$

where $f_{\mathcal{H}i}$ is given by (3).

In the frequency domain this definition reads

$$F_A(\boldsymbol{u}) = F(\boldsymbol{u})(1 - i\operatorname{sign}(u)\operatorname{sign}(v)) \; .$$

The spectrum of f_A according to Def. 2 is shown in Fig. 1. It does not vanish anywhere such that property 1 from Sect. 2 is not satisfied by this definition. A common approach to overcome this fact can be found e.g. in [7]. This definition starts with the construction in the frequency domain. While in 1D the analytic signal is achieved by suppressing the negative frequencies, in 2D one half-plane of the frequency domain must be set to zero. It is not immediately clear how negative frequencies can be defined in 2D. However, it is possible to introduce a direction of reference defined by the unit vector $\hat{e} = (\cos(\theta), \sin(\theta))$. A frequency $\boldsymbol{u}$ with $\hat{e} \cdot \boldsymbol{u} > 0$ is called positive while a frequency with $\hat{e} \cdot \boldsymbol{u} < 0$ is called negative. The 2D analytic signal can then be defined in the frequency domain.

Definition 3. *Let f be a real 2D signal and F its Fourier transform. The Fourier transform of the analytic signal is defined by:*

$$F_A(\boldsymbol{u}) = \begin{cases} 2F(\boldsymbol{u}) & \text{if } \boldsymbol{u} \cdot \hat{e} > 0 \\ F(\boldsymbol{u}) & \text{if } \boldsymbol{u} \cdot \hat{e} = 0 \\ 0 & \text{if } \boldsymbol{u} \cdot \hat{e} < 0 \end{cases} = F(\boldsymbol{u})(1 + \operatorname{sign}(\boldsymbol{u} \cdot \hat{e})) \; . \tag{4}$$

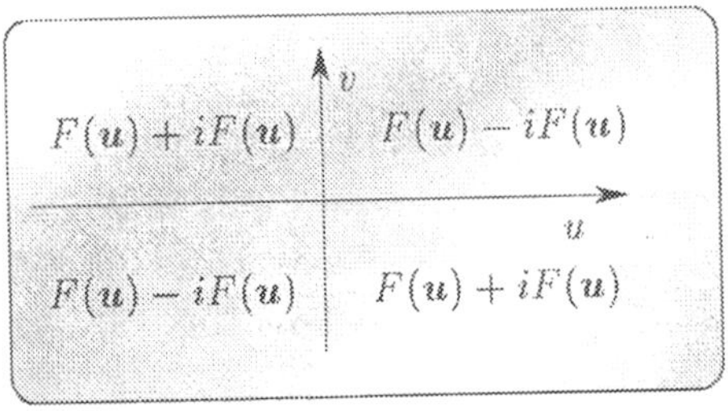

$$F(u) + iF(u) \qquad F(u) - iF(u)$$

$$v$$

$$u$$

$$F(u) - iF(u) \qquad F(u) + iF(u)$$

Fig. 1. The spectrum of the analytic signal according to Def. 2.

Please note the similarity of this definition with (2). In the spatial domain (4) reads

$$f_A(\boldsymbol{x}) = f(\boldsymbol{x}) ** \left(\delta(\boldsymbol{x} \cdot \hat{e}) + \frac{i}{\pi \boldsymbol{x} \cdot \hat{e}} \right) \delta(\boldsymbol{x} \cdot \hat{e}_\perp) \ . \tag{5}$$

The vector $\hat{e}_\perp$ is a unit vector which is orthogonal to $\hat{e} : \ \hat{e} \cdot \hat{e}_\perp = 0$.

According to this definition the analytic signal is calculated line-wise along the direction of reference. The lines are processed independently. Hence, Def. 3 is intrinsically 1D, such that it is no satisfactory extension of the analytic signal to 2D. Another definition of the 2D analytic signal was introduced by Hahn [8][1].

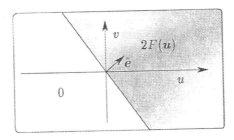

Fig. 2. The spectrum of the analytic signal according to Def. 3.

Definition 4. *The 2D analytic signal is defined by*

$$f_A(\boldsymbol{x}) = f(\boldsymbol{x}) ** \left(\delta(\boldsymbol{x}) + \frac{i}{\pi x} \right) \left(\delta(y) + \frac{i}{\pi y} \right) \tag{6}$$

$$= f(\boldsymbol{x}) - f_{\mathcal{H}i}(\boldsymbol{x}) + i(f_{\mathcal{H}i_1}(\boldsymbol{x}) + f_{\mathcal{H}i_2}(\boldsymbol{x})) \ , \tag{7}$$

where $f_{\mathcal{H}i}$ is the Hilbert transform according to Def. 2 and $f_{\mathcal{H}i_1}$ and $f_{\mathcal{H}i_2}$ are the so called partial Hilbert transforms, which are the Hilbert transforms according to Def. 3 with $\hat{e}^\mathsf{T} = (1,0)$ and $\hat{e}^\mathsf{T} = (0,1)$, respectively:

$$f_{\mathcal{H}i} = f ** \frac{1}{\pi^2 xy}, \quad f_{\mathcal{H}i_1} = f ** \frac{\delta(y)}{\pi x} \quad \text{and} \quad f_{\mathcal{H}i_2} = f ** \frac{\delta(x)}{\pi y} \ . \tag{8}$$

---

[1] Hahn avoids the term "analytic signal" and uses "complex signal" instead.

The meaning of Def. 4 becomes clearer in the frequency domain: Only the frequency components with $u > 0$ and $v > 0$ are kept, while the components in the three other quadrants are suppressed (see Fig. 3):

$$F_A(\boldsymbol{u}) = F(\boldsymbol{u})(1 + \text{sign}(u))(1 + \text{sign}(v)) \ .$$

A main problem of Def. 4 is the fact that the original signal is not reconstructible

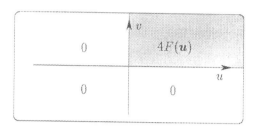

Fig. 3. The spectrum of the analytic signal according to Hahn [8] (Definition 4).

from the analytic signal, since due to the Hermite symmetry only one half-plane of the frequency representation of a real signal is redundant. For this reason Hahn proposes to calculate not only the analytic signal with the spectrum in the upper right quadrant but also another analytic signal with its spectrum in the upper left quadrant. It can be shown that these two analytic signals together contain all the information of the original signal. Thus, the complete analytic signal consists of two real parts and two imaginary parts or, in polar representation, of two amplitude- and two phase-components which makes the interpretation, especially of the amplitude, difficult.

4 The Quaternionic Fourier Transform

Since our definition of the analytic signal is based on the quaternionic Fourier transform (QFT), we will briefly review this transform. The QFT was recently introduced in [2, 3, 1] and [5], independently. The QFT of a 2D signal $f(\boldsymbol{x})$ is defined as

$$F^q(\boldsymbol{u}) = \int\limits_{-\infty}^{\infty} \int\limits_{-\infty}^{\infty} e^{-i2\pi ux} f(\boldsymbol{x}) e^{-j2\pi vy} d^2\boldsymbol{x} \ , \tag{9}$$

where i and j are elements of the algebra of quaternions $\mathbb{H} = \{q = a + bi + cj + dk \,|\, a, b, c, d \in \mathbb{R}, i^2 = j^2 = -1, ij = -ji = k\}$. Note that the quaternionic multiplication is not commutative. The magnitude of a quaternion $q = a + bi + cj + dk$ is defined as $|q| = \sqrt{qq^*}$ where $q^* = a - bi - cj - dk$ is called the conjugate of q.

The 1D Fourier transform separates the symmetric and the antisymmetric part of a real signal by transforming them into a real and an imaginary part, respectively. In real 2D a signal splits into four symmetry parts (symmetric and antisymmetric with respect to each argument). These four symmetry components are decoupled by the QFT and mapped to the four algebraic components of the quaternions [3].

The phase concept can be generalized using the QFT: In 1D the phase of a frequency component is represented by one real number. In the 2D quaternionic frequency domain a triple of real numbers can be defined, which can be regarded as the generalized phase in 2D [4, 1].

The operation of conjugation in $\mathbb{C}$ is a so-called algebra involution, i.e. it fulfills the two following properties: Let $z, w \in \mathbb{C} \Rightarrow (z^*)^* = z$ and $(wz)^* = w^* z^*$. In $\mathbb{H}$ there are three nontrivial algebra involutions:

$$\alpha : q \mapsto -iqi, \quad \alpha(q) = a + bi - cj - dk,$$
$$\beta : q \mapsto -jqj, \quad \beta(q) = a - bi + cj - dk \quad \text{and}$$
$$\gamma : q \mapsto -kqk, \quad \gamma(q) = a - bi - cj + dk \ .$$

Using these involutions we can extend the definition of Hermite symmetry: A function $f : \mathbb{R}^2 \to \mathbb{H}$ is called quaternionic Hermitian if:

$$f(-x, y) = \beta(f(x, y)) \quad \text{and} \quad f(x, -y) = \alpha(f(x, y)) \ , \tag{10}$$

for each $(x, y) \in \mathbb{R}^2$. The QFT of a real 2D signal is quaternionic Hermitian!

5   The Quaternionic 2D Analytic Signal

Using the QFT we can follow the arguments of Hahn [8] and keep only one of the four quadrants of the frequency domain. Since the QFT of a real signal is quaternionic Hermitian (see Sect. 4) we do not lose any information about the signal in this case (see Fig. 4).

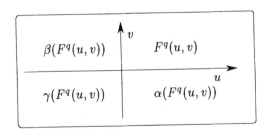

Fig. 4. The quaternionic spectrum of a real signal can be reconstructed from only one quadrant.

Definition 5. *In the frequency domain we define the quaternionic analytic signal of a real signal as*

$$F_A^q(\boldsymbol{u}) = (1 + sign(u))(1 + sign(v)) F^q(\boldsymbol{u}) \ ,$$

where F^q is the QFT of the real two-dimensional signal $f(\boldsymbol{x})$ and F_A^q denotes the QFT of the quaternionic analytic signal of f.

Definition 5 can be expressed in the spatial domain as follows:

$$f_A^q(\boldsymbol{x}) = f(\boldsymbol{x}) + \boldsymbol{n} \cdot \boldsymbol{f}_{\mathcal{H}i}(\boldsymbol{x}) \ , \tag{11}$$

where $\boldsymbol{n} = (i, j, k)^{\top}$ and $\boldsymbol{f}_{\mathcal{H}i}$ is a vector which consists of the partial and the total Hilbert transforms of f according to (8):

$$\boldsymbol{f}_{\mathcal{H}i}(\boldsymbol{x}) = (f_{\mathcal{H}i_1}, \ f_{\mathcal{H}i_2}, \ f_{\mathcal{H}i})^{\top} \ . \tag{12}$$

Note that, formally, (11) resembles (1).

6 Comparison of the Different Approaches

The 2D analytic signal according to all approaches is most easily constructed in the frequency domain. While the first approach Def. 2 does not suppress any parts of the frequency domain, all other approaches do have this analogy to the 1D case. According to Def. 3 one half of the frequency domain is suppressed. According to Def. 4 and Def. 5 only one quarter of the frequency domain is kept while the rest is suppressed. The difference between the last two definitions is that Def. 4 uses the complex frequency domain while Def. 5 uses the quaternion-valued frequency domain of the QFT.

The requirement that the original signal be reconstructible from the analytic signal is fulfilled is by all approaches except for Def. 4. The main requirement is that the magnitude of the analytic signal (the instantaneous amplitude of the original signal) should be the envelope of the original oscillating signal. We demonstrate the instantaneous amplitude according to the four definitions of the 2D analytic signal on an example image containing oscillations in all directions of the image plane. In Fig. 5 we show the magnitudes of the different analytic

Fig. 5. The instantaneous amplitude according to the different definitions of the analytic signal. From left to right: Original image, real envelope, inst. amplitude according to Def. 2, Def. 3 (oriented along the x-axis), Def. 4, and Def. 5, respectively.

signals of a test-images containing oscillations in all orientations of the image plane. Definition 3 is applied with $\hat{e} = (1, 0)$. Obviously Def. 2 is not successful in providing the envelope of the test-image. The quality of Def. 3 depends strongly on the orientations of the local structure. If it is orthogonal to the chosen $\hat{e}$ the envelope is constructed well, while the Def. 3 fails as soon as the local structure is parallel to $\hat{e}$. Definition 4 looses information about the signal and so yields the envelope only for structures that correspond to frequencies in the upper right quadrant. Also Def. 5 does not yield the perfect envelope. However, compared to the other approaches this one is the most satisfactory.

7 Conclusion

The quaternionic analytic signal combines in itself the earlier approaches in a natural way: The vector $f_{\mathcal{H}i}$ in (12) is not constructed artificially. The quaternionic analytic signal is rather defined in a simple way in the frequency domain and everything falls in place automatically. In our opinion the successful definition of the analytic signal in 2D, which can only be obtained using the QFT, shows also that the QFT is a reasonable definition of a 2D harmonic transform.

As mentioned in the introduction the analytic signal is the theoretical foundation of Gabor filter techniques. These techniques are widely used in image processing. Based on the quaternionic analytic signal introduced here it is possible to define quaternion-valued Gabor filters instead of complex-valued ones. These filters have already been successfully applied in image processing. Based on quaternion-valued image representations the concept of the local phase has been extended and used for local structure analysis as well as for disparity estimation [4, 1].

References

1. Th. Bülow. *Hypercomplex Spectral Signal Representations for Image Processing and Analysis*. PhD thesis, University of Kiel, 1999. to appear.
2. Th. Bülow and G. Sommer. Das Konzept einer zweidimensionalen Phase unter Verwendung einer algebraisch erweiterten Signalrepräsentation. In *E. Paulus, F. M. Wahl (Eds.), 19. DAGM Symposium Mustererkennung, Braunschweig*, pages 351–358. Springer-Verlag, 1997.
3. Th. Bülow and G. Sommer. Multi-dimensional signal processing using an algebraically extended signal representation. In *G. Sommer, J.J. Koenderink (Eds.), Proc. Int. Workshop on Algebraic Frames for the Perception-Action Cycle, Kiel, LNCS vol. 1315*, pages 148–163. Springer-Verlag, 1997.
4. Th. Bülow and G. Sommer. Quaternionic Gabor filters for local structure classification. In *14th International Conference on Pattern Recognition, August 17-20 1998, Brisbane, Australia*, 1998. 808-810.
5. T.A. Ell. Quaternion Fourier transforms for analysis of 2-dimensional linear time-invariant partial-differential systems. In *Proc. 32nd IEEE Conf. on Decision and Control, San Antonio, TX, USA, 15-17 Dec.*, pages 1830–1841, 1993.
6. D. Gabor. Theory of communication. *Journal of the IEE*, 93:429–457, 1946.
7. G.H.Granlund and H. Knutsson. *Signal Processing for Computer Vision*. Kluwer Academic Publishers, 1995.
8. S. L. Hahn. Multidimensional complex signals with single-orthant spectra. *Proc. IEEE*, 80(8):1287–1300, 1992.
9. H. Stark. An extension of the Hilbert transform product theorem. *Proc. IEEE*, 59:1359–1360, 1971.

Shift Detection by Restoration

Herbert Suesse, Klaus Voss, Wolfgang Ortmann, and Torsten Baumbach

Friedrich-Schiller-University Jena, Department of Computer Science
Ernst-Abbe-Platz 1-4
D-07743 Jena, Germany
{nbs,nkv,noo,nbu}@uni-jena.de
http://pandora.inf.uni-jena.de

Abstract. In this paper an approach is presented for robust shift detection of two given images. The new unifying idea is that we determine
a shifted delta impulse using some well-known restoration techniques,
e.g. the Wiener filtering, constraint restoration, entropy restoration, and
Baysian restoration. The used restoration techniques imply the robustness of the presented method. Our approach is a generalization of the
matched filtering approach. Additionally, we describe in the paper the
problem of calculating an evaluation measure of the restored delta impulse image. This measure is the basis for the uncertainty of the detected shift. The unifying approach of shift detection by restoration
(SDR-method) could be tested successfully, for example using a series
of fundus image pairs which are of practical interest and which contain
also small rotations, scalings and even deformations.

1 Introduction

For the sake of simplicity all formulas refer to discrete functions of a single
variable. Let be given two images f and f'. Let be δ the delta impuls and $\mathbf{S}_d$
the shift operator with a shift d. Then follows

$$f' = \mathbf{S}_d f = \mathbf{S}_d f * \delta = f * \mathbf{S}_d \delta = f * \delta_d \ . \tag{1}$$

Thus, the cyclic shift $x \to x-d$ is reduced to the convolution of f with the shifted
delta impulse. If we apply to (1) the convolution theorem, then we receive the
shift theorem ($\alpha_k(f)$ are the discrete Fourier coefficients of f):

$$\alpha_k(f') = \alpha_k(f) \cdot e^{2\pi i \frac{kd}{N}} \ . \tag{2}$$

The shift theorem is the basis of all phase-based techniques, see ([3, 4, 14,
15]). But because the shift theorem works only with one frequency k, the integrating effect of the deconvolution (i.e. taking into account many frequencies) is
lost. And nothing is changed substantially if one uses different resolution levels
and operates gradually and hierarchically.

The local Fourier transform with different window functions uses only the
space-dependence of the problem. As with the block matching to a given block

a reference block is then looked up, which possesses the same phase of only one Fourier coefficient. Weng shows in [14] that Gabor filters are even more problematic than the use of the local Fourier transform with rectangle windows, since Gabor filters do not even guarantee "quasi-linearity " of the phase of one resolution level. However as with the block matching a search strategy must be executed, often root finding techniques are used in the case of small disparities.

The robustness of cepstrum techniques is already well-known for a long time, since they use actually also the convolution, see e.g. [6, 7]). Contrary to (1) one uses the equation

$$h = f + \mathbf{S}_d f = f * (\delta + \mathbf{S}_d \delta) . \tag{3}$$

But this method is only then meaningful if only the overlay image h is given. Are however as in most cases f and f' well-known images, then we lose information, which expresses itself in the following disadvantages of the cepstrum method

- There is always a "peak" in the origin, so that small shifts cannot be detected surely,
- The actual "peaks" are symmetrical with respect to the origin, therefore only relative shifts can be detected,
- The "peaks" overlay the cepstrums of f and $\mathbf{S}_d f$ respectively, which we do not know, therefore generally the "peak" detection can be difficult.

In the following a new method is to be stated, which not only uses the integrating effect of the deconvolution of model (1), but beyond that the image intensity disturbances and noise are also added to the model. First results are given in [13]. It seems in such a way that our approach is similar to the classical matched filtering random signal processing technique, see [11]. However, we don't maximize the special correlation signal-to-noise ratio, our unifying approach is the restoration of the shifted delta impuls. Determining the image shift by any restoration method we call our approach "shift detection by restoration" (SDR-approach).

2 Shift Detection by Restoration (SDR Approach)

If we regard the model (1), then the similarities are noticeable to the field of image restoration, whereby the restoration problem is limited at first only to the pure deconvolution. Therefore, the basic idea is that we interpret f' as the given image, f we interpret as a given "point spread function" (PSF) and the shifted delta impulse δ_d is the original image which has to be restored. As in the image restoration and the matched filtering approach, noise occurs also here, which is strengthened by a pure deconvolution (e.g. by pseudoinverse filtering). Therefore we must consider in the model also the problem of noise suppression.

Let two images f and f' be given that the model (1) is correct, then we presuppose the cyclic shift. But the cyclic shift is for practical images not satisfied. Some structures disappear from the image and other structures emerge. Often one tries to correct these "wrap around effects" or "boundary effects" with window functions. We want to add these effects as additive disturbances to the model. Therefore, we extend the model (1) with the general random field N to

$$f' = f * \delta_d + N \ . \tag{4}$$

Therefore the field N contains noise, objectively caused grey level modifications and the wrap around effects. Thus N is neither signal independent nor uncorrelated and it is strongly space-dependent due to the wrap around effects. This implies that N depends itself from the shift d.

Phase-based techniques as well as the normalized coefficient of correlation are invariant against linear contrast-and brightness modifications. Therefore we have to add this linear grey level modification $g' = a \cdot g + b$ to the model:

$$f' = f * (a \cdot S_d \delta) + b + N \ . \tag{5}$$

The contrast factor a causes only an upsetting or a scaling of the delta impulse and the offset b influences only the zeroth Fourier coefficient α_0. Therefore we can use without loss of generality equation (4) and this equation forms the basis for the shift estimation for a pair of images. Thus, we can use any restoration technique to restore the shifted delta impulse.

Now first of all, we elect a minimum mean square error (MMSE) estimation technique using the Wiener filter. We have to solve the problem

$$E\left(||\delta_d - f' * w||^2\right) \to minimum \tag{6}$$

for the unknown LSI-filter w. We remember that the LSE-estimation x

$$||g - f * x||^2 \to minimum \tag{7}$$

of the linear convolution equation $f * x = g$, whereby f and g are given, satisfies the convolution equation

$$f^r * f * x = f^r * g \ , \tag{8}$$

f^r is the reflected function of f. We use this knowledge to solve our MMSE problem. From this equation we obtain

$$E\left\{(f * \delta_d + N)^r * (f * \delta_d + N)\right\} * w = E\left\{(f * \delta_d + N)^r\right\} * \delta_d \ . \tag{9}$$

In contrast to the classical Wiener approach we have shift dependent noise $N(d)$ because the random field N depends on the shift d. But we can assume that the shift d affects only the variance and not the expectation value of the random field $N(d)$. Therefore, an assumption is in the following only the restriction $E\{N(d)\} = 0$ for all d. This assumption implies that some terms vanish, and we can derive the convolution equation

$$[f^r * f + E\{N^r * N\}] * w * f' = f^r * f' \ . \tag{10}$$

In the frequency domain then we get with $g = w * f'$

$$\alpha_k(g) = \frac{\overline{\alpha_k(f)} \cdot \alpha_k(f')}{|\alpha_k(f)|^2 + E\{|\alpha_k(N)|^2\}} \ . \tag{11}$$

For the MMSE-solution we need the spectrum of the autocorrelation of noise. If we assume white noise then we have the solution

$$\alpha_k(g) = \frac{\overline{\alpha_k(f)} \cdot \alpha_k(f')}{|\alpha_k(f)|^2 + \beta} \ . \tag{12}$$

The assumption that the noise is white, which is reasonable over the matched image regions, clearly does not hold on the unmatched part. We have received with the Wiener filtering the same expressions (11) and (12) as in the matched filtering approach [11].

To compare the Wiener solution with other restoration techniques we consider for example the constraint restoration technique. For the random field N we only assume that the "mean value" $E(||N||^2)$ is given. Then we form the following optimization problem

$$||\mathbf{L}g||^2 \ \rightarrow \ minimum \ subject \ to \ ||f' - f * g||^2 = E(||N||^2) \ . \tag{13}$$

Here, g is the solution of the problem as an approximation for δ_d. The operator $\mathbf{L}$ is a LSI operator which can be given in such a manner that we select the most favorable solution from all admissible solutions for our problem. Due to the LSI characteristic we can assign to $\mathbf{L}$ a discrete function l. Then the general solution of (13) is:

$$[f^r * f + \beta \cdot l^r * l] * g = f^r * f' \ . \tag{14}$$

Here, β is the regularization parameter which is to be determined as a function of the (unknown) random field N. We solve the convolution equation (14) once more in the frequency domain:

$$\alpha_k(g) = \frac{\overline{\alpha_k(f)} \cdot \alpha_k(f')}{|\alpha_k(f)|^2 + \beta|\alpha_k(l)|^2} \ . \tag{15}$$

The equation contains still two unknown parameters: The regularization parameter β has to be estimated dependently of the given practical problem. Furthermore, we have to specify the operator $\mathbf{L}$ resp. the function l or the Fourier coefficients of l.

a) Firstly, we choose β and $\alpha_k(l)$ in (15) or $E\{|\alpha_k(N)|^2\}$ in (11) in such a way that $\sum (constant - |\alpha_k(g)|)^2 \rightarrow minimum$ using the a priori knowledge $|\alpha_k(\delta_d)| = constant$. Thus, we obtain directly from (15) or (11)

$$\alpha_k(g) = \frac{\overline{\alpha_k(f)} \cdot \alpha_k(f')}{|\alpha_k(f)| \cdot |\alpha_k(f')|} \quad , if \ |\alpha_k(f')| > |\alpha_k(f)|$$

$$\alpha_k(g) = \frac{\alpha_k(f')}{\alpha_k(f)} \ , \qquad\qquad if \ |\alpha_k(f')| \le |\alpha_k(f)| \tag{16}$$

The first part of the expression (16) was mentioned in the literature by Kuglin and Hines [1975] in [5] and is called *phase correlation*. The phase correlation is a special case in our constraint restoration method with a LSI-operator determined by a priori knowledge that we have only one delta-impuls. Additionally, the phase correlation is a special case in the Wiener filtering method using a special chosen autocorrelation function of noise. Additionally, if we have more than one delta-impulse then the choice of the LSI-operator l and the autocorrelation function is not correct.

b) Secondly, the operator $\mathbf{L}$ can be defined in such a way that the norm of $\mathbf{L}g$ is approximately the same as the norm of the shifted δ-impuls. The δ-impulse itself has a minimal norm, therefore we choose the identity operator $l = \delta$, i.e. we receive with $\alpha_k(\delta) = constant$ the expression

$$\alpha_k(g) = \frac{\overline{\alpha_k(f)} \cdot \alpha_k(f')}{|\alpha_k(f)|^2 + \beta} . \tag{17}$$

Surprisingly, this is the same expression as in the Wiener filtering (12) and the matched filtering approach, but the main difference is the interpretation of the parameter β. This parameter is now a regularization parameter β and has nothing to do with the case of white noise. Surprisingly, the expression (17) is not symmetric, i.e. g' with

$$\alpha_k(g') = \frac{\overline{\alpha_k(f')} \cdot \alpha_k(f)}{|\alpha_k(f')|^2 + \beta} \tag{18}$$

is not the reflected image of g. One possibility is the symmetrizing of the expression (17), but then we have a loss of information.

In Fig 1 the restoration results for different values of the regularization parameter are represented. The smaller the overlapping area of two blocks, the stronger is the mean value of the random field N. The regularization parameter therefore depends on the shift. The regularization parameter β expresses for our model the "distance" from g to the shifted delta impulse. For $\beta \to 0$ we obtain from (17) the "pure deconvolution" and for $\beta \to \infty$ we get the "pure cross correlation" between f' and f. It can be seen in Fig. 1 that the correct shift is obtained with a "middle" β, even in the case of drastic pertubations without using window functions. Since a motion compensation algorithm runs often hierarchically, the parameter β can be decreased monotonously from hierarchy level to hierarchy level.

c) Thirdly, we can use nonlinear techniques to restore the shifted delta impulse. For example, it can be used the entropy restoration technique:

$$-\sum \frac{g_i}{\sum g_i} ln \left(\frac{g_i}{\sum g_i} \right) \to minimum \ \ subject \ to \ ||f' - f * g||^2 = E\left\{ ||N||^2 \right\} \tag{19}$$

The solution is given by a transcendental equation which is to be solved in dependence of the regularization parameter.

In the next step we would have to find an objective measure which evaluates the restored image in comparison to the ideal delta impulse. Additionally, this objective measure can be used to find the best regularization parameter β during an optimization process.

3 Peak Detection and Evaluation

The idea now is to evaluate the restored delta implulse image (using the cross correlation this image is called correlation surface). The derived quality measure can be used for the optimization process to select the optimal β. The restored δ-impulses approximate the ideal δ-impulse more or less. The main peak is often asymmetrical and flat. Furthermore more or less significant subpeaks occur (see also Fig. 1). The determination of the position(s) of the peak(s) cannot take place therefore with the well-known methods of the cluster analysis, since these relatively well presuppose separable peaks. We can manage however differently. In the case of a "good" approximation of the δ-impulse the position of the global maximum corresponds to our looked up shift.

Now a probability or a quality measure is to be only determined as a function of features of this maximum and the other image areas, which says about it, how safe this certain shift is. In other words, a measure is to be found that describes the "distance" of the restored image from the ideal δ-impulse. Simple measures, like the height of the peak (see also [12]), are however unsuitable for it. Thus a simple contrast modification already influences the peak height.

We have to take into account the following considerations:

a) As long as the quantization effects do not produce large grey level plateaus, a reliable positioning is both with flat, and with pointed peaks possible. Therefore the expansion of a peak area alone has no relevance for the quality measure.

b) The reliability of the peak detection is strongly worsened with a flat peak by noise, against what it is preserved with a pointed peak.

c) If the value of the largest maximum differs only around a small (of the noise dependent) amount from other local maxima in the picture, then a high uncertainty exists concerning the correctness of the found peak position. If only very small maxima exist apart from the largest maximum, then a high agreement with our model exists.

From this catalog of demands one cannot derive an accurate measure e.g. that the entropy, but only an algorithmic description of the measure. For the algorithmic description and numerical experiments, see Baumbach in [1].

4 Experimental Investigations

In the following Fig. 1 one can see the extreme robustness of this method. Additionally, in the Fig. 1 is displayed the optimization process to determine the correct regularization parameter β using the developed evaluation measure. In the most cases of practical applications a "middle" parameter β is sufficient to determine the correct shift. For extreme situations it is helpful to know both restored images g (17) and g' (18). Is the shift symmetric then we assume a correct shift. If this is not the case then we choose the shift with the best evaluation measure.

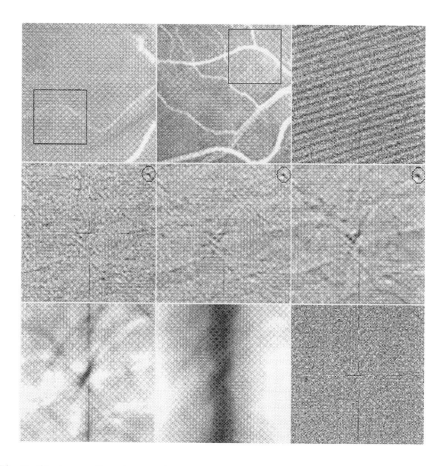

Fig. 1. First row: Two images (an extreme situation with a shift, 5 degree rotation and grey level disturbances, note that corresponding structures are to find only in the marked frames) and the restoration by "pure deconvolution". Second row: Restored images with encircled delta impulses for $\beta = 10^2$, the optimal $\beta = 10^3$, and $\beta = 10^4$. Third row: Restored images for $\beta = 10^6$, the "pure cross correlation", and last not least the estimator (16) called a modified "phase correlation".

5 Rotations and Scalings

Rotations and scalings can be detected using the log-polar-representation of the translation-invariant power spectras of two images (see for example [2][8][9][10]).

The practical difficulty is situated however in the fact that one must have an extremely robust method for shift detection since the log-polar spectra indicate various disturbances. The SDR-approach could be tested successfully in a series of fundus images to detect the rotations and scalings.

References

1. Baumbach T.: Bewegungskompensation von Fundus-Bildern. Technical Report FSU Jena and Carl Zeiss Jena, Jena 1998.
2. De Castro E., Morandi C.: Registration of translated and rotated images using finite Fourier transform. IEEE Trans. PAMI 9 (1987) 700-703
3. Fleet D.J., Jepson A.D.: Computation of component image velocity from local phase information. IJCV 5 (1990) 77-104
4. Fleet D.J., Jepson A.D., Jenkin M.R.M.: Phase-based disparity measurement. CVGIP Image Understanding 53 (1991) 198-210
5. Kuglin C.D., Hines D.C.: The phase correlation image aligment method. Proc. IEEE on Cybernetics and Society, New York 1975, pp. 163-165
6. Lee D.J., Mitra S., Krile T.F.: Analysis of sequential complex images, using feature extraction and two-dimensional cepstrum techniques. JOSA A-6 (1989) 863-870
7. Lehmann T., Goerke C., Schmitt W., Repges R.: Rotations- und Translationsbestimmung durch eine erweiterte Kepstrum-Technik. Proceedings 17. DAGM-Symposium Bielefeld 1995. Springer 1995, S. 395-402
8. Marcel B., Briot M., Murrieta R.: Calcul de translation et rotation par la transformation Fourier. Traitement du Signal 14 (1997), No.2
9. Messner E.R.A., Szu H.H.: An image processing architecture for real time generation of scale and rotation invariant patterns. CVGIP 31 (1985)
10. Murrieta R., Briot M., Marcel B., Gonzales H.: Aspectos dinámicos de la visión: Seguimiento de objetos no rígidos y estimación de la rotación de una cámara. Memorias Visión Robótica, Primer Encuentro de Computación ENVC'97, Querétaro/México, Sept. 1997, pag. 144-152
11. Pratt W.: Digital Image Processing. John Wiley, New York 1978
12. Vlachos T., Thomas G.: Motion estimation for the correction of twin-lens telecine flicker. Proc. ICIP 1996, pp. 109-112
13. Voss K., Ortmann W., Süße H.: Bildmatching und Bewegungskompensation bei Fundus-Bildern. Proc. 20. DAGM-Symposium,439-446, Stuttgart 1998, Germany
14. Weng J.J.: Image matching using the windowed Fourier phase. IJCV 11 (1993) 211-236
15. Xiong Y., Shafer S.A.: Hypergeometric filters for optical flow and affine matching. IJCV 24 (1997) 163-177

Edge Preserving Probabilistic Smoothing Algorithm

Bogdan Smolka   and   Konrad W. Wojciechowski

Dept. of Automatics Electronics and Computer Science
Silesian University of Technology
Akademicka 16 Str, 44-101 Gliwice, Poland
bsmolka@peach.ia.polsl.gliwice.pl

Abstract In the presented paper a new probabilistic approach to the problem of noise reduction has been presented. It is based on the concept of a virtual particle performing a random walk on the image lattice, with transition probabilities derived from the median distribution.
The probabilistic smoothing algorithm combined with a cooling procedure, known from the simulated annealing optimization method, constitutes a new powerful technique of noise suppression, capable of preserving edges and other image features.

1 Introduction

The reduction of image noise without the degradation of the image structure is one of the major problems of the low-level image processing. A whole variety of algorithms has been developed, but none of them can be seen as a final solution of the noise problem.

In this paper, a new probabilistic approach to the problem of noise reduction has been presented. It is based on the concept of a virtual particle performing a random walk on the image lattice, with transition probabilities derived from the median distribution.

This work is divided into three parts. In the first section, a brief overview of the smoothing operations is presented. The second part introduces the model of a virtual walking particle and shows how the probabilistic smoothing operator is constructed. The third part contains the results of noise attenuation achieved using some of the methods described in the paper, in comparison with the effects obtained with the new probabilistic method.

2 Short Overview of Noise Reduction Filters

The noise suppression of digital images has been a topic of considerable interest in the past decades, due to its importance in numerous applications in various fields of computer vision.

The most frequently used noise reducing transformations are the linear filters, which are based on the convolution of the image matrix with the filter kernel. This kind of filtering replaces the initial value $B(i,j)$ of the pixel (i,j) with the weighted mean of the gray scale values of its neighbours. The simplest linear filter is the moving average (box filter) and the Gauss filter. Linear filters are simple and fast, especially when they are separable, like the box and Gaussian filter, but their major drawback is that they cause blurring of the edges and strongly attenuate small details.

The blurring effect can be reduced using the median filter, which replaces the value of the centre pixel (i, j) by the median of the gray scale values of the pixels lying in a predefined filter window. The median is very effective in removing of the impulse noise, however its efficiency in suppressing the Gaussian or uniform distribution noise is rather poor, as the filter wipes out such important features as lines, corners and small details.

The blurring caused by smoothing can be also diminished choosing an appropriate adaptive filter kernel, which performs the averaging in a selected neighbourhood [1]. The term *adaptive* means, that the filter kernel coefficients change according to the image structure, which is to be smoothed and as a result the kernel becomes a function $H(i, j; k, l)$

$$J(i, j) = \frac{1}{Z} \sum_{(k,l) \in W(i,j)} H(i, j; k, l) \, B(k, l), \tag{1}$$

where $H(i, j; k, l)$ are weighting coefficients depending on the values of $B(i, j)$, $B(k, l)$ and $W(i, j)$ denotes the filter window with a centre in (i, j).

Different kinds of edge and structure preserving filter kernels has been proposed [2–6]. If $H(i, j; k, l) = 1 - |B(i, j) - B(k, l)|$ then [1, 7–10]

$$J(i, j) = \frac{1}{Z} \sum_{(k,l) \in W(i,j)} (1 - |B(i, j) - B(k, l)|) B(k, l) \,, \quad Z = \sum_{(k,l) \in W(i,j)} 1 - |B(i, j) - B(k, l)|, \tag{2}$$

where Z is the filter normalizing factor. A simple version of this filter assigns to the filter coefficients $H(i, j; k, l)$ from (2) the values [1]

$$H(i, j; k, l) = 1 \text{ if } |J(i, j) - B(k, l)| \leq \gamma \text{ and } 0 \text{ otherwise}, \tag{3}$$

where γ is a constant threshold or the standard deviation of the gray scale values of the neighbourhood pixels or of the whole image. This filter takes with greater weighting coefficients the pixels of the neighbourhood, whose gray tones are close to the intensity of the centre pixel. Another very effective procedure [11] is defined by

$$J(i, j) = \frac{1}{Z} \sum_{(k,l) \sim (i,j)} \exp \left\{ -\frac{r^2}{2\sigma^2} - \frac{[B(k, l) - B(i, j)]^2}{t^2} \right\} B(k, l), \tag{4}$$

where $\sim$ denotes the neighbourhood relation, $r = \sqrt{(k - i)^2 + (l - j)^2}$, σ controls the scale of spatial smoothing and t is a brightness threshold.

Similar properties has the gradient inverse weighted operator, which forms a weighted mean of the pixels belonging to a filter window. Again, the weighting coefficients depend on the difference of the gray scale values between the central pixel and its neighbours [8, 10]

$$T(i, j) = \frac{1}{Z} \sum_{(k,l) \in W(i,j)} \frac{1}{\max\{\frac{1}{2}, |B(k, l) - B(i, j)|\}} B(k, l), \tag{5}$$

The Lee's local statistics filter [9] estimates the local mean $\mu(i,j)$ and variance $\sigma^2(i,j)$ of the intensities of pixels belonging to the window $W(i,j)$ of the size $Z = (2n+1)(2m+1)$

$$\mu(i,j) = \frac{1}{Z} \sum_{(k,l)\in W(i,j)} B(k,l), \quad \sigma^2(i,j) = \frac{1}{Z} \sum_{(k,l)\in W(i,j)} [B^2(k,l) - \mu^2(k,l)], \quad (6)$$

and assigns to the pixel (i,j) the value $J(i,j)$

$$J(i,j) = B(i,j) + [1 - \alpha(i,j)]\mu(i,j), \quad \alpha(i,j) = \max\left\{0, \frac{\sigma^2(i,j) - \sigma_n^2}{\sigma^2(i,j)}\right\}, \quad (7)$$

where σ_n^2 is the estimate of the image noise. If $\sigma(i,j) \gg \sigma_n$ then $\alpha(i,j) = 1$ and $J(i,j) = B(i,j)$ and no change occurs, but for $\sigma(i,j) \ll \sigma_n$ then $\alpha(i,j) = 0$ and $J(i,j) = \mu(i,j)$. In this way, the filter smooths with a local mean when the noise is not very intensive and leaves the pixel value unchanged when a strong signal activity is detected. The major drawback of this filter, is that it leaves noise in the vicinity of edges, lines and other image details.

Good results of noise reduction can usually be obtained by performing the σ-filtering [12, 13, 4]. This procedure computes a weighted average over the filter window, but only those pixels, whose gray values do not deviate too much from the value of the center pixel are permitted into the averaging process

$$J(i,j) = \frac{1}{Z} \sum_{(k,l)\in W(i,j)} H(k,l)B(k,l), \; \{(k,l) : |B(k,l) - B(i,j)| \leq \sigma\}, \quad (8)$$

where Z is the normalizing factor, σ is the standard deviation of all pixels belonging to $W(i,j)$ or the value of the standard deviation estimated from the whole image and $H(k,l)$ values can be taken from the box or Gaussian filter.

The image noise can be also reduced applying a filter, which substitutes the gray scale value of a pixel, by a gray tone from the neighbourhood, which is closest to the average μ of all points in the filter window $W(i,j)$ [7, 2, 3].

$$T(i,j) = B(k,l), \quad (k,l) = \arg \min_{(k,l)\in W(i,j)} \{ |B(k,l) - \mu| \} , \quad (9)$$

An important role in the noise suppression play the rank transformations defined with the use of an ordering operator, the goal of which is the transformation of the set of pixels lying in a given filter window $W(i,j)$ of the size $n_1 \times n_2$ with the centre at (i,j) into a monotonically increasing sequence

$$\{B(i-k,j-l)\}, \, k = -n_1, \ldots, n_1, \, l = -n_2, \ldots, n_2\} \rightarrow \{B_1, \ldots, B_N\},$$
$$B_k \leq B_{k+1}, \, k = 1, 2, \ldots, N-1 , \quad (10)$$

where $N = (2n_1 + 1)(2n_2 + 1)$ is the number of pixels belonging to $W(i,j)$.

In this way the rank operator is defined on the ordered values from the set $\{B_1, B_2, \ldots, B_N\}$ and has the form $T(i,j) = \frac{1}{Z} \sum_{k=1}^{N} \varrho_k I_k$, where once again Z is the normalizing constant and ϱ_k are the weighting coefficients. Taking appropriate ranking coefficients allows the defining of a variety of useful operators [5, 3, 2].

3 Probabilistic Noise Reduction Algorithm

3.1 Probabilistic Smoothing Operator

Let (i, j) denote the initial position of a virtual particle performing a classical random walk [14] on the image lattice and let $(i, j)_n$ denote its position after n steps. If the probability, that the walking particle starting from the point (i, j) reaches in n steps the point (k, l) is denoted as $P[n, (i, j), (k, l)] = P[(i, j)_n = (k, l)]$, then the probabilistic smoothing transformation U can be defined as

$$U(i, j) = \sum_{(k,l) \sim (i,j)} P[1, (i, j), (k, l)]\, B(k, l) = \sum_{(k,l) \sim (i,j)} P_{ij,kl}\, B(k, l) \qquad (11)$$

where $\sim$ denotes the neighbourhood relation of the image points.

3.2 Edge Preserving Smoothing Operator

Let us now assume the 8-neighbourhood system and let the probability $P_{ij,kl}$ of a transition from (i, j) to (k, l) be derived from the median distribution.

$$P_{ij,kl} = \frac{\exp\{-\beta |B(i, j) - B(k, l)|\}}{Z}, \quad Z = \sum_{(m,n) \in W(i,j)} \exp\{-\beta |B(i, j) - B(m, n)|\} \quad (12)$$

is now the statistical sum, equivalent to the previously used normalizing factor and $\beta = T^{-1}$ is the reciprocal of the temperature of a statistical system.

The properties of a such defined operator depend strongly on the value of the temperature parameter β. At low β, a weighted average over the neighbourhood is taken. The increase of β causes, that the value of the central pixel (i, j) is taken with growing weighting coefficients. When $\beta \to \infty$ then U becomes an identity operator and the pixel value remains unchanged.

The experiments have shown, that if an appropriate value of β is chosen, then the smoothing operator (12) is very efficient at reducing the uniformly distributed noise. Difficulties arise however, when the image is corrupted with a strong Gaussian or impulse noise. In such a case, it is better not to allow the particle to stay at its current position, which means that it must jump to one of its neighbour. In this way

$$P_{ij,kl} = \frac{\exp\{-\beta |B(i, j) - B(k, l)|\}}{\sum_{(m,n) \sim (i,j)} \exp\{-\beta |B(i, j) - B(m, n)|\}}, \quad (i, j) \neq (k, l), \qquad (13)$$

For low β the weighted mean is assigned to the central pixel (i, j) (the central pixel is not involved in the averaging) and at high β, the pixel whose value is closest to the value of the central pixel is assigned to (i, j). In this way, the transformation (13) seems to be a tradeoff between the median, box and the agglomerating filter (9).

Unfortunately, the iterative application of this filter leads to a gradual image blurring, which is also a drawback of the smoothing operators defined by (1), (4) and (5).

Fig. 1. Test images (left) and images enhanced using the new method ($\beta(0) = 15$, $\kappa = 1.25$, 5 iterations).

To avoid the blurring effect, a "cooling schedule", similar to the one used in simulated annealing [15], has been proposed. The first smoothing of the noisy image with the probabilistic filter is performed at low β (high temperature in the pseudo - Gibbsian (median) distribution) and the successive iterations are being applied with increasing β (the system is being cooled down). The experiments were performed using the cooling schedule $\beta(n+1) = \kappa\,\beta(n)$ with experimentally chosen $\kappa = 1.25$ and $\beta(0) = 15$.

The iterations terminate, when the image becomes "frozen' 'and the next iterations does not change the image pixel values, which requires about 5 iterations. This procedure of increasing of the value of β prevents blurring and leads to an image with suppressed noise and preserved edges.

Figures 1, 2 and 3 show the results achieved using the introduced filtering procedure. The last two images depict the comparison with the results obtained using the algorithms described in the first part of this paper. The quality of noise suppression was evaluated using the root of the mean square error. The ability of preserving edges was tested using the standard Sobel edge detector.

4 Conclusions

The results obtained using the probabilistic approach to the image smoothing confirm its good properties. Only the new method, together with the method of Smith [11] and Lee [9] performed well, when applied to the test images corrupted by strong Gaussian noise. Although the rms value obtained when using the new method was higher than the respective value of Smith's algorithm, the application of the cooling procedure increased the sharpness of the image edges and therefore this algorithm can be applied as a preliminary step of edge detection.

References

1. **Nagao M.   Matsuyama T.** : Edge preserving smoothing, Computer Graphics & Image Processing, **9**, 394-407, 1979
2. **Klette R.   Zamperoni P.** : Handbuch der Operatoren für die Bildverarbeitung, Vieweg Verlag, Braunschweig, Wiesbaden, 1992
3. **Zamperoni P.** : Methoden der digitalen Bildsignalverarbeitung, Vieweg, Braunschweig, 1991
4. **Yaroslavsky L.   Murray E.** : Fundamentals of Digital Optics, Birkhäuser, Boston, 1996
5. **Pratt W.** : Digital Image Processing, New York, John Willey & Sons 1991
6. **Gonzalez R.C.   Woods R.E.** : Digital Image Processing, Reading MA, Addison-Wesley, 1992
7. **Scher A.   Dias V.   Rosenfeld F.R.** : Some new image smoothing techniques, IEEE Trans. on SMC, **10**, 153-158, March 1980
8. **Wang D.   Vagnucci A.H.   Li C.C.** : Gradient inverse smoothing scheme and the evaluation of its performance, Computer Graphics & Image Processing, **15**, 167-181, 1981
9. **Lee J.S.** : Digital image enhancement and noise filtering by use of local statistics, PAMI **2**, 165 -168, 1980
10. **Wang D.   Wang Q.** : A weighted averaging method for image smoothing, Proceedings of the 8th. ICPR, 981-983, Paris, 1988
11. **Smith S.M.   Brady J.M.** : SUSAN - a new approach to low level image processing, Int. Journal of Computer Vision, **23**, 1, 45 -78, 1997
12. **Lee J.S.** : Digital image smoothing and the sigma filter, Computer Vision Graphics and Image Processing, **24**, 255 - 269, 1983
13. **Lohmann G.** : Volumetric Image Analysis, John Wiley and Teubner, 1988
14. **Spitzer F.** : Principles of Random Walk, D. van Nostrand Company, Princeton, New Jersey, 1975
15. **van Laarhoven P.J.M.   Aarts E.A.** : Simulated Annealing: Theory and Application, Kluver Academic Publishers, Dordrecht, 1989

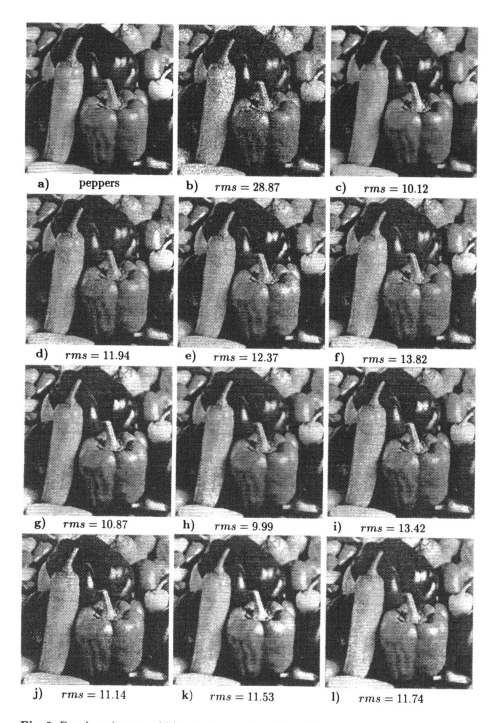

Fig. 2. Results: **a)** test and **b)** noisy image ($\sigma = 30$), **c)** image smoothed with the new method, **d)** box filter, **e)** Gauss filter, **f)** median filter, **g)** filter (2), **h)** filter (4), **i)** filter (5), **j)** filter (7), **k)** filter (8), **l)** filter (9).

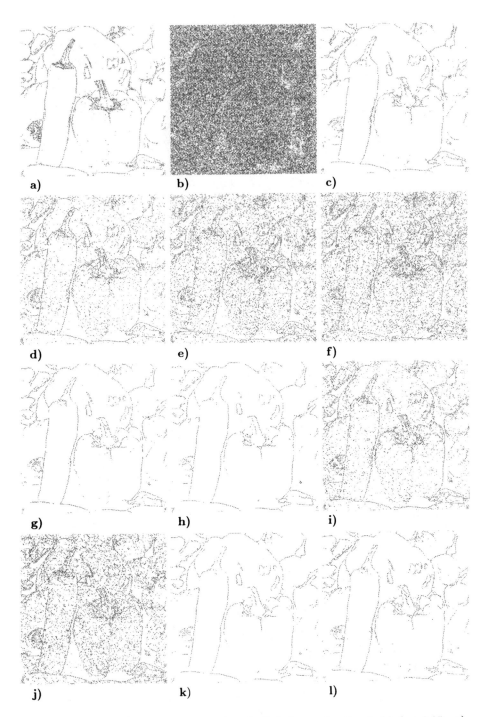

Fig. 3. The results of edge detection using the Sobel operator thresholded at 0.05 : **a)** test image, **b)** noisy image ($\sigma = 20$), **c)** image smoothed with the new method, **d)** box filter, **e)** Gaussian, **f)** median, **g)** filter (2), **h)** filter (4), **i)** filter (5), **j)** filter (7), **k)** filter (8), **l)** filter (9).

Evaluating the Complexity of Databases for Person Identification and Verification*

G. Thimm, S. Ben-Yacoub, J. Luettin

IDIAP, CP 592, CH-1920 Martigny, Switzerland

Abstract. Databases play an important role for the development and evaluation of methods for person identification, verification, and other tasks. Despite this fact, there exists no measure that indicates whether a given database is sufficient to train and/or to test a given algorithm. This paper proposes a method to rank the complexity of databases, respectively to validate whether a database is appropriate for the simulation of a given application. The first nearest neighbor and the mean square distance are validated to be suitable as minimal performance measures with respect to the problems of person verification and person identification.
Keywords: Person identification, person verification, database evaluation

1 Introduction

Computer vision systems are more and more based on statistical or heuristic methods. It is therefore important to compare alternative algorithms in order to evaluate their individual performance. As this is impossible in an analytical way, comparison is usually done by means of experiments. This requires the use of an identical database and an identical test protocol. The database, especially the test set, should be as close as possible to the real world conditions. If the test set is to easy, the algorithm will be overestimated, if it is too difficult, it will be underestimated. What one normally wants is to estimate the performance for the actual condition.

In practice, real-world data are often unavailable for legal and/or practical reasons. On the other hand, artificial datasets are often insufficient due to limited size, artifacts, similar illumination, similar position of subjects relative to the camera, the same background, similar facial expressions, or alike. Different databases have been used in face recognition which does not allow an objective comparison of results.

We describe therefore a procedure to rank databases. In the following, the term *complex* is used to compare two databases, although we are aware, that a proper, non-subjective definition in the mathematical sense can not exist. For example, the complexity of two databases is compared by means of the error rate yielded by a nearest neighbor algorithm.

* This work has been performed with financial support from the Swiss National Science Foundation under Contract No. 21 49 725 96 and the Swiss Office for Science and Education in the framework of the European ACTS-M2VTS project.

2 Ranking Databases

Before a database for training and testing is created, the recording and definition of the test protocol has to be well designed. Parameters of a dataset like noise, number of classes, items per class, the amount of artifacts, control of illumination, head position, and so on, are important in this context and determine in some way the complexity of a database. It is surprising that only two recently registered databases define a test protocol: the Extended M2VTS database [13] and the FERET[1] [18] database.

Another, often neglected but still important, question concerns the reliability of obtained results. This reliability depends on the statistical significance of the test (see [21]) and how similar the evaluation database is compared to real world data. It is therefore desirable to rank datasets according to their complexity for a given task or problem P.

Let P be problem defined for some class of objects. A computer can not directly act on physical objects, but on digital or analog signals. Such signals are obtained from applying transformations $T_i \subset \{t_1, t_2, \ldots, t_p\}$ to the real objects, giving databases D_i (such transformations include for example the projection of the object to an image, as well as filtering in the computer). The transformations in T_i used for the production of D_i are directly related to the complexity of problem P. Consequently, the goal is to rank the databases D_i according to a measure that reflects how well they incorporate the transformations that influence the complexity of a specific problem. However, datasets can not be ranked easily:

- The ranking depends on the problem P.
- It is often unknown which transformations were applied to the objects and how well they reflect the transformation encountered in a real application.
- The assignment of a "degree of complexity" to specific transformations and to determine how complexities add up is difficult.
- Transformations are often continuous, implying different, continuous valued, degrees of complexity.

Given these unknown parameters, we propose to test a database my means of the performance of a gauge algorithm A. The performance achieved by this algorithm on a particular database D_i is then used as the complexity measure of D_i with respect to T and P.

3 Face Identification and Verification

In the context of this paper, the problems are $P_1 = \textit{person identification}$ and $P_2 = \textit{person verification}$ by means of faces (*i.e.* T operates on faces). The set of possible transformations that increase the complexity of problems $P_{1,2}$ are a

---

[1] The FERET database is not publicly available. The authors have been unable to obtain it.

subset of $\mathcal{T} = \{rotation,\ illumination\ change,\ scaling,\ facial\ expression, \ldots\}$. The gauge algorithms are chosen to be the $\mathcal{A}_1 = nearest\ neighbor\ classifier$, respectively the $\mathcal{A}_2 = mean\ square\ distance$ (applied to zero-mean normalized image vectors). Obviously, both algorithms are not robust against illumination changes, translations, rotations, or scaling. Although other choices for $\mathcal{A}_{\{1,2\}}$ are possible, the chosen methods are most suitable for the following reasons:

- No free parameters have to be defined. Algorithms based on approaches like neural networks, genetic algorithms, and so on, require some parameters to be standardized (learning rate, network topology, crossover ratios,...).
- The first nearest neighbor algorithm and the mean square distance are well known and easily implemented and therefore cause only little work overhead.
- The complexity of the algorithms is reasonably low.

4 Experiments

The aim is (1) to evaluate the complexity of commonly used databases, (2) to compare identification and verification performance, (3) to compare the performance of the nearest neighbor algorithm, respectively the mean square distance, with published algorithms, and (4) to estimate for which evaluation tasks these datasets are appropriate. Scanning papers concerned with face recognition based on frontal views ([3, 4, 6–8] and others) reveals that many different datasets are used (table 1) - some not even publicly available - which prevents the comparison of published results. Some databases are available via [9]. Sometimes, mixtures of databases from different independent sources were used, in the aim to increase the significance of an evaluation [12, 23].

Name of the database	#		
		M2VTS [17]	4
FERET [18]	11	Weizmann [14]	2
Private or unspecified databases	11	Yale Face Database [5]	1
ORL [19]	6	Bern [1]	1
Mixtures of other databases	5	MIT [22]	1

Table 1. Databases used for person identification or verification and the number of times (column #) used in literature. The FERET database is often used in parts only.

Four datasets are used in this report (for sample images see figure 1). Only the Extended M2VTS database includes a well defined training and testing procedure:

1. The **Weizmann Institute of Science database** (subjects: 28, images per subject: 30) [14]. The images show the head, the neck, and some amount of background. The images are scaled to 18×26 pixels. The database was split twice into 50 pairs of training and test sets. The first 50 training sets included 8 images of each identity with the same, randomly chosen head positions and illumination. The second set of training sets contains also 8 images per identity, but not necessarily the same shots.

2. The **Bern database** (subjects: 30, images per subject: 10) [1]. The position and orientation of the faces is controlled, but the faces are neither centered nor scaled. As the rotation angles of the head positions are smaller as compared to the Extended M2VTS database, we decided to "crop" the images. In this operation, first top rows and columns with a high amount of background are removed (the hair remained mainly). Then, the lower part of the image is cut/extended, in order to obtain an image that has a height/width ratio of 2/3. Finally the images are scaled to 20×30 pixel.

   Two experiments were performed, each using 20 pairs of training and test sets. Each training set contains 4 randomly chosen images of each person. However, during the first experiments, the training sets contained always the same shots of each person.

3. The **ORL database** from the Olivetti Research Laboratory (subjects: 40, images per subject: 10) [19]. The faces in this database are already centered and show only the face. For the experiments, the images are scaled to 23×28 pixels, the database was split into 10 training and test sets. Each identity is represented 4 times in each training set.

4. The **Extended M2VTS database** (subjects: 295, images per subject: 8) [13]. The faces in this database are neither equally positioned, nor scaled. The faces are detected using the Eigenface algorithm [22], and the eyes are located using again the Eigenface approach. Then, the positions of the eyes are used to normalize the scale, to rotate the head into an upright position, and to define the region of interest. The region of interest is extracted, scaled, and stored as grey level image of the size 24×35 pixels. In a small percentage of the images the eyes were hand-labeled, as the head or eyes were not properly detected. The experiments were performed with six different pairs of training and test sets, each containing 4 images from two sessions.

Two tests were performed with these databases:

1. Person identification using a first nearest neighbor classifier with a mean square distance measure. The performance measure is the correct classification rate (see table 2).
2. Person verification using the mean square distance. The performance measure is the equal error rate (see table 3).

As the Weizmann and Bern databases are controlled (head position, illumination direction, and facial expression in the Weizmann database, the head position in the Bern database), two experiments were performed for each dataset. In the first experiment, with the results documented in the second column of table 2 and 3, the same shots of each identity were included in the training set. In the second experiment, all shots were selected randomly. It could, for example, occur that the training set for one identity includes only views of the left side of the face, whereas for another identity only frontal views are included.

The different performances show that small details in the configuration of an experiment may result in large changes of the error rate. Remarkable is also the high variance of the equal error rate when the training sets include always the

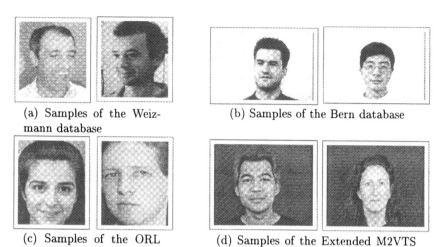

(a) Samples of the Weizmann database

(b) Samples of the Bern database

(c) Samples of the ORL database

(d) Samples of the Extended M2VTS database

Fig. 1. Samples from the four databases compared in this study.

Database	Average correct Identification		Number of Subjects
	similar shots	random shots	
Weizmann	85%	55%	28
Bern	85%	80%	30
ORL	—	92%	40
Ext. M2VTS	—	56%	295

Table 2. Percent of correct identification of the nearest neighbor classifier for face recognition (for the similar/random shots for all identities in the test set, if applicable).

same shot for each person. Equal error rates in the range of 5 to 13% for the Bern database and 9 to 33% for the Weizmann database have been observed. This demonstrates the importance of defining a common test protocol.

The high discrepancy of the identification rate and equal error rate, although not very intuitive, is explainable; see figure 2. Generally, this discrepancy is likely to occur when the inner- and inter-class distances are similar.

The results show, that, in terms of person identification, the ORL database has a lower complexity than the other three databases. The fact, that the number of persons contained in this database is higher than in the Weizmann and Bern database, supports the hypothesis, that the size of the database is not necessarily an indicator for the complexity of a database. Overall, the complexity does not increase with the size of the data set (as it would be expected).

For the problem of person verification, one would expect, that the average equal error rate is almost independent of the size of the data set (or slightly increasing with it). This is not true for the Weizmann and the Extended M2VTS

Database	Average Equal Error Rate similar shots	Average Equal Error Rate random shots	Number of Subjects
Weizmann	16%	18%	28
Bern	10 %	11%	30
ORL	—	7%	40
Ext. M2VTS	—	11%(*)	295

Table 3. Equal error rate using the mean square distance for face verification (for the similar / random shots for all identities in the test set, if applicable). (*) Using the protocol described in [13] (without imposter accesses by clients), the equal error rate is 14.8%.

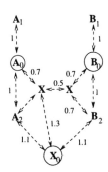

Consider the classes **A**, **B**, and **X** in a two dimensional space, with their elements distributed as shown. The dashed arrows indicate the euclidian distance between the elements, and circles indicate the test set. It can be easily seen that a nearest neighbor classifier has a 0% recognition rate. The equal error rate is 33% for a threshold of 1.2: in 6 tests, only A_0 and B_0 are accepted falsely for class **X**; in 3 tests, only X_0 is falsely rejected. Note that the tree classes can easily be separated vertical lines.

Fig. 2. The performance of a identification and a verification system can be very different.

database: the latter includes a factor of 10 more identities, but the observed equal error rate is considerably lower.

It can be observed that the complexity of databases depends on the defined problem (*i.e.* for the Ext. M2VTS database). A database with a high complexity with respect to classification, is not necessarily complex with respect to verification, and vice versa. According to the experiments, the Extended M2VTS database is the most challenging of the four examined database for person identification and the Weizmann database for person verification.

5   Comparison with Other Publications

The nearest neighbor classifier obtains a considerable performance (table 4), when compared to other methods. Note that the ranking may change slightly due to different, in the respective papers often unspecified, test protocols. Unfortunately, the authors could not find any publications using one of these databases in the context of person verification.

In real applications, the nearest neighbor algorithm can in the presence of, for example, rotation and illumination changes **not** be expected to perform better

Weizmann database	
100%	Elastic matching [23]
85%	**Nearest neighbor**
84%	Eigenfaces [23]
~80%	Garbor-like filters [2]
41%	Auto-Association and Classification networks [23]

Bern database	
93%	Elastic matching [23]
87%	Eigenfaces [23]
85%	**Nearest neighbor**
43%	Auto-association and Classification networks [23]

ORL database	
96.2%	Convolutional neural networks [11]
95%	Pseudo-2D HMMs [20]
92%	**Nearest neighbor**
90%	Eigenfaces [22]
87%	HMMs [19]
84%	HMMs [16]
84%	Point matching and 3D modelization [10, 15]
80%	Eigenfaces [23]
80%	Elastic matching [23]
20%	Auto-association and Classification networks [23]

Table 4. Recognition rates reported in other publications.

than more sophisticated methods that take advantage from *a priori* knowledge. From the high performance of the nearest neighbor method, it can be concluded that the ORL database and probably the Bern database are insufficient for realistic tests of person identification applications.

6 Conclusion

It is argued and supported by experiments that it is necessary to rank databases used for the development and the comparison of classification and verification tasks. This helps to prevent a gross over- or underestimation of a system due to an inappropriate database. In consequence, a simple method is proposed that ranks databases according to their complexity prior to their usage. The argumentation is supported by experiments using four datasets and the nearest neighbor classifier for person identification, respectively a mean square distance measure for identity verification.

Among the four examined databases, the Extended M2VTS database is the most challenging database for person identification and the Weizmann database for person verification.

The first nearest neighbor method is shown to perform better than several other methods for person identification. Similarly, the mean square distance performs rather well for person verification on some databases. These outcomes and the simplicity of both approaches suggest to use these two methods as a minimal performance measure for other algorithms in their respective domains.

References

1. B. Ackermann: Bern data base (1995). Anonymous ftp: `iamftp.unibe.ch/pub/Images/FaceImages/`.

2. Y. Adini, Y. Moses, and S. Ullman: Face recognition: The problem of compensating for changes in illumination direction, IEEE Trans. on Pattern Analysis and Machine Intelligence 19 (July 1997) 721–732.
3. IEEE Int. Conf. on Automatic Face and Gesture Recognition, Killington, Vermont, IEEE (October 14-16, 1998).
4. IEEE Proc. of the Second Int. Conf. on Automatic Face and Gesture Recognition, Nara, Japan, IEEE (April 14-16, 1998).
5. P. N. Belhumeur and D. J. Kriegman: The Yale face database (1997). URL: http:// giskard.eng.yale.edu/yalefaces/yalefaces.html.
6. J. Bigün, G. Chollet, and G. Borgefors, eds.: Audio- and Video-based Biometric Person Authentication (AVBPA'97), Lecture Notes in Computer Science 1206, Crans-Montana, Switzerland, Springer (March 1997).
7. H. Burkhardt and B. Neumann, eds.: Computer Vision - ECCV'98, II of Lecture Notes in Computer Science 1406, Freiburg, Germany, Springer (June 1998).
8. IEEE Computer Society Conf. on Computer Vision and Pattern Recognition (CVPR-96), San Francisco, California (June 18-20, 1996).
9. P. Kruizinga: The face recognition home page. URL: http://www.cs.rug.nl/ ~peterkr/FACE/face.html.
10. K.-M. Lam and H. Yan: An analytic-to-holistic approach for face recognition on a single frontal view, IEEE Trans. on Pattern Analysis and Machine Intelligence 20 (July 1998) 673–689.
11. S. Lawrence, C.L. Giles, A.C. Tsoi, and A.D. Back: Face recognition: a convolutional neural-network approach, IEEE Trans. on Neural Networks 8 (1997) 98–113.
12. S.Z. Li and J. Lu: Generalized capacity of face database for face recognition, in [4] 402–405.
13. K. Messer, J. Matas, J. Kittler, J. Luettin, and G. Maitre: XM2VTSDB: The extended m2vts database, in Proc. Second Int. Conf. on Audio- and Video-based Biometric Person Authentication (AVBPA'99) (1999). http://www.ee.surrey.ac. uk/Research/VSSP/xm2vts
14. Y. Moses: Weizmann institute database (1997). Anonymous ftp: ftp.eris. weizmann.ac.il/pub/FaceBase.
15. A.R. Mirhosseini, H. Yan, K.-M. Lam, and T. Pham: Human face image recognition: An evidence aggregation approach, Computer Vision and Image Understanding 71 (August 1998) 213–230.
16. A. V. Nefian and M. H. Hayes III: Hidden markov models for face recognition, in ICASSP'98 5, IEEE (1998) 2721–2724.
17. S. Pigeon and L. Vandendorpe: The M2VTS multimodal face database, in [6].
18. P. Phillips, H. Wechsler, J. Huang, and P. Rauss: The FERET database and evaluation procedure for face recognition algorithms. To appear in: Image and Vision Computing Journal (1998).
19. F. Samaria and A. Harter: Parameterization of a stochastic model for human face identification, in Proc. of 2nd IEEE Workshop on Applications of Computer Vision, Sarasota, FL (1994). URL: http://www.cam-orl.co.uk/facedatabase.html.
20. F. S. Samaria: Face Recognition using Hidden Markov Models. PhD thesis, Trinity College, University of Cambridge, Cambridge (1995).
21. W. Shen, M. Surette, and R. Khanna: Evaluation of automated biometrics-based identification and verification systems, Proc. of the IEEE 85 (September 1997) 1464.
22. M. Turk and A. Pentland: Eigenfaces for recognition, Journal of Cognitive Neuroscience 3:1 (1991) 71–96. ftp: whitechapel.media.mit.edu/pub/images/.
23. J. Zhang, Y. Yan, and M. Lades: Face recognition: Eigenface, elasic matching, and neural nets, Proc. of the IEEE: Automated Biometric Systems 85 (1997) 1423–1435.

Efficient Shape Retrieval by Parts

S. Berretti, A. Del Bimbo, and P. Pala

Dipartimento di Sistemi e Informatica, Università di Firenze, via S. Marta 3,
50139 Firenze, Italy
{berretti, delbimbo, pala}@dsi.unifi.it

Abstract. Modern visual information retrieval systems support retrieval
by directly addressing image visual features such as color, texture, shape
and spatial relationships. However, combining useful representations and
similarity models with efficient index structures is a problem that has
been largely underestimated. This problem is particularly challenging in
the case of retrieval by shape similarity.
In this paper we discuss retrieval by shape similarity, using local features
and metric indexing. Shape is partitioned into tokens in correspondence
with its protrusions, and each token is modeled by a set of perceptually salient attributes. Two distinct distance functions are used to model
token similarity and shape similarity. Shape indexing is obtained by ar-
ranging tokens into a M-tree index structure. Examples from a prototype
system are expounded with considerations about the effectiveness of the
approach.

1 Introduction

In the last few years, visual information retrieval (VIR) has become a major area
of research, due to the ever increasing rate at which images are being generated in
many application fields. VIR is an extension of traditional information retrieval,
so as to deal with visual information: given a collection of images or videos, the
purpose of VIR is to retrieve those images or videos which are relevant to a
visual query.

Modern VIR systems support retrieval by visual content by directly addressing image perceptual features such as colors [1] [4], textures [5] [6] [7],
shapes [2] [8] [9], spatial relationships [13] [14] [15]. At archiving time, Image
Processing and Pattern Recognition techniques are used to extract content descriptors from database images. At retrieval time, the user's query is expressed
by means of visual examples which contain the prominent visual features of
the images that are searched for. Content descriptors of database images are
compared with the user's query descriptors and a similarity matching score is
associated with each image.

Color, texture and shape features are commonly represented through multidimensional feature vectors, whose elements correspond to significant parameters
that model the feature. Feature vectors correspond to points in the multidimensional feature space and close points are considered as representatives of similar
features.

One fundamental problem in VIR is that of combining useful representations and similarity models with index structures which provide efficient access to images in large databases. Most systems have focussed only on feature representation or on modeling the correspondence with human similarity judgment, disregarding the dimensionality course that occurs in large databases. Other systems, have exploited the fact that the feature vector model and metric distances permit to build index structures according to classical point access methods (PAMs), developed for spatial data [13]. In order to adapt feature vector dimensionality (which is usually large) with PAM requirements, vectors are mapped in a low–dimensional distance–preserving space, according to appropriate transformations.

The combination of useful representations with index structures is particularly challenging in the case of retrieval by shape similarity. On the one hand, there is little knowledge about the way in which humans perceive shape similarity. However, there is enough evidence that metric distances are not suited to model shape similarity perception. On the other hand, shape descriptions commonly used to model the perceptual appearance of a shape are usually complex and cannot be easily combined with effective indexing.

A number of solutions have been proposed in the literature to deal with shape representation. Among representations that don't follow the feature vector model, the relational approach [8], and the shape–through–transformation approach [2] [9] [10] are the most distinguished ones.

Representations following the feature vector model, are more suitable for effective indexing. Solutions which employ global feature descriptors, among which the QBIC [1] system, use attributes which characterize the form of the shape considered as a whole. However, it is demonstrated that metric distances (such as the Euclidean) between feature points do not match closely the human judgement of similarity. This makes ineffective the combination of this representation with PAM indexing. Moreover this model is not able to support similarity evaluation in the case of partially occluded objects, which is a relevant aspect in shape perception.

Alternative solutions employ local feature descriptors - such as edges and corners of boundary segments - and are essentially based on a partitionment of the shape boundary into perceptually significant tokens. This approach is suited for the evaluation of similarity in the case of partially occluded shapes. Local boundary features have been used by Grosky and Mehrotra [11] and Mehrotra and Gary [12].

2 Shape Representation

In our approach minima of the shape curvature are used to partition the shape into tokens. Partitioning of the curve along such points, determines tokens that correspond to *protrusions* of the curve. Protrusions can be used as signatures to identify the curve. The way in which different protrusions are arranged along the curve can be used to carry out a description of the curve.

Mathematically, a planar, continuous curve can be parameterized with respect to its arc-length t, and expressed as $c(t) = \{x(t), y(t)\}$, $t \in [0,1]$. Let $\gamma(t)$ be the curvature of $c(t)$ and $P = \{p_i\}_{i=1}^{N}$ the set of minima of the curvature $\gamma(t)$. If we assume to deal with curves such that their curvature is continuous, between two consecutive minima p_k, p_{k+1} there is always a maximum of $\gamma(t)$, namely m_k.

A feature that significantly affects the perception of a shape token is its width (narrow or wide tokens). The maximum value of token curvature m_k can be used to represent the token width: narrow tokens are associated with high values of m_k, while wide tokens are associated with low values of m_k.

The visual appearance of a shape is also related to the way in which narrow and wide tokens are arranged. Different arrangements of similar tokens result into completely different shapes. In our approach token arrangement is retained by considering the orientation of each token in the 2D space. Given a token τ_k, identified by two consecutive points of minima, p_k and p_{k+1}, the orientation θ_k is the orientation, in polar coordinates, of the vector linking the median point of the segment $p_k - p_{k+1}$ with the point where the token curvature is maximum, p_{m_k}. This feature preserves the tokens organization by reducing the degrees of freedom allowed for tokens arrangement in the 2D space.

Each token is therefore represented by two features (m_k, θ_k) and a generic closed curve c, with N partition points, is represented by the set: $T(c) = \{(m_k, \theta_k)\}_{k=1}^{N}$. As an example, in Fig. 1 it is shown the shape of a horse, its tokens and the description of each token.

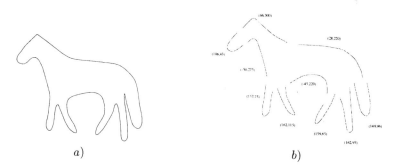

$a)$ $b)$

Fig. 1. a) Shape of a horse; b) The shape is partitioned in correspondence with minima of the curvature function. Each token τ_k is described through the features (m_k, θ_k).

3   Shape Similarity Measure

Since shapes are represented by a finite set of token descriptors, two measures of distance are defined: a *token distance* and a *shape distance*. Token distance is used to provide a measure of the similarity between two tokens. Shape distance is obtained by combining token distances in order to derive a global measure of shape similarity.

Given two generic tokens τ_1 and τ_2, their distance is computed according to Proc. 1 of Fig. 2. Curvature and orientation distances are combined in a convex

---

Dist(Token τ_1, Token τ_2)
{

        compute $d_{curvature} = |curvature(\tau_1) - curvature(\tau_2)|$
        compute $d_{orientation} = |orientation(\tau_1) - orientation(\tau_2)|$
        compute $d_{ij} = \alpha d_{curvature} + (1 - \alpha)d_{orientation}$

}

Fig. 2. Procedure 1 to compute the distance between two tokens. α is the weight to combine the curvature distance and the orientation distance.

form, through a parameter α, to compute the token distance. By changing the α value in the range $[0, 1]$, the contribute of curvature and orientation can be weighted differently. Since the token distance fulfills the triangular inequality, the token space (m_k, θ_k) is a metric space.

Given two generic shapes A and B, let n and m be the number of tokens of shape A and B respectively, with $n < m$ (if this is not the case, A and B are exchanged). The distance between A and B is computed by combining the similarity of shape tokens according to Proc. 2 shown in Fig. 3.

4   Shape Indexing

Since a generic token is represented by using two features, each token is a point in a two-dimensional space and a generic shape can be represented as a set of points in the 2D space of curvature and orientation. In order to have effective indexing, shapes are broken out into tokens and tokens are indexed. In this way, the dimensionality of the feature space is kept low and the metric property of the distance between index entries is fulfilled.

Each entry in a tree node has the following format:

$$entry(O_i) = [\, O_i,\ ptr(T(O_i)),\ r(O_i),\ d(O_i, P(O_i)) \,]$$

where O_i are the feature values of the indexed object, (if the node is a leaf node), or an identifier of the indexed object, (if the node is an internal node); $ptr(T(O_i))$ is a pointer to the root of the sub-tree $T(O_i)$, which is composed of elements O_j such that the distance from O_i is less than the covering radius $r(O_i) > 0$:

$$d(O_i, O_j) \leq r(O_i) \ \forall\, O_j \ in \ T(O_i)$$

The term $d(O_i, P(O_i))$ represents the distance between the node entry and its parent object $P(O_i)$. For leaf nodes, no covering radius is needed and the field

```
Distance( Shape A, Shape B )
{
        if (number of tokens in A > number of tokens in B)
            exchange A and B

        for each token a_i in A do
            for each token b_j in B do
                compute d_{ij}=Dist(Token a_i, Token b_j)

        for each token a_i in A do
            associate a_i to the nearest token b_j in B
                that is not yet assigned to tokens of A.

        if (∀ token a_i in A the token distance d_{ij} = d_i
            is <= of a predefined threshold)
            then compute d(A,B) = Σⁿᵢ₌₁ d_i / n

}
```

Fig. 3. Procedure 2 to compute the distance between two shapes.

$ptr(.)$ is replaced with the identifier of the object or with the object itself. Given a query point Q and a range r, a range query selects all the database objects O_i such that: $d(O_i, Q) \leq r$. Distances $d(O_i, P(O_i))$ and $r(O_i)$ are precomputed and stored in the M-tree nodes, so that the number of accessed nodes and distance computations are reduced at the search time.

The tree traversing algorithm reduces the number of accessed nodes as well as the number of distance computations. Distances $r(O_i)$ are used to prune a sub-tree from a search path using these two results:

$$if \quad d(O_r, Q) > r + r(O_r) \Rightarrow d(O_i, Q) > r \quad \forall \, O_i \; in \; T(O_r) \qquad (1)$$

$$if \quad |d(O_p, Q) - d(O_r, O_p)| > r + r(O_r) \Rightarrow d(O_r, Q) > r + r(O_r) \qquad (2)$$

Equation (1) states the condition to prune the subtree $T(O_r)$ from the search path. Equation (2) avoids the distance computation $d(O_r, Q)$. Both these relations derive from the M-tree definition and the triangle inequality applied to the distance function $d(.)$.

5 Experimental Results

In the following, some retrieval examples are presented for a database of over 400 shapes taken from 20th century paintings. Each shape is sampled at 100 equally spaced points and descriptions of shape tokens are automatically extracted from each shape. Shape descriptions are then organized into a M-tree index structure, following concepts expounded in Sect. 4.

The system interface allows the user to tune the value of a parameter (the *trust factor*) which triggers the degree of selectiveness of the indexing process (in the range $[0, 100]$). Retrieved images are presented in decreasing order of similarity on the right part of the interface. Retrieval examples with the system are presented in Figs. 4 – 5.

In Fig. 4, the sketch roughly represents the contour of a horse. The system selects not only those shapes which closely resemble the sketch, but also shapes that represent a similar object with different postures. Neither the difference in the number of legs of the animals, nor the absence of some occluded parts affect the performance of the system. Fig. 5 shows an example of retrieval in presence

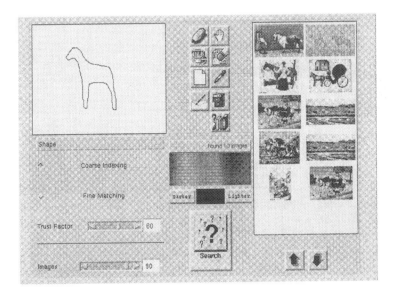

Fig. 4. Sketch representing a horse and the corresponding output of the indexing process. If two different shapes in the same image match the query shape, then the image appears twice in the result set.

of occlusions. The query shape represents only a portion of a horse. Nevertheless the best ranked images all represent horse shapes.

System performance has been analyzed in terms of *retrieval accuracy*. Retrieval accuracy is concerned with the effectiveness of shape retrieval at a *quantitative* level. Precision and recall measures are computed and performance figures obtained from the experiments are reported.

To assess the retrieval accuracy of the proposed technique, precision and recall figures are reported. For a given query, let T be the total number of relevant items available, R_r the number of relevant items retrieved, and T_r the total number of retrieved items. Then the *precision* $\mathcal{P}$ is defined as R_r/T_r and the *recall* $\mathcal{R}$ as R_r/T. In order to obtain values of $\mathcal{P}$ and $\mathcal{R}$, database shapes are first classified through human annotation. In this way, given a generic query

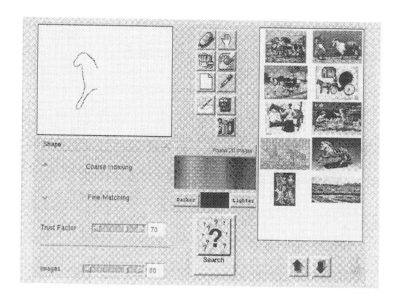

Fig. 5. Sketch representing a part of a horse shape and the corresponding output of the indexing process.

shape, the exact number of similar database shapes T is known. In Fig. 6(a), the values of $\mathcal{R}$ and $\mathcal{P}$ are shown as a function of the *trust factor*.

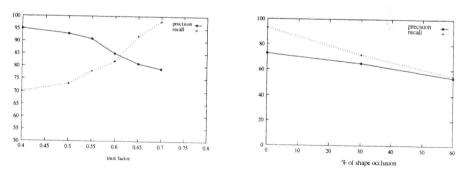

Fig. 6. a) Recall and precision values as a function of the trust value. b) Recall and precision for increasing degrees of occlusion

System robustness in case of partial object occlusion has been estimated by feeding the system with a query representing a partially occluded contour. To analyze the extent by which the system is able to cope with occlusions, some reference shapes were partially occluded. By progressively occluding a shape, a set of shapes is devised, each shape representing the contour of the original shape with different degrees of occlusion. Experiments were carried out using a set of 20 shapes representing the contours of several types of objects, such as bottles, horses, busts and vases. Each shape was subject to three degrees of occlusions: 0% (the original shape), 30% and 60% of the shape contour length.

System robustness has been measured by evaluating how much the presence of an occlusion affects the accuracy of retrieval in terms of precision and recall. Fig. 6(b) shows system precision and recall figures for increasing degrees of occlusion.

References

1. Faloutsos, C., Flickner, M., Niblack, W., Petkovic, D., Equitz, W., Barber, R.: The QBIC Project: Efficient and Effective Querying by Image Content. Res. Report 9453, IBM Res. Div. Almaden Res.Center, (1993).
2. Del Bimbo, A., Pala, P.: Visual image retrieval by elastic matching of user sketches. IEEE Trans. on Pattern Analysis and Machine Intelligence, Vol. 19, no. 2, (1997) 121-132.
3. Patella, M., Ciaccia, P., Zezula, P.: M-tree: An efficient access method for similarity search in metric spaces. In: Proc. of the International Conference on Very Large Databases (VLDB). Athens, Greece, (1997).
4. Del Bimbo, A., Mugnaini, M., Pala, P., Turco, F.: Visual Querying by Color Perceptive Regions. Pattern Recognition, Vol. 31, (1998) 1241–1253.
5. Tamura, H., Mori, S., Yamawaki, T.: Texture Features Corresponding to Visual Perception. IEEE Trans. on Systems, Man and Cybernetics. Vol. 6, no. 4, (1976) 460–473.
6. Liu, F., Picard, R.W.: Periodicity, directionality, and randomness - Wold features for image modeling and retrieval. IEEE Trans. on Pattern Analysis and Machine Intelligence. Vol. 18, no. 7, (1996) 722–733.
7. Picard, R.: A Society of Models for Video and Image Libraries. MIT Media Lab Perceptual Computing Section T.R. no. 360, (1995).
8. Shapiro, L.G., Haralick, R.M.: Structural Descriptions and Inexact Matching. IEEE Trans. on Pattern Analysis and Machine Intelligence. Vol. 3, no. 5, (1981) 504–519.
9. Pentland, A., Picard, R.W., Sclaroff, S.: Photobook: Tools for Content-Based Manipulation of Image Databases. In: Proc. SPIE Storage and Retrieval for Image and Video Database. San Kose CA, (1994).
10. Sclaroff, S.: Deformable Prototypes for Encoding Shape Categories in Image Databases. Pattern Recognition, Vol. 30, no. 4, (1997) 627–642.
11. Grosky, W.I., Mehrotra, R.: Index–Based Object Recognition in Pictorial Data Management. Computer Vision, Graphics and Image Processing. Vol. 52, (1990) 416–436.
12. Mehrotra, R., Gary, J.E.: Similar-Shape Retrieval in Shape Data Management. IEEE Computer, (1995) 57–62.
13. Samet, H.: Hierarchical Representations of Collections of Small Rectangles. ACM Computing Surveys, Vol. 20, no. 4, (1988) 271–309.
14. Chang, S.K., Shi, Q.Y., Yan, C.W.: Iconic Indexing by 2–D Strings. IEEE Transactions on Pattern Analysis and Machine Intelligence. Vol. 9, no. 3, (1987) 413–427.
15. Egenhofer, M.J., Franzosa, R.: Point-set Topological Spatial Relations. International Journal of Geographical Information Systems. Vol. 9, no. 2, (1992).

Curvature Scale Space for Shape Similarity Retrieval Under Affine Transforms

Farzin Mokhtarian and Sadegh Abbasi

Centre for Vision, Speech, and Signal Processing
Department of Electronic and Electrical Engineering
University of Surrey, Guildford, England GU2 5XH, UK

F.Mokhtarian@ee.surrey.ac.uk
http://www.ee.surrey.ac.uk/Research/VSSP/imagedb/demo.html

Abstract

The maxima of Curvature Scale Space (CSS) image have already been used to represent 2-D shapes in different applications. The representation has showed robustness under the similarity transformations. In this paper, we examine the performance of the representation under affine transformations. Since the CSS image employs the arc length parametrisation which is not affine invariant, we expect some deviation in the maxima of the CSS image under shear. However, we show that the locations of the maxima of the CSS image do not change dramatically even under large affine transformations.

Applying transformations to every object boundary of our database of 1100 images, we construct a large database of 5500 boundary contours. The contours in the database demonstrate a great range of shape variation. The CSS representation is then used to find similar shapes from this prototype database. We observe that for this database, 95% of transformed versions of the original shapes are among the first 20 outputs of the system. This provides substantial evidence of stability of the the CSS image and its contour maxima under affine transformation.

Keywords: Affine transformation, Image databases, Shape Similarity

1 Introduction

A number of shape representations have been suggested to recognise shapes even under affine transformation. Some of them are the extensions of well-known methods such as Fourier descriptors [1] and moment invariants [2][8]. The methods are then tested on a small number of objects for the purpose of pattern recognition. In both methods, the basic idea is to use a parametrisation which is robust with respect to affine transformation. The arc length representation is not transformed linearly under shear and therefore is replaced by *affine length* [3]. The shortcomings of the affine length include the need for higher order derivatives which results in inaccuracy, and inefficiency as a result of computation complexity.

In all cases, the methods are tested on a small number of objects and therefore the results are not reliable. Moreover, almost the same results can be achieved with the conventional methods without modifications [1][8]. Affine invariant scale space is introduced in [6]. It generalises the definition of curvature and introduces *affine curvature*. This curve evolution method is proven to have similar properties as curvature evolution [4], as well as being affine-invariant.

We have already used the maxima of Curvature Scale Space (CSS) image to represent shapes of boundaries in similarity retrieval applications [5]. The representation is proved to be robust under similarity transformation which include translation, scaling and changes in orientation. In this paper, we examine the robustness of the representation under general affine transformation which also includes *shear*. As a result of shear, the shape is deformed and therefore the resulting representation may change. We will show that the changes in the maxima of the CSS image are not severe even in the case of severe deformations.

The following is the organisation of the remainder of this paper. In section 2, CSS image is introduced and the CSS matching is briefly explained. Section 3 is about the affine transformation and the way we create large databases consisting of 1100 original and 4400 transformed shapes. In section 4, we study the movement of the CSS maxima under affine transformation and conclude that it is not considerable and therefore the measure of similarity between a shape and its transformed version is large according to the CSS matching. As a result, in shape similarity retrieval, the method can easily find the transformed versions of an input query and retrieve them. Concluding remarks are presented in section 5.

2 Curvature Scale Space image and the CSS matching

The formula for computing the curvature function can be expressed as:

$$\kappa(u) = \frac{\dot{x}(u)\ddot{y}(u) - \ddot{x}(u)\dot{y}(u)}{(\dot{x}^2(u) + \dot{y}^2(u))^{3/2}} \quad . \tag{1}$$

Equation (1) is for continuous curves and there are several approaches in calculating the curvature of a digital curve [7]. We use the idea of *curve evolution* which basically studies shape properties while deforming in time.

If $g(u, \sigma)$ is a 1-D Gaussian kernel of width σ, then $X(u, \sigma)$ and $Y(u, \sigma)$ represent the components of *evolved* curve,

$$X(u, \sigma) = x(u) * g(u, \sigma) \quad Y(u, \sigma) = y(u) * g(u, \sigma)$$

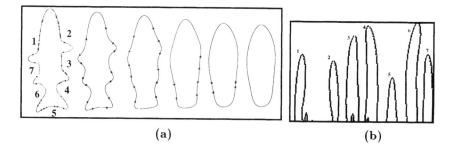

Figure 1: a)Shrinkage and smoothing of the curve and decreasing of the number of curvature zero crossings during the evolution, from left: $\sigma = 1, 4, 7, 10, 12, 14$. b)The CSS image of the shape.

according to the properties of convolution, the derivatives of every component can be calculated:

$$X_u(u, \sigma) = x(u) * g_u(u, \sigma) \qquad X_{uu}(u, \sigma) = x(u) * g_{uu}(u, \sigma)$$

and we will have a similar formula for $Y_u(u, \sigma)$ and $Y_{uu}(u, \sigma)$. Since the exact forms of $g_u(u, \sigma)$ and $g_{uu}(u, \sigma)$ are known, the curvature of an evolved digital curve is given by:

$$\kappa(u, \sigma) = \frac{X_u(u, \sigma)Y_{uu}(u, \sigma) - X_{uu}(u, \sigma)Y_u(u, \sigma)}{(X_u(u, \sigma)^2 + Y_u(u, \sigma)^2)^{3/2}} \qquad (2)$$

As σ increases, the shape of Γ_σ changes. This process of generating ordered sequences of curves is referred to as the evolution of Γ. If we calculate the curvature zero crossings of Γ_σ during evolution, we can display the resulting points in (u, σ) plane, where u is the normalised arc length and σ is the width of Gaussian kernel. For every σ we have a certain curve Γ_σ which in turn, has some curvature zero crossing points. As σ increases, Γ_σ becomes smoother and the number of zero crossings decreases. When σ becomes sufficiently high, Γ_σ will be a convex curve with no curvature zero crossing, and we terminate the process of evolution. The result of this process can be represented as a binary image called CSS image of the curve (see Figure 1). Black points in this image are the locations of curvature zero crossings during the process of evolution. The intersection of every horizontal line with the contours in this image indicates the locations of curvature zero crossings on the corresponding evolved curve Γ_σ.

The algorithm used for comparing two sets of maxima, one from the input and the other from one of the models, has been described in [5]. The algorithm first finds any possible changes in orientation which may have been occurred in one of the two shapes. A circular shift then is applied to one of the two sets to compensate the effects of change in orientation. The summation of the Euclidean distances between the relevant pairs of maxima is then defined to be the matching value between the two CSS images.

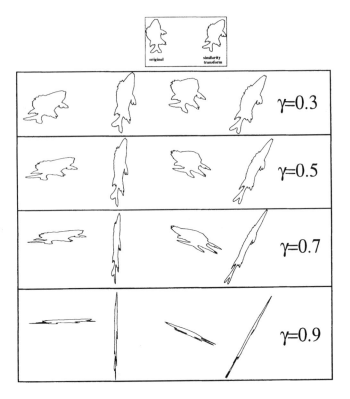

Figure 2: Affine transformation can be considered as a change in camera viewpoint. The original is in top left, top right shows a similarity transform with $a = 20, b = 10, c = -10$ and $d = 20$. The first, second, and third rows show four different affine transformed shapes with $\gamma = 0.5$, $\gamma = 0.7$ and $\gamma = 0.9$ respectively.

3   Affine transformation

The affine transformed version of a shape can be represented by the following equations.

$$\begin{cases} x_a(u) = ax(u) + by(u) \\ y_a(u) = cx(u) + dy(u) \end{cases} \tag{3}$$

where $x_a(u)$ and $y_a(u)$ represent the coordinates of the transformed shape. Now if we choose:

$$a = d, \quad and \quad c = -b,$$

then the transformation is just a similarity transform. Any deviation from these constraints will cause more shape deformation.

We may consider γ as a measure of the degree of deformation, where each parameter such as a, can move from γa to $(1+\gamma)a$. For example, if we choose $\gamma = 0.5$, we can create affine transformed versions of our existing shape contours. Five different sets of affine parameters in this case are selected as follows.

$$A_1 = \begin{pmatrix} 20 & 10 \\ -10 & 20 \end{pmatrix} A_2 = \begin{pmatrix} 30 & 15 \\ -5 & 10 \end{pmatrix} A_3 = \begin{pmatrix} 10 & 5 \\ -15 & 30 \end{pmatrix}$$

$$A_4 = \begin{pmatrix} 30 & 5 \\ -15 & 10 \end{pmatrix} \; A_5 = \begin{pmatrix} 10 & 15 \\ -5 & 30 \end{pmatrix}.$$

A_1 represents a similarity transform and the others represent four different transformations. Note that each parameter is chosen in extreme higher or lower limit of the above mentioned region. By choosing larger values for γ, e.g. 0.7, we can create more deformed shapes. Figure 2 shows an example which shows how a shape is deformed under each of these transformations. Four different values, 0.3, 0.5, 0.7 and 0.9 are considered for γ.

By choosing four different values for γ, we created four different databases, each consisting of 5500 shapes from which 1100 were our original database of marine creatures and 4400 were the transformed versions of the shapes. Different experiments were carried out on these databases to investigate the performance of the CSS representation under affine transformation.

4 Locations of the CSS maxima under affine transformation

In this section, we first show that the curvature zero crossing points are preserved under affine transformation. As a result, the overall configuration of the CSS image is also preserved. This can also be verified by recalling the fact that the number and orders of the shape segments remain unchanged under general affine transformation. We then verify these facts through several examples.

From equation 3 we have:

$$\dot{x}_a(u) = a\dot{x}(u) + b\dot{y}(u) \qquad\qquad \dot{y}_a(u) = c\dot{x}(u) + d\dot{y}(u)$$

$$\ddot{x}_a(u) = a\ddot{x}(u) + b\ddot{y}(u) \qquad\qquad \ddot{y}_a(u) = c\ddot{x}(u) + d\ddot{y}(u)$$

And therefore from equation 1:

$$\kappa_a(u) = \frac{(a\dot{x}(u) + b\dot{y}(u))(c\ddot{x}(u) + d\ddot{y}(u)) - (a\ddot{x}(u) + b\ddot{y}(u))(c\dot{x}(u) + d\dot{y}(u))}{((a\dot{x}(u) + b\dot{y}(u))^2 + (c\dot{x}(u) + d\dot{y}(u))^2)^{3/2}}$$

$$\kappa_a(u) = \frac{(ad - bc)(\dot{x}\ddot{y} - \dot{y}\ddot{x})}{((a\dot{x}(u) + b\dot{y}(u))^2 + (c\dot{x}(u) + d\dot{y}(u))^2)^{3/2}} \tag{4}$$

Now compare the numerators of equations 1 and 4 and observe that the curvature zero crossings of the original and the transformed shapes have a one to one correspondence.

Now we look at several examples of the CSS images. The shapes of Figure 2 have been shown in Figure 3 together with their corresponding CSS images. The effects of change in orientation have not been reflected in the CSS images of this Figure so that comparisons can be made more easily. All CSS contours experience the same circular shift as a result of change in orientation which is easily detected during the matching algorithm. As this Figure shows, in all cases, even those with severe deformations, the numbers and orders of the CSS contours are preserved. Even the height of the corresponding CSS contours are not very different. However, small changes in the locations of maxima are inevitable. Note for example that the largest maxima in the CSS image of the left column are different from the largest maxima of the other columns. Note also the last row of this Figure which shows the shapes and their corresponding CSS images for $\gamma = 0.9$. The deformations of shapes are noticeably severe, however, even small contours of the CSS images are preserved and they are quite similar to the CSS image of the original shape. This demonstrates the robustness of the CSS image under affine transformation. As a result of the changes in the CSS maxima, the matching value between a shape and its transformed version is not zero. We have already used the CSS representation in shape similarity retrieval [5] and learned that a matching value between zero and 0.5 reflects a very good and between 0.5 and 1.0 a reasonably good measure of similarity between the two shapes. Therefore, we expect that in response to a query, the system should be able to identify the transformed versions of the query and retrieve them among the other similar shapes.

In order to verify this expectation, we measured the matching value of each shape and its transformed versions. Having 1100 shapes and 4 transformed versions for each shape, we obtained 4400 values, each related to an original shape and one of its transformed versions. We observed that for $\gamma = 0.3$, in 98.3% of cases the matching value was less than 0.5. This can be seen in Figure 4a, where accumulative histograms of these values are presented for different values of γ. For $\gamma = 0.5$, although the deformation is considerable, the changes in matching CSS maxima and the corresponding matching values are not dramatic. In 91.1% of cases, the matching value is less than 0.5. This figure is 80.0% and 70.66% for $\gamma = 0.7$ and $\gamma = 0.9$, respectively.

In conclusion, we can observe the robustness of the CSS maxima under affine transformation, even in presence of severe deformations.

4.1 Matching a shape with its transformed version

In order to evaluate the performance of the CSS representation under affine transformation, we performed another experiment with our databases. Every original shape was selected as the input query and the first n outputs of the system were observed. We found out that for $\gamma = 0.3$, almost all transformed shapes appear in the first 10 outputs of the system. This figure is 90% for $\gamma = 0.5$. The results are presented in Figure 4b, where n is the number of outputs and *success rate* for a query is defined as:

$$success \quad rate = 100 \times \frac{k}{k_{max}}$$

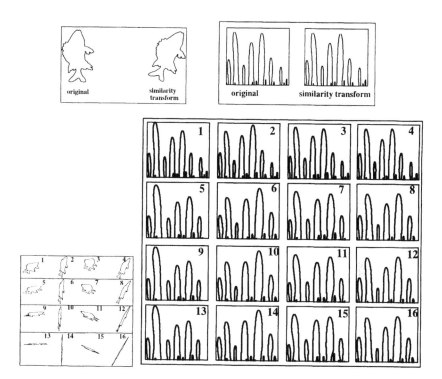

Figure 3: Affine transformed shapes and their associated CSS images. Above: Similarity transform does not change the CSS image. Below: Affine transform causes minor changes in the CSS images. Note that the CSS images are shifted to compensate the effects of orientation changes.

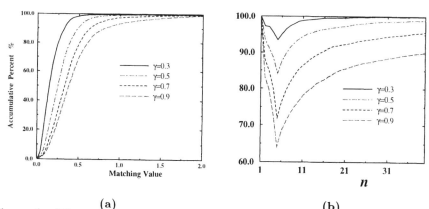

Figure 4: a)The matching value between a shape and its transformed versions is small. b)The average success rate for retrieval of the transformed versions of a shape among the first n outputs.

where k is the number of transformed versions of the input query which are in the first n outputs. k_{max} is the maximum possible value of k which is 5 if $n >= 5$ and is equal to n if $n < 5$.

5 Conclusion

A number of different experiments on our large database of marine creatures demonstrated that the maxima of the Curvature Scale Space image experience little change under general affine transformation. We showed that the curvature zero crossings as well as the order and number of shape segments are preserved. As a result, the order and number of the CSS contours remain unchanged. We observed that the horizontal coordinates of the CSS maxima remain almost unchanged under general affine transforms. We tested our method on a large database of shapes and observed that the method assigns a small matching value to a shape and its transformed version. This clearly shows that the method identifies them as similar shapes. Interestingly though, we observed that the method is able to identify and retrieve the transformed versions of a shape, even if they are hidden among a large number of potential candidates.

Acknowledgements Sadegh Abbasi is on leave from the University of Guilan, Rasht, Iran. He is grateful to the Ministry of Culture and Higher Education of Iran for its support during his research studies.

References

[1] K. Arbter et al. Applications of affine-invariant fourier descriptors to recognition of 3-d objects. *IEEE Trans. Pattern Analysis and Machine Intelligence*, 12(7):640–646, July 1990.

[2] J. Flusser and T. Suk. Pattern recognition by affine moment invariants. *Pattern Recognition*, 26(1):167–174, January 1993.

[3] H. W. Guggenheimer. *Differential Geometry*. McGraw-Hill, New York, 1963.

[4] B. B. Kimia and K. Siddiqi. Geometric heat equation and nonlinear diffusion of shapes and images. *Computer Vision and Image Understanding*, 64(3):305–332, 1996.

[5] F. Mokhtarian, S. Abbasi, and J. Kittler. Robust and efficient shape indexing through curvature scale space. In *Proceedings of the seventh British Machine Vision Conference, BMVC'96*, volume 1, pages 53–62, Edinburgh, September 1996.

[6] G. Sapiro and A.Tannenbaum. Affine invariant scale space. *International Journal of Computer Vision*, 11(1):25–44, 1993.

[7] D. Tsai and M. Chen. Curve fitting approach for tangent angle and curvature measurement. *Pattern Recognition*, 27(5):699–711, 1994.

[8] A. Zhao and J. Chen. Affine curve moment invariants for shape recognition. *Pattern Recognition*, 30(6):895–901, 1997.

FORI-CSDR - A New Approach for Context Sensitive Image Data Reduction

Dr. Uwe Strohbeck[1], Prof. Dr. Uwe Jäger[1,2], Dr. Alastair Macgregor [3]

[1]Fachhochschule Heilbronn,
University of Applied Sciences
Department of Electronics
Max-Planck-Strasse 39
74081 Heilbronn
Germany
Phone: +49 7131 504 419
Fax: +49 7131 252470
EMail: elstro@fh-heilbronn.de

[2]Steinbeis Transferzentrum Heilbronn
Robert-Bosch-Strasse 32
74081 Heilbronn
Germany
Phone: +49 7131 507 892
Fax: +49 7131 507 894

[3]University of Northumbria at Newcastle upon Tyne,
3, Ellison Terrace
Newcastle upon Tyne
NE1 8ST
U.K.

Abstract. This paper presents the theory of the FORI-CSDR algorithm (Focus On Regions of Interest - Context-Sensitive Data Reduction). The FORI-CSDR algorithm is able to reduce the amount of image data in dependence of the image context. Within the important image regions nearly no image data will be reduced. The image data of the unimportant image regions are reduced in a very strong way. The data reduction factors for the important and the unimportant image regions can be chosen independently. The FORI-CSDR (focus on regions of interest - context sensitive data reduction) algorithm is able to increase the total data compression rate by varying the image quality settings block-wise, depending on a determined DOI (degree of interest) value.

Introduction

Many image data compression tasks are performed by the use of the well-known JPEG algorithm. This algorithm achieves good results for multimedia applications but not for industrial image processing tasks. An 'industrial image processing disadvantage' of the JPEG algorithm is, that JPEG uses the same image quality settings for every pixel block, independently of whether the pixel block contains interesting or non-interesting parts of an object. Also for the JPEG algorithm some research projects are existing. Adaptive quantization within the JPEG algorithm is mainly used for the improvement of the PSNR of an image and not to obtain the maximum image quality for the important image regions. These algorithms are optimised for multimedia use [1], [2], [3], [4], [6]. An other huge difference to the presented FORI-CSDR algorithm is, that those algorithms are not able to vary their image quality parameters for each pixel block. They examine all pixel blocks at the beginning and then calculate the optimal image quality parameters for all pixel blocks. The FORI-CSDR (focus on regions of interest - context sensitive data reduction) algorithm is able to increase the total data compression rate by varying the image quality settings depending on a DOI (degree of interest) value.

Theory

The reason for the 'industrial image processing disadvantage' of the JPEG method is that for every pixel block the same quantization table is used. JPEG uses the same quantization table for the interesting parts within one frame (e.g. the objects) and also for the non interesting parts (e.g. the background). This is problematic when we want to preserve the full information in the interesting parts, e.g. preserving the edges within the frame for object measurement. JPEG smoothes the edges on high reduction rates, therefore we cannot use the JPEG image, e.g. for a subpixel algorithm in a measurement task. If we want to focus on the interesting parts within the frame, we have to modify the JPEG method to the FORI-CSDR method. Similar to the JPEG algorithm the FORI-CSDR algorithm consist also of a data reduction and data compression part. Therefore, the total data compression rate of the FORI-CSDR is determined by the data reduction rate (quantization unit) multiplied by the data compression rate (VLC unit). The FORI-CSDR method changes the data reduction rate of a pixel block dependent on the degree of interest of the pixel block (DOI value). Every pixel block within one frame will be classified as whether it contains interesting parts or not. Depending on this classification, the FORI-CSDR method changes the quantization tables of the quantization unit. Only the quantization unit is responsible for the data reduction rate (the following units in the JPEG method are only performing lossless data compression). Therefore it is possible to vary the data reduction rate of each block and focus on the interesting parts within one frame by changing the quantization table for each block. The resulting FORI-CSDR image contains now high resolution and low resolution areas.

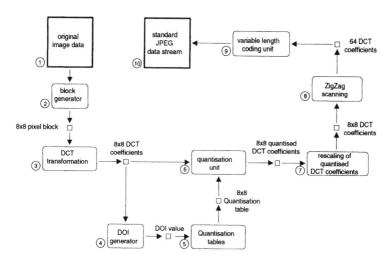

Fig. 1. This figure shows the complete FORI-CSDR algorithm.

In order to allow different image quality factors, different quantization tables (5) have to be defined and chosen by the DOI generator (4). The quantization unit (6) is responsible for the data reduction rate of the FORI-CSDR. The main task of the quantization unit is to reduce the image quality of each 8x8 pixel block by a given image quality factor. In order to make it possible to decode a FORI-CSDR encoded image with a standard JPEG decoder we have to rescale the quantized DCT coefficients. The rescaling unit (7) makes it possible to implement the FORI-CSDR functionality as a standard JPEG data stream (10) without changing the format.

DOI ranges

The FORI-CSDR method is able to handle different quantization tables for different DOI ranges (=image quality ranges). The developed FORI-CSDR software is able to handle five different quantization tables and DOI ranges. Table 1 shows the five different DOI ranges with the suitable image quality factors.

Table 1. different FORI ranges

DOI-range	image block content is...	image quality factor
0	... very important	100%
1	... important	70%
2	... average	50%
3	... less important	30%
4	... unimportant	10%

It is also necessary to define four DOI thresholds in order to assign each block to one of these five DOI-ranges. These thresholds, related to the image quality factors are very important for the total data reduction rate of the FORI-CSDR algorithm. The thresholds and the image quality factors are the only necessary a priori knowledge and have to be determined heuristically.

Rescaling of the quantized DCT data

A big problem of the described context sensitive data reduction method was the incompatibility to the standard JPEG method. Therefore it was not possible to decode the generated data stream by a standard JPEG decoder. This incompatibility is caused by the block-wise change of the quantization table. The JPEG algorithm allows only a frame-wise change of the quantization table - not a block-wise change. In order to decode the FORI-CSDR data stream a special FORI-CSDR decoder is necessary, which is able to block-wise change the de-quantization tables. This requires also an additional data field in the non-standard JPEG data stream, in order to tell the FORI-CSDR decoder which quantization table it has to use for the current block. Such an additional data field results in a standard JPEG decoder not being able to interpret the FORI-CSDR data stream. In order to enable the use of standard JPEG decoders to decode the FORI-CSDR data stream it is necessary to modify the FORI-CSDR encoder so that the FORI-CSDR data is in the standard JPEG data stream format. This can be achieved by ensuring the data is presented to the VLC unit in the same form as that used by the JPEG standard. Of course it is desired, that the increased data reduction rate, which is produced by the FORI-CSDR, is not cancelled by a decreasing data compression rate of the VLC- unit (variable length coding) which may be caused by this slightly change. Therefore it is also necessary to have a closer look to the VLC unit in order to understand, which properties of the data statistics influence the data compression rate of the VLC unit.

Four properties of the quantised DCT data are important in order to reach an adequate data compression rate by the JPEG VLC-unit.

- **bit-length of the quantized DCT values**.
  This property has a *direct* effect on the amount of coded data (data compression ratio).
- **data statistics**
  The huffman coder inside the VLC unit uses data statistics in order to compress the data.
- **repetition of same values.**
  Repeating values in the quantized data stream can be easily compressed by the run-length coding, which is implemented in the VLC.
- **number of trailing zero values**
  This property is most important for achieving high data compression ratios. If the quantized 8x8 DCT block consists of only 10 non-zero values, that are followed by trailing 54 zero values, then only these 10 values are coded by the VLC unit and sent together with an EOB marker (End of block) to the decoder. Trailing zero values are completely ignored.

In order to take care about these requirements the FORI encoder uses a trick. After the quantization of the DCT coefficients a rescaling unit is inserted. This rescaling unit is responsible for rescaling the quantization results back to the previous data range, independent of which image quality factor is used.

The original data range of the DCT coefficients is scaled down by the quantization unit to a new data range. The rescaling unit stretched the data range back to the original DCT data range, independently of the quality factor is used. In general we can say, that the quantization step is reversed but not completely cancelled. The reason is, that two of the described four properties (see above) are untouched by the rescaling unit. These untouched properties are a) repetition of the same values and b) number of trailing zero values. These are the two most important properties of the data stream in order to reach an adequate data compression rate. The bit-length property is compensated by the rescaling unit, because, after the rescaling step the DCT coefficients are containing the same data range as before. In order to reduce the negative effect on the data statistics, caused by the rescaling unit, it may be necessary to modify the corresponding Huffman tables.

FORI-CSDR compared with JPEG

JPEG is one of the best still image data reduction algorithm available at the moment. Therefore it is important to compare the results of FORI-CSDR algorithm with the results of the JPEG algorithm. It is important to remember that the JPEG algorithm is optimised for multimedia purpose and the FORI-CSDR algorithm was developed for industrial purpose. It is desired that the FORI-CSDR algorithm produces better results than the JPEG algorithm for edges in images so that further industrial image processing steps, such as subpixeling, are more accurate. The following experiments show the results of the FORI-CSDR and JPEG algorithm in different forms. The first part compares the results of both algorithms referring to the maximum image quality. This means, that both algorithms are configured, to preserve the same maximum image quality. In the second part the results are compared referring to the same data reduction rate. The JPEG images were encoded by standard JPEG techniques and the FORI-CSDR images by the FORI-CSDR encoder software. The experiment shows images, recovered by standard JPEG decoder software. The error images show the differences of the original image to the decoded image as grey-scale values. In order to improve the visibility of the errors the error images are inverted and a standard histogram equalisation technique was performed.

Maximum image quality
The JPEG image quality was set to 99%. The FORI-CSDR image quality was set to 99% for the important and 5% for the unimportant pixel blocks.

The result of the FORI-CSDR algorithm is shown in figure 2. The show image presents some visible distortions at the unimportant pixel blocks. Some very obvious differences to the original image are marked (see marker 1- 3).

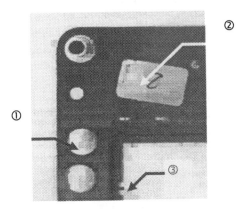

Fig. 2. This figure shows the result of the JPEG decoding unit. The orginal image was encoded by a FORI-CSDR encoder with 99% / 5% image quality setting

Fig. 3. The FORI-CSDR error image shows the differences of the greyscale values between the original image and the FORI-CSDR decoded image

Table 2. JPEG/FORI-CSDR data reduction factors

	file size	total reduction factor[1]
JPEG 99% image quality	20.407 Bytes	2,36
FORI-CSDR 99% / 5% image quality	9.364 Bytes	5,14

Table 2 shows the comparison of the total data reduction factors of the JPEG algorithm against the FORI-CSDR algorithm. The table shows, that the FORI-CSDR algorithm reaches a better data reduction factor without any visible errors at the important pixel blocks. Thus a factor of over two improvement in data reduction is achieved without degradation of the image in the important regions.

---

[1] Original file size 48.142 Bytes

Data reduction rate

In order to achieve the same file size for the coded JPEG data as for the FORI-CSDR data, the JPEG image quality factor is set to 91%. The FORI-CSDR image quality settings remain at 99% / 5% image quality (see table 3).

Table 3. JPEG/FORI-CSDR data reduction factor

	file size	reduction factor[2]
JPEG 91% image quality	9.184 Bytes	5,24
FORI-CSDR 99% / 5% image quality	9.364 Bytes	5,14

The following images show the results of 91% image quality setting at the JPEG algorithm.

Fig. 4. JPEG error image

A decoded JPEG data (coded with 91% image quality setting) offers no visible difference compared to the original image. Figure 4 shows some errors at the important pixel blocks (object edges). Slight concentrations of darker pixel values in the error images are located at the object edges. This indicates more differences compared to the original image at the important pixel blocks as at the unimportant.

These differences are also visible in the following diagram which presents the different pixel data at the object edge. In contrast to a image quality setting of 99% JPEG now offers different pixel data at the important pixel blocks . The measurement line in this figure shows the difference between the FORI (= original) and the JPEG edge data. This difference will cause an unrecoverable error in subpixel measurement tasks.

---

[2] Original file size 48.142 Bytes

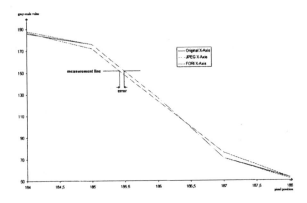

Fig. 5. pixel comparison at the object edge

Conclusion

The FORI-CSDR algorithm is able to preserve the original image quality at the important image regions. In a direct comparison of FORI-CSDR with JPEG the results show a clear advantage of the FORI-CSDR for the use in industrial image processing tasks. Figure 5 shows the difference of the original object edge and the JPEG object edge. This difference is too poor for a 5-time subpixeling algorithm, which is state of the art in industrial image processing measurement tasks.

The presented algorithm is easy to handle and needs little a priori knowledge.

References

1 Crouse, M and Ramchandran, K, Joint thresholding and quantizer selection for transform image coding: entropy-constrained analysis and applications to baseline JPEG, Proceedings of the society of photo-optical instrumentation engineers (SPIE) VOL. 2847 (1996), 356-364

2 Mitchell J L and Pennebaker, W B, MPEG Video: Compression Standard (Digital Multimedia Standard Series), in Chapman & Hall USA (1996)

3 Ortega, A and Vetterli, M, Adaptive quantization without side information, Int'l. Conf. on Image Proc. ICIP '94; Austin / Texas (Oct. 1994)

4 Ortega, A and Vetterli, M, Adaptive scalar quantization without side information, IEEE Transactions on Image Processing (1997)

5 Strohbeck, U and Jäger, U and Macgregor, A E, Context-sensitive image data reduction by FORI, Proceedings IWK '98, 43rd International Scientific Colloquium; Technical University of Ilmenau / Germany (1998), 372-377

6 Wu, S.W.; Gersho, A.: Rate-constrained picture-adaptive quantization for JPEG baseline coders; ICASSP 93; IEEE international conference on acoustics, speech and signal processing; VOL 1-5; Pages 389-392;1993

Image Compression by Approximated 2D Karhunen Loeve Transform

Władysław Skarbek[1] and Adam Pietrowcew[2]

[1] Department of Electronics and Information Technology
Warsaw University of Technology
Nowowiejska 15/19, 00-665 Warszawa, Poland
[2] Department of Informatics, Technical University of Bialystok
tel:(48 22)6605315, fax: (48 22)8255248, email:Skarbek@ire.pw.edu.pl

Abstract. Image compression is performed by 8×8 block transform based on approximated 2D Karhunen Loeve Transform. The transform matrix W is produced by eight pass, modified Oja-RLS neural algorithm which uses the learning vectors creating the image domain subdivision into 8×1 blocks. In transform domain, the stages of quantisation and entropy coding follow exactly JPEG standard principles. It appears that for images of natural scenes, the new scheme outperforms significantly JPEG standard: at the same bitrates it gives up to two decibels increase of PSNR measure while at the same image quality it gives up to 50% lower bitrates. Despite the time complexity of the proposed scheme is higher than JPEG time complexity, it is practical method for handling still images, as C++ implementation on PC platform, can encode and decode for instance LENA image in less than two seconds.

1 Introduction

In JPEG standard (conf. [5]), the image is subdivided into 8×8 blocks. Each block U is transformed into DCT (Discrete Cosine Transform) block V using a DCT matrix C :

$$V = CUC^T \tag{1}$$

The image compression scheme proposed here replaces the DCT matrix by a matrix W^T :

$$V = W^T U W \tag{2}$$

where W is a stochastic approximation of KLT matrix for a vector data sequence $X = (x_0, \ldots, x_{L-1})$ obtained from the image by the partition of the image domain into 8×1 vectors. The columns of KLT matrix create a maximal set of unit length independent eigenvectors (ordered by non increasing eiegenvalues) of the covariance matrix R_X :

$$R_X \doteq \frac{1}{L} \sum_{i=0}^{L-1} (x_i - \overline{x})(x_i - \overline{x})^T \tag{3}$$

where $\overline{x}$ is the average vector of the sequence X (compare Jain [4]).

The feasibility of Oja-RLS algorithm for image compression was verified by Cichocki, Kasprzak, and Skarbek in [2]. However, the authors considered there only 1D KLT transform for $k = 8 \cdot 8 = 64$ dimensional vectors and no attempts were made to join quantisation and binary coding stages. Here, we apply Modified Oja-RLS algorithm which contrary to Oja-RLS algorithm has the proved stochastic convergence. Moreover, we make a fair comparison with the existing standard, by replacing only JPEG's modelling stage while preserving other coding details.

2 Neural approximation of KLT matrix

The proposed neural algorithm is based on classical Oja algorithm [6] for principal component stochastic approximation. We have used the recurrent least square approach to estimate gains in the learning scheme. This idea was suggested independently by a number of authors: by Diamantaras, Kung in 1995 [3], by Bannour, Azimi-Sadjadi in 1995 [1], and by Cichocki, Kasprzak, Skarbek in 1996 [2] who gave the name Oja-RLS rule for this new scheme.

In this work we use Modified Oja-RLS rule introduced by Sikora and Skarbek in [8] where modification of the gain is made after the modification of the weight vector w (in Oja-RLS the order of modifications is opposite).

At step $n = 0, 1, \ldots$
  choose randomly x_n from X and compute:
  if $n = 0$ then $w_0 = x_0/\|x_0\|$, $\eta_0 = 2\max_{x \in X} \|x\|^2$ $\qquad$ (4)
  $y_n = x_n^T w_n$
  $w_{n+1} = w_n + y_n(x_n - y_n w_n)/\eta_n$
  $\eta_{n+1} = \eta_n + y_n^2$

Thorough analysis of stochastic convergence to the principal eigenvector e_1 for the above scheme is given in [8]. Authors describe there also certain preliminary compression results for still images. However, the complete description of the image coder and more extensive experimental evidence for the advantage of the proposed scheme over JPEG standard is given here.

The next eigenvector of R_X is found by a kind of Gramm-Schmidt orthogonalization process for the learning sequence X which gives the current X for the next stage of the algorithm.

In order to get an efficient compression algorithm while applying Modified Oja-RLS rule to approximate KLT matrix W we tune here this scheme in few points:

1. the number of steps is equal to L, i.e. to the number of elements in the training sequence;
2. at the n-th step the n-th element of the training sequence is chosen, i.e. the randomness is dropped and the algorithm becomes one epoch neural algorithm;
3. as the matrix R_X can be singular or near singular the number of its significant eigenvectors is controlled by the variance of the current learning sequence X.

The actual algorithm used to build the matrix W for the given training sequence $X = (x_0, \ldots, x_{L-1})$ $(x_i \in R^N)$ and the variance threshold τ has the form:

$$
\begin{aligned}
&\text{for } i = 1 \text{ to } N \text{ while variance of } X \text{ is greater than } \tau \text{ do:} \\
&\quad w_0 = x_0/\|x_0\|; \ \eta = 2\max_{x \in X}\|x\|^2; \\
&\quad \text{for } n = 0, 1, \ldots, L-1 \\
&\qquad y = x_n^T w; \\
&\qquad w\mathbin{+}= y(x_n - yw)/\eta; \\
&\qquad \eta\mathbin{+}= y^2; \\
&\quad w = w/\|w\|; \\
&\quad \text{set column } i \text{ of } W \text{ to } w \\
&\quad \text{for } n = 0, 1, \ldots, L-1 \\
&\qquad y = x_n^T w; \\
&\qquad x_n = x_n - yw \\
&\text{if } i \le N \text{ fill columns } i \text{ to } N \text{ with zeros;}
\end{aligned}
\tag{5}
$$

Fig. 1. Original image LENA (left) with 8[bpp] and its reconstructions after JPEG (center) with 0.53[bpp] and after KLT (right) with 0.33[bpp].

3   Experimental comparative study

The image processing in our method was very similiar to JPEG standard. At the modelling stage for each $k \times k$ ($k = 8$ as JPEG implies) image block X, we calculate the coefficient block $Y = W^T X W$. At the quantisation stage Y is quantised using the same quantisation matrix as JPEG uses (cf. [5]). The symbols for entropy coding are created using zig-zag scan, run length and value level encoding scheme, just in the same way as JPEG does.

Table 1. Comparison of bitrates at the same quality for image Lena.

Quant. scale JPEG	Bitrate (bpp) JPEG	PSNR JPEG	Quant. scale KLT	Bitrate (bpp) KLT	PSNR KLT
0.80	0.58	33.65	1.75	0.41	33.70
1.00	0.50	32.99	2.25	0.35	32.96
1.20	0.44	32.38	2.75	0.31	32.37
1.30	0.42	32.09	3.00	0.30	32.09

The KLT transform approximation for image blocks of size 8×8 is obtained by using the algorithm (4) on the learning set which is obtained by subdivision of the given image domain into blocks of size 8×1. The extracted vectors are mean shifted by subtraction the average vector of the learning set. Therefore for the image with N pixels we have about $N/8$ learning vectors.

Fig. 2. Original image LENA (left) with 8[bpp] and its reconstructions after JPEG (center) with 0.80[bpp] and after KLT (right) with 0.49[bpp].

Because for natural images N is rather large, the advantage of the neural scheme in calculating KLT matrix is rather obvious. In the standard technique, where we must establish the covariance matrix the number of operations is much greater.

The idea of using KLT for compression is not new. Actually DCT transform was invented to approximate KLT optimal energy compaction feature. It was an obvious need in seventies when computer resources such as time and memory was very limited. The technological progress gives now a chance to improve the compression efficiency using optimal KLT transform which gives better quality of

Table 2. Compression results for image Lena at the same quantisation scale.

Quant. scale	Bitrate (bpp) JPEG	PSNR JPEG	Bitrate (bpp) KLT	PSNR KLT
0.60	0.73	35.06	0.87	37.25
1.00	0.50	32.99	0.62	35.74
1.20	0.44	32.38	0.55	35.21
1.50	0.38	31.53	0.48	34.54
1.80	0.34	30.88	0.43	33.98

the reconstructed image and less bitrate. Moreover, the adaptive and incremental structure of modified Oja-RLS algorithm gives a chance to design effective compression schemes for 3D signals (i.e. time or space slices).

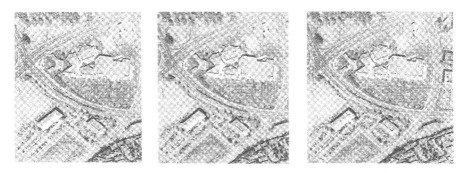

Fig. 3. Original image AERIAL (left) with 8[bpp] and its reconstructions after JPEG (center) with 1.04[bpp] and after KLT (right) with 0.63[bpp].

Figures 1–3 show visual results for compression of three different classes of natural images. The image LENA is rather smooth, BARBARA includes textures, while AERIAL contains a lot of details formed of regular lines.

Experiments were conducted on many different images. The compression efficiency is measured by the bitrate in bits per pixel (bpp) which equals the output bit stream size divided by the number of pixels. The quality of the reconstructed (decoded) image is measured by PSNR (Peak Signal to Noise Ratio) in decibels (dB) which expresses the quotient (in logarithmic scale) of the energy for the range maximal signal to the energy of the difference image between the original and the decoded images.

The results show a significant increase of the image quality for $k = 8$. The slight decrease of bitrate in this case follows from the fact that the matrix A must be send by the encoder to the output stream.

The tables 1, 3 and 5 shows an attempt to find similar level of PSNR for JPEG and KLT in order to observe the difference in bitrates between these

Table 3. Comparison of bitrate at the same image quality level for image Barbara.

Quant. scale JPEG	Bitrate (bpp) JPEG	PSNR JPEG	Quant. scale KLT	Bitrate (bpp) KLT	PSNR KLT
0.80	0.88	30.13	1.60	0.69	30.24
1.00	0.77	29.18	1.95	0.60	29.35
1.20	0.69	28.41	2.40	0.53	28.43
1.30	0.65	28.11	2.55	0.51	28.21

Table 4. Results of compression for image Barbara at the same quantisation scale.

Quant. scale	Bitrate (bpp) JPEG	PSNR JPEG	Bitrate (bpp) KLT	PSNR KLT
0.6	1.06	31.42	1.21	33.61
1.0	0.77	29.18	0.93	32.49
1.2	0.69	28.41	0.82	31.20
1.5	0.59	27.47	0.72	30.46
1.8	0.52	26.78	0.64	29.70

Table 5. Comparison of bitrate at the same image quality level for image Aerial.

Quant. scale JPEG	Bitrate (bpp) JPEG	PSNR JPEG	Quant. scale KLT	Bitrate (bpp) KLT	PSNR KLT
0.80	1.15	29.89	1.50	0.92	29.82
0.90	1.07	29.51	1.65	0.88	29.52
1.00	1.01	29.19	1.80	0.83	29.23
1.20	0.90	28.57	2.20	0.73	28.58

Table 6. Results of compression for image Aerial at the same quantisation scale.

Quant. scale	Bitrate (bpp) JPEG	PSNR JPEG	Bitrate (bpp) KLT	PSNR KLT
0.6	1.37	30.94	1.58	33.20
1.0	0.96	29.19	1.18	31.26
1.2	0.90	28.57	1.06	30.61
1.5	0.78	27.80	0.92	29.85
1.8	0.69	27.20	0.83	29.25

two methods. It was made by the manipulation of the quantisation scale which multiplies each element of the standard quantisation matrix for JPEG and by modifying the block size k for KLT.

The tables 2, 4 and 6 were prepared for the same block sizes in JPEG and KLT $k = 8$. They contain results of applying different quantisation scale coefficients for quantisation matrices in both methods.

The table 7 includes comparison of PSNR at the same bitrates for image Barbara.

Table 7. Comparison of image quality after reconstruction between JPEG and KLT for image Barbara. Block size $k = 8$.

Quant. scale JPEG	Bitrate (bpp) JPEG	PSNR JPEG	Quant. scale KLT	Bitrate (bpp) KLT	PSNR KLT
0.80	0.88	30.13	1.05	0.87	31.89
0.90	0.82	29.66	1.20	0.82	31.40
1.00	0.77	29.18	1.30	0.77	31.07
1.20	0.69	28.41	1.60	0.69	30.24
1.35	0.64	27.96	1.80	0.65	29.75

4 Conclusion

Comparing to the JPEG standard, the developed algorithm exploiting a neural approximation of 2D KLT transform in place of the DCT, allows to obtain much better results. From the one side, at the same bitrates, we could achieve incrase of PSNR measure almost by two decibels. From the other, preserving comparative quality of the image after reconstruction we could decrease bitrates up to 50% in described above method.

Though 2D KLT transform is of higher complexity than 2D DCT, the resulting quality is worth of such effort. Moreover, the algorithm stays practical to use in today home computing to compress still images. Implemented with C++ on PC platform, it can encode images 512×512 pixels in less then two seconds.

Acknowledgments: This work was in part supported by Dean's research grant no. 503/034/394/8 of Faculty of Electronic and Information Technology, Warsaw University of Technology.

References

1. Bannour S., Azimi-Sadjadi M.R.: Principal component extraction using recursive least squares learning. IEEE Transactions on Neural Networks. **6** (1995)457–469

2. Cichocki A., Kasprzak W., Skarbek W.: Adaptive learning algorithm for Principal Component Analysis with partial data. Proceedings of 13-th European Meeting on Cybernetics and Systems Research. Austrian Society for Cybernetics Studies, Vienna, Austria (1996) 1014–1019
3. Diamantaras K. I., Kung S. Y.: Principal component neural networks – Theory and applications. John Wiley & Sons, Inc. (1995)
4. Jain A.K.: Fundamentals of digital image processing. Prentice-Hall International, Englewood Cliffs NJ (1994)
5. Mitchell, Pennebaker: The JPEG standard (1995)
6. Oja, E.: A simplified neuron model as a principal component analyzer. Journal of Mathematical Biology. **15** (1982) 267–273
7. Sikora, R., Skarbek, W.: On stability of Oja-RLS algorithm. Fundamenta Informaticae. **34** (1998) 441–453
8. Skarbek, W., Sikora, R., Pietrowcew, A.: Modified Oja-RLS Algorithm – Stochastic Convergence Analysis and Application for Image Compression. Fundamenta Informaticae. (to appear in 1999)

A New Watermarking Method Using High Frequency Components to Guide the Insertion Process in the Spatial Domain

Iñaki Goirizelaia, Juanjo Unzilla, Eduado Jacob, Javier Andiano

University of the Basque Country, Electronics and Telecommunications Departament
Alameda Urquijo s/n - 48013 - Bilbao, (Spain)
{jtpgoori | jtpungaj | jtpjatae @ bicc00.bi.ehu.es}, {jtaanazj@bipt106.bi.ehu.es}

Abstract. This paper presents a new method for signing digital images using high frequency components to guide the way digital watermarks are inserted in the image. Although this is not a new idea, we implement it working in the spatial domain, which means a more efficient algorithm and less processing time. We use a user's configurable line segment chain to define the watermark and line segments that can be found in high frequency components of an image to guide the process of inserting the watermark. Therefore, this method offers a new way of finding specific zones of the image where watermark is to be inserted. It also presents a distributed system for signing digital images using watermarks and delivering watermarked images over Internet. This system provides secure access using Java applets and a Certification Authority.

1 Introduction

A digital watermark can be defined as a digital image that can be embedded in the original image and therefore be used to prove who the owner of the image is. As it is generally accepted [1], a watermark should be invisible to human perception and it should not affect image quality. The retrieved watermark should unambiguously identify the owner. It should not be detected by statistical proofs and it should be difficult to remove. The watermark should be robust to different image processing algorithms like filters, compression, scaling, etc.. During the last years digital watermarking has become a very active research field and lot of new methods has been proposed. As it is described in [1], most of these methods are applied in the frequency domain, using DCT or another image transform to determine where to embed the signal (watermark).

Our work is based on the idea that all the images have inside information that can be used to guide the process of inserting the watermark in the image. The method that we propose is based on locating high frequency components of the image in the spatial domain and the embedding operation is also performed in this domain. After that we applied a line segment detection algorithm that produces several line segment chains, that are used to guide the way we insert the watermark in the image. So our algorithm is image adaptive, because these areas of the image are strongly dependent of the image itself. Previous works proposed for working in spatial domain were based on less significant bit (LSB) modification of the pixels located in several areas selected by pseudo-random sequence number [2]. These methods are very sensible to common signal processing operations, and don't resist lossy compression. Other proposals insert the watermark in image contours or texture, but using frequency domain processing [3].

2   Digital image processing techniques

2.1   Edge detection using gradient operator

Transition detection is done using different enhancement edge operators. We used the gradient operator because it allows us to detect those areas of the image that will be used as a guide in the process of watermarking. Gradient calculation process is based on matrix operations, as it's well known in the image processing literature [4].

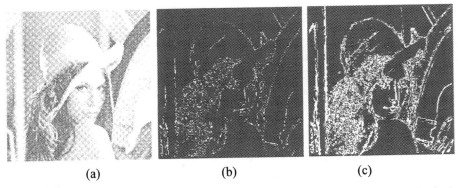

(a)   (b)   (c)

Fig. 1. Original image (a), and processed image with high gradient threshold (b) and with low gradient threshold (c).

Results of applying gradient operator to an image (figure 1 (a)) are shown in figures 1 (b) and 1 (c). The image representation shown in these figures is obtained after calculating and applying a **gradient threshold**. Those pixels with a gradient module higher than the gradient threshold are set to 255 (white) and the rest are set to 0 (black). Using different values of the gradient threshold implies different results in the edge detection process (see figures 1 (b) and 1 (c)). In our implementation, the value of the gradient threshold is defined by the average value of luminance of the image and it is calculated by analyzing its histogram. So the value of gradient threshold depends on the image and it is obtained dynamically, making the algorithm more flexible in that way.

2.2 Image Segmentation

Once we define all of the pixels in the image where luminance transitions take place by using gradient operator, we code them generating a list of connected segments according to a previous chain code definition. Figure 2(a) shows all the direction codes considered for encoding segments found in images. Therefore this image segmentation process allows us to go from local pixels to line segment chains. We can define a line segment chain as a set of connected segments where each segment is composed of pixels where the intensity variation has the same direction.

To obtain line segment chains we take pixel (x,y), for which we know its gradient module and argument. If module is higher than threshold, it means that around that pixel there may be a color transition, and therefore pixels in that area may belong to an image contour. The argument tells us the color transition direction. In this direction we analyze the module value of next pixel. If this value is higher than threshold, that pixel also belongs to the same image contour. This line segment finishes when we find a pixel where the gradient module is not higher than the threshold value. We codify all the pixels considering that all the pixels a line segment have to be connected, they should have the same chain code and a new line segment should be created as soon as there is a change in the direction. Each line segment is defined by its starting point, chain code and number of pixels, as is shown in figure 2 (b). As we discuss later, watermarks will be defined in the same way.

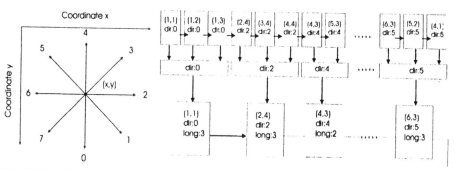

Fig. 2. Directions considered for encoding line segments (a) and chain code example (b)

2.3 Enhanced segmentation algorithm

We use an enhanced segmentation algorithm, because experimental results of segmentation algorithm described previously lead us to small line segments, less than 10 pixels average in almost 90% of them. We look for point M of the list where the distance to the segment defined by p_0 and p_n is the highest (figure 3 (a)). If distance from point M to segment defined by p_0 and p_n, is lower than predefined value called tolerance, we model the set of pixels p_0 ... p_n like line segment (figure 3 (b)), in the other case, we call the algorithm recursively taking as arguments two new lists, (p_0, M) and (M, p_0). Applying this modification to our segmentation method, we get longer line segments and lower number of them, that reduces considerably processing time.

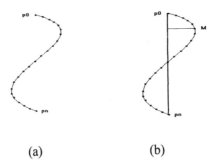

(a)   (b)

Fig. 3. Set of pixels (a), and line segment approximation model (b).

3   Watermark definition

The watermark will be described in the same way that the image itself after segmentation, i.e. using chain codes. Watermarks are generated using author, customer and Trusted Third Party (TTP) identification data, a timestamp and a specific code for each operation. In our tests we use numerical codes, but any data should be used applying a hash function [5]. This information is coded in five blocks, each one consisting of four bytes, so watermark is 20 bytes long. The coding scheme uses BCD code to represent author, customer and TTP data. This means that we can use values between 0 and 99.999.999. Timestamp is defined by using a Unix format, represented by 32-bit integer (four bytes).

Once these 20 bytes are defined, they must be converted to a chain code in terms of direction and length. We use 3 bit to code the length of line segment (0 to 7) and other 3 bits to code its direction. This process is developed considering the whole watermark data as a bit sequence, grouping them in three bit sets. The following example shows the process; for an input like this: 011 101 000 111 011 010 .. the chain code, in pairs (direction-length), will be: 3-5; 0-7; 3-2; ..

Inserting a timestamp in the watermark and using TTP certificates [4] we can address the problem of several watermarks of different people in the same image, discussed in [6].

4 Watermark insertion algorithm

4.1 Carrier selection algorithm

The criteria that we use to insert the watermark is based on the two data structures that we have: the watermark and the line segments found in the image. Both of these data structures are defined as a chain of line segments. To insert the watermark in the image we take the first line segment of the watermark and the first one of the line segment chain. We begin analyzing the segmens of the original image trying to find the first segment in the chain of the original image that has the same chain code and a length greater or equal than the one selected from the watermark. Once we locate it, we find the central point of this segment and from this point and in a perpendicular direction a matrix of pixels is selected. The size of this matrix (n x n) and the distance (d) from the center point (see figure 4) can be user's configurable. We repeat this process until all the segments of the original image have been analyzed and all the segments of the watermark have been used to mark different areas of the image.

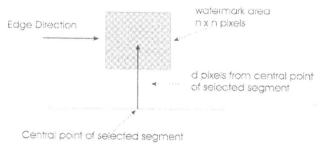

Fig. 4. Line segment chain and selected zone to insert watermark

If one segment fits with watermarking conditions, before it is watermarked we analyze whether it has been previously marked, then it is skipped and we continue analyzing the next segment. If it hasn't been marked yet, it is added to a list (carrier) where we put marked segments. Graphically, the process is shown in figure 5.

By using this implementation, pixels are watermarked only once, and we require that direction sequence defined by watermark must be equal to the direction sequence of the image contour.

4.2 Pixel modification

Pixel modification is also image dependent. In the selected areas we consider pixel luminance values, increasing this value for N pixels in the selected matrix and decreasing in the same quantity other N pixels, where N is the size of the square area to be modified. This variation is done in one of the three RGB (red, green, blue) colors, the less significant one in that image. The N increased pixels are the N lowest ones, and the decreased pixels are the N highest ones.

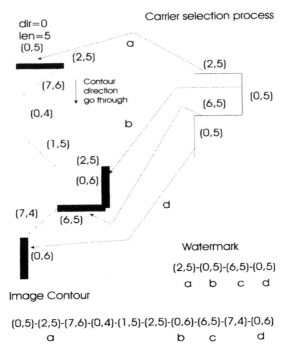

Fig. 5.: Carrier selection.

The quantity to add or to reduce is the average value of the difference between V_{max} (the maximum value of luminance in this area) and V_{min}, (the minimum one). To avoid saturation, i.e. for any pixel the new value be higher than 255 or lower than 0, a previous calculation is required. In this case, the value to add/reduce is divided by 2 successively until the new values fits between the limits. If more than N pixels have the same luminance value, we choose pixels to add/reduce its luminance following the left to right and up to down criteria. With this modification the luminance average value in this matrix is not modified. During detection process we look for an incremental variation and not absolute values, which makes the algorithm more robust and tolerant to signal processing operations.

5   Watermarking detection algorithm

The proposed method establishes two different ways to detect if a watermark is present or not in a image. In both of them we need either watermarking process information or the original image. First of the ways establishes that the watermarked image should be processed in the same way as during insertion process, i.e. using gradient operator, segmentation and line segment encoding. With watermark definition data, we verify that correspondent areas have the luminance modification previously defined. In this case we need all the data that defines the watermark, but the original image is not needed. The second way is based on comparing pixel luminance values in the selected areas. When we insert the watermark, we save in a file the position of carrier pixels. By using this file and watermarked image we can select those pixels in the watermarked image where a luminance modification is expected. Luminance values of the watermarked image in the selected areas and values of the original image after applying our modification algorithm are compared.

Based on the results, we can deduce if the image has a watermark or not. Three options are possible: image has the watermark (more than 90% of position coincidence), the image has the watermark but it has been manipulated (values between 90% and 30%) and the image hasn't it (values less than 30%). Other proposed techniques don't need the original image to watermarking extraction process. Original image or information of watermarking process could be requiered as another proof of ownership.

6 Distributed system for validation

Validation distributed system elements are the server, the Database, the Certification Authority (CA), and the Client (see figure 6). Image distribution over Internet is done by watermark insertion in real time, providing access to them in a secure way using Java applets. There are two operation modes depending on client features. In the first case (unidentified client) the client must know the CA public key (1) in order to encrypt data that only CA could decrypt. CA signs with his private key server's public key (2). The server sends this signed key to the client (3). So, client gets server's public key in a secured way (he can verify it using CA public key). That is, client has a proof of server's identity. The client creates a random session key and sends it to the server encrypted with the server's public key (4). Server decrypts it and from now on they have an encrypted channel between them. The client requests home page from server through this channel (5). The server creates a random session key that is sent to the client in secured way and the applet that will manage the connection with database and watermark application. The server sends this random session key to the database application in a secured way (6).

The applet that establishes the connection with the database begins a connection protocol sending client identification data (for example IP address, a machine name, a code) encrypted with the session key created by the server (7). Database application confirms the reception of this identifier assigning a code that client must use to request images from database (8). This code forms part of the watermark. The client selects an image from the database (9) and it is watermarked in real time with author, customer and TTP identifiers, the code previously defined and a time stamp. The data related with the operation is stored in the database (10).

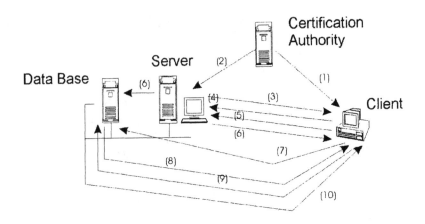

Fig. 6. Watermarked image distribution process.

In the second case (identified client), the client should have a certificate (a public key signed by one CA). With it, the server can prove client identity and optionally use two different session keys (one in each way). Mutual identification and watermarking processes follow a similar mechanism to the one explained previously.

7   Experimental results

Experimental tests have proven that our algorithm is very robust to JPEG compression, colours and resolution modifications if this variation is not visible, spatial filters that don't produce significant modifications in the original image and invisible random noise addition. Figure 7 shows the original image (a), the same image after applying gradient operator and segmentation algorithm (b), the selected areas where we insert the pixel modification for a given identification data (c). Image areas where we insert the watermark are very similar to the ones obtained using frequency domain processing, as is shown in [7].

8   Conclusions

The new algorithm that we present, allows us to insert watermarks using spatial domain image processing techniques, which provide us faster processing times. We prove that it is possible to watermark digital images using high frequency components to guide the process of digital image watermarking in the spatial domain. The implemented carrier selection procedure guarantees robustness against usual processing techniques and common compression algorithms because the watermark is inserted in important information areas of the image. This work describes a new algorithm that uses the same method to define the watermark, and to describe and select those areas of the image where the watermark is to be inserted.

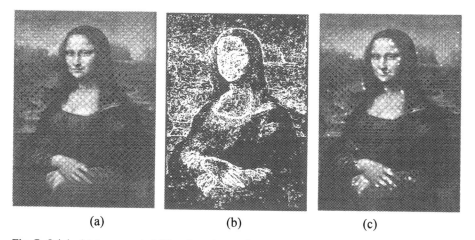

(a)   (b)   (c)

Fig. 7. Original (a), segmented (b), selected areas for watermark insertion (c) and watermarked images (d).

A distributed arquitecture is proposed for watermarked image distribution over Internet, using Java applets that allows us to work with almost any kind of platform. CA interaction provides us a way to certificate ownership, and develop high secure transactions.

9 Acknowledgments

This work has been partially supported by UPV/EHU (University of the Basque Country) under research projects UPV 147.345-TA044/95 and 147.345-TA119/97, and also supported by GV/EJ (Basque Country Government) under the research project PI95/82. Implementation was done in the Iberdrola's classroom of the School of Engineering in Bilbao (University of the Basque Country).

References

[1] I. Cox, Kilian, J, Leighton, T. Shamonn. 1997. "Secure Spread Spectrum Watermarking for Multimedia". IEEE Transactions on Image Processing, vol 6, no. 12, pp. 1673-1687. (1997).
[2] S. Steve Walton. "Image Authentication for a slippery new age". Dr. Dobb's Journal, pp 18-26. (1995).
[3] J. Zhao. "Look, It's not there". Byte International, pp. 7-12. (1997).
[4] R. C. Gonzalez, R. E. Woods. "Digital Image Processing". Addison-Wesley. (1996).
[5] W. Ford, M. S. Baum. "Secure Electronic Commerce". New Jersey: Prentice Hall (1997).
[6] S. Craver, N. Memon, B. Yeo, M. Yeung. "Can Invisible Watermarks Resolve Rightful Ownerships?". IBM Research Report. 96-07. (1996).
[7] R. Wolfgang, C. Podilchuck, E. J. Delp. "The effect of matching watermark and compression transforms in compressed color images". International Conference on Image Processing. Vol 1, MA-11.04.(1998).

Adventurous Tourism for Couch Potatoes

Luc Van Gool[1,2], Tinne Tuytelaars[1] and Marc Pollefeys[1]

[1] University of Leuven, Kard. Mercierlaan 94, 3001 Leuven, Belgium
[2] Inst. Kommunikationstechnik, ETH, Gloriastr. 35, CH-8092 Zürich, Switzerland

Abstract. Two tourist guides are described. One supports a virtual tour through an archaeological site, the other a tour through a real exhibition. The first system is based on the 3D reconstruction of the ancient city of Sagalassos. A virtual guide, represented by an animated mask can be given commands using natural speech. Through its expressions the mask makes clear whether the questions have been understood, whether they make sense, etc. Its presence largely increases the intuitiveness of the interface. This system is described only very concisely.

A second system is a palmtop assistant that gives information about the paintings at the ongoing Van Dijck exhibition in the Antwerp museum of fine arts. The system consists of a handheld PC with camera and Ethernet radio link. Images are taken of paintings or details thereof. The images are analysed by a server, which sends back information about the particular painting or the details. It gives visitors more autonomy in deciding in which order to look at pieces and how much information is required about each. The system is based on image database retrieval, where interest points are characterised by geometric/photometric invariants of their neighbourhoods.

1   Introduction

The amalgam of more powerful computer vision techniques, faster computers, and wireless communications coupled to progressing miniaturisation yields technology that will soon pervade our lifes in many, so far mostly unimaginable ways. One area of applications probably is cultural tourism. One can imagine systems that allow us to visit sites only virtually and yet create a feeling of really being there. Alternatively, we may get on that plane and actually visit other places. But rather than spending our time on searching for the things we want to see and trying to get some information about them, virtual guides could accompany us and supply information 'just-in-time'.

In this paper, we describe two systems under development that can assist future tourists to get more information with less effort. Section 2 describes a system that supports virtual tours through the archaeological site of Sagalassos. Section 3 describes a palmtop guide to lead visitors through a real exhibition. The 1999 Van Dijck exposition in the Antwerp Museum of Fine Arts was taken as a show case. Section 4 concludes the paper.

2   Virtual Sagalassos

The first system integrates a diversity of techniques to create virtual tours through the ancient city of Sagalassos. This old Greek city was destroyed by an earthquake in the 6th century. The goal of the demonstrator is to show this archeological site in 3D to a virtual visitor. He/She is accompanied by a virtual guide, who can be spoken to. The visitor can e.g. ask to go to a certain building or ask questions about finds that have been made at different places. The system integrates techniques for the 3D reconstruction of the site and its buildings from uncalibrated video sequences [6], techniques for the recognition of fluent speech (both Dutch and English, developed by our colleagues of speech recognition) and techniques to animate the face of the virtual guide. Missing parts of the buildlings are replaced by CAD-models composed by archeologists.

The guide makes the interaction more intuitive. A nod of the head shows that the system has "understood" the instruction, otherwise the eyebrows will be raised to prompt a repetition of the question. The visitor gets a 'no' if the instruction doesn't make sense at the current stage of the tour, e.g. if one asks to show a statue where there is none. Fig. 1 shows some example views during such a virtual visit. In the near future we plan to also let the guide's face talk, to initiate a real conversation with the visitor.

Fig. 1. *Holiday pictures from a trip to virtual Sagalassos.*

Such systems in a way yield an experience that is more complete then if one were on-site. It is e.g. possible to show the city in its diverse stages of development and to see the city grow towards its most prosperous age. Finds that are now stored in a museum can be put on their original locations. Ruins can be shown in their original context.

This sytem allows us remote access to a site. The system described next is meant as a guide to a real exhibition with the visitor on-site.

3 Palmtop pictures of an exhibition

3.1 System overview

It is quite common for museums to provide tape or CDROM players as a replacement for human guides. Although they offer a kind of individual service in terms of the time a visitor can spend on different exhibits, it simultaneously is also more homogeneous, as no questions can be asked to get additional information. It would be better if a visitor could interact more dynamically with the virtual guide than pushing the 'next' or 'feedforward' button. An example would be pointing a camera at a piece to solicit information about it, and zooming in on parts if the visitor is intrigued by certain details. Such a system is the subject of this section.

The Museum of Fine Arts in Antwerp has for 1999 organised an important exhibition about the work of Antoon Van Dijck, in remembrance of the painter's 400th birthday. We have used this exhibition as a show case.

The basic setup is a handheld PC with built-in camera (e.g. Sharp's HC 4600 A) or a separate digital camera connected to the handheld pc via an infrared link. The digital image is then sent to a server via a wireless ethernet connection, where it is processed such that the correct information can be returned to the user. Figure 2 gives an overview. The server recognizes the painting the visitor

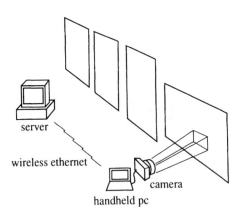

Fig. 2. *Overview of the system's architecture.*

is pointing the camera at. The visitor can also select details for more specific

information. In fact, the server performs 'database retrieval'. Once the content of the picture has been recognized, the appropriate information is returned to the user over the same wireless ethernet connection.

As the visitor cannot be expected to always stand right in front of a painting and as illumination conditions may change (e.g. shadows of other visitors), the extracted image descriptors should remain invariant under such variability. Also, other visitors may occlude parts of the piece. This calls for the use of local features, i.e. features that are based on relatively small parts of the images. Section 3.2 describes the image processing steps in somewhat more detail.

An advantage of such a system - apart from its intuitive user interface - is that it can derive from the picture what the user is actually interested in and adapt the returned information accordingly. This is difficult to achieve with e.g. electronic tour guides that use the viewer's location as primary input (e.g. using infrared transmitters placed in the ceiling above each piece to be described) [1]. The visitor is also free to roam the museum and ask for information in whatever order, which is difficult with tape and CD-ROM players. Also, since all that is needed is a picture of the piece, no external adaptations have to be made. This makes the basic approach equally suited for less controlled environments like a moving exhibition, an open air show, or even a city tour.

Probably the most similar system is the Dynamic Personal Enhanced Reality System (DyPERS) of Jebara et al. [2], which has (among other applications) also been evaluated in a museum-gallery scenario. However, they recognize objects based on multidimensional receptive field histograms, which is probably less robust to changes in viewpoint than our system.

3.2 Recognizing a painting or a detail thereof

The recognition process should be sufficiently robust, so that our camera–empowered visitor doesn't have to push aside other visitors to get shadow-free, frontal views. People may be walking in front of the camera causing partial occlusions, lighting conditions may change, etc. To deal with these problems retrieval is based on local, invariant features.

In a first step interest points are selected. For the moment these are only corners, detected with the Harris corner detector (i.e. points with a high degree of information content, as they are surrounded by intensity profiles that show large changes in different directions). Then, around each of these points a number of neighbourhoods are selected. The goal is to arrive at neighbourhoods that cover the same parts of the painting, irrespective of the viewpoint. This implies that the neighbourhood should deform as the observer moves around. Finally, from these neighbourhoods invariant features are extracted and used for interest point recognition in the database.

The system combines geometric and photometric invariance. Not only can the viewpoint change, but also the illumination conditions may vary or the spectral responses of the camera or scanner with which the database was produced may differ from those of the visitor's camera. Both the construction of the neighbour-hoods and the extraction of features from these neighourhoods must yield the same results irrespective of such changes. For this sytem, we consider invariance under 2D affine deformations

$$\begin{pmatrix} x' \\ y' \end{pmatrix} = \begin{pmatrix} a & b \\ c & d \end{pmatrix} \begin{pmatrix} x \\ y \end{pmatrix} + \begin{pmatrix} e \\ f \end{pmatrix}$$

and linear changes in the 3 colour bands, where each band is subject to its own scaling and offset.

$$\begin{pmatrix} R' \\ G' \\ B' \end{pmatrix} = \begin{pmatrix} s_R & 0 & 0 \\ 0 & s_G & 0 \\ 0 & 0 & s_B \end{pmatrix} \begin{pmatrix} R \\ G \\ B \end{pmatrix} + \begin{pmatrix} o_R \\ o_G \\ o_B \end{pmatrix}$$

The crux of the matter are the invariant neighbourhoods, i.e. the definition of neighbourhoods that depend on the underlying image structure in such a way that their content is not altered by a change in viewpoint. The selected neighbourhoods correspond to viewpoint dependent parallelograms with corners as one of their vertices. When changing the position of the camera the neighbourhoods in the image change their shape so that they cover the same physical part of the object or scene. Figure 3 illustrates this. The image on the right is part of the database image "Ages of Man" (also shown in figure 5). The image on the left is part of a query image taken from a different viewpoint. Although the illumination conditions have changed drastically and the image is distorted due to the change in viewpoint, the affinely invariant neighbourhoods found in both images correspond, i.e. they are deformed in such a way that they still have the same physical content in spite of the change in viewpoint. Note that these neighbourhoods are constructed solely on the basis of each image separately.

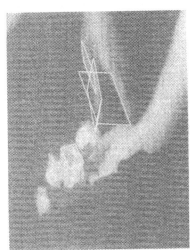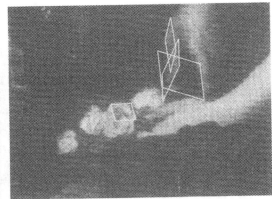

Fig. 3. *A detail of the database image "The Ages of Man"(right) and of the query image (left) with corresponding affinely invariant neighbourhoods.*

This approach is reminiscent of the work of Schmid and Mohr [5] who also used interest points and their neighbourhoods, but in their case invariance is under Euclidean motions. Pritchett and Zisserman [4] also used parallelogram-shaped regions to find stereo correspondences under wide baseline conditions. In their case these were explicitly present in the image and not constructed from general surface textures as is the case here.

The steps in the construction of the affinely invariant neighbourhoods are all invariant under the forementioned geometric and photometric changes. More information on this construction is given elsewhere [7]. The geometric/photometric invariants used to characterize the neighbourhoods are moment invariants. For a complete classification of moment invariants of lower orders, we refer to [3]. The moments are so-called 'Generalized Colour Moments'. These better exploit the multispectral nature of the data. They contain powers of the image coordinates *and* the intensities of the different colour channels. They yield a broader set of features to build the moment invariants from and, as a result, moment invariants that are simpler and more robust to image noise.

Once such local, invariant descriptions have been derived, the image can easily be matched with the images in the database using a voting mechanism. For each neighbourhood, a feature vector is composed from moment invariants. The closest matching neighbourhood in the database is retrieved using the Mahalanobis distance. The painting with the maximum number of closest matches is selected. The efficiency of this voting mechanism is enhanced using hashing techniques. This renders the retrieval procedure less dependent on the image database size.

Each correspondence also gives an approximation for the affine transformation linking both images. This can be exploited in two ways. Firstly, this allows to reject false matches (i.e. matches that give a different affine transformation than the other matches). Secondly, it allows the server to precisely define what part of the database image the user is actually interested in, as the borders of the query image can be backprojected into the database image. The latter use is illustrated in the next section.

3.3 Experimental results

Due to organisational problems as well as security considerations, we were in the end not able to evaluate our system at the real Van Dijck exhibition. Instead, we created a small mock-up gallery in our lab based on $50 \times 70 \ cm^2$ reproductions. It is important to note that this has complicated rather than simplified matters. The database images were scanned form a catalogue with high quality reproductions. Several posters in the mock-up gallery were of much lower quality, however, leading to a loss of colour.

Images of these posters were matched to high-resolution (1000×1500 pixels) digital images of 20 of Van Dijck's paintings stored in our database. An overview of the database is given in figure 4. In total, 12.978 affinely invariant neighbourhoods were found in these 20 images, i.e. an average of 650 regions per image.

Fig. 4. *An overview of our database.*

As a first example, look at the painting called "The Ages of Man", shown in figure 5. At the top, the database image of the painting is shown with some possible user queries just below. Note the large differences in colour and the affine image deformations. Not visible in this rescaled figure is the difference in scale, which varied in between 80 and 130 percent. Nevertheless, the correct painting was retrieved from our database for each of the three query images. On top of the database image, parallellograms are added. These represent the corresponding image part as computed by the server, through the backprojection method mentioned earlier.

Similarly, a second example based on the painting "Maria Louisa de Tassis" is shown in figure 6.

Fig. 5. *Given the query images shown below, the system was each time able to retrieve the correct database image (i.e. to identify the painting out of the list of 20) and to highlight the corresponding part therein (top).*

Fig. 6. *Given the query images shown at the right, the system was each time able to retrieve the correct database image and to highlight the corresponding part in the database image (left).*

4   Conclusion

Two tourist guides were described. The first one supports virtual tours through the archaeological site of Sagalassos based on a 3D reconstruction of the site. Mid-term goal is to fully integrate immersive visualisation with a natural dialogue with the animated virtual guide.

The second system describes a real tour through an exhibition. A visitor simply takes a picture of a piece s/he wants to learn more about. This picture is then transferred to a server via a wireless ethernet connection, where it is matched to the images in the database based on the notion of local, affinely invariant neighbourhoods. Based on the corresponding regions found, the correct information can be returned to the visitor. Advantages of the system are its easy and intuitive user-interface, its ability to respond to different levels of detail and its flexibility (no external actions required). Although the experimental results presented in this paper are limited to images of 2D objects, the method may also be applied to 3D objects such as buildings or sculptures, as long as the surfaces can locally be approximated by planes [7].

Acknowledgements: Tinne Tuytelaars gratefully acknowledges an FWO grant of the Flemish Fund for Scientific Research. Implementations were done in TargetJr. The authors gratefully acknowledge the support by the IUAP 4/24 project IMechS, financed by the Belgian OSTC.

References

1. B. Bederson *Audio Augmented Reality: A prototype Automated Tour Guide*, ACM Human Computer in Computing Systems conference (CHI'95), pp. 210-211, 1995.
2. T. Jebara, B. Schiele, N. Oliver, A. Pentland *DyPERS: Dynamic Personal Enhanced Reality System* MIT Media Laboratory, Perceptual Computing Technical Report nb 463.
3. F. Mindru, T. Moons, L. Van Gool *Color-based Moment Invariants for the Viewpoint and Illumination Independent Recognition of Planar Color Patterns*, to appear at ICAPR, Plymouth, 1998.
4. P. Pritchett, A. Zisserman *Wide baseline stereo matching*, Proc. International Conference on Computer Vision (ICCV '98), pp. 754-759, 1998.
5. C. Schmid, R. Mohr *Local Greyvalue Invariants for Image Retrieval*, PAMI Vol. 19, no. 5, pp 872-877, may 1997.
6. M. Pollefeys, R. Koch and L. Van Gool, *Self-calibration and metric reconstruction in spite of varying and unknown internal camera parameters*, Int. Conf. on Computer Vision, pp. 90-95, 1998.
7. T. Tuytelaars, L. Van Gool *Content-based Image Retrieval based on Local Affinely Invariant Regions*, International Conference on Visual Information Systems, Visual99, 1999.

Linear vs. Quadratic Optimization Algorithms for Bias Correction of Left Ventricle Chamber Boundaries in Low Contrast Projection Ventriculograms Produced from Xray Cardiac Catheterization Procedure

Jasjit S. Suri[1], Robert M. Haralick[1], and Florence H. Sheehan[2]

[1] Intelligent Systems Laboratory, Department of Electrical Engineering,
University of Washington, Seattle, WA 98105, USA
[2] Cardiovascular Research and Training Center, Division of Cardiology,
University of Washington Medical Center,
University of Washington, Seattle, WA 98195, USA

Abstract. Cardiac catheterization procedure produces ventriculograms which have very low contrast in the apical, anterior and inferior zones of the left ventricle (LV). Pixel-based classifiers operating on these images produce boundaries which have systematic positional and orientation bias and have a mean error of about 10.5 mm. Using the LV convex information, comprising of the apex and the aortic valve plane, this paper presents a comparison of the linear and quadratic optimization algorithms to remove these biases. These algorithms are named after the way the coefficients are computed: the identical coefficient and the indepen-

left ventriculograms have poor image quality. The main reasons of the poor quality in the LV apex is as follows: (i) The contrast agent is unable to reach the apex zone of the LV. This is partially due to the curling of the catheter which is necessary to avoid irritation to the patient. (ii) Large LV size with respect to the catheter outlet source. (iii) Abnormality of the LV shape also contributes to the poor propagation of the contrast agent towards apex. (iv) The dynamics of blood-mixing with the contrast agent is not homogeneous in the LV chamber. This is because of the muscle resistivity. Some boundary muscle tissue are thick which resists the agent to penetrate towards the apex. Besides apex, the inferior wall of the LV chamber is also of poor quality because of the superposition of diaphragm over the LV. The projection of the ribs over the LV in the LVgrams is another cause of the poor contrast. The motion artifacts and noise due to the scattering of the X-ray radiation by tissue volumes which is not related to the LV also contribute towards the low quality LVgrams.

In an attempt to automatically estimate the accurate boundaries of the LV chamber, several researchers have tried proposing their models in ventriculograms and echocardiograms. Image processing techniques applied to these two sets of images fall broadly in many classes but we will highlight the major and directly related once. Van Bree et al. [2] estimated the LV borders using a combination of probability surfaces and dynamic programming in LVgrams. Cootes et al. [3] attempted using an active shape model to infer the position of boundary parts where there was missing data (top of the ventricle). Cootes et al. used the knowledge of the expected shape combined with information from the areas of the image where good evidence of the wall could be found. The least squares method was used. Cootes et al. used weighted algorithm for final shape estimation where the weights were proportional to the std. deviation of the shape parameter over the training set. Lee [4] used a pixel-based Bayesian approach for the LV chamber boundary estimation where the gray scale values of the location throughout the cardiac cycle was taken as a vector. The above methods do produce boundaries but due to the reasons stated above, the boundaries fall short, have jaggedness, over estimation, under-estimation, irregularities and are not close to the ground truth boundaries [4] thereby making the system incomplete and unreliable. In the inferior wall region, the papillary muscles have a non-uniform structure unlike the anterior wall region. This non-uniformity causes further variation in the apparent boundary during the heart cycle. Because of this, the initial boundary position of the inferior walls are sometimes over-estimated. In an attempt to correct the initial image processing boundaries, Suri et al. [5] first attempted the two linear calibration algorithms to estimate the LV boundaries without taking the apex information. For further improving the LV boundary error, Suri et al. [6] presented a Greedy LV boundary estimation algorithm which fused the boundaries coming from two different techniques. To reduce further error in the apex zone, Suri et al. [7] developed ES apex estimation technique using the ED apex, so called dependence approach. This estimated ES apex was then used in ES boundary estimation process. Using the WLS algorithm, Suri et al. [8] developed the apex estimation using the ruled surface model. Though the apex error did reduce but the inferior wall zone in many subjects could not be controlled due to the large classifier error, which was due to the overlap of the LV and the diaphragm. This was validated by Suri et al. [9] using a training based system which utilized the LV gray scale images and the estimated boundaries. Suri et al. [10] then developed the surface fitting approach and mathematical morphology to identify the diaphragm edges and separate the LV from diaphragm. Then classifier boundaries was merged with edge-based LV and better of the two was taken for calibration. By penalizing the apex vertex in linear calibration, Suri et al. [11] then forced the estimated curve to pass through the apex by drastically reducing the apex error given the mean system error to 2.89 mm. Recently a further improvement was made by Suri, Haralick and Sheehan to develop a Quadratic calibration scheme [12]. This paper compares the linear and quadratic calibration schemes for LV boundary quantification. This approach is different from the earlier approaches is that we use a training based optimization procedure over the initial boundaries to correct its bias. This bias correction can be thought as a calibration procedure where the boundaries produced by the image processing techniques are corrected using the global shape information gathered from the ideal database.

2 Two Linear and Two Quadratic Calibrators

This section gives the mathematical statements of the two calibration methods used for estimation of the LV boundaries. Ground truth boundaries refer to the hand delineated boundaries drawn by the cardiologist. Raw boundaries refer to the boundaries produced by the initial pixel-based automated classification procedure [4]. In the *identical coefficient* (IdCM), each vertex is associated with a set of coefficients. The calibrated x-coordinate for that vertex is computed as the linear combination of raw x-coordinates of the LV boundary using the coefficients associated with that vertex. The calibrated y-coordinate of that vertex is similarly computed as the *same* linear combination of raw y-coordinates of the LV boundary. In the *independent coefficient* (InCM), the calibrated x-coordinate is computed as the linear combination of raw x-and raw y-coordinates of the LV boundary, using the coefficients associated with that vertex. The calibrated y-coordinate of that vertex is computed with a *different* linear combination of raw x-and y-coordinates. The problem of calibration then reduces to a problem of determining the coefficients of the linear combination. This can be accomplished by solving a regression problem. Since input raw and ground truth LV boundaries are initially in a 100 vertices polygon format with unit dimensions in mm, we therefore resample and interpolate each of these polygons into equally spaced vertices before it undergoes the calibration procedure discussed below.

Identical Coefficient Method for any Frame of Cardiac Cycle: Let g_n' and h_n' be the row vectors of x-coordinates and y-coordinates respectively for the ground truth boundaries for patient n. Let r_n' and s_n' be the row vectors of x-coordinates and y-coordinates respectively for the classifier boundary for any patient n, where $n = 1, ..., N$. For the calibrated boundary estimation in ventriculograms using the IdCM, we are:

- **Given**: Corresponding pairs of ground truth boundary matrix $\mathbf{R}$ $[2N \times P]$, and the classifier boundary matrix $\mathbf{Q}$ $[2N \times (P + k)]$, respectively as:

$$\mathbf{R} = \begin{pmatrix} g_1' \\ h_1' \\ \cdots \\ \cdots \\ g_N' \\ h_N' \end{pmatrix} \qquad \mathbf{Q} = \begin{pmatrix} r_1'' \\ s_1'' \\ \cdots \\ \cdots \\ r_N'' \\ s_N'' \end{pmatrix}$$

where for linear calibration, r_n'' and s_n'' are given as: $r_n' \ \underbrace{1 \ u_{1n} \ u_{2n}}_{LV} \ \underbrace{p_n}_{AoV} \ \underbrace{}_{Apex}$ and

$s_n' \ \underbrace{1}_{} \ \underbrace{v_{1n} \ v_{2n}}_{AoV} \ \underbrace{q_n}_{Apex}$. For quadratic calibration, r_n'' and s_n'' are:

$\underbrace{r_n'}_{LV} \ \underbrace{1 \ u_{1n} \ u_{2n}}_{AoV} \ \underbrace{p_n}_{Apex} \ \underbrace{u_{1n}^2 \ u_{2n}^2 \ u_{1n}v_{1n} \ u_{1n}v_{2n} \ u_{2n}v_{1n} \ u_{2n}v_{2n}}_{Quadratic-Terms}$ and

$\underbrace{s_n'}_{LV} \ \underbrace{1 \ v_{1n} \ v_{2n}}_{AoV} \ \underbrace{q_n}_{Apex} \ \underbrace{v_{1n}^2 \ v_{2n}^2 \ u_{1n}v_{1n} \ u_{1n}v_{2n} \ u_{2n}v_{1n} \ u_{2n}v_{2n}}_{Quadratic-Terms}$ where (u_{1n}, v_{1n}), and

(u_{2n}, v_{2n}) are the anterior and inferior aspect of the AoV plane for patient n. (p_n, q_n) represents the apex coordinates of the LV for patent n. Note $k=4$ for linear identical coefficient method (L-IdCM) and $k=10$ for quadratic identical coefficient method (Q-IdCM).

- Let $\mathbf{A}$ $[(P + k) \times P]$ be the unknown coefficient matrix.
- **The problem is to estimate the coefficient matrix $\mathbf{A}$, to minimize $\| \mathbf{R} - \mathbf{Q} \mathbf{A} \|^2$.** Then for any classifier boundary matrix $\mathbf{Q}$, the calibrated vertices of the boundary are given by $\mathbf{Q}\hat{\mathbf{A}}$, where $\hat{\mathbf{A}}$ is the estimated coefficients.

Note that from the problem formulation, the coefficients that multiply g_n' also multiply h_n', hence the name *identical coefficient*. Also note that the new x-coordinates for the nth boundary only depend on the old x-coordinates from the nth boundary, and the new y-coordinates from the nth boundary only depend on the old y-coordinates from the nth boundary.

Independent Coefficient Method for any Frame of Cardiac Cycle: As before, let g_n' and h_n' be the row vectors of x- and y- coordinates for any patient n. Let r_n' and s_n' be the row vectors of x-and y-coordinates of the classifier boundary. For the calibrated boundary estimation in ventriculograms using the *independent coefficient* method, we are:

- **Given**: Corresponding ground truth boundary matrix $\mathbf{R}$ $[N \times 2P]$, classifier boundary matrix $\mathbf{Q}$ $[N \times (2P + k)]$ respectively as:

$$\mathbf{R} = \begin{pmatrix} g_1' \ h_1' \\ \cdots \\ \cdots \\ g_N' \ h_N' \end{pmatrix} \qquad \mathbf{Q} = \begin{pmatrix} r_1'' \ s_1'' \\ \cdots \\ \cdots \\ r_N'' \ s_N'' \end{pmatrix}$$

where, for linear calibration, $(r_n'' s_n'') = (r_n' s_n' c_n')$, $c_n' = \underbrace{1 \, u_{1n} \, u_{2n} \, v_{1n} \, v_{2n}}_{AoV} \, \underbrace{p_n, q_n}_{Apex}$.

For quadratic calibration, $(r_n'' s_n'') = (r_n' s_n' c_n')$, $c_n' = \underbrace{1 \, u_{1n} \, u_{2n} \, v_{1n} \, v_{2n}}_{AoV} \, \underbrace{p_n, q_n}_{Apex} \, \underbrace{b_n'}$,

where,

$b_n' = \underbrace{u_{1n}^2 \, u_{2n}^2 \, v_{1n}^2 \, v_{2n}^2 \, u_{1n}v_{1n} \, u_{1n}v_{2n} \, u_{2n}v_{1n} \, u_{2n}v_{2n}}_{Quadratic-Terms}$, (u_{1n}, v_{1n}), and (u_{2n}, v_{2n})

are the coordinates of the anterior aspect and inferior aspect of the AoV plane of the LV. (p_n, q_n) are the apex coordinates of the left ventricle for patent n. Note $k=7$ for L-InCM and $k=15$ for Q-InCM.

- Let $\mathbf{A}$ $[(2P + k) \times 2P]$ be unknown coefficient matrix.
- The problem is to estimate the coefficient matrix $\mathbf{A}$, to minimize $\| \mathbf{R} - \mathbf{Q} \mathbf{A} \|^2$. Then for any classifier boundary matrix $\mathbf{Q}$, the calibrated vertices of the boundary are given by $\mathbf{Q}\hat{\mathbf{A}}$ where $\hat{\mathbf{A}}$ is the estimated coefficients.

Note that the new (x, y)-coordinates of the vertices of each boundary is a *different* linear combination of the old (x, y)-coordinates for the polygon, hence the name *independent coefficient* method.

Classifier Matrix: The above two methods are different in the way the calibration model is set up. In linear calibration (L-IdCM) $\mathbf{Q}$ is of size $2N \times (P + 4)$, while in L-InCM is $N \times (2P + 7)$. The classifier boundary matrix $\mathbf{Q}$ in Q-IdCM is of size $2N \times (P + 10)$ while in Q-InCM is of size $N \times (2P + 15)$.

Number of Coefficients: In linear calibration (L-IdCM), the number of coefficients estimated in the $\hat{\mathbf{A}}$ matrix is $(P+4) \times P$, while the number of coefficients in L-InCM are: $(2P + 7) \times 2P$, while for Q-IdCM, the number of coefficients estimated in the $\hat{\mathbf{A}}$ matrix is $(P + 10) \times P$. For Q-InCM, the number of coefficients estimated is $(2P + 15) \times 2P$. Thus the *independent coefficient* method requires around four times the number of coefficients of the *identical coefficient* method to be estimated, and this difference represents a significant factor in the ability of the technique to *generalize* rather than *memorize* for our data size.

3 Training and On-Line System

Suri [11] presented a two stage training-based cardiovascular imaging system for clinical cardiology and its research as shown in fig. 1 (left). LV classification boundaries are produced on-line (right half) using the training parameters produced off-line. Stage-II (lower half) is the calibration process which calibrates out the bias errors of the raw boundaries using the training based system. The training parameters are applied on-line to estimate the accurate boundaries (right half). Generalizing for any frame t, the minimizing $\hat{\mathbf{A}}_{tr}$ and estimated boundaries $\hat{\mathbf{R}}_{te}$ on the test set $(\mathbf{Q}_{te})$ as:

$$\underbrace{\hat{\mathbf{A}}_{tr} = (\mathbf{Q}_{tr}^T \mathbf{Q}_{tr})^{-1} \mathbf{Q}_{tr}^T \mathbf{R}}, \, \underbrace{\hat{\mathbf{R}}_{te} = \mathbf{Q}_{te} \hat{\mathbf{A}}_{tr}} \tag{1}$$

Singular value decomposition was used for solving for $\hat{A}_{tr}$. The calibration procedure discussed in this paper removes all systematic position, orientation, and shape errors from the initial classifier boundaries by taking the convex information of LV $\{(u_1, v_1), (u_2, v_2), (p_1, q_1)\}$.

Cross-Validation Optimization Algorithm: The CV technique is used in estimating the accuracy of the calibration procedure that takes a database of N patients and partitions it into K equal sized subsets. Then for all K *choose* L combinations, we estimate the calibration transformation using L subsets. Using the estimated transformation on the remaining $K-L$ subsets, we estimate the mean error of the transformed boundary. We employ two different calibration techniques: the *identical coefficient* and the *independent coefficient* . Each method produces estimates for the vertices of the polygon bounding the LV. Because of the small number of patient studies, $N=377$ and large number of parameters (about 200 times N) in the transformation, there is a danger of memorization rather than *generalization* in the estimation of the *transformation parameters*. Therefore, the number of vertices, P in the LV polygon must be carefully chosen. As P decreases, the generalization will be better but the representation of the true shape will get worse, thereby causing higher error with respect to the ground truth. As P increases, generalization will be lost but the representation of the true shape will get better. With the other parameters K, L and N fixed, there will be an *optimal* number of boundary vertices balancing the representation error with the memorization error. Our protocol finds this *optimal number*. Rotating through all L *choose* K combinations, we measure the accuracy of the results on the remaining $K-L$ subsets using the *polyline distance metric*. The mean and standard deviation of the resulting set of $N \times F \times P \times \frac{(K-1)!}{(K-L-1)! \, L!}$ numbers is then used to estimate the overall performance.

4   Polyline Distance Measure

The polyline distance $D_s(B_1 : B_2)$ between two polygons representing boundary B_1 and B_2 is symmetrically defined as the average distance between a vertex of one polygon to the boundary of the other polygon. To define this measure precisely, first requires having defined a distance $d(v, s)$ between a point v and a line segment s. The distance $d(v, s)$ between a point v having coordinates (x_o, y_o), and a line segment having end points (x_1, y_1) and (x_2, y_2) is:

$$d(v, s) = \begin{cases} \min\{d_1, d_2\}; & if \ \lambda < 0, \lambda > 1 \\ |d^\perp|; & if \ 0 \le \lambda \le 1, \end{cases} \tag{2}$$

where

$$
\begin{aligned}
d_1 &= \sqrt{(x_0 - x_1)^2 + (y_0 - y_1)^2} \\
d_2 &= \sqrt{(x_0 - x_2)^2 + (y_0 - y_2)^2} \\
\lambda &= \frac{(y_2 - y_1)(y_0 - y_1) + (x_2 - x_1)(x_0 - x_1)}{(x_2 - x_1)^2 + (y_2 - y_1)^2} \\
d^\perp &= \frac{(y_2 - y_1)(x_1 - x_0) + (x_2 - x_1)(y_0 - y_1)}{\sqrt{(x_2 - x_1)^2 + (y_2 - y_1)^2}}
\end{aligned}
\tag{3}
$$

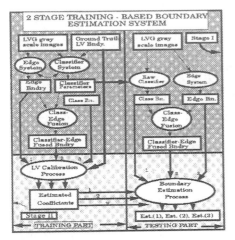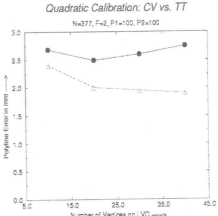

Fig. 1. Two stage system for LV boundary estimation. Stage I (upper half, shown black in color) consists of three approaches: Pixel-classification approach, edge detection approach and the classification-edge fusion approach. Stage II (lower half, shown white in color), consists of the calibration stage which smoothes the raw boundaries produced in stage I. Left Half: Off-line training system, Right Half: On-line boundary estimation Right: The *cross validation* protocol consists of calibration parameters: $N=377$, $K=188$, $L=187$, $F=2$, $P_1=100$, $P_2=100$. Note the optimized points 20 are interpolated back to 100 points for polyline metric computation.

The distance $d_b(v, B_2)$ measuring the polyline distance from vertex v to the boundary B_2 is defined by:

$$d_b(v, B_2) = \min_{s \in \; sides \; B_2} d(v, s) \tag{4}$$

The distance $d_{vb}(B_1, B_2)$ between the vertices of polylgon B_1 and the sides of polygon B_2 is defined as the sum of the distances from the vertices of the polygon B_1 to the closest side of B_2.

$$d_{vb}(B_1, B_2) = \sum_{v \in \; vertices \; B_1} d(v, B_2)$$

On reversing the computation from B_2 to B_1, we can similarly compute $d_{vb}(B_2, B_1)$. Using Eq. 4, the polyline distance between polygons, $D_s(B_1 : B_2)$ is defined by:

$$D_s(B_1 : B_2) = \frac{d_{vb}(B_1, B_2) + d_{vb}(B_2, B_1)}{(\#vertices \in B_1 + \#vertices \in B_2)} \tag{5}$$

Using the above definition, the overall mean error e_{NFP}^{poly} of the calibration system can be given as:

$$e_{NFP}^{poly} = \frac{\sum_{t=1}^{F} \sum_{n=1}^{N} D_s(G_{nt}, C_{nt})}{F \times N} \tag{6}$$

where, $D_s(G_{nt}, C_{nt})$ is the polyline distance between the ground truth G_{nt} and calibrated polygons C_{nt} for patient study n and frame number t.

5 Quadratic vs. Linear Optimization Results

Suri et al. showed the linear calibration with convex information [5] [6] [7] [9] [10] for different data sets (N) ranging from 245 to 377. Using the *cross validation* protocol discussed in section 3, the polyline mean error in linear calibrator for InCM was **3.47** mm when N=291. Corresponding estimated boundaries are shown in fig. 2. We here show results for the quadratic calibration under InCM conditions. The optimization curve for cross-validation is shown in fig. 1 (right) with a dip when P=20. The corresponding mean error e_{NFP}^{poly}=**2.49** mm. The calibration parameters were: N=377, F=2, K=188, P_1=100, P_2=100. The mean error when patient boundary lies both in training and testing data (TT case) condition is below **2** mm. The mean error for InCM technique under 4 conditions are: Without apex: **4.09** mm, with apex alone: **3.59** mm, with apex and AoV (linear): **2.97** mm, with apex and AoV (quadratic): **2.49** mm.

6 Diagnostic Clinical Acceptance and Discussions

The mean error over ED and ES frames using a cross validation protocol and polyline distance metric was **2.49** mm over the database of **377** patient studies. The goal of the diagnostic system was **2.5** mm.

InCM vs. IdCM: We also observed that when the training data was less (around 245 patient studies) then the IdCM technique was performing better than the InCM technique, and when the data was larger than 291 patient studies then the InCM technique was performing better than the IdCM technique. One reason of large error in IdCM with low data was due to the large number of coefficients it had to compute. Also we used the singular value decomposition to evaluate the classifier matrix, **Q**, which is very critical in inverse computations, a full rank matrix in InCM was more superior to a full rank matrix in IdCM with large data size.

Dynamics and Apparatus Design: Though we are able to obtain a goal on the cost of the heavy data processing, it may be worth while to discuss if computer processing of huge data is the only approach to handle the poor quality data sets (LVgrams). If careful analyzation is done, we find that there can be less complexity in computer processing (LV classification, edge detection, calibration) if the LV chamber would receive enough contrast agent (dye). How can we improve the apparatus setup to inject dye in apical zone ? One way would be to bring a change in curvature of the catheter for handling variability of the LV's. If the LV is more longitudinal we should be able to change the curvature of the catheter to let dye flow towards the apex. On the contrary if the heart is very wide or huge, can a dual catheter facing opposite walls be a good choice ? One catheter can be used for filling the anterior side while the other catheter can be used to fill the inferior side. Another possibility would be to look over the lateral movement of the catheter during the motion of the LV . If we can detect the crests and troughs of the LV border muscles, we can then fill the apical zone using computer control. Careful design is feasible by controlling the fluid dynamics inside the LV chamber to improve the quality of the LVgrams.

7 Conclusions and Acknowledgements

We compared the two training-based linear and quadratic calibration algorithms: the *identical coefficient* and the *independent coefficient*. We showed that the mean boundary error under quadratic calibration is better than the linear calibration with convex information of the LV. The mean error over end diastole and end systole frames using a cross validation protocol and polyline distance metric is **2.49** mm over the database of **377** patent studies. The goal of the diagnostic system is **2.5** mm. The software runs on PC and SUN work stations and written in **C** language. The authors thank Professors Linda G. Shapiro, Dean Lytle, Arun K. Somani, D. D. Meldrum, Werner Stuetzle, and Dr. Ajit Singh, Siemens for motivations.

References

1. Dumesnil et al.: LV Geometric Modeling and Quantification, Circulation (1979).
2. VanBree and Pope: Improving LV Border Recognition Using probability surfaces, IEEE Computers in Cardiology, pp. 121-124 (1989).
3. T.F. Cootes, C.J. Taylor, D.H. Cooper and J. Graham: Active Shape models: Their Training and Applications, Computer Vision and Image Understanding, Vol. 61, No. 1, January, pp. 38-59 (1995).
4. C. K. Lee: Ph.D Thesis, Department of Electrical Engineering, University of Washington, Seattle (1994).
5. Jasjit S. Suri, R. M. Haralick and F. H. Sheehan: Two Automatic Calibration Algorithms for Left Ventricle Boundary Estimation in X-ray Images, Proceedings of IEEE-Engineering in Medicine and Biology, Oct 31-Nov. 3 (1996).
6. Jasjit S. Suri, R. M. Haralick and F. H. Sheehan: Systematic Error Correction in automatically produced boundaries in Low Contrast Ventriculograms, International Conference in Pattern Recognition, Vienna, Austria, Vol. IV, Track D, pp. 361-365 (1996).
7. Jasjit S. Suri, R. M. Haralick and F. H. Sheehan: Accurate Left Ventricle Apex Position and Boundary Estimation From Noisy Ventriculograms, Proceedings of the IEEE Computers in Cardiology, Indianapolis, pp 257-260 (1996).
8. Jasjit S. Suri, R. M. Haralick and F. H. Sheehan: Left Ventricle Longitudinal Axis Fitting and LV Apex Estimation using a Robust Algorithm and its Performance: A Parametric Apex Model, International Conference in Image Processing, Santa Barbara, 1997, Volume III of III, IEEE ISBN Number 0-8186-8183-7, pp 118-121 (1997).
9. Jasjit S. Suri, R. M. Haralick and F. H. Sheehan: A General technique for automatic Left Ventricle Boundary Validation: Relation Between Cardioangiograms and Observed Boundary Errors, Journal of Digital Imaging, Vol. 10, No.2, Suppl 1, May, pp 1-7 (1997).
10. Jasjit S. Suri, R. M. Haralick and F. H. Sheehan: Effect of Edge Detection, Pixel Classification, Classification-Edge Fusion Over LV Calibration, A two Stage Automatic system, 10th Scandinavian Conference on Image Analysis (SCIA '97), June 9-11, Finland (1997).
11. Jasjit S. Suri, R. M. Haralick and F. H. Sheehan: Two Automatic Training-Based Forced Calibration Algorithms for Left Ventricle Boundary Estimation

in Cardiac Images, Proceedings of IEEE-Engineering in Medicine and Biology, Chicago, Oct 31-Nov.2, ISBN 0-7803-4265-9(CD-ROM) SOE 9710002, pp. 528-532 (1997).

12. Jasjit S. Suri, R. M. Haralick and F. H. Sheehan: Automatic Quadratic Calibration for Correction of Pixel Classification Boundaries to an Accuracy of 2.5 Millimeters: An Application in Cardiac Imaging, Proceedings of Int. Conference in Pattern Recognition, Brisbane, Austrialia, pp. 30-33, Aug. (1998).

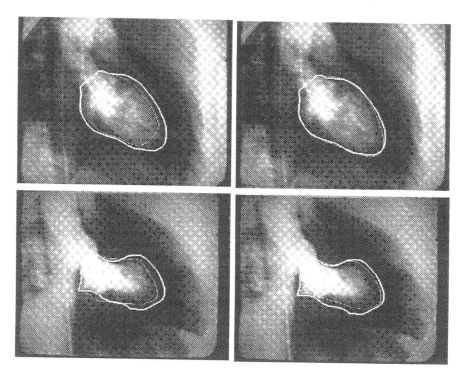

Fig. 2. Upper (left & right): ED Input and output to the Quadratic Calibration System along with the ground truth boundary. **Bottom**: ES Input and output. Calibration Parameters: $N=377$, $K=188$, $L=187$, $F=2$, $P_1=100$, $P_2=20$, Mean error (e_{NFP}^{poly}) = **2.49** mm.

Registration and Composition of Stacks of Serial Optical Slices Captured by a Confocal Microscope

Martin Čapek

Institute of Physiology, Academy of Sciences of CR,
Vídeňská 1083, Praha 4, 142 20, Czech Republic
Tel. +420 2 475 2334, fax +420 2 444 72269
Capek@biomed.cas.cz

Abstract. This article deals with image and volume registration of stacks of serial optical slices from a large biological tissue specimen captured by a confocal microscope. Due to the limited depth of observation and the restricted field of view of the confocal microscope the oversized specimen has to be sliced into smaller physical sections and scanned individually. The composition of the stacks of optical slices, which is based on data registration, is achieved in two steps. First, sub-volumes are created by volume registration of overlapping stacks of optical slices (volumes) captured from individual physical section. Second, image registration of peripheral images of sub-volumes of neighboring physical slices makes possible to compose 3D image of the whole specimen. Both registrations are based on similarity measures, such as the sum of absolute valued differences, normalized correlation coefficient, and mutual information. Data registration requires optimization of the search for the global extreme of a similarity measure over a parametrical space. Therefore, optimization strategies—n-step search, adaptive simulated annealing and stochastic approach—are used, and their optimal set-up is presented. The composition of stacks enables us to visualize and study a large biological specimen in 3D in high resolution.

1   Introduction

To register data means to determine coefficients of the chosen type of geometrical transformation between two data sets so that the corresponding elements of both data sets are mapped adequately. This can be expressed by the equation,

$$T' = \arg\min_{T} F(u(.),v(T(.)))\,, \tag{1}$$

where T', T stand for geometrical transformations; F for measure of quality of data registration; $u(.)$ for an element of a reference data set; $v(T(.))$ for an element of a data set to be registered (that is related to the corresponding element of the reference data set u by transformation T). As most of optimization strategies search for global minimum of the registration quality measure, "min" is put in (1).

We register images and volumes captured by confocal laser-scanning microscope (CLSM, Bio-Rad MRC 600) from a large biological tissue specimen. Due to the limited depth of observation and the restricted field of view of CLSM the specimen has to be sliced into smaller physical sections. These sections are then scanned individually. Another problem arises from the restricted field of view of the microscope, which is compensated by scanning a physical section by parts. Thus, we obtain an ensemble of partially overlapping stacks of serial optical slices (volumes) that has to be composed. After their composition, alignment of neighboring physical sections is carried out by registration of their peripheral optical slices. Thus, we get a whole individual volume that represents the oversized specimen in electronic way enabling us to render and visualize it in 3D in high resolution (see Fig. 1).

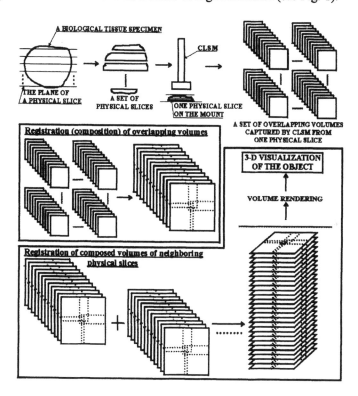

Fig. 1 A process of 3D visualization of a large biological tissue specimen.

2 Data

Scanning a physical section by parts results in an ensemble of overlapping stacks of optical slices. These slices are gray-scale 768x512 pictures, and their stack represents a volume. Adjacent volumes of the same physical section are captured by CLSM using object stage controlled by stepping motor; therefore, we know a priori mutual overlap and approximate position of the volumes. Registration will be achieved only by small linear shifts in x, y and z axe without rotation. To compose the volumes, we need to specify the position for which the overlapping parts of the volumes attain maximum similarity. The similarity is highest at registration because of similar conditions of data acquisition. However, a global shift of brightness aimed at better visibility of observed objects in volumes may cause some problems, and a measure of registration quality should be robust enough to handle such a matter.

Alignment of the neighboring physical sections requires registration of their peripheral optical slices, and re-computing one stack of optical slices (volume) according to the obtained coefficients of geometrical transformation. The peripheral optical slices, however, are not identical, but only similar in contents, because there is some spatial distance between them, see Fig. 2. Also, physical sectioning can mechanically deform objects in these slices. Misregistration of the peripheral optical slices is due to general rotation and shifts along x and y axes restricted to approx. ±150 pixels. The optical slices can be mutually shifted regarding brightness scale too.

As biological specimens are used terminal villi of human placenta studied in connection with morphology of their capillary bed.

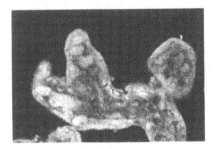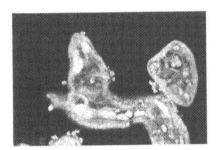

Fig. 2 Example of optical slices with mutual distance of approx. 10 μm.

3   Methods and Results

Since we attempt at an automatic procedure for registration and composition of stacks of optical slices that can differ to some degree, the boundary-based and point-mapping registration methods are of little use for our purposes. Therefore, we focus on similarity-based methods maximizing certain similarity function of data sets. To be able to cope with relatively great image diversity of peripheral optical slices we look for a robust but computationally fast enough criterion of data similarity. We have already experimented [1], [2] with criteria such as the sum of absolute valued differences (*SAVD*) [3], normalized correlation coefficient (*NCC*) [3] and mutual information (*MI*) [4]. On the basis of practical experiments we came to the conclusion that *SAVD* is a very fast criterion, but it is not suitable for registration of data sets that differ in brightness scale. *NCC* works much better in this case and its computational demands are still acceptable. *MI* is the most computationally demanding criterion, but it is the most robust criterion managing even non-linear dependencies between elements of data sets. Also, *MI* gave the most exact results concerning registration of distant peripheral optical slices in our pilot tests, which proved opportunity to register optical slices up to min. 10 μm. Characteristic shapes of the criteria in a neighborhood of registered position of distant optical slices for shifts ±15 pixels are depicted in Fig. 3. *SAVD* and *NCC* have broader extremes when compared with *MI* whose global extreme is very sharp, narrow and masked by local ones. For the sake of better comparison all graphs in Fig. 3, *NCC* and *MI* are depicted multiplied by -1 because of their positive maximums.

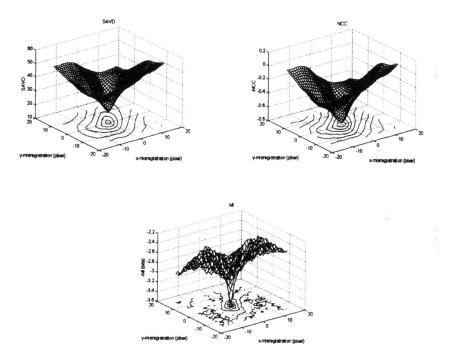

Fig. 3 Characteristic shapes of criteria in surroundings of registered position of distant optical slices.

Data registration requires finding the global extreme of a similarity measure in a parametrical space, which can be hidden among local ones. Therefore, an optimization strategy as robust and fast as possible is desirable. From the point of view of an applied optimization strategy, the broader the shape of the criterion extreme, the easier is to localize it. On the other hand, the sharper the shape of the extreme, the more precise registration is achievable.

The most simple optimization strategy is represented by the so-called "full-search". It is based on exhaustive searching through the whole parametrical space, and it is rather time-consuming. Consequently, we studied [5] more sophisticated strategies such as genetic algorithm [6], downhill simplex method (ameba) [7], Powell's method [7], and adaptive simulated annealing (ASA) [8]. We applied the strategies [5] to registration of serial (distant) optical slices (which is equivalent to registration of peripheral optical slices). We obtained the best results with ASA in combination of stochastic approach. While ASA is initialized by a starting point, the stochastic approach enriches it by multiple re-starts with random starting points evenly distributed in a parametrical space. ASA[1] (also called "very fast simulated re-annealing") is an optimization strategy whose application on data registration has not probably been reported yet. It is fast—its speed depends on chosen predefined accuracy—and robust. Therefore, it is suitable for data registration.

Table 1 illustrates the number of successful localizations of global extreme—whose position we knew—when ameba, Powell's method, and ASA with combination of stochastic approach (i.e. stochastic ameba, stochastic Powell's method, stochastic ASA) were applied to the registration of serial (distant) optical slices (see Fig. 2). As similarity criterion $SAVD$, NCC and MI were used. The number of repeated re-starts was set to 30. As one can see, the worst result for all optimizations strategies were obtained in case of MI, because it has the narrowest and the worst detectable global extreme. The highest number of successful registrations was reached in case of stochastic ASA for each of similarity criteria. The lowest number of successful registrations we get in case of stochastic Powell's method which also requires the highest number of criterion evaluations. The number of criterion evaluations qualifies an optimization strategy from the aspect of its speed. The latter number serves for an assessment of strategy's speed. The lower it is, the faster the strategy convergence to an extreme of the optimized criterion. From this point of view ASA is slower than stochastic ameba but—as mentioned above—more robust.

Table 1. Numbers of successful localizations of global extreme during registration of serial (distant) optical slices for different optimization strategies and similarity criteria (30 re-starts); bracketed numbers represent numbers of function evaluations necessary for registration.

similarity criterion	stochastic ameba	stochastic Powell's method	stochastic ASA
SAVD	14 (5014)	9 (15 373)	**19 (8220)**
NCC	9 (8287)	8 (17 335)	**19 (8220)**
MI	1 (3985)	0 (9569)	**6 (8220)**

---

[1] Source code of ASA in C is available via anonymous ftp-servers [11], [12].

On the basis of the aforementioned experiments we suggested algorithms for data registration that are optimal from the point of view of speed, robustness and reliability.

Registration (composition) of adjacent partially overlapping volumes can be solved simply. We know approximate position of both volumes and the size of the overlap. Thus, registration requires only small shifts of the first volume with respect to the second one according to x, y and z axe. It is sufficient to apply shifts along x and y axes in extents ±20 voxels and along z-axis only ±1 voxel[2]. One possibility how to find the global extreme of a criterion is to use "full-search" with multi-resolution approach that lowers the size of data sets [9]—there is no need of more sophisticated optimization strategies, since the parametrical space is rather small. Our proposal is to replace full-search by "n-step search" [10]. This technique hinges on searching the parametrical space with large step first, and when an approximate position of the extreme is found, the step is decreased to its half. The extreme position is refined only in vicinity of the previous result, and the step is decreased to its quarter, etc. "n" represents number of the partitioning of the step. To achieve the sub-voxel accuracy we set "n" equal to four (4-step search) with searching in the first step by four voxels, in the second step by two voxels, in the third step by one voxel, and, finally, in the fourth one by the half of voxel. By this way we reached certain simplification of the previously reported procedure [9], preserving low computational demands and increasing registration accuracy, because data set of low resolutions is used[3]. Our registration procedure requires about 750 criterion evaluations. But it should be noted that this approach is not quite suitable[4] for optimizing criteria that have narrow global extreme masked by local ones, e.g. *MI;* then "full-search" should be applied.

Alignment of neighboring physical sections is a more complicated task. It is based on registration of peripheral optical slices that are generally mis-positioned and that are only similar in contents. Because of the very large parametrical space to be investigated we applied the multi-resolution approach. In case of this, it is vitally important to detect the global extreme on the level having the lowest number of data elements (pixels) as a prerequisite of correct registration of the original images. Therefore, robust, but fast enough optimization strategy should be applied. From all the strategies we tested the most suitable is stochastic ASA. The chosen number of repeated re-starts depends on the criterion used. It got proven that *SAVD* and *NCC* needed about twenty re-starts, but *MI* required at least thirty re-starts. One possibility how to continue searching on the next levels is to apply ASA again, but only once for a point of the extreme of the previous levels. Alternatively, n-step search—described above—can be applied too. Both approaches are approximately equivalent as far as the convergence speed (expressed by number of criterion evaluations) and accuracy of registration are concerned.

Registration ends by specifying position of the global extreme of images of original resolution. Fig. 4 shows registration of 10-μm distant optical slices accomplished by procedure described and with *NCC* as a similarity criterion.

---

[2] We assume good stability in z-direction when a physical slice is studied by CLSM.

[3] When small objects, like grains or stars, are presented in images, after significant lowering resolution of the images they are practically eliminated.

[4] If the global extreme is narrow with regard to the size of step, "n-step search" can miss it.

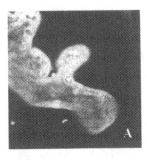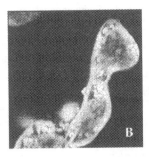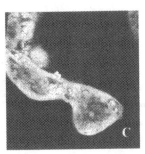

Fig. 4 Registration of distant optical slices (10 µm; 512x512) accomplished by the above described algorithm with *NCC*, A), B) are slices to be registered and C) is slice B) after registration; global extreme was found in position x=-9,63 (pixel), y=-50,08 (pixel), α=97,03° for *NCC*=0,55425 with 4242 criterion evaluations.

4 Conclusions

This work is devoted to the composition of stacks of serial optical slices captured by CLSM, which is based on data registration. The task consists of two steps. First, it is necessary to compose adjacent overlapping volumes captured from single physical section. This is based on slightly modified procedure that utilizes "n-step search", which enables us to achieve higher accuracy and easier implementation. Second, image registration is applied to create a volume from neighboring physical slices. Image registration makes use of multi-resolution approach combined with adaptive simulated annealing and stochastic approach. This combination proved to be robust, fast and general enough for our purposes and it represents an original modification of the image registration procedure. Also, we paid attention to similarity criteria whose application depends on distortion of transfer function (linear, non-linear distortions) of both data sets. Selection of suitable criterion is in hands of an operator that judges the character of the distortion.

Our future aim is to develop a software package for automatic or semi-automatic registration and composition of optical slice stacks in order to enable physicians and biologists to study human and biological specimens in high resolutions. The package should be based on the results of this practical study.

5 Acknowledgements

This study was supported by grant of Ministry of Education of Czech Republic No. VS 97033 "Laboratory of Computer Supported Medical Diagnostics", grant of Ministry of Education of Czech Republic supporting Czech-Slovenia cooperation KONTAKT: ME 256/1999 grant of Agency of the Academy of Sciences of the Czech Republic No. A 5011810, and grant of the Czech Ministry of Health No. 4178-3.

By this I would like to express thanks to Dr. Lucie Kubínová for kind provision of optical slices of biological tissues and to Dr. Ivan Krekule for inspirational ideas concerning this work.

References

1. Čapek, M., Krekule, I., Kubínová, L.: Practical Experiments with Automatic Image Registration of Serial Optical Slices of Thick Tissue. Conf. Proc., 8-th Intern. IMEKO Conf. on Measurement in Clinical Medicine, Dubrovnik, Croatia, p. 10-6–10-9, September 1998.
2. Čapek, M., Krekule, I.: Alignment of Adjacent Picture Frames Captured by a CLSM. IEEE Transaction on Information Technology in Biomedicine, 1999. In press.
3. Brown, L. G.: A Survey of Image Registration Techniques. ACM Computing Surveys, Vol. 24, No. 4, p. 325-376, December 1992.
4. Maes, F., Collignon, A., Vandermeulen, D., Marchal, G., Suetens, P.: Multimodality Image Registration by Maximization of Mutual Information. IEEE Trans. Med. Imag., Vol. 16, No. 2, p. 187-198, April 1997.
5. Čapek, M.: Optimisation Strategies Applied to Global Similarity Based Image Registration Methods. Conf. Proc., 7-th Intern. Conf. in Central Europe on Computer Graphics, Visualization and Interactive Digital Media'99, Plzeň, Czech Republic, February 1999. In press.
6. Parker, J. R.: Algorithms for Image Processing and Computer Vision. John Wiley & Sons, Inc., New York, 1997.
7. Press, W. H., et al.: Numerical Recipes in C: the Art of Scientific Computing. 2nd ed., Cambridge University Press, New York, 1992.
8. Ingber, L.: Simulated Annealing: Practice versus Theory. Mathl. Comput. Modelling, Vol. 18, No. 11, p. 29-57, 1993.
9. Dani, P., Chaudhuri, S.: Automated Assembling of Images: Image Montage Preparation. Pattern Recognition, Vol. 28, No. 3, p. 431-445, 1995.
10. Tekalp, A., M.: Digital Video Processing. Prentice Hall, New York, 1995.
11. ftp://ringer.cs.utsa.edu:/pub/rosen/vfsr.zip
12. ftp://ftp.caltech.edu:/pub/ingber/ASA.tar.gz

Tracking Articulators in X-ray Movies of the Vocal Tract*

Georg Thimm

IDIAP, CP 592, CH-1920 Martigny, Switzerland

Abstract. Tongue, lips, palate, and throat are tracked in X-ray films showing the side-view of the vocal tract. Specialized histogram normalization techniques and a new tracking method that is robust against occlusion, noise, and spontaneous, non-linear deformations of objects are used. Although the segmentation procedure is optimized for the X-ray images of the vocal tract, the underlying tracking method can be used in other applications.
Keywords: contour tracking, edge template, joined forward-backward tracking

1 Introduction

Although speech and speaker recognition systems achieve nowadays high performances in laboratory conditions, their performance is still unsatisfying under real-life conditions. Many researchers advocate, that more knowledge about the speech production process (*e.g.* co-articulation, dynamics, inter/-intra speaker differences) lead to improved feature extraction methods. Therefore, we attempted to extract the position and shape of articulators in the ATR X-ray films [6], showing a side-view of the vocal tract. We have investigated contour modeling techniques proposed in the literature [5, 2, 7], but obtained only insufficient results. Furthermore, we observed that the optical flow fields obtained with be best performing algorithms in [1] do not give good indications on the motion of the articulators. Another approach to segment the ATR X-ray films is described in [4]. But no final results were published and lips or jaws were not tracked. Note that other methods to gain quantitative data on the motion of articulators exist: MRI and tags are used in [3], and ultra sound in [8, 9].

2 Overview of the Segmentation

Some articulators are more distinct in the images than others. Therefore, these are located first, and then, using the benefits of image normalization and constraints on relative positions, other articulators are tracked (compare figure 1):

---

* This work has been performed with financial support from the Swiss National Science Foundation under Contract No. 21 49 725 96.

1. The X-ray images are filtered with a Gaussian filter and their histograms are then zero-normalized (section 3).
2. The upper front teeth are located by the means of a pattern matching algorithm using distorted grey-level histograms (section 3).
3. Similarly, a reference point in the rear upper teeth is tracked. Teeth fillings are very robust objects for this purpose.
4. The images are normalized for the position of the upper teeth and their grey-levels.
5. and 6. The position of the lower teeth is determined.
7. The images obtained in 4. are filtered with a *Canny edge detector*.
8., 9., and 10. The edges corresponding to the lips and the rear throat are tracked in the edge images extracted in 7.
11. A modified Canny filter, that neglects negative gradients in x-direction, is applied to the images obtained in 4.
12. The front throat is tracked in the edge images obtained in 11.
13. Image parts representing the upper and lower jaw with the tongue in an advanced and lowered, as well as in a back and high position, are subtracted.
14. These two series of images are filtered with a Canny edge detector.
15. The tongue is tracked. Possible positions are restricted by temporal information as well as the location of the front throat.

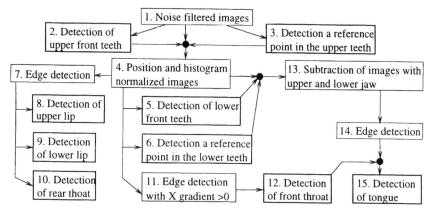

Fig. 1. Tracking of teeth, lips, front and rear throat, as well as the tongue

3 Illumination Normalization

The X-ray films are affected by a variable illumination, caused by either an instable X-ray energy or a varying shutter speed of the camera. This effect can not be eliminated by standard linear histogram normalization. Standard pattern matching algorithms yield therefore unsatisfactory results.

A first step to overcome this problem is to remove parts of the histogram: black parts with gray-values smaller than some value g_0 correspond to noise and parts of the image that are of no interest. Furthermore, gray-values in the interval $[g_0, g]$ are almost not occurring, as shown in figure 2(a). Consequently, all pixels with gray-values smaller than g are set to g and the image is normalized to cover the whole range of gray-levels. Although modified images have a higher contrast, the main benefit is to obtain the same brightness for an object in all images. The cut-off value g is chosen for each image according to the formula:

$$g = \max\left\{ K \mid \sum_{i=g_0}^{K} \text{hist}(i) < N \cdot q \quad \text{and } K \in [g_0, 255] \right\}. \tag{1}$$

In this formula, $\text{hist}(i)$ is the number of pixels with gray-value i, N is the number of pixels, g_0 is chosen close to zero in the left of the nearly empty region of the gray-level histogram, and q is the fraction of pixel values that are allowed in the interval $[g_0, g]$. Figure 2(b) shows the histogram of the resulting image.

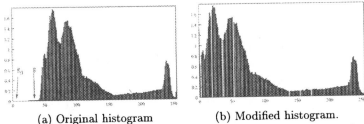

(a) Original histogram   (b) Modified histogram.

Fig. 2. The lower part of the histogram of the images is removed.

The histogram of the images are also subject to non-linear distortions. We compensated for this in the pattern matching process: the histogram of the template is modified during a comparison with some image location by the means of a flexible histogram mapping function (for more details see [11]).

4   Tracking Using Edge Templates

4.1   Edge-Based Template Matching

This section describes the basic matching procedure for edge templates with an edge image. The approach assumes that the object (*e.g.* the tracked articulatory feature)   1. is exactly once present in an image,   2. it is invariably at the same place and has the same orientation and size, respectively all possible places, orientations, and sizes of the object are represented in the training data.   The procedure is, however, robust against **small** deformations, translations, and rotations, as well as occlusion and noise. It does not require that the edges corresponding to the object are connected.

The matching procedure uses edge images, as produced by a Canny edge detector (figure 3(a)). In a first step, edges are detected in all normalized images. From these edge images, representative edges are extracted (figure 3(b)). Such images are called *state images* in the following. These state images are inverted and blurred by a Gaussian filter, resulting in so-called *matching images* S_i (figure 3(c)). Both images are further associated with the same state which is proper to them.

(a) The Canny-filtered image.   (b) A selected edge: the tongue.   (c) A matching image.

Fig. 3. Creation of a state (image). Image (a) shows a side view of the vocal tract with lips (right), jaws, and throat (left).

The matching images S_i (the figure background is encoded as 0, the foreground as positive values) are used in the matching procedure. The score of a matching image S_i with respect to an image X is calculated as

$$\text{score}(S_i, X) = \sum_{x,y} X(x,y) * S_i(x,y) \qquad (2)$$

The matching image S_i with $i = \text{argmax}_k(\text{score}(S_k, X))$ is defined as the optimal state and written as $S(X)$. Although equation (2) evokes a rather high computational complexity (per frame n multiplications of matrices in the size of the images, if n is the number of possible states), an implementation can be efficient: only non-zero parts of the matching image need to be considered in equation (2) which permits considerable optimizations. Furthermore, the tracking procedure described in section 4.3 limits the number of matching images for which the score has to be calculated to a small subset.

4.2 Selecting State Images

In order to obtain good results with the matching procedure, the edges used for the state images should be selected consistently. In particular, the size of the selected edges and cut-off points should be similar. Example choices are given in figure 4 by the bold lines.

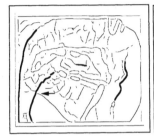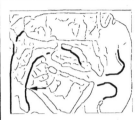

Fig. 4. The bold lines are examples for selected edges. From the points marked by the flash downwards, the edge of the tongue is also the edge of the front throat.

4.3 Adding Temporal Information to the Tracking Procedure

The performance of the basic matching procedure described in section 4.1 can be improved by using temporal information. *I.e.*, if the deformation and/or displacement of a feature is small between consecutive frames, a prediction of possible next states can be done on the base of the current state. Whether or not certain transitions are possible, is here estimated by calculating a heuristic distance between contours and then permitting only transitions corresponding to the smallest distances.

These distances would ideally be defined as the mean traveling distance of contour points. As such a strong correspondence between contour points can not be determined, the distance $D_{i,j}$ between two edges i and j is defined as the ratio of surface delimited by the splines to the mean length of the splines approximating the contours. Note, that generally the endpoints of the splines do not correspond to the same points of the feature. The splines to be used for this calculation are therefore only parts of the splines which approximate the edges (compare [10]).

Then, for each state S_i, the element $T_{i,j}$ of the state transition matrix, is defined as 1, if $D_{i,j}$ is among the p percent smallest elements of vector D_i and 0 otherwise. Furthermore, T is augmented by a row $T_{0,j} = 1$, corresponding to an initial state S_0. For this project, $p = 30\%$ was chosen. A transition from state i to state j is possible if, and only if, $T_{i,j} = 1$.

4.4 Tracking a Feature

The tracking procedure is an iterative process, in which the selection of a set of possible next states by means of the transition matrix T alternates with the calculation of the optimal state with respect to this selection. More precisely, the score of S_i with respect to matching image X_t and the optimal state $\overrightarrow{S}(X_{t-1})$ for the previous frame is calculated using the following formula:

$$\overline{\text{score}}(S_i, X_t) = \begin{cases} \text{score}(S_i, X_t) & \text{if } T_{j,i} = 1 \text{ with } \overrightarrow{S}(X_{t-1}) = S_j \\ 0 & \text{otherwise.} \end{cases} \qquad (3)$$

Whereas for the image X_1, the preceding state is defined as S_0. Then, the optimal state $\overrightarrow{S}(X_t)$ is defined as the state with the maximal score.

5 Joined Forward-Backward Tracking

One important assumption in section 4.3 is, that objects move slowly. Although this is true most of the time, there are exceptions: tongue and lips can move so fast so that they assume almost opposite extreme positions in consecutive frames.

However, before and after those spontaneous, high-velocity movements, the velocity and acceleration of the respective articulators are low or even zero and fulfill for a certain time the assumption of slow movements. The following approach exploits these observations and reduces the tracking errors that are due to fast movements.

1. Calculate the forward tracking sequence $\overrightarrow{S}$ as in section 4.3.
2. Calculate the backward tracking sequence $\overleftarrow{S}$ in a similar manner, but using a score that restricts the states in a backward manner.
3. Join state sequences $\overrightarrow{S}$ and $\overleftarrow{S}$ to form the forward-backward sequence $\overleftrightarrow{S}$:

$$\overleftrightarrow{S_i} = \begin{cases} \overrightarrow{S_i} & \text{if } \overline{\text{score}}(\overrightarrow{S_i}) \geq \overleftarrow{\text{score}}(\overleftarrow{S_i}) \\ \overleftarrow{S_i} & \text{otherwise} \end{cases} \tag{4}$$

This approach can be used for the lips as well as for the rear and front throat.

5.1 Tracking the Tongue

The tongue is often hidden by the jaws, especially the teeth, which means that its contour is not or only hardly visible. Sometimes even a human observer is unable detect the precise location of the tongue. The tracking procedure is consequently augmented by a specialized version of background subtraction. As the upper jaw is not moving, its image can be directly subtracted (figure 5(a)). The background image with the lower jaw has to be oriented according to the current position of the jaw, which is known from the tracking of the lower teeth. Furthermore, two background images of the lower jaw with different tongue positions are required (figures 5(b) and 5(c)): as no background image without the tongue is available, the contour of the tongue will disappear if the tongue in the image is at the same position as in the subtracted background image. Therefore, according to the position of the more easily tracked front throat, one of the two different background images of the lower jaw are used.

Figure 6 shows an example: the jaws are subtracted from the original image 6(a), resulting in image 6(b). It can be seen that the image region corresponding to the tongue is more uniform and the fillings in the upper teeth disappeared. In consequence, the edge of the tongue in the region of the mouth is nicely detected by a Canny edge detector (image 6(c)).

  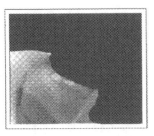

| (a) The upper jaw | (b) The lower jaw with the tongue advanced | (c) The lower jaw with the tongue drawn back |

Fig. 5. Background images subtracted from images before the edge detection for tongue.

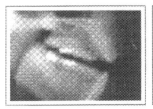  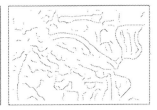

| (a) A part of a X-ray image. | (b) Background images subtracted. | (c) The resulting edge image. |

Fig. 6. An example for the subtraction of the jaws followed by an edge detection used for the tracking of the tongue.

The tracking procedure is similar, with exception that possible state tran. tions are also limited by the estimated horizontal distance of the contours of front throat and the tongue. The cut-off values for *tongue–tongue* and *tongu front throat* distances are tuned in order to restrict the number of states ima, that are actually matched with an image to approximately 30%.

6 Results

.The film Laval 43 with 3944 frames showing the vocal tract was completely a. lyzed and the results are made available on the WWW page of the IDIAP vis' group (URL: http:www.idiap.ch/vision). The data includes the position the front teeth, some position in the rear upper and lower teeth, as well as cc tours in the form of splines for the lips, tongue, and throat. We assume that precision of the tracking procedure is sufficient for speech analysis purposes, though a quantization of the error is infeasible in practice. However, we estim: that a human would have located the contours of lips and throat in more tl 98% very similarly. For the position of the tongue we estimated this figure to above 95%. Further, the maximal error for the position of the teeth is likely be below 2mm in real world dimensions.

7 Conclusion

We proposed a contour tracking algorithm that can be applied to objects of which the general position is known (or limited to a small number of positions), but that are subject to non-linear, spontaneous deformations. The approach is very robust against noise and occlusion, and is based on the assumption that deformations, with the exception of rarely occurring spontaneous deformations, are slow. The approach associates contours with states, and motion as well as object deformations with state transitions. During the tracking procedure, the state transitions are restricted to those associated with small movements. Spontaneous deformations are dealt with by joining the state sequences obtained by tracking the image sequence forward and backward in time.

With respect to the obtained results, the low quality of the X-ray database, and the difficulties proper to this type of data, the method can be considered to be very robust.

We expect the obtained results for the film Laval 43 of the ATR database to be sufficient for speech research purposes.

References

1. J. Barron, S. Beauchemin, and D. Fleet: On optical flow, in Int. Conf. on Artificial Intelligence and Information-Control Systems of Robots, (1994) 3–14.
2. T. Cootes, A. Hill, C. Taylor, and J. Haslam: Use of active shape models for locating structures in medical images, Image and Vision Computing 12 (1994) 355–365.
3. E.P. Davis, A.S. Douglas, and M. Stone: A continuum mechanics representation of tongue deformation, in Proc. of Int. Conf. on Spoken Language Processing (Bunnell and Idsardi, eds.) 2, New Castle, Delaware, Citation Delaware (1996) 788–792.
4. Y. Laprie and M. Berger: Towards automatic extraction of tongue contours in x-ray images, in Proc. of Int. Conf. on Spoken Language Processing 1, Philadelphia, USA (1996) 268–271.
5. J. Luettin and N.A. Thacker: Speechreading using probabilistic models, Computer Vision and Image Understanding 65:2 (1997) 163–178.
6. K. Munhall, E. Vatikiotis-Bateson, and Y. Tokhura: X-ray film database for speech research, J. Acoust. Soc. Am. 98:2 (1995) 1222–1224.
7. L.H. Staib and J.S. Duncan: Boundary finding with parametrically deformable models, IEEE Trans. on Pattern Analysis and Machine Intelligence 14 (1992) 1061–1075.
8. M. Stone and E. Davis: A head and transducer support system for making ultrasound images of tongue/jaw movement, J. Acoust. Soc. Am. 98:6 (1995) 3107–3112.
9. M. Stone and L. Lundberg: Three-dimensional tongue surface shapes of english consonants and vowels, J. Acoust. Soc. Am. 99:6 (1996) 1–10.
10. G. Thimm: Segmentation of X-ray image sequences showing the vocal tract, IDIAP-RR 1, IDIAP, CP 592, CH-1920 Martigny, Switzerland (1999).
11. G. Thimm and J. Luettin: Illumination-robust pattern matching using distorted color histograms, in Lecture Notes in Computer Science (5th Open German-Russian Workshop on Pattern Recognition and Image Understanding), Springer Verlag (1998). To appear.

Modeling Morphological Changes During Contraction of Muscle Fibres by Active Contours

Aleš Klemenčič, Franjo Pernuš, Stanislav Kovačič

Faculty of Electrical Engineering
University of Ljubljana
Tržaška 25, 1000 Ljubljana, Slovenia
Tel. +386 (61) 176 84 67, Fax +386 (61) 126 46 30
{alesk,feri,stanek}@fe.uni-lj.si

Abstract. An active contour model with expansion "balloon" forces was used as a tool to simulate the changes in shape and increase in cross-sectional area, which occur during the contraction of isolated muscle fiber. A polygon, imitating the boundaries of the relaxed muscle fiber cross-section, represented the initial position of the active contour model. This contour was then expanded in order to increase the cross-sectional area and at the same time intrinsic elastic properties smoothed the contour. The process of expansion was terminated, when the area of the inflated contour surpassed the preset value. The equations that we give, lead to a controlled expansion of the active contour model.

1 Introduction

Skeletal muscles, which are exquisitely tailored for force generation and movement, are characteristically composed of muscle fibers with different structural and functional properties. Besides the prevailing contractile proteins organized into cylindrical myofibrils, a fiber's interior contains other organelles essential to cellular function. Among these mitochondria, lipid droplets, and glycogen provide the metabolic support needed by active muscle. Externally a muscle fiber is bounded by the sarcolemma, an elastic material, composed of plasma membrane, an external basal lamina, and a collagen layer having together a thickness of about 0.1 μm.

In transverse section the profiles of the muscle fibers are usually polygonal because endomysium, the bounding connective tissue, enforces tight apposition of the fibers as they grow (Fig. 1). The endomysium is the site of metabolic exchange between muscle and blood, and capillaries run, together with small nerve branches, in this layer.

When a motor neuron is excited, an action potential is propagated along the axon and its branches to all of the muscle fibers that it supplies and elicits a twitch contraction. If a second nervous impulse is delivered before the muscle fibers relax, the fibers contract again, building the tension to a higher level. Because of this mechanical summation the higher the impulse frequency, the more force is produced.

Recently Taylor and colleagues [7,10], found that an isolated fiber's cross-sectional area increased during isometric tetanus. The magnitude of the increase varied from 1% to 40% along the fiber with the mean of 20-25%. Active force development along a fiber's axis thus evidently produces outwardly directed radial forces.

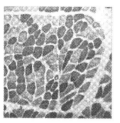

Fig. 1. A sample of muscle fibers. The profiles of muscle fibers are polygonal.

The purpose of this study was to simulate changes in cross-sectional area and shape, due to contraction of isolated muscle fibers. We have explored the idea of active contour models or snakes, introduced by Kass *et al.* [4], which have proved to be a useful technique for detecting object contours and shape changes [5,9]. The energy functional was formulated in such a way that only the contour's intrinsic elastic properties and outwardly directed radial forces governed the behavior of an active contour model. No other external forces were thus present.

The paper is organized in the following way. In section 2 we review the classical active contour models and give the equations that control their well-known shrinking effect. In section 3 we give equations that enable controlled interaction of internal elasticity and inflation forces. In section 4 we give the results of the muscle fiber contraction simulations and in conclusions, section 5, we evaluate the results of the research and give ideas, how these results can be of use in future work.

2 Active Contour Models

An active contour model is a planar curve defined by a set of connected points, which iteratively move from an initial position to its final state. The evolution is driven by minimization of the energy functional E_{snake}, which is a function of internal and external energies [4]:

$$E(v)_{snake} = \int_0^1 (E_{int}(v(s)) + E_{ext}(v(s))) ds , \tag{1}$$

where the position of the snake is represented parametrically by $v(s)=(x(s),y(s))$, along contour s, $0 \le s \le 1$. The internal energy term E_{int}, controls the elastic properties of the snake and is expressed by:

$$E_{int}(v(s)) = \frac{1}{2}\left(\alpha(s)\left|\frac{\partial v(s)}{\partial s}\right|^2 + \beta(s)\left|\frac{\partial^2 v(s)}{\partial s^2}\right|^2 \right) , \tag{2}$$

where the parameter $\alpha(s)$ controls the amount of stretching the snake is willing to undergo and $\beta(s)$ controls the amount of flexing it will allow. The external energy E_{ext} comes from the image and/or higher-level processes and is responsible for the deformation and moving of a snake.

Using the calculus of variations, minimizing the energy functional gives rise to an Euler-Lagrange equation which, when discretized and solved iteratively, gives [4]:

$$\mathbf{v}_k = (\mathbf{A} + \gamma \mathbf{I})^{-1}(\gamma \, \mathbf{v}_{k-1} - \mathbf{f}(\mathbf{v}_{k-1})), \tag{3}$$

where $\mathbf{v}_k = \{v_{1,k}, v_{2,k}, \ldots, v_{N,k}\}$ are the positions of N control points at iteration k, $\mathbf{A}$ is a pentadiagonal matrix, depending on the elasticity parameters α and β, γ is the step size parameter, and $\mathbf{f}(\mathbf{v}_{k-1})$ are the external forces at control points of the discretized snake. Internal forces provide regularization through the inverse of the positive definite matrix $(\mathbf{A} + \gamma \mathbf{I})$.

In the absence of external forces, i.e. $\mathbf{f}(\mathbf{v}_{k-1}) = 0$, active contour models tend to shrink into a point. To evaluate the shrinking effect, Gunn [2] used a discrete circular contour defined as:

$$\mathbf{v}_{i,0} = r_0 \, \mathbf{c}_{i,0} + \mathbf{t}_0 = r_0 \begin{pmatrix} \cos\left(\dfrac{2\pi i}{N}\right) \\ \sin\left(\dfrac{2\pi i}{N}\right) \end{pmatrix} + \begin{pmatrix} x_0 \\ y_0 \end{pmatrix}, \tag{4}$$

where $\mathbf{v}_{i,0}$ is the initial position of a point i on the circular contour, which is defined by N points $\mathbf{c}_{i,0}$, $i = \{1,\ldots,N\}$, r_0 is the radius of the initial contour and $\mathbf{t}_0$ is the vector indicating the circle's center. If the contour $\mathbf{v}_0$ is exposed only to internal forces, it shrinks uniformly and the amount of contraction in one iteration, expressed as Δr, $\Delta r = r_{k-1} - r_k$, depends on the values of elasticity parameters α and β, and the step size parameter γ as follows [4]:

$$\Delta r \approx \frac{r_k(\alpha 4\pi^2 + \beta 16\pi^4)}{\gamma}, \tag{5}$$

provided that the number of points $N \gg \pi$. On the other hand, for preselected elasticity parameters α and β and radius r_{k-1} defined, any desired decrease of radius Δr (except $\Delta r = 0$) can be achieved by setting the value of step size parameter γ to:

$$\gamma \approx \frac{(r_{k-1} - \Delta r)(\alpha 4\pi^2 + \beta 16\pi^4)}{\Delta r}. \tag{6}$$

The parameter Δr may thus control the step size parameter γ, i.e. the speed of contraction. When the value of Δr is small, the contraction is smaller and as the value of Δr increases, the amount of contraction of the snake increases accordingly. Fig.2. shows the evolution of the active contour model in five iterations from the same initial position, a square, using different values of Δr. Equation 3 with external forces $\mathbf{f}(\mathbf{v}_{k-1})$ set to zero was used to define the position of points after each iteration. A smaller Δr, and a correspondingly higher number of iterations would eventually lead to the same final contour position as using higher value of parameter Δr and lower number of iterations.

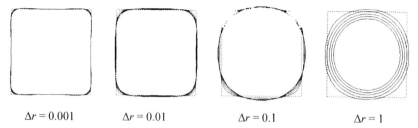

$\Delta r = 0.001$ $\Delta r = 0.01$ $\Delta r = 0.1$ $\Delta r = 1$

Fig. 2. The initial position of the active contour model is indicated with a dotted line, which has a shape of a square. Five iterations were executed, using different values of Δr.

Fig.3 shows how the values of elasticity parameters α and β influence the evolution of the active contour model, when step size parameter γ is calculated following equation (6). The values of parameters α and β were changed in the range from 10^{-6} to 10^{+6}, while parameter Δr was set to $\Delta r = 1$. Results were similar for most combinations, except the combination, when parameter α had a significantly higher value than parameter β, $\alpha=10^{+6}$ and $\beta=10^{-6}$. This shows that elasticity doesn't depend on the absolute values of parameter α and β, but on their ratio, i.e. when parameter α is bigger than parameter β the contour is more rigid.

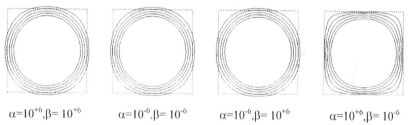

$\alpha=10^{+6},\beta=10^{+6}$ $\alpha=10^{-6},\beta=10^{-6}$ $\alpha=10^{-6},\beta=10^{+6}$ $\alpha=10^{+6},\beta=10^{-6}$

Fig. 3. Different values of elasticity parameters α and β, despite their large variability, return almost identical results, showing that parameters α and β do not significantly influence the behavior of a snake.

Due to its internal energy, the active contour model always tends to shrink. The amount of shrinking can be controlled by a suitable value of Δr. To achieve an expansion of the active contour model, strong enough external forces have to be applied. Inflation forces, which were proposed by Cohen [1], are pushing the points of the snake outwards, in the direction perpendicular to the contour. The snake acts as if air was introduced inside and these kinds of active contour models are also known as "balloons". The external forces, as defined in [1], are:

$$\mathbf{f}(v_{i,k}) = f_{bf}\mathbf{n}_{i,k}, \tag{7}$$

where $\mathbf{n}_i$ is the normal unit vector to curve at points v_i in iteration k, and f_{bf} is the "balloon force" amplitude. The normal unit vector at discrete control points of an active contour model was defined according to the solution proposed by Lobregt and

Viergever [8]. Fig. 4 illustrates the effect of internal, shrinking, and external, inflating forces.

3 Simulation of Muscle Fiber Contraction

Active contour models with expanding forces were used as a tool to simulate changes in shape and cross-sectional area during contraction of isolated muscle fibres. The internal forces of the active contour model were used to simulate the elements of the cytoskeleton that normally restrict lateral expansion of an intact fiber. The expansion of muscle fiber boundaries was achieved by applying a pressure force pushing outside in every point $v_{i,k}$, in the direction normal to the contour. This simulated the outwardly directed radial forces that are produced by active force development during fiber contraction.

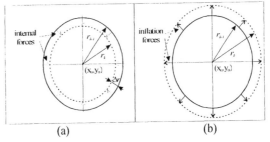

(a)   (b)

Fig. 4. The amount of shrinking of a circular contour (a) can be supervised by the step size parameter γ, which is based on the value of Δr (equation 6). In order to expand the contour, inflation forces, strong enough to overweight the contraction, have to be applied (b).

The resemblance between muscle fibers and Voronoi polygons was first observed by Honda [3]. Based on that observation we selected a 'Voronoi-like' polygon to imitate the boundaries of a muscle fiber and at the same time represent the initial position of the active contour model. This initial contour was then expanded in order to increase the cross-sectional area, and at the same time intrinsic elastic properties smoothed the contour. After every iteration k the active contour model area A_k was calculated and checked for the stopping criterion. The process of expansion was terminated, when the area of the expanded contour surpassed the preselected value $A_{contracted}$:

$$A_k \geq A_{contracted} , \qquad (8)$$

which represented the cross-sectional area of muscle fiber in contracted state.

This simulation is an iterative process and, in order to get a good approximation, the increase of area in each iteration should be small, to gradually approach the desired value of cross-sectional area $A_{contracted}$. This calls for carefully tailored inflation forces which, on one hand, balance the contraction forces and, on the other hand, expand the active contour model for a small amount δA. Based on user selected value of δA we have derived the necessary inflation forces. The expansion of a circular contour, as defined in equation (4), in every iteration increases the contour's area for:

$$\delta A = \pi(r_k^2 - r_{k-1}^2).$$ (9)

Because the shrinking effect can be made insignificant by setting Δr in equation (6) to a small enough value, the size of inflation forces also defines the size of expansion of the contour and therefore $r_k = r_{k-1} + f_{hf}$. Inserting $r_k = r_{k-1} + f_{hf}$ in equation (9) and expressing f_{hf} leads to the amplitude of inflation forces:

$$f_{hf} = -r_{k-1} + \sqrt{r_{k-1}^2 + \frac{\delta A}{\pi}}.$$ (10)

Inflation forces are added to every active contour model point and they act perpendicularly to the contour in the direction toward outside. The complete evolution equation for controlled expansion of the snake is:

$$\mathbf{v}_k = (A + \gamma I)^{-1} \gamma (\mathbf{v}_{k-1} + f_{hf} \mathbf{n}_{k-1}).$$ (11)

Internal forces act uniformly on each contour point only in case of a circular contour with evenly spaced control points. Equations (9)-(11) are derived on the assumption that we are dealing with such a contour, as is defined in equation (4). These conditions are not exactly met in the case of contours representing the muscle fiber boundaries, but nevertheless the differences are not too excessive.

4 Experimental Results

As we have shown in Fig.3, the values of parameters α and β do not significantly influence the behavior of the active contour model, especially when the value of Δr is small. Therefore we have set them to constant values ($\alpha=\beta=1$). The values of parameter Δr varied from 10^{-2} to 10^{-6}. Neering et al. [7], have found that the cross-sectional area of an isolated frog's fiber increased from 1-40% from the relaxed to the contracted state. Following these figures we studied the increase of area form its initial size for 0.1%, 1%, 10% and 40%. Results of the experiments, using a polygonal initial shape, different values of parameter Δr, and different area changes ΔA are shown in Fig.5.

In our experiments the parameter Δr could not take values greater than 10^{-2}. Inflation forces k_{hf} are calculated based on the assumption that the shrinking effect is negligible. If the values of Δr are to big, than this no longer holds, and the inflation forces are not strong enough to overcome the shrinking effect. Nevertheless, the allowed range of parameter Δr leaves enough room to simulate the desired changes in shape and size of cross-sectional area during contraction of isolated muscle fibers.

Fig.5 shows that, in order to get a rounder final shape, the increase of area should not be too small. In case, when the area increased for 0.1%, the smoothing effect was limited, even with higher value of parameter Δr. Vertices were softened, but the overall shape of the contour didn't change much.

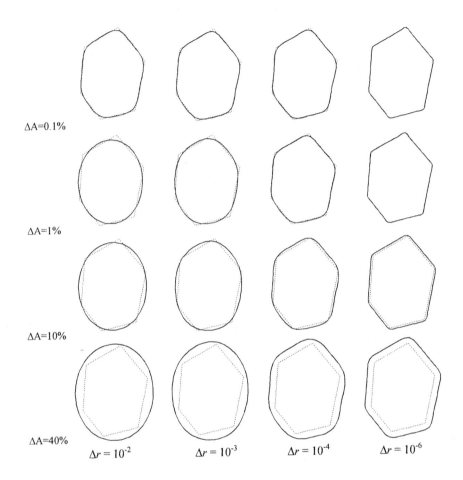

ΔA=0.1%

ΔA=1%

ΔA=10%

ΔA=40%

$\Delta r = 10^{-2}$ $\Delta r = 10^{-3}$ $\Delta r = 10^{-4}$ $\Delta r = 10^{-6}$

Fig. 5. The dotted line indicates the 'Voronoi-like' polygon, representing the initial position of the snake. Desired increase of contour's area was set to 0.1%, 1%, 10% and 40% (following the results of Neering *et al.*[7]) and the results obtained with different values of Δr parameter are shown in the first, second, third and forth row, respectively.

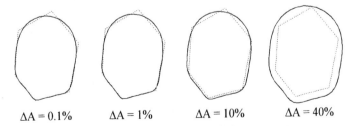

ΔA = 0.1%   ΔA = 1%   ΔA = 10%   ΔA = 40%

Fig. 6. Simulation of different flexibility conditions along the contour. Parameter Δr is changing from 10^{-3} at the top to 10^{-6} at the bottom of the contour.

In another experiment we studied the changes in cross-sectional area and shape by presuming that different segments of the contour have different elastic properties. We have simulated these conditions by changing the value of parameter Δr along the contour. The value of parameter Δr a the top of the polygon in Fig. 6 is $\Delta r = 10^{-3}$, allowing a greater flexibility, while at the bottom of the contour $\Delta r = 10^{-6}$, decreasing the flexibility.

5 Conclusions

The present study is a preliminary one. The experimental design involved isolated muscle fibers. In living muscle in situ, where extracellular space is restricted, the amount of cross-sectional area change may be much smaller and the contracted fibers are most probably much less circular than the contracted isolated fibers. Evidently, any attempt to extrapolate the magnitude of cross-sectional area and shape changes simulated in this study to the other situation should be taken with caution. This research proved that active contour models with addition of inflation forces are a promising tool for the simulation of muscle fiber contraction. All these data can now be further used to simulate the interaction of a number of active contour models corresponding to fibers in a fascicle or the whole muscle.

References

1. L.D. Cohen, "On Active Contour Models and Balloons", *CVGIP: Image Understanding*, vol. 53/2, pp. 211-218, March 1991.
2. S.R. Gunn, "Dual Active Contour Models for Image Feature Extraction", *Ph.D. Thesis*, Faculty of Engineering and Applied Science, Department of Electronics and Computer Science, University of Southampton, 10th May 1996.
3. H. Honda, "Description of cellular patterns by Dirichlet domains: The two-dimensional case" *J Theor Biol* 72:523-543, 1978.
4. M. Kass, A. Witkin, D. Terzopoulus, "Snakes: Active Contour Models", *International Journal of Computer Vision*, vol. 1/4, pp. 321-331, 1988.
5. A. Klemenčič, S. Kovačič, F. Pernuš, "Automated Segmentation of Muscle Fiber Images Using Active Contour Models", *Cytometry 32*, pp. 317-326, August 1998.
6. R.L. Lieber, "Skeletal Muscle Structure and Function: Implications for Rehabilitation and Sports Medicine", *Williams & Wilkins*, Baltimore, 1992.
7. I.R. Neering, L.A. Quesenberry, V.A. Morris, S.R. Taylor, "Nonuniform volume changes during muscle contraction", *Biophys. J.*, vol. 59, pp. 926-932, 1991.
8. S. Lobregt, M.A. Viergever, "A Discrete Dynamic Contour Model", *IEEE Transactions on Medical Imaging*, vol. 14/1, pp. 12-24, 1995.
9. T. McInerney, D. Terzopoulos, "Deformable models in medical image analysis: A survey", *Medical Image Analysis*, vol. 1, pp. 91-108, 1996.
10. S.R. Taylor, I.R. Neering, L.A. Quesenberry, V.A. Morris, "Volume changes during contraction of isolated frog muscle fibers", In: Excitation-Contraction Coupling in Skeletal, Cardiac, and Smooth Muscle, (Ed.: G.B. Frank), pp. 91-101, Plenum Press, New York, 1992.
11. K. Trombias, P. Baatsen, J. Schreuder, G.H. Pollac, "Contraction-induced movements of water in single fibers of frog skeletal muscle", *J. of Muscle Research and Cell Motility*, vol. 14, pp. 573-584, 1993.

Fuzzy Similarity Relations for Chromosome Classification and Identification

M. Elif Karsligil[1], M. Yahya Karsligil[1]

[1]Yildiz Technical University, Computer Sciences and Engineering Department, Yildiz, 80750, Istanbul - TURKEY

{elif, yahya}@ce.yildiz.edu.tr

Abstract. This paper presents a new approach to the classical chromosome classification and identification problem. Our approach maps distinctive features of chromosomes, e.g. length, area, centromere position and band characteristics, into fuzzy logic membership functions. Then fuzzy similarity relations obtained from the membership functions are used to classify and identify the chromosomes. This method has several advantages over classical methods, where usually a prebuilt single-criteria of template chromosomes is used to compare the unknown chromosome as to make a decision about its identity. First the formulation of chromosome characteristics using fuzzy logic better compensates for the ambiguities in the shape or band characteristics of chromosome in the metaphase images, second the use of all the characteristics of the chromosomes produce a more fail-safe method. As a preparatory step to the actual identification process we divide chromosomes according to their fuzzy similarity relation based on length and area into groups. To recover from the situations where a chromosome may be misgrouped because of its disconfirmity to ideal definitions, we refine the grouping of chromosome by applying fuzzy similarity relations which represent the relative centromere positions of chromosomes. Then the band characteristics of each chromosome in a group is correlated with the band characteristics of the chromosomes in the same group of a preprocessed template to obtain identity of the chromosome. The templates used at this step are updated each time when a chromosome is identified, so the system has an adaptive decision algorithm.

1 Introduction

Cytogenetics deals with the analysis of chromosomes (Fig. 1), which carry genetic information. There are normally 46 chromosomes in a human cell, which are arranged as 22 pairs of autosomes (numbered from 1 to 22) and 2 genosomes. Each chromosome consists of two chromatids which are joined together at the centromere, that forms the primary constriction of a chromosome. The two chromatides of a chromosome are called the p-arm and q-arm, with p-arm always being the shortest. Therefore the location of the centromere classify a chromosome as metacentric, submetacentric or acrocentric(Table 1). [6] arranges the chromosomes into seven groups(A-G) according to their descending order of size and centricity.

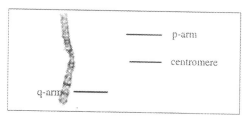

Fig. 1. A typical human chromosome

Besides the shape characteristics of a chromosome, there is another distinctive feature of each chromosome, which is called a band. A band is a region along the chromosome, that is apparently darker or lighter in intensity compared with the adjacent regions. Bands are the result of special staining procedures which are applied to the chromosome during their metaphase stage.

Table 1. Metacentric, submetacentric and acrocentric chromosomes

Centromere Group	Chromosome Number
Metacentric	1 3 16 19 20
Submetacentric	2 4 5 6 7 8 9 10 11 12 X 17 18
Acrocentric	13 14 15 21 22 Y

Manual identification of chromosomes is a time consuming and challenging task, since it requires the evaluation of lengths, centromere positions and also band characteristics of the chromosomes in at least 20 cells for each patient to reach at satisfying results. Also the fact that cytogenetics is very important in medical cases where time is the primary restricting factor like the identification of chromosomal disorders and/or prenatal diagnosis the necessity of automatic classification and identification systems for chromosomes with high accuracy cannot be argued at.

An analysis of current classification and identification systems reveals the fact that these systems usually employ a template-matching algorithm, where the template is constructed by the combination of one or more chromosome characteristics. The weak point of this approach is the fact that a chromosome not matching a template, e.g. length template may be misidentified because of a random match in another template, e.g. band template. Therefore we decided to include all of the characteristics of a chromosome in the classification and identification process and to design the decision algorithm as similar as a professional in cytogenetics would do, by formulating fuzzy logic functions of chromosome characteristics.

2 Fuzzy similarity groups for length and area characteristics

Test conducted on many chromosome samples showed that a chromosome supposed to be smaller than another chromosome according to [6] may be larger than this chromosome in some metaphases. This fact shows that length and area characteristics are necessary but not sufficient for automatic chromosome identification, so we decided to use these characteristics to formulate fuzzy similarity functions. The use of these functions will group the chromosomes based on their length and area as a preparatory step in actual identification. The formulation of the aforementioned fuzzy functions is as follows:

If we denote the length of the chromosome i, with L_i, then the total length of N chromosomes is given by

$$L_T = \sum_{i=1}^{N} L_i \tag{1}$$

Then we can express the relative length of each chromosome with

$$l_i = \frac{L_i}{L_T} \tag{2}$$

Using the relative lengths of the chromosomes i and j, their membership degree can be given by the formula

$$\mu_{ij}(l_i) = \begin{cases} \dfrac{l_j}{l_i} & l_i \geq l_j \\ \dfrac{l_i}{l_j} & l_j > l_i \end{cases} \tag{3}$$

which satisfies the following three conditions:

1. each chromosome is identical to itself, $\mu_{ii} = 1$
2. the membership degrees μ_{ij} and μ_{ji} for two chromosomes i and j are equal,
3. the membership degree function is defined within the lower and upper limits of the independent variable l

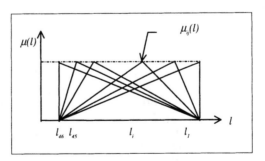

Fig. 2. fuzzy membership degree relation for the lengths of 46 chromosomes

For 46 chromosomes sorted in descending order by length $\mu_{ij}(l_i)$ $i=1..46$, $j=1..46$ (Fig. 2) can be arranged in a matrix as shown in Equation 4.

$$M_R = \mu(L) = \begin{bmatrix} \mu_{1,1} & \cdots & \mu_{1,46} \\ \vdots & \vdots & \vdots \\ \mu_{46,1} & \cdots & \mu_{46,46} \end{bmatrix} \tag{4}$$

This matrix represents the similarity relation for 46 chromosomes, where each value is between 0 and 1 inclusive. 0 means totally dissimilar, 1 means identical. In this form M_R is both reflexive and symmetric, so it is called a tolerance fuzzy relation. What we are seeking for is the transitivity of M_R so we apply repeated max-min compositions as in Equation 5 until the condition $M_{R^{N+1}} = M_{R^N}$ is satisfied.

$$M_{R^{N-1}} = \prod_{i=1}^{N-1} M_R \qquad (5)$$

According to Warshall algorithm the condition $M_{R^{N+1}} = M_{R^N}$ is satisfied in at most $(N-1)$ steps each consisting of N^3 logic multiplications. In our case $\varphi = 46^3 * 45$ calculations is required for each membership degree matrix to reach the steady state. Therefore we developed an algorithm called, successive power taking, which reduces the number of calculations drastically. If we denote

$$T_1 = M_R * M_R$$

then it follows that

$$T_2 = T_1 \circ T_1$$

$$T_r = T_{r-1} \circ T_{r-1}.$$

This equality can be expressed as

$$T = (((M_R \circ M_R)^2)^2 \cdots)^r$$

This formulation reduces the total number of calculations by $(N - 1) / \log_2 (N - 1)$ steps.

The resulting matrix is the equivalence relation for 46 chromosomes. An α-cut operation is performed on the equivalence relation matrix to defuzzify the relation. At this step each value of 1 in the ith row of the defuzzified relation matrix shows the similar chromosomes to the ith chromosome. In this way we obtained the relative length similarity groups (G_l). Applying the same technique to the areas of the chromosomes we obtain the relative area similarity groups (G_a) for the chromosomes.

3 Fuzzy similarity groups for centromere positions

The length l of a chromosome can be expressed as:

$$l = p + q \qquad (6)$$

where p and q correspond to the lengths of individual arms of the chromosome. Then the relative centromere position of a chromosome is given by

$$c = \frac{p}{p+q} \qquad (7)$$

The calculated values for the relative centromere position of metacentric, submetacentric and acrocentric chromosomes are given in Table 2.

Table 2. Chromosomes and their relative centromere positions

Chromosome Number	Relative Centromere Position	Chromosome Number	Relative Centromere Position
1	.45-.50	13	.13-.22
2	.35-.42	14	.13-.22
3	.44-.50	15	.13-.22
4	.24-.30	16	.41-.45
5	.24-.30	17	.28-.37
6	.34-.42	18	.23-.33
7	.34-.42	19	.42-.50
8	.33-.38	20	.41-.50
9	.32-.40	21	.22-.30
10	.30-.37	22	.22-.30
11	.35-.45	X	.36-.41
12	.24-.30	Y	.28-.34

As can be seen from Table 2, the majority of relative centromere positions for submetacentric chromosomes are in the range 0.30-0.42, but chromosomes 4, 5, 12 and 18 in this group disobey this distribution and have values between 0.23 and 0.33. The same behavior can be observed in the group of acrocentric chromosomes too. The only exception is the metacentric group where all the chromosomes have values between 0.41 and 0.50. Therefore we decided to group the chromosomes into four groups (Table 3) based on their centricity in contrast to the three groups defined in [6].

Table 3. New arrangement of chromosomes in to four groups by their centromere position.

Centromere Group	Relative Centromere Position	Chromosome Number
S1	.42-.50	1 3 16 19 20
S2	.30-.41	2 6 7 8 9 10 11 X 17 18 21 22
S3	.28-.34	4 5 12 Y
S4	.13-.22	13 14 15

If we denote the lower limit of a group by c_{kl} and the upper limit by c_{ku}, then the fuzzy membership function to give the similarity among to chromosomes based on their centricity can be written as:

(8)

$$\mu_{ij}(c_k) = \begin{cases} 1 & c_{kl} < c_j < c_{ku} \\ \dfrac{c_j}{c_{ku}} & c_j < c_{kl} \\ \dfrac{c_{ku}}{c_j} & c_j > c_{ku} \end{cases}$$

The same technique given in the previous section applied to $\mu_{ij}(c_k)$ $k=1..4$, $i=1..46$, $j=1..46$ gives the relative centromere position similarity membership degree groups (G_s).

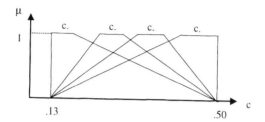

Fig. 3. Fuzzy membership degrees for the relative centromere positions

The intersection of G_l, G_a. G_s groups is calculated to divide the chromosomes into seven groups which correspond to the A-G groups [6]. The next step involves the actual identification of each chromosome within the seven groups using their band characteristics.

4 Identification of chromosomes based on their band characteristics

Each chromosome shows a different band pattern, but external conditions, staining procedure or the chromosome itself can influence the resulting band pattern, so that some bands may be missing or other bands become visible. It is therefore mandatory to use only characteristic band information in the calculations to make the method more resilient to such changes in band patterns. To extract characteristic band information we run histogram equalization on the gray level chromosome images and reduced the gray levels such that only transitions from white to gray and from gray to black remained. Then we calculated a template band pattern for each chromosome in each group. To identify a chromosome in a group, this chromosomes band pattern is correlated with the band patterns of the chromosomes in the template group using the following formula

$$r(x) \;=\; \frac{E\{(x - \eta_x)(y - \eta_y)\}}{\sqrt{E\{(x - \eta_x)^2\} * E\{(y - \eta_y)^2\}}} \tag{9}$$

which results in $-1 \leq r(x) \leq 1$, where values between 0 and 1 shows the degree of similarity. So we write the following equation to obtain a fuzzy membership function based on $r(x)$

$$\mu(x) \;=\; \frac{1}{2} * r(x) + \frac{1}{2} \tag{10}$$

The highest $\mu(x)$ value gives us the identity of the chromosome. This template scheme is designed to be adaptive in the sense that the template values are adjusted and recalculated each time when a new chromosome is identified correctly.

5 Conclusion

In this paper we presented an original and new method to the classification and identification problem of chromosomes based on fuzzy logic membership functions. This method, at the first step divides the chromosomes into preliminary groups using similarity relations resulting from fuzzy membership degrees. Second it rearranges these preliminary groups according to their centromere position similarity. At last step it identifies each chromosome of a group by correlating its band characteristics with the band characteristics of template chromosomes of the corresponding group. The system adapts itself as chromosomes are identified by updating the templates of the chromosomes.

This method has been successfully applied to 150 metaphase images with an average rate of 98% correct grouping of chromosomes and 94% correct identification of chromosomes within these groups.

References

1. Carothers, A., Piper, J.: Computer-Aided Classification of Human Chromosomes: A Review. Statistics and Computing, vol.4, no.3 (1994) 161-171
2. Fukunaga, K.: Introduction to Statistical Pattern Recognition, Academic Press (1990)
3. Graham, J., Piper, J.: Automatic Karyotype Analysis. Methods in Molecular Biology, vol.29 (1994) 141-183
4. Ji., L.: Fully Automatic Chromosome Segmentation. Cytometry 17 (1994) 196-208
5. Johnston, D., Tang, K., Zimmerman, S.: Band Features as Classification Measures for G-banded Chromosome Analysis. Comput. Biol. Med., vol.23, (1993) 115-129
6. Mitelman., F.: An International System for Human Cytogenetic Nomenclature. Karger, Switzerland (1995)
7. Papoulis, A.: Probability Random Variables and Stochastic Process, Mc Graw Hill Inc., New York (1965)
8. Rooney, D.E., Czepulkowski, B.H.: Human Cytogenetics, Oxford University Press, Volume I, Oxford (1992)
9. Ross, T.J.: Fuzzy Logic with Engineering Applications, Mc Graw Hill Inc., New York (1995)

Enhanced Neural Networks and Medical Imaging[*]

Luis F. Mingo[1], Fernando Arroyo[1], Carmen Luengo[1], and Juan Castellanos[2]

[1] Dept. de Lenguajes, Proyectos y Sistemas Informáticos. E.U. de Informática, Universidad Politécnica de Madrid, Carretera de Valencia km. 7, 28031 Madrid Spain. Phone: +34913367867. Fax: +34913367526. E-mail: lfmingo@eui.upm.es
[2] Dept. de Inteligencia Artificial. Facultad de Informática, Universidad Politécnica de Madrid, Campus de Montegancedo, 28660 Boadilla del Monte, Madrid Spain.

Abstract. This paper shows that the application of Enhanced Neural Networks when dealing with classification problems is more powerfull than classical Multilayer Perceptrons. These enhanced networks are able to approximate any function $f(x)$ using a n-degree polinomial defined by the weights in the connections. Also, the addition of hidden layers in the neural architecture, increases the degree of the output equation associated to output units. So, surfaces generated by these networks are really complex and theoretically they could classify any pattern set with a number n of hidden layers. Results concerning medical imaging, breast cancer diagnosis, are studied along the paper. The proposed architecture improves obtained results using classical networks, due to the implicit data transformation computed as part of the neural architecture.

Keywords. Neural Networks, Breast Cancer, Approximation Theory.

1 Introduction to Enhanced Neural Networks

Since 1962 Rosenblatt published the Perceptron architecture, most of the applications and theorical researches in the field of Neural Networks are based on the signal propagation from the input to the output and the adjustment of the weights in the connections.

The most powerful characteristic of *NN* is the generalization of data not presented to the net. *MLPs* with no lineal activation function have demonstrated [12, 5] to be able to solve most of the problems that have been stated, provided that the data pre-processing is carefully done. That is, neural networks with one hidden layer are universal approximators [5, 6, 3], the main problem is to choose the right number of hidden units [4].

Recent approaches take into account that the weights do not need to be static for all the patterns in the training and test sets. Among them, Jordan [8] and Jacobs [7] introduced Modular Neural Networks (*MNN*) with a different learning algorithm for each module and the possibility of having context units. Vapnik and

---

[*] Supported by INTAS 96-952.

Bottou combined local algorithms [2] and some statistical methods [1] in order to have different weight sets. In 1992, Pican [11] proposed a similar approach. Pican defined an *OWE (Orthogonal Weight Estimator)* model in which each weight is computed by an *MLP* using as input a specific context information.

Multilayer Perceptrons *MLP* are based on the fact that the addition of hidden layers increases the performance of the Perceptron. This way, data sets with a no linear separation can be divided by the neural network and a more complex geometric interpolation can be achieved [13], provided that the activation function of the hidden units is not a linear one.

The proposed neural networks *ENNs* are characterized for having different weights for each different pattern which is introduced to the net. Such mechanism could be thought as a local interpolation in some specific points of the function $f(\bar{x})$ that the pattern set defines instead of the global interpolation that *MLP* networks provide.

In order to reach this property two neural networks have been used. The assistant network computes the weights of the main network depending on the input pattern. That is, each pattern produces a set of weights that is employed in the main network to output the desired response. The main and assistant networks share the inputs.

Such behaviour seems very logic from a neuropsicological point of view. Barbizet proposes connections among the neurons, these connections propagate the input signal to different neurons increasing or decreasing the power of the signal, depending on the features of the connections.

Mathematically, the previous idea could be expressed as: $w_{ji} = o_k$, being w_{ji} a weight of the main network and o_k the output of a neuron in the assistant network. Such idea permits to introduce quadratic terms in the output equations of the network where they were lineal equations using Backpropagation Neural Networks *BPNN* with linear activation functions.

2   Aproximation Capabilities

The initial idea of Artificial Neural Networks was very disapointed in their earliest stage, since the Perceptron architecture does not classify in the appropiate class the *XOR* boolean function, in which a lineal separation of classes was not possible. Besides, and more important, there were linear activation functions in the units. This fact makes that the addition of new hidden layers does not improve the performance of the network since the architecture could be reduced to another one without hidden layers.

In the proposed architecture *ENN*, the addition of new hidden layer with lineal activation functions increases the computational performance concerning *MSE*, time, and generalization. This is one of the most important advantages with respect to the Perceptron.

Next figures show the scheme of an *ENN* without any hidden layers and with one hidden layer and with lineal activation function. These networks are not equivalent since the equations of the output neuron is different for each network. Lineal *MLP* does not have this property.

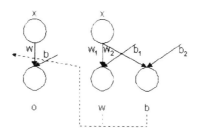

Fig. 1. Lineal ENN without hidden layer.

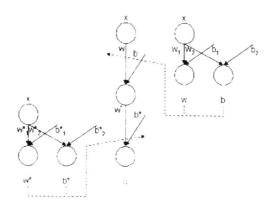

Fig. 2. Lineal ENN with $n = 1$ hidden layers.

Equations (2) and (1) show the expression of the output unit with and without hidden layer respectively. The activation function is the identity function, that is, $f(x) = x$. These equations verify that the addition of hidden layers increases the computational capacity of *ENN*, concerning the geometric separation of the patterns.

$$
\begin{aligned}
o_1 &= wx + b = \\
&= (w_1 x + b_1)x + w_2 x + b_2 = \\
&= w_1 x^2 + (b_1 + w_2)x + b_2
\end{aligned}
\tag{1}
$$

$$
\begin{aligned}
o_2 &= w^*(wx + b) + b^* = \\
&= (w_1^* x + b_1^*)[(w_1 x + w_2 + b_1)x + b_2] + w_2^* x + b_2^*
\end{aligned}
\tag{2}
$$

The addition of new hidden layer in the main network increases the degree of the output polynomial expression $P(x)$, that is, the expression that the output unit of the network generates $o(\bar{x}) = P(x)$. If there are not hidden layers then the degree of the polynomial is two, that is a quadratic polynomial in the output of the network. The feature of being able to increase the degree of the output polynomial, adding more hidden layers, makes this kind of neural networks a powerfull tool against the *MLP* neural networks.

A function could be approximated with a certain error, previously fixed, using an n-degree polynomial $P(x)$. The achieved error using this polynomial is bound by a mathematical expression, in such a way that if you want to approximate a function with an error $< \epsilon$ then you only have to compute the sucessive derivates of $f(x)$ until a certain degree and to generate the polynomial $P(x)$.

$$|e(x)| \leq \frac{1}{(n+1)!} M |x - a|^{n+1} \tag{3}$$

The proposed neural network *ENN* behaves as universal n-degree approximators, depending on the number of hidden layers of the main network. To achieve this goal, activation functions of the units do not need to be no lineal functions.

Figures 1 and 2 and equations (1) and (2) associated to each one of the neural networks, show that the addition of one hidden layer in the main network increases the degree of the output polynomial $P(x)$. So, as the number of hidden layers increases, the degree of the output polynomial will increase too.

The enhanced neural network architecture has the property that the output equation of an output unit is a n-degree polynomial if the activation function is a lineal one. This output can be compared with the *Taylor* approximation polynomials since the both methods are very similar. A set of data can be approximated by *ENNs*, computing a n-degree polynomial as the network output. But, the activation function can be a sinusoidal, instead being a lineal one. With this function, the approach of *ENN* is similar to *Fourier* series decomposition. This way, the activation function can be changed in order to get a better approximation than in the case of *MLPs*.

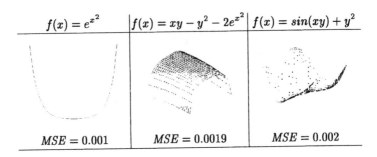

$f(x) = e^{x^2}$	$f(x) = xy - y^2 - 2e^{x^2}$	$f(x) = sin(xy) + y^2$
$MSE = 0.001$	$MSE = 0.0019$	$MSE = 0.002$

Fig. 3. Data approximation computed by an *ENN*. Surfaces computed.

Figure 3 shows that the proposed network is able to learn very difficult surfaces with a low *MSE*. This is mainly due to the special architecture of the net. The input to the net affects to the weights in the connection, and even changes them in order to optimize the error achieved by the net [10].

3 Breast Cancer Diagnosis

This paper covers just one aspect of this field, mammographic pattern recognition, which is concerned with developing computers programs that are able to interpret patterns in digitized mammographic images. Problems which required complex interpretive decisions to be made are often formulated mathematically as systems for classifying patterns. The traditional solution of these systems involves a branch of mathematical statistics known as statistical pattern recognition.

Artificial Neural Networks techniques show excellent behaviour in being able to overcome the limitations of presently used computer methods to predict carcinoma. Because these networks can be trained to recognize complex relationships that exist among inputs and outputs. These subtle relationships in the data are automatically recognized by the network, even if they are unkown to clinicians. *NN* can normally be expected to perform at least as well as, and usually better, than other modelling technique.

In order to be able to detect breast cancer, it is necessary to pay attention to four mammographies, two ones per each breast (two frontal shots and two shots from above). The cancer could be detected if some of the following situations [9] is into the mammography: *asimmetry between the shape of right and left mammographies, asimmetry between the tissue in both mammographies, architecture disorders concerning the global structure of the breasts, nodules or tumour masses, microcalsifications, and spiculating images.*

Image processing is one of most important fields that must be taken into account when designing an automatic system to breast cancer diagnosis. It is necessary to provide adaptive filters [9] to avoid changing situations and robust subsystems, such as neural networks, due to the importance of the global system, concerning moral and human decisions.

4 Neural Networks and Breast Cancer Diagnosis

Two different methods could be employed when computing a diagnosis from a given image, depending on the stage where neural networks are introduced, or depending on the input to the neural networks. Two both methods are valid ones, and they can be used together in order to obtain or to reinforce the diagnosis made by the radiologist.

The first one consists on extracting some features from the image, via filtering process, and to feed them into the neural network. This way, the *NN* will be able to learn the specific features that lead a cancer diagnosis. This method will consider only a few inputs to the net, and the most important phase is not done by the *NN*.

Table 1. Performance in breast cancer diagnosis using first method

Hidden Layers	Features	Patterns	MSE	Performance	Validation
1	10	200	0.07	92%	85%
1	12	200	0.05	95%	90%
1	16	200	0.02	98%	97%
2	10	200	0.04	92%	89%
2	12	200	0.02	97%	92%
2	16	200	0.01	99%	98%
1	16	500	0.01	99%	98%
2	16	500	0.01	99%	99%

Among the different features that can be obtained from the mammographies are: *density, number of nodules, number of microcalsifications, simmetry, % of bran, etc.* for each one of mammography views and data concerning the patient: *age, previous diagnosis, history, etc.*. These data are fed into an *ENN* with two outputs in order to compute a diagnosis: *cancer, no cancer and other*, the network only establish a relationship among different inputs to make a regression analysis. The most important result using *ENN* is that the regression analysis could be made using a n-degree polynomial depending on the number of hidden layers. Table 1 shows results concerning this first method. All feature data have been obtained from *CBMR - Center for Biomedical Modelling Research*.

The second method consists on pre-processing the image associated to the mammography and feed it, or some part of it, to the network. This way, the net will be the most important part in the diagnosis process. Even, an incorrect pre-processing will not affect the overall behaviour of the system.

Along this paper, the *ENN* architecture has shown that it is very suitable when dealing with any problem. Decision surfaces generated by the net are complex enough to represent any data set. The powerfull of these nets is in the number of hidden layers, that is, in the degree of the output polynomial associated to one output unit.

Table 2 shows *MSE* using different network architectures, and varying the % of filtering, that is, the number and kind (adaptive, fuzzy, etc.) of filters applied to the initial mammography. Results are very similar to table 1, but this method has the advantage of being an unattended process.

Table 3 shows a comparison between the *MSE* of an Enhanced Neural Network and a classical Back Propagation, both dealing the same problem, that is,

Table 2. Performance in breast cancer diagnosis using second method

Hidden Layers	% Filtering	Patterns	MSE	Performance	Validation
1	2	200	0.07	92%	85%
1	4	200	0.04	95%	91%
1	6	200	0.02	98%	97%
2	2	200	0.04	92%	89%
2	4	200	0.02	97%	92%
2	6	200	0.01	99%	98%
1	6	500	0.01	99%	98%
2	6	500	0.01	99%	99%

breast cancer diagnosis. In both cases, all parameters have been set to the same value in the two nets.

Table 3. Enhanced Neural Networks vs. Back Propagation Nets

ENN	0.5	0.1	0.08	0.06	0.04	0.01	0.008	0.006	0.005	0.004
BP	0.5	0.2	0.1	0.09	0.07	0.05	0.02	0.01	0.009	0.008

Previous results confirm the power of *ENN* against classical network models. The architecture does not introduce computational complexity, it only modifies ,in a light sense, the concept of varying weights from biology point of view. This improvement permits to approximation a function $f(x)$ using a n-degree polynomial represented by the net, and therefore, the application of this net to a given problem, breast cancer diagnosis, obtains good results.

5 Conclusions

This paper has presented to methods in order to ellaborate a breast cancer diagnosis. Both methods are based on the application of a special neural network architecture, known as *Enhanced Neural Network*. This architecture permits to approximate any data set using a n-degree polynomial, depending on the number of hidden layers. Therefore, the decision surface generated by the *ENN* is more powefull, in the sense of complexity, than the surface generated by classical *NN*, see back propagation. These decision surfaces permit to approximate with a low mean squared error any pattern set, so as to approximate with no error any n-boolean function due to the special architecture of the net. In some sense, *ENNs* are very similar to the *Taylor* series, in the field of function approximation. All theory and examples are based on lineal activation functions but, if a sinusoidal or sigmoid function is choosen as activation function, then results will vary in good sense, that is, the *MSE* will decrease considering the same architecture, and even the net architecture could be simplify to get a similar *MSE*. Concerning the breast cancer diagnosis problem, both methods are valid in order to support radiologist diagnosis. First one depends on the features extracted from mammographies while the other depends on the *ENN* process. The performance using both methods with Enhanced Neural Networks is really good enough to be considered as a second opinion in the diagnosis process.

References

1. Bottou, L. & Gallinari, P.: *A Framework for the Cooperation of Learning Algorithms*. Laboratoire de Recherche en Informatique. Université de Paris XI. 91405 Orsay Cedex. France.
2. Bottou, L.; Cortes C. & Vapnik V.: *On the Efeectiveness VC Dimension*. AT&T Bell Laboratories, Holmdel NJ 07733. USA. (1994).
3. Funahashi, K: *On the Approximate REalization of Continuous Mappings by Neural Networks*. Neural Networks, Vol. 2. Pp. 183-192. (1989).
4. Hecht-Nielsen, R.: *Kolmogorov´s Mapping Neural network Existence Theorem*. IEEE First International Conference on Neural Networks. San Diego. Vol III. Pp. 11-14. (1987).
5. Hornik, K.; Stinchombe, M. & White, H.: *Universal Approximation of an Unknown Mapping and its Derivates Using Multilayer Feedforwar Networks*. Neural Networks, Vol. 3. Pp. 625-633. (1990).
6. Hornik, K.: *Approximation Capabilities of Multilayer Feedforward Networks*. Neural Networks, Vol. 2. Pp. 551-560. (1991).
7. Jacobs, R. A.; Jordan M. I. & Barto A. G.: *Task Decomposition through Competition in a Modular Connectionist Architecture: The what and where vision tasks*. MIT. COINS Technical Paper 90-27. (1990).
8. Jordan M. I. & Rumelhart D. E.: *Forward Models: Supervised Learning with a Distal Teacher*. MIT Center for Cognitive Science. Occasional Paper #40.
9. Mingo, L.F.; Castellanos, J.; Giménez, V. & Villarrasa, A.: *Hierarchical Neural Network System to Breast Cancer Diagnosis*. 9th. European Signal Processing Conference. EUSIPCO-98. Signal Procesing IX. Theories and Applications. Rhodes, Greece, 8-11 September. Pp. 2457-2460. (1998).
10. Mingo, L.F.; Arroyo, F.; Luengo, C. & Castellanos J.: *Learning HyperSurfaces iwth Neural Networks*. 11th Scandinavian Conference on Image Analysis. SCIA'99. Kangerlussuaq, Greenland. June 7-11. (1999). SUBMITTED
11. Pican, N.: *An Orthogonal Delta Weight Estimator for MLP Architectures*. ICNN'96 Proceedings, Washington DC. June 2-6. (1996).
12. Stinchombe, M. & White, H: *Multilayer Feedforward Networks are Universal Approximators*. Neural Networks, Vol. 2. Pp. 551-560. (1989).
13. Wieland, A. & Leighton, R.: *Geometric Analysis of Neural Network Capabilities*. IEEE First International conference on Neural Networks, Vol. 3. Pp. 385-392. (1987).

Learning Behavior Using Multiresolution Recurrent Neural Network

Satoru Morita

Image Sciences Institute, Univeristy Hospital Utrecht
Heidelberglaan 100, 3584 CX, Utrecht, NL
Faculty of Engineering, Yamaguchi University, Tokiwadai Ube 755, Japan

Abstract. We propose the multiresolution recurreint neural network to learn behavior based on view and action. Recurrent neural network structure has the multiresolution channel to establish between the view and the action. It is difficult to learn action using only a image generally. We solve this problem by using the 3 kinds of image on the frequency. We control the multiresolution vision using Genetic Algorithm. The action sequences is acquired by the pan-tilt camera.

1 Introduction

One of the purposes of robot study is to realize robots in the real world. Many researchers started to study the robot action [1] after success of the intelligence based on the action of Brooks [2] in the field of the robot study. However, a behavior-based robot system is controlled based on the rule sets which are designed by humans. If the object, task and environment are complex, it is difficult to design the robot system.

Gibson discusses that we get the information given the worth "affordance" based on our action without the physical stimulus [3]. The discussion concerns the relations between view and action essentially. Wilson discusses a model where humans have 4 kinds of filters for analyzing images, and use there filters [4][5]. When a human moves from a place to another place, he uses the recognized map based on relative space arrangements is gotten from the usual visual and active experiments [6].

We propose a method which acquires actions by getting the recognized map, while we control the multiresolution vision. We use a recurrent neural network to establish the relation between behavior and vision. Recurrent neural network has been introduced by Rummelhart [7]. This kind of neural network is suited for learning of time sequences. But if some kinds of output exist in a input, the pattern is not learned using the recurrent neural network [7]. In this paper, we propose a method which acquires the behavior by controlling the view to avoid some kinds of output. So we propose a recurrent neural network that relates view, behavior and the multiresolution channel. We control the multiresolution vision using a genetic algorithm.

In section 2, we discuss the multiresolution vision using 3 channels. In section 3, we discuss the neural network that relates view and behavior. In section 4, we discuss the control of multiresolution vision using a genetic algorithm. In section 5, we discuss the algorithm of behavior acquisition. We evaluate the effectiveness of this method by applying this method for the learning of action sequence for a pan-tilt camera.

2 Multiresolution Vision Using 3 Channels

In this section, we generate the multiresolution vision using 3 channels to get the view needed to learn the action sequences. Scale-space is proposed to analyze signal sets with several scale [8][9]. We use the filtering to generate 3 kinds of the image on multiresolution.

Assuming a parametric form of a surface $f(u, v) = (x(u, v), y(u, v), z(u, v))$ for the pixel value z on the (x,y) coordinate in an image, the point on the derived surface is the result of the convolution $\phi(u, v, \sigma) = f(u, v) * g(u, v, \sigma)$. The image convolved Gaussian satisfies with the following diffusion equation.

$$\frac{\partial^2 \phi}{\partial u^2} + \frac{\partial^2 \phi}{\partial v^2} = \frac{1}{\sigma}\frac{\partial \phi}{\partial \sigma} \tag{1}$$

This equation is approximated by a difference equation.

$$\phi_{i0}(t + \Delta t) = \phi_{i0}(t) + \frac{\Delta t}{n}\Sigma_{k=1}^{n}\frac{1}{l_{ik}^2}(\phi_{ik}(t) - \phi_{i0}(t)) \tag{2}$$

where $\phi_{i0}(= (x_{i0}, y_{i0}, z_{i0}))$, ϕ_{ik} and l_{ik} are respectively ith sample point, neighbor samples and the distance between $i0$ and ik.

As equation (1) is approximated by a difference equation, if scale σ become big, the value of the curvature is diffused such that the curvature of x direction, y direction and z direction become the constant value.

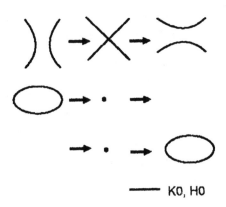

—— K0, H0

Fig. 1. The topology changes of KH-image.

We show the algorithm generating the description from the camera in the following. We get the intensity image $z = f(x, y)$ using camera. We transform the intensity image $z = f(x, y)$ to the parametric representation $\phi(u, v) = (x(u, v), y(u, v), z(u, v))$ for the pixel value z on the (x,y) coordinate. The surface is smoothed using the filter of scale σ. The parametric surface $\phi_{\eta\zeta}(u, v, \sigma) = (x(u, v, \sigma), y(u, v, \sigma), z(u, v, \sigma))$ is classified into elements using Mean curvature H and Gaussian curvature G. Each element is discriminated by negative and positive of KH. The boundaries of the elements are the zero-crossings of Mean curvature and Gaussian curvature (K=0(K0), H=0(H0)). We define the image as the KH-image on the scale σ. We use 3 channel multiresolutions ($\sigma = 90, \sigma = 270, \sigma = 540$). Where σ is the iteration number of the diffusion equation. In this experiments, we use the multiresolution vision of these 3 channels. The resolution $r(t)$ selected from 3 channels is defined as 0, 1, 2 for σ=90, 270, 540.

The KH-image is transformed to the image topology which describes the neighbor relation between elements. We use the toppology change to classify images which have a similar property. Topology changes are classified into 3 types that K0 lines or H0 lines appear and vanish and two H0 lines and two K0 lines contact (Figure 1). This theory is based on the Morse theory [10]. Because the cross sections between 3 and 4 K0 and H0 lines exist actually, many topology changes exist. The changes are approximated using 3 types of topology changes. If the image can be generated from another image with one or two topology changes, we define the two images as the same classification. If the image is generated from another image using two topology changes, we define the two images as the same classification. We use the label of the image topology in the sensor input of the recurrent neural network.

3 Recurrent Neural Network Based on Behavior and View

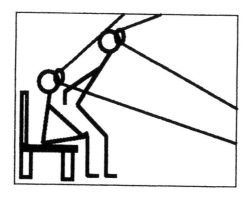

Fig. 2. The relation between action and view.

When a man who was sitting in a chair, stands up, his viewpoint changes (Figure 2). The change of viewpoint is determined by the behavior. We use the view as the input of the neural network and the output is the behavior. The output behavior is also feedback on the input. As it is important to control the view, we propose a recurrent neural network where a multiresolution channel is connected to the output and is connected to the feedback input. The input layer has $n + m$ which are n sensor inputs and m feedback inputs. The hidden layer has $n + m$ units and output layer has m units.

- Sensor input $x(t) = \{x_i(t)(i = 1, 2, ..., n)\}$
- Feedback input $y'(t) = \{y_i'(t)(i = 1, 2, ..., m)\}$
- Unit input of hidden layer $J(t) = \{J_i(t)(i = 1, 2, ..., m + n)\}$
- Unit output of hidden layer $zhid(t) = \{zhid_i(t)(i = 1, 2, ..., m + n)\}$
- Unit input of output layer $I(t) = \{I_i(t)(i = 1, 2, ..., n)\}$
- Behavior (output) $y(t) = \{y_i(t)(i = 1, 2, ..., m)\}$

If the parts move, the description of the action is 1. If the parts do not move, the description of the action is 0. We use 3 different parts which are ZOOM, PAN and TILT in this experiments. If the parts rotate in the direction $(+)$ and $(-)$, the description is 1. If the parts do not rotate, the description is 0.

The image on sensor input is classified using the topology of image. We discuss the method which generate the topology of image in section 2. The structure of recurrent neural network proposed is shown in figure 3(A).

- Sensor Input $x(t) = \{x_i(t)(i = 1, 2, ..., 8)\}$, IMAGE TOPOLOGY(Image topology)
- Feedback Input $y'(t) = \{y_i'(t)(i = 1, 2, ..., 9)\}$, RESOLUTION CHANNEL(Resolution channel), ZOOM(+), ZOOM(−), PAN(+), PAN(-), TILT(+), TILT(−)
- Behavior (output) $y(t) = \{y_i(t)(i = 1, 2, ..., 9)\}$, ZOOM (+), ZOOM (−), PAN (+), PAN (−), TILT (+), TILT (−), RESOLUTION CHANNEL (Resolution channel)

We discuss the learning of action sequences in this section. We define the input of recurrent neural network $z_j(t)(i = 1, .., n + m)$ according to:

$$z_j(t) = \begin{cases} x_j(t) & (j = 1, ..., n) \\ y_{j-n}'(t) & (j = n + 1, ..., n + m) \end{cases} \tag{3}$$

Output for units of hidden layer is defined as follows:

$$zhid_i(t) = s_i(I_i(t))(i = 1, ..., m + n) \tag{4}$$

$$I_i(t) = \Sigma_{j=1}^{m+n} w1(t - 1)_{ij} z_j(t - 1)(i = 1, ..., m + n) \tag{5}$$

Output for units of output layer is defined in the following.

$$y_i(t) = s_i(J_i(t))(i = 1, ..., m) \tag{6}$$

$$J_i(t) = \Sigma_{j=1}^{m+n} w2(t-1)_{ij} zhid_j(t-1)(i=1,...,m) \qquad (7)$$

where is $s_i(u)$ as the sigmoid function in this paper. Input for the units of the input layer is defined by:

$$y_i'(t) = y_i(t-1)(i=1,...,m) \qquad (8)$$

We get the input sequences and output sequences in the action by acting the behavior. By repeating the input sequences $x = \{x(0),...,x(stop-1)\}$ and the output sequences $y = \{y(1),...,y(stop)\}$, we learn the weight $w2_{ij}(i = 1,...,m, j = 1,...,n+m)$ between the hidden layer and the input layer, and the weight $w1_{ij}(i = 1,...,m+n, j = 1,...,n+m)$ between the hidden layer and the output layer. The action sequences are trained by changing the weight $w2_{ij}(i = 1,...,m, j = 1,...,n+m)$, $w1_{ij}(i = 1,...,m+n, j = 1,...,n+m)$ to decrease the average square error of $y(t+1)$ for $x(t)$. In figure 3(B) , we show the flow of the learning the action sequences using the recurrent neural network with a multiresolution channel.

The stop condition for the training of the neural network is defined by the square error function $error(t)$. If the value of the function is under the threshold, the learning is stopped.

$$\begin{aligned} error~(t) = \\ \Sigma_{i=1,j=1}^{m,n+m}(w2(t)_{ij} - w2(t-1)_{ij})^2 \\ +\Sigma_{i=1,j=1}^{n+m,n+m}(w1(t)_{ij} - w1(t-1)_{ij})^2 \end{aligned} \qquad (9)$$

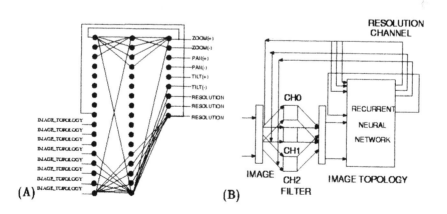

Fig. 3. (A)The recurrent neural net structure is related with resolution, view and action.(B)The figure shows the recurrent neural network structure which has the multiresolution channel.

4 Control of Multiresolution by Genetic Algorithm

We assume that the sequences of multiresolution channel are $r = \{r(1), ..., r(stop)\}$ and $stop$ is the number of the action sequence. We evaluate the input sequences and output sequences obtained from the given resolution channel. The evaluation is high, as the trained pattern is small in the case that the different relations between behavior and view are small. Otherwise, the evaluation is low, because it is difficult to learn the pattern in the case that the contradiction between behavior and view is large. The number of situations that have different output in spite of the same input as $g(i, j)$ are determined in the following.

$$F(i, j) = \begin{cases} 1 & x(i) = x(j) \text{ and} \\ & y(i+1) = y(j+1) \\ 0 & \text{otherwise} \end{cases} \tag{10}$$

$$f1 = \Sigma_{i=0}^{i<stop-1} \Sigma_{j=0}^{j<stop-1}[k - F(i, j)] \tag{11}$$

This is the function that evaluates the number of different relations between behavior and view. k is a constant value to prevent the evaluation to become negative. In this experiment, we use $k = 1$.

$$G(i, j) = \begin{cases} 1 & x(i) = x(j) \text{ and} \\ & y(i+1) \neq y(j+1) \\ 0 & \text{otherwise} \end{cases} \tag{12}$$

$$g1 = \Sigma_{i=0}^{i=stop-1} \Sigma_{j=0}^{j=stop-1} G(i, j) \tag{13}$$

This is the function that evaluates the contradiction between behavior and view.

$$h1 = \Sigma_{i=1}^{i=stop} r(i) \tag{14}$$

This is the function that gives the priority for the low-pass filtered image because the cost of the image processing is low.

$$Min_{r(i)(i=1,...,stop)}[a \cdot f1 + b \cdot g1 + c \cdot h1] \tag{15}$$

We define the evaluation rate between the functions as a, b, c. In this experiments, $c << a << b$. We define the final evaluation function using the previous 3 evaluation functions. In this experiments, we use $c = 1.0, a = 50.0, b = 100.0$.

We define the sequences of the multiresolution channel $r = \{r(1), ..., r(stop)\}$ such that the evaluation function is minimal. It is necessary to get the global optimal solutions which are good for the actual problem though the solutions are not exact. A genetic algorithm is a good technique to get the global optimal solution. So we solve the problem using the genetic algorithm. We define the genetic code as the sequences of multiresolution channel $r = \{r(1), ..., r(stop)\}$. The length of the genetic code is $stop*2$. We use the equation (13) to calculate the fitness value of a gene. In this experiment, the crossing rate is 0.8, the mutation rate is 0.005 and the number of gene is 30.

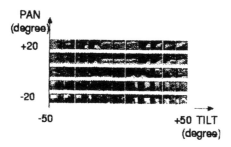

Fig. 4. Changes of view are shown, when pan moves from +20 to −20 and tilt moves from +50 to −50. 55 viewpoint images are shown.

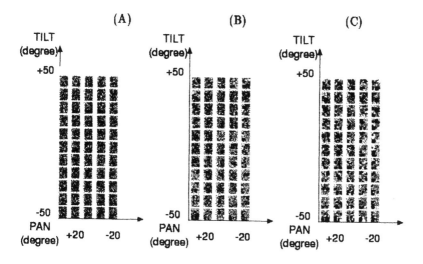

Fig. 5. Multiresolution visions of three channels are shown ((A)σ=90, (B)σ=270, (C)σ=540)

5 Algorithm of Behavior Acquisition

Behavior acquisition algorithm using the control of multiresolution vision and the recurrent neural network is discussed in the following.

- We record the view and action after the behavior.
- The resolution sequences are defined using a genetic algorithm(section 4).
- The neural network is trained to relate the behavior and the view, obtained from the gotten resolution (section 3).

The training is continued until the average square error of the weight which is shown in equation (9) becomes less than a threshold. If the number of training

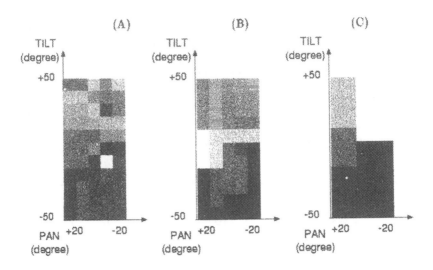

Fig. 6. Image topologies generated from multiresolution visions using the clustering method are shown when the pan of the camera moves from +20 degree to –20 degree and the tilt of the camera moves from +50 degree to –50 degree. Same image topology is represented by the same color. ((A)σ=90, (B)σ=270, (C)σ=540)

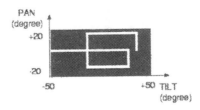

Fig. 7. The example of an action sequence. A white line means the route where the camera moves on.

reaches 100, and the error still is larger than the threshold, we define the learning as impossible. If the initial view and behavior is given, the learned action sequences are done in the case that the learning is finished.

In this experiments, we use a pan-tilt camera controlled by SUN Workstation. We fixed the pan-tilt camera on the desk and trained the motion of the pan-tilt camera by the computer. The range of the pan-tilt camera is tilt (–70, +70) and pan (–20, +20). Figure 4 is shown the change of view as pan(+20, –20) and tilt (+50, –50) change. Figure 5 shows the change of 3 channel multiresolution vision. Figure 6 is the image topology classification determined according to section 3. The topology is distinguished using the gray level.

The results are discussed below. We tried to make the pan-tilt camera learn the 25 action sequences. The movement of pan, tilt and zoom is shown in the

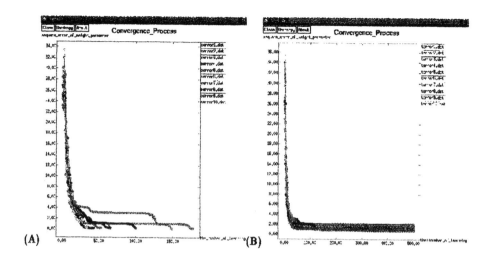

Fig. 8. (A)Convergence process for the learning of the action in figure 7 is shown when the resolution sequence is decided using this method. (B)Convergence process for the learning of the action in figure 7 is shown when the resolution sequence is decided using the random.

following. $y = \{$ (10, 0, 0), (10, 0, 0), (10, 0, 0), (10, 0, 0), (10, 0, 0), (10, 0, 0), (10, 0, 0), (10, 0, 0), (10, 0, 0), (0, 10, 0), (0, 10, 0), (-10, 0, 0), (-10, 0, 0), (-10, 0, 0), (-10, 0, 0), (-10, 0, 0), (0, -10, 0), (0, -10, 0), (0, -10, 0), (0, -10, 0), (10, 0, 0), (10, 0, 0), (10, 0, 0), (10, 0, 0), (0, 10, 0), (0, 10, 0), (0, 10, 0)$ \}$ In this example, the zoom is fixed. The white line of figure 7 is the tracks of the camera movement. The pan-tilt camera moves according to the movements y from the initial position of (0, −50, 0). The relations between behavior and the 3 channel multiresolution vision are recorded. We get the multiresolution sequences using the genetic algorithm based on the results. The multiresolution sequences of 3 channels are $r = \{$ (3), (3), (3), (2), (3), (3), (3), (2), (1), (3), (3), (3), (3)$\}$ $r = \{$ 0, 1, 1, 1, 1, 1, 1, 1, 1, 1, 0, 0, 0, 0, 1, 2, 1, 0, 1, 0, 1, 1, 1, 0, 1, 1$\}$. The weights of the recurrent neural network are trained by repeating the behavior. If we give the initial parameter (0, −50, 0) to the pan-tilt camera, the pan-tilt camera moves according to the acquired behavior when the training is finished.

If the resolution calculated by the genetic algorithm is given, the possibility of the training is 100 %. The training requires a number of iterations to learn relation between behavior and view. 75 % need the 100 iterations and the other 25 % need the 200 iterations. If the resolution calculated by the random numbers is given, the possibility of the learning is 5 %. The 5 % need more than 300 iterations. Figure 8(A) shows the relation between the average square error shown in equation (7) and the number of the iterations in the case that we chose the resolution using this method. Figure 8(B) shows the relation between the average square error shown in equation (7) and the number of the iterations in the case that the resolution is determined using the random numbers. 8 trials need the 100 iterations, the other 2 trials need 200 iterations in 10 trials in this method. The square error did not reach the threshold in the 10 trials in the case that we define the resolution using the random numbers. Even if the initial view given for the pan-tilt camera is based on 1000 iterations, the behavior could not be realized. We conclude that the training of the action sequences is realized without many iterations by using the feedback input of the multiresolution.

6 Conclusions

We proposed a recurrent neural network relates view and multiresolution vision to acquire the behavior by controlling the view using 3 channels of the multiresolusion vision. The action sequences are acquired by the pan-tilt camera.

Acknowledgements

I thanks Dr. Theo van Walsum for improving the English of this paper and Prof. J. J. Koenderink for interesting discussions. The author thanks the member of University Hospital Utrecht, Image Sciences Institute, Dr. Bart ter Haar Romeney and Prof. Max. A. Viergever.

References

1. L. Steels,: The Artificial Life Roots of Artifitial Intelligence, Journal of Artificial Life, Vol. 1, No.1/2, pp. 75-110, MIT Press (1994)
2. R. Brooks,: A Robust Layered Control System for a Mobile Robot, IEEE J. Robotics and Automation, Vol. RA-2, pp.14-23 (1986)
3. J. J. Gibson,: The ecological approach to visual perception, Houghton Mifflin, (1979)
4. H. R. Wilson and S. C. Giese,: Threshold visibility of frequency gradient patterns, Vision Research, 17, pp. 1177-1190 (1977)
5. H. R. Wilson and J. R. Bergen,: A four mechanism model for spatial vision, Vision Research, 19, pp. 19-32 (1979)
6. R. M. Downs and D. Stea,: Maps in mind, New York: Harper & Row, (1977)
7. D. E. Rumelhart, G. E. Hinton and R. J. Williams,: Learning internal representations by error propagation, in eds D.E. Rumelhart D.E. and J.L. McClelland, Parallel Distributed Processing, MIT Press, pp. 318-362 (1986)
8. J. J. Koenderink,: The structure of images, Biological Cybernetics, Vol. 50, pp. 363-370 (1984)
9. A. P. Witkin,: Scale-space filtering, Int. Joint Conf. Art. Intelligence, pp. 1019-1021 (1983)
10. R. Thom,: Structural Stablity and Morphogenesis, W. A. Benjamin, Inc. (1975)
11. D. Marr,: Vision, W.H.Freeeman and Company, (1982)

Automatic Classification System of Marble Slabs in Production Line According to Texture and Color Using Artificial Neural Networks

Juan Martínez-Cabeza-de-Vaca-Alajarín[1] and Luis-Manuel Tomás-Balibrea[2]

Grupo de Visión y Robótica, E.T.S.I. Industriales, Universidad de Murcia
Paseo Alfonso XIII, 48, 30203 Cartagena (Murcia), SPAIN,
Phone: +34.968.325477   Fax: +34.968.325433
E-mail: [1]jcmcv@plc.um.es, [2]lmtb@plc.um.es

Abstract. This article describes the algorithms and the mechatronic system developed for the clustering and classification of marble slabs on production line according to their texture. The method used for the recognition of textures is based on the Sum and Difference Histograms, a faster version of the Co-occurrence Matrices, and the classifier has been implemented by using an LVQ neural network. For each pattern (a marble slab color image), a set of statistical, texture-dependant, parameters is extracted. The input of the classifier (the LVQ network) is the set of parameters calculated before, normalized in the range [0,1], which forms a vector that characterize the pattern shown to the net; the desired output of the network is the class where the pattern belongs to (supervised learning). In our tests, seven different color spaces were used, each one with three different neighbourhoods of pixels. The selected samples chosen for testing the algorithms have been marble slabs of "Crema Marfil Sierra de la Puerta" type. The neural network has been implemented by using MATLAB.

1   Introduction

Marble products have their principal application in covering surfaces for decoration, using small pieces of marble (slabs). The requirements needed for the quality of these slabs do not refer only to technological parameters (such as endurance or polish rate), but also to aesthetic appearances, such as color homogeneity, texture, or spots. However, as a natural material, marble is heavily heterogeneous and therefore, there are not two marble slabs alike.

These requirements make necessary the classification of marble slabs into homogeneous classes or groups. But it is not feasible to intervene in the extraction of the raw material, and thus, the classification process is carried out at the end of the production line, where human experts evaluate the product according to visual parameters. This kind of classification presents two major problems: a) the subjective criterion of the operator (even different operators), and b) the visual fatigue after a period of time. This generates poor results from an aesthetic point of view.

To solve these two problems, we present in this article an automatic system for the inspection and classification of marble slabs in production line according to their texture, by using artificial color vision and neural networks. In it, a vector of texture features is extracted for each pattern (marble slab image), and then classified into the correct class with an LVQ neural network.

2 System Overview

In our system, the marble slabs on the production line pass under a CCD color camera, which obtains the color image for the entire slab, located perpendicular to the slab plane. The camera is placed in a closed box (Fig.1) with a specific illumination to avoid influences of external lighting and thus assure that the slab does not receive any reflects from other materials. In this way, the image obtained does not need any subsequent treatment to overcome the effects of irregular light condition.

The color approach used in our system, instead of monochrome, is necessary since the slabs present different tones in their surfaces. The number of grey levels that the image presents is 256 for each one of the three channels of the color space.

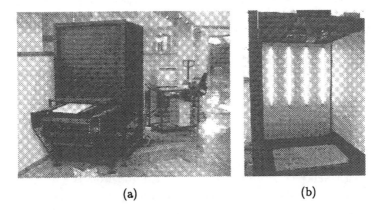

(a)   (b)

Fig. 1. a) Prototype of the mechatronic system (the black box) used for capturing images of marble slabs. b) Detail of the inside of the black box, where the CCD color camera and the illumination system are placed

After the image has been captured with the image acquisition board, it is transmitted to the computer and transformed from RGB-color space to other different color spaces: XYZ, IHS, DIF (using the differences between components, R-B, R-G and G-B) [1], UVW, YIQ and KL (Karhunen-Loève transform) [2].

Fig. 2. Images of marble slabs of 2 different classes

Then a texture recognition algorithm called *Sum and Difference Histograms* (SADH) [3] is used for extracting texture-dependant features from the slab image. This algorithm extracts for each pattern a vector of statistical parameters as texture features. Once the network is trained, the feature vector is utilized as input to the network and the output of the net will give the class to where that pattern belongs to.

2.1 Characteristics of a Marble Slab

The proposed system has been designed to work with any type of marble slabs, but for testing the classification algorithm we have used slabs of the type "Crema Marfil Sierra de la Puerta", which is the type of mayor production in the Region of Murcia. This kind of slabs presents small gradient of colors and hardly-distinguishable veins in the surface (Fig.2).

As a natural product, in a marble slab we can not find any repeated structures. For this reason, we have used statistical methods to extract texture features, instead of structural methods, that are not of application here.

3 Feature Extraction

The approach used in this article to classified marble slabs into groups has been the analysis of their texture. When the human expert examines the slab, the classification he or she makes is based on the visual appearance of color tone and veins distribution of the slab. For this reason, it is necessary to utilize a method capable of analyzing the texture of the image. Previous approaches using only the first-order histogram (color tone) [1] achieved unsatisfactory results.

The SADH algorithm implemented in our system for texture recognition is an alternative to the Co-occurrence matrices (COM) [4–6], based on the Spatial Grey Level Dependence Method (SGLDM). Because of each of the three channels of the color image presents 256 grey levels, the requirements of memory-storage

Neighbourhood Directions			Matrix Image	Co-occurrence Matrices	Sum and Difference Histograms
$(-1,1)$	$(0,1)$	$(1,1)$	$\begin{bmatrix} 0&0&1&1 \\ 0&1&2&3 \\ 1&0&1&2 \\ 1&2&2&3 \end{bmatrix}$	$\begin{bmatrix} 1&3&0&0 \\ 1&1&3&0 \\ 0&0&1&2 \\ 0&0&0&0 \end{bmatrix}$	SH: $[\,0\;2\;1\;3\;1\;4\;1\,]$
$(-1,0)$	*	$(1,0)$			DH: $[\,0\;0\;1\;3\;8\;0\;0\,]$
$(-1,-1)$	$(0,-1)$	$(1,-1)$		matrix of N_G^2 elements	two vectors of $[2N_G - 1]$ elements

(a)   (b)

Fig. 3. a) Neighbourhood directions of a cell. b) Example of Co-occurrence Matrices and Sum and Difference Histograms for a 4 × 4 image with 4 grey levels. The results have been calculated for the horizontal nearest neighbour (1,0)

and time-consumption with COM were huge (processing of three 256 × 256-matrices). The SADH method offers in this way a good alternative to the usual COM used for texture analysis, with experimental results showing that SADH are as powerful as COM for texture discrimination, decreasing computation time and memory storage.

Figure 3 shows a four-grey-level image in matrix form, where the grey levels have been represented with numbers from 0 to 3. Here we can see both the COM and the SADH for the image proposed. For a large number of grey levels (usually 256), the two vectors of the SADH require less storage space than the matrix of the COM, and so, the processing is much faster because (1) the number of data to process is smaller, and (2) the double sums for the COM statistical parameters calculated are transformed into single sums for the SADH ones.

The algorithm is as follows: let be a discrete texture image defined on a $K \times L$ rectangular grid, and denoted by $\{y_{k,l}\}$, where $\{k = 1, \ldots K; \ l = 1, \ldots L\}$. Let $G = \{1, 2, \ldots, N_G\}$ be the set of the N_G quantized grey levels for each one of the three channels of the color image. Let us consider two picture elements separated by the vector distance $(d_1, d_2) \in D$ (where D specifies the texture region to analyze):

$$\begin{cases} y_1 = y_{k,l} \\ y_2 = y_{k+d_1, l+d_2} \end{cases} \tag{1}$$

Both the sum and difference vectors contain $s_{k,l}$ and $d_{k,l}$ elements respectively, defined as follows:

$$s_{k,l} = y_{k,l} + y_{k+d_1, l+d_2} \tag{2}$$

$$d_{k,l} = y_{k,l} - y_{k+d_1, l+d_2} \ . \tag{3}$$

Table 1. Statistical features used for classification with SADH

FEATURE	FORMULA
mean	$\mu = \frac{1}{2}\sum_i i\, P_s(i)$
energy	$\sum_i P_s(i)^2 \sum_j P_d(j)^2$
entropy	$-\sum_i P_s(i)\log\{P_s(i)\} - \sum_j P_d(j)\log\{P_d(j)\}$
contrast	$\sum_j j^2 P_d(j)$
homogeneity	$\sum_j \frac{1}{1+j^2} P_d(j)$

The SADH uses two normalized $[2N_G - 1]$-dimensional vectors, formed each one with $P_s(i)$ and $P_d(j)$ elements, defined as follows:

$$P_s(i) = h_s(i)/N; \quad (i = 2, \ldots, 2N_G) \tag{4}$$

$$P_d(j) = h_d(j)/N; \quad (j = -N_G + 1, \ldots, N_G - 1) \tag{5}$$

with

$$h_s(i; d_1, d_2) = h_s(i) = \mathrm{Card}\{(k,l) \in D, \; s_{k,l} = i\} \tag{6}$$

$$h_d(j; d_1, d_2) = h_d(j) = \mathrm{Card}\{(k,l) \in D, \; d_{k,l} = j\} \tag{7}$$

$$N = \mathrm{Card}\{D\} = \sum_i h_s(i) = \sum_j h_d(j) \; . \tag{8}$$

Thus, for each color image there are six 511-dimensional vectors: P_{s1}, P_{d1}, P_{s2}, P_{d2}, P_{s3} and P_{d3}. In our case, $N_G = 256$.

Once the six histogram vectors have been calculated, we extract statistical information from them, in order to reduce the dimension of the set of characteristics used for describing the texture. Table 1 gives the list of features [3] used in our case.

4 Classifier

The classifier used in our application has been a supervised neural network, in particular the LVQ1 algorithm [7] that learns to classify input vectors into target classes chosen by the user. The LVQ network consists of two weight layers. The first layer is a competitive layer (CL), which learns to classify input vectors forming subclasses. The second layer (linear) transforms the competitive layer's subclasses into target classifications defined by the user (target classes).

The Kohonen learning rule assures that competitive neurons move towards input vectors which belong to their class, and away from input vectors that belong to other classes. Thus, the advantages that this network presents in comparison to the unsupervised algorithm are two: (1) the target classes are chosen by the user, and (2) the convergence of the algorithm is faster.

In our case, the input to the network is a 15-elements vector, normalized in the range [0,1]: for each color channel we have used the mean, energy, entropy, contrast and homogeneity parameters. The output layer of the network is a binary-type one ('1' for the winning neuron and '0' for the rest), with 10 neurons (one neuron per target class, 10 in our case). The number of neurons in the CL should be greater than the number of neurons in the output layer for that several subclasses (several $S1$-neurons) could be combined into only one target class (an $S2$-neuron).

5   Experimental Results

The classification software was executed on a 166 MHz PC computer, and the image acquisition board used was the Matrox Genesis. The LVQ network was implemented with the Neural Network Toolbox of MATLAB [8], and the image processing and feature extraction software were developed in C++ under MSDOS. For the LVQ network, the design parameters were the following: 50 neurons in the CL, 5000 epochs for training, and 0.05 for learning rate.

Seven color spaces were tested for classification: RGB, XYZ, IHS, DIF, UVW, YIQ and KL. For each one, three training series were made, according to the values of (d_1, d_2) (nearest neighbours directions): (1) M_1, using the mean of the statistical parameters for the directions $(1,0)$ and $(0,1)$, (2) M_2, the same as the former for directions $(1,1)$ and $(1,-1)$, and (3) M_3, corresponding to the mean of the statistical parameters for the four directions $(1,0)$, $(0,1)$, $(1,1)$ and $(1,-1)$. The four remaining neighbourhood-directions, $(-1,0)$, $(0,-1)$, $(-1,-1)$ and $(-1,1)$, are symmetrical to the four former, and thus, produce COM equals to that produced with M_3, so there is no need to consider them.

For each series, 4 training sets were made, each one containing different patterns: for each class, all but one patterns were used for training and one for testing. The patterns used for testing were rotating for the different training sets, so in this way, for the 4 trainings, all the patterns were used as both training patterns and test patterns. Therefore, 84 trainings (7 color spaces × 3 neighbourhood directions × 4 training sets) were made for the totality of the color spaces. A total of 40 patterns (4 patterns per class) were used for training and testing the system (3 patterns for training and 1 pattern for testing).

Table 2 shows the percentage of patterns classified correctly. As it can be seen, the best results have been achieved with the RGB and the UVW color spaces, respectively, for M_1 and M_2, with 72.5% of patterns classified correctly. However, the average results achieved with the RGB color space are better than those obtained with the UVW color space. Thus, for this reason, and for the simplicity of computation (the image acquisition board provides an RGB image), the RGB color space is preferred. Other good results correspond to the YIQ and XYZ color spaces. However, the IHS, DIF and KL color spaces achieved very poor classification rates (less than 53%), maybe due to the approximation used in the case of the KL transform. Table 3 shows the confusion matrix for the results obtained with the RGB color space.

Table 2. Results obtained for the classification with the LVQ network, in %. M_1, M_2 and M_3 are the average of the calculated parameters for directions $[(1,0),(0,1)]$, $[(1,1),(1,-1)]$, and $[(1,0),(0,1),(1,1),(1,-1)]$ respectively

Color	Direction			Mean
Space	M_1	M_2	M_3	
RGB	72.5	67.5	67.5	69.2
XYZ	62.5	65.0	55.0	60.8
IHS	50.0	52.5	50.0	50.8
DIF	45.0	35.0	47.5	42.5
UVW	60.0	72.5	67.5	66.7
YIQ	65.0	67.5	62.5	65.0
KL	45.0	45.0	35.0	41.2
Mean	57.1	57.9	55.0	

6 Conclusion

By using the automatic inspection system proposed, we may reduce costs and increase the quality control in the classification of marble slabs in production line, with a significantly higher performance when it is compared to the traditional classification system. The algorithms used for feature extraction and classification perform high-speed processing and low memory-storage requirements. The result is that marble slabs are classified according to their color texture features, thus achieving an objective, uniform-through-time classification of marble slabs, which derives in an homogeneous classification criterion.

Table 3. Confusion matrix for results obtained with RGB color space (Row: Experts labelling, Column: Neural network results)

	Class 1	Class 2	Class 3	Class 4	Class 5	Class 6	Class 7	Class 8	Class 9	Class 10
Class 1	9	0	1	1	0	0	0	0	1	0
Class 2	0	12	0	0	0	0	0	0	0	0
Class 3	4	0	5	0	0	0	0	0	3	0
Class 4	0	0	0	7	0	0	0	2	3	0
Class 5	0	0	0	0	8	3	1	0	0	0
Class 6	0	0	0	0	0	9	3	0	0	0
Class 7	0	0	0	0	1	5	6	0	0	0
Class 8	0	0	0	0	0	0	0	12	0	0
Class 9	0	0	3	6	0	0	0	0	3	0
Class 10	0	0	0	0	0	0	0	0	0	12

Although the number of patterns used in this case for training the neural network is very small (3 patterns per class plus 1 pattern for testing) the results obtained are promising. We hope to improve these results when we have more patterns available. At the same time, the introduction of fuzzy technics could also improve the classification results. However, in spite of the errors committed in the classification, the classes assigned in those cases were very similar to the correct ones, so the classification can still be considered homogeneous, even better than the one obtained with the human expert.

Finally, the processing time to make a decision is sufficient in our case. The software implemented is faster enough to perform the classification on-line with an image size of 204 × 294 pixels. However, once the prototype system is completely finished and the whole system works correctly, the final goal in this project is to implement the texture-feature extraction system and the neural network classifier on Very Large Scale Integration (VLSI) circuits.

Acknowledgment

This work has been supported in part by the Spanish R&D Plan (Grant PTR95-0210-OP) and with the collaboration of the Technological Center of Marble of the Region of Murcia (Spain). The authors would like to thank them for their technical support about practical aspects of marble classification.

References

1. Garcerán-Hernández, V., García-Pérez, L. G., Clemente-Pérez, P., Tomás-Balibrea, L. M., Puyosa-Piña, H. D.: Traditional and neural networks algorithms: applications to the inspection of marble slabs. IEEE International Conference on Systems, Man and Cybernetics; 0-7803-2559-1, Vancouver (Canada) (1995) 3960–3965
2. Van de Wouwer, G.: Wavelets for multiscale texture analysis. University of Antwerpen (1998)
3. Unser, M.: Sum and difference histograms for texture classification. IEEE Trans. on PAMI Vol.PAMI-8 No.1 (1986) 118–125
4. Haralick. R. M., Shanmugam, K., Dinstein, I.: Textural features for image classification. IEEE Trans. on SMC Vol.SMC-3 No.6 (1973) 610–621
5. Haralick, R. M.: Statistical and structural approaches to texture. Proc. IEEE Vol.67 No.5 (1979) 786–804
6. Gotlieb, C. C., Kreyszig, H. E.: Texture descriptors based on co-occurrence matrices. Computer Vision, Graphics and Image Processing 51 (1990) 70–86
7. Kohonen, T.: The self-organizing map. Proc. IEEE Vol.78 No.9 (1990) 1464–1480
8. Demuth, H., Beale, M.: Neural network toolbox user's guide. The Mathworks Inc. (1994)

Polygonal Approximation Using Genetic Algorithms[1]

Peng-Yeng Yin

Department of Information Management, Ming Chuan University, 250 Sec. 5,
Chung Shan N. Rd., Taipei 111, Taiwan
pyyin@mcu.edu.tw

Abstract. In this paper, three polygonal approximation approaches using genetic algorithms are proposed. The first approach approximates the digital curve by minimizing the number of sides of the polygon and the approximation error should be less than a prespecified tolerance value. The second approach minimizes the approximation error by searching for a polygon with a given number of sides. The third approach, which is more practical, determines the approximating polygon automatically without any given condition. Moreover, a learning strategy for each of the proposed genetic algorithm is presented to improve the results. The experimental results show that the proposed approaches have better performances than do the existing methods.

1. Introduction

Shape representation is an important process in image processing and pattern recognition since two-dimensional shapes in an image usually appear as the projection of three-dimensional objects in the real world. Approximations of such shapes not only provide compact description with less memory requirement, but also facilitate feature extraction for further image analysis. Polygonal approximation is one of the approaches which can provide good approximations of two-dimensional shapes with any desired accuracy. Many polygonal approximation methods can be found in the literature. They can be classified into four categories as sequential approaches [5, 10], split-and-merge approaches [4, 6], dominant points detection approaches [2, 9], and K-means based approaches [3, 11]. These methods are designed to solve three types of problems: (1) For a given tolerance of error, the objective is to minimize the number of approximating line segments. (2) For a given number of approximating line segments, the objective is to minimize the error norm. (3) Approximate the digital curve according to its own characteristics without any prespecified restriction.

The solutions of all of the three types of problems involve a searching process in a complex parameter space. The scheme of this searching process leads to a local or global optimal solution. Sato [7] used dynamic programming technique to solve Type 2 problem but consuming too much time. Genetic algorithms (GAs) are based on stochastic search theorems which simulate the biological model of evolution [1]. GAs perform parallel search in complex parameter space and have been found many advantages over traditional searching methods. In this paper, we employ GAs to obtain near-optimal solutions for all of the three types of problems. We also present a

---

[1] This research was supported by the National Science Council of R.O.C., under contract NSC-88- 2213-E-130-002

learning strategy to improve the results of GA search. The performance of the proposed approach is compared to those of other existing methods.

2. Genetic Algorithms to Polygonal Approximation

GAs are search algorithms based on natural genetic systems [1]. Each solution of the application is usually coded as a binary string of 0's and 1's, called chromosome. A collection of such strings forms a population. GAs start with a randomly generated population of a fixed size. This initial population evolves to the next population with the same size using three genetic operators: selection, crossover, and mutation. The evolution process is iterated until a near-optimal solution is obtained or a prespecified number of generations is reached. Selection operation mimics the natural survival of the fittest creatures. The probability of each string to be selected is proportional to its fitness. Crossover is a process where each individual can interchange information with its mate chosen randomly. Since the highly fit strings take a large proportion of the population, they will receive increasing trials of crossover in subsequent generations and extend the exploration in "good" candidate solutions. Mutation is an occasional alteration of a bit. It provides sufficient diversity of the population to prevent the unwilling premature convergence and it also guarantees a non-zero probability of the search to any feasible string.

Assume the input digital curve has n points as $S = \{ x_0, x_1, ..., x_{n-1} \}$. Polygonal approximation can be viewed as a searching process of the vertices from S. The aim is to meet the different goals such as minimizing the number of approximating segments or minimizing the error between the polygon and S. Since GAs have a high probability of tendency to a global optimal solution, we employ GAs for solving all of the three kinds of polygonal approximation problems.

2.1 Genetic Algorithm for Solving Type 1 Problem (GA-Type-1)

The aim is to find the minimal subset of S as the vertices of the polygon such that the error norm between S and the polygon is less than a prespecified tolerance value ε. The error norm can be measured in different ways, such as integral square error, maximal perpendicular distance error, and the area deviation error. In all of our methods presented in the following, we use integral square error (which will be referred to as ε_2) as the error measure since it is appropriate for experimental data.

String Representation and Fitness Function. Each solution of Type 1 problem can be encoded into a binary string as $\alpha = a_0 a_1 \cdots a_{n-1}$, where $a_i = 1$ if x_i is a vertex of the polygon and $a_i = 0$ otherwise. The number of the vertices of the polygon is equivalent to the sum of all bit values. Thus, Type 1 problem can be formulated as minimizing the cost function

$$C(\alpha) = \sum_{i=0}^{n-1} a_i + P(\alpha),\qquad(1)$$

where $P(\alpha)$ denotes the penalty function and is equal to 0 if $\varepsilon_2 < \varepsilon$, and ε_2 otherwise. The fitness of each string grows inversely with its cost and is given by

$$f(\alpha) = k - C(\alpha),\qquad(2)$$

where k is a constant. To prevent an outstanding string with a very high fitness value taking over a significant proportion of the next population, which is a leading cause of the premature convergence, the fitness values are properly scaled [8].

The standard genetic operators are used in GA-Type-1. However, the crossover and the mutation probabilities are adaptive to preserve the diversity of the population.

Learning Strategy. Since GAs improve the fitness of each string by exchange the information with its mate chosen blindly, the string may not have the chance to meet its "best" mate. It is infeasible to consider any mate of any string due to expensive CPU time. Nevertheless, the current best string of each generation might improve its fitness by copying substrings from a randomly chosen individual. In this work, the implementations of the learning strategy for various problems are different depending on the string representations. For Type 1 problem, it is conducted as follows.

Let $\alpha = a_0 a_1 \cdots a_{n-1}$ be the current best string, and $\beta = b_0 b_1 \cdots b_{n-1}$ be a randomly chosen string. We define the set of delimiters as $D(\alpha, \beta) = \{ i \mid$ if $a_i = b_i = 1$, $0 \le i \le n\text{-}1 \}$. Let $D(\alpha, \beta) = \{d_0, d_1, ..., d_m\}$, where $d_0 < d_1 < ... < d_m$. We can improve the fitness of α by checking the fitness values of the new strings of copying the substring of β between d_i and d_{i+1}, $0 \le i \le m - 1$, to α, respectively, and retain the best one, i.e., α can improve its fitness by learning from β. The proposed learning strategy will increase the probability of reaching to the global optimal solution and does not affect the behavior of the original GAs.

2.2 Genetic Algorithm for Solving Type 2 Problem (GA-Type-2)

The aim is to minimize the value of ε_2 between S and the approximating polygon with the restriction that the number of segments is equal to a prespecified value c (which will be referred to as the c-segment constraint).

String Representation and Fitness Function. Each solution of Type 2 problem can be encoded into a string as $\beta = b_0 b_1 \cdots b_{c-1}$, where $b_i \ne b_j$ if $i \ne j$, and each b_i has $\log_2 n$ bits with a value lying in the range of [0, n-1]. String β means $\{ x_{b_0},$

$x_{b_1}, \ldots, x_{b_{c-1}}$ } are the c vertices of the polygon. Then Type 2 problem can be realized by minimizing the cost function

$$C(\beta) = \sum_{i=0}^{c-1} \sum_{j=b_i}^{b_{i+1}} d_{ij}^{2}(x_j, \overline{x_{b_i} x_{b_{i+1}}}), \qquad (3)$$

where $b_c \equiv b_0$ and $d_{ij}(x_j, \overline{x_{b_i} x_{b_{i+1}}})$ means the perpendicular distance between point x_j and segment $\overline{x_{b_i} x_{b_{i+1}}}$. And the fitness function is defined by

$$f(\beta) = k - C(\beta). \qquad (4)$$

A special crossover operator which is similar to the partially matched crossover (PMX) [1] is proposed here to prevent violating the c-segment constraint. Let strings α and β be selected and $\alpha = a_0 a_1 \cdots a_{c-1}$ and $\beta = b_0 b_1 \cdots b_{c-1}$. First, generate a random number s between $[0, c-2]$, two strings α' and β' are obtained by swapping the substrings of α and β after position s, i.e., $\alpha' = a_0 a_1 \cdots a_s b_{s+1} \cdots b_{c-1}$, and $\beta' = b_0 b_1 \cdots b_s a_{s+1} \cdots a_{c-1}$.

However, α' and β' may have repeating characters and violate the c-segment constraint. Let the character a_i from the substring $a_0 a_1 \cdots a_s$ be identical to another character b_j from the substring $b_{s+1} b_{s+2} \cdots b_{c-1}$. This problem can be solved by replacing a_i with a_j which is swapped by b_j in the previous step. This operation should be done iteratively for both α' and β' until there is no repeating character in both strings.

Learning Strategy. Let $\alpha = a_0 a_1 \cdots a_{c-1}$ be the current best string, and $\beta = b_0 b_1 \cdots b_{c-1}$ be another randomly chosen string, the character positions can be used as the natural delimiters, i.e., $D(\alpha, \beta) = \{ i \mid 0 \le i \le c-1 \}$. The fitness of α can be improved by checking the results of the new strings of copying b_i of β, $i \in D(\alpha, \beta)$, to α, respectively, and retain the best one.

2.3　Genetic Algorithm for Solving Type 3 Problem (GA-Type-3)

The aim is to approximate S by a polygon with respect to any objective without prespecified restrictions. After inspecting Type 1 and Type 2 problems, we know that the general goal of approximation is to maximize the compression ratio and minimize the approximation error as much as possible. To join the two factors, the fitness function can be defined as

$$f(\alpha) = R^w \Big/ \varepsilon_2 ,$$ (5)

where R is the compression ratio measured as $R = n \Big/ c$, and w is the weighting parameter. The string representation is as same as that of GA-Type-1 method. In our experiments, the approximation error changes more rapidly than does the compression ratio during the evolution; hence, we set $w = 4$ in our proposed method. For simplicity of computation, the fitness function can be replaced by

$$f'(\alpha) = k - \varepsilon_2 * c^4, \qquad \text{if } 3 \le c \le n-1;$$ (6)
$$= k - \varepsilon_2 * c^4 - (\varepsilon_2 + c^4), \quad \text{otherwise,}$$

since n is a constant and can be ignored, and $\varepsilon_2 + c^4$ is the penalty to decrease the fitness of invalid polygonal approximation.

The genetic operators and the learning strategy used for GA-Type-3 method are as same as those for GA-Type-1 method since they have the same string representation.

3. Experimental Results and Performance Analysis

The proposed methods (GA-Type-1, GA-Type-2, and GA-Type-3) and three other existing methods (Wall-Danielsson's method [10], Phillips-Rosenfeld's method [3], and Ray-Ray's method [5]) are implemented for comparison. Three curves as shown in Fig. 1 are tested and they are broadly used in the literature [2, 5, 6, 9, 11].

For Type 1 problem with the same given error bound, GA-Type-1 method always produces fewer number of vertices than that of Wall-Danielsson's method (see Table 1). For Type 2 problem with the same given number of approximating line segments, GA-Type-2 method always produces smaller errors than those of Phillips-Rosenfeld's method (see Table 2). For Type 3 problem which is more practical since no input parameter is required, GA-Type-3 method tends to produce higher compression ratios than those of Ray-Ray's method (see Table 3). Moreover, for the digital curve of Fig. 1(a), GA-Type-3 method not only has a higher compression ratio but also yields smaller error. Hence, GA-Type-3 method considers more correctly based on the tradeoff between the compression ratio and the approximation error. Fig. 2 shows the approximation results determined by the proposed GA-Type-3 method automatically.

4. Conclusion
In this paper, we have proposed three GA-based approaches for solving different types of polygonal approximation problems. We conclude the features of the proposed methods as follows. (1) GA-Type-1 and GA-Type-2 methods outperform the existing methods in compression ratio and approximation error. (2) GA-Type-3 method adopts superior factors for Type 3 problem to determine the "optimal"

approximating polygon automatically. (3) The existing methods are designed to solve one of the problems, while the proposed methods are provided for all kinds of problems. (4) The proposed methods can be speeded up by the parallel implementation.

Table 1 The results of Wall-Danielsson's method and the 3-run average results of the proposed GA-Type-1 method for Type 1 problem.

Algorithms	Fig. 1(a)		Fig. 1(b)		Fig. 1(c)	
	ε_2	c	ε_2	c	ε_2	c
Wall-	30	9	150	19	60	15
Danielsson's	10	12	90	25	25	18
method	6	36	20	44	15	25
GA-Type-1	30	7	150	16	60	13
	10	10	90	17	25	17
method	6	15	20	23	15	23

Table 2 The 3-run average results of Phillips-Rosenfeld's method and the proposed GA-Type-2 method for Type 2 problem.

Algorithms	Fig. 1(a)		Fig. 1(b)		Fig. 1(c)	
	c	ε_2	c	ε_2	c	ε_2
Phillips-	6	55.01	17	71.0	15	67.53
Rosenfeld's	8	34.97	19	63.54	17	62.83
method	10	23.37	21	49.26	19	33.29
GA-Type-2	6	46.05	17	44.37	15	25.99
	8	18.02	19	29.38	17	14.75
method	10	12.9	21	20.24	19	12.76

Table 3 The results of Ray-Ray's method and the 3-run average results of the proposed GA-Type-3 method for Type 3 problem.

Algorithms	Fig. 1(a)		Fig. 1(b)		Fig. 1(c)	
	ε_2	c	ε_2	c	ε_2	c
Ray-Ray	7.67	14	16.43	26	16.33	19
GA-Type-3	6.74	12	18.56	19	15.54	19

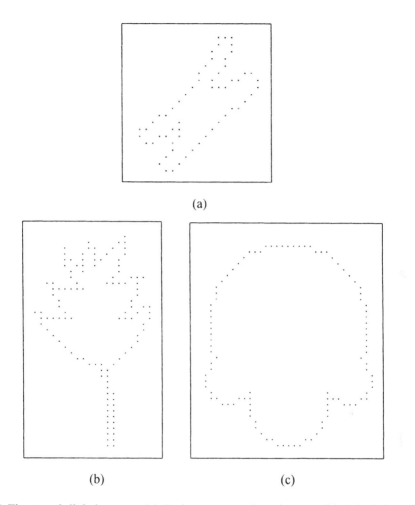

(a)

(b)   (c)

Fig. 1 The tested digital curves. (a) A chromosome-shaped curve. (b) A leaf-shaped curve. (c) A curve with four semicircles.

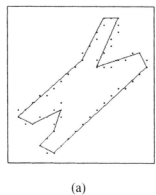

(a)

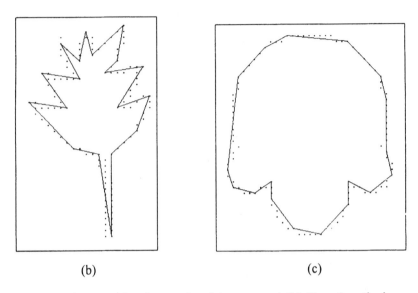

(b)   (c)

Fig. 2 The approximation results of the proposed GA-Type-3 method.

References

1. Goldberg, D.E.: Genetic Algorithms: Search, Optimization and Machine Learning. Addison-Wesley, Reading, MA, 1989.
2. Held, A., Abe, K., Arcelli, C.: Towards a hierarchical contour description via dominant point detection. IEEE Trans. Syst., Man, Cybern. **24** (1994) 942-949.
3. Phillips, T.Y,. Rosenfeld, A.: An ISODATA algorithm for straight line fitting. Pattern Recognition Lett. **7** (1988) 291-297.
4. Ramer, U.: An iterative procedure for the polygonal approximation of plane curves. Computer Graphics and Image Processing **1** (1972) 244-256.
5. Ray, B.K., Ray, K.S.: Determination of optimal polygon from digital curve using L_1 norm. Pattern Recognition **26** (1993) 505-509.
6. Ray, B.K., Ray, K.S.: A new split-and-merge technique for polygonal approximation of chain coded curves. Pattern Recognition Lett. **16** (1995) 161-169.
7. Sato, Y.: Piecewise linear approximation of plane curves by perimeter optimization. Pattern Recognition **25** (1992) 1535-1543.
8. Singh, M., Chatterjee, A., Chaudhury, S.: Matching structural shape descriptions using genetic algorithms. Pattern Recognition **30** (1997) 1451-1462.
9. Teh , C.H., Chin, R.T.: On the detection of dominant points on digital curves. IEEE Trans. Pattern Anal. Machine Intell. **11** (1989) 859-872.
10. Wall, K., Danielsson, P.E.: A fast sequential method for polygonal approximation of digitized curves. Computer Vision, Graphics, and Image Processing **28** (1984) 220-227.
11. Yin, P.Y.: "Algorithms for straight line fitting using K-means," Pattern Recognition Lett. **19** (1998) 31-41.

Computation of Symmetry Measures for Polygonal Shapes *

Stanislav Sheynin, Alexander Tuzikov, and Denis Volgin

Institute of Engineering Cybernetics
Academy of Sciences of Republic Belarus
Surganova 6, 220012 Minsk, Belarus
{sheynin,tuzikov,volgin}@mpen.bas-net.by

Abstract. In this paper we propose efficient algorithms for computation of symmetry degree (measure of symmetry) for polygonal shapes. The algorithms are based on turning function representation and the approach developed in [1] for comparing polygonal shapes. These algorithms allow computation of reflection and rotation symmetry measures which are invariant to translations, rotations and scaling of shapes. Using normalization technique we show how it is possible to apply them for computation of skew symmetry measures as well.

Keywords: symmetry measures, polygons.

1 Introduction

Symmetry of objects is one of their fundamental features [11]. Therefore it is not surprising that the problem of object symmetry identification is of great interest in image analysis and recognition, computer vision and computational geometry.

Recently a number of publications devoted to computation of symmetry degree (symmetry measure) of shapes and images was published [2, 7, 8, 10, 9, 12]. Naturally, we mentioned only several publications dealing with different kinds of symmetries, namely, central, rotation, reflection or skew ones. In this paper like to [10] we employ contour representation of shapes for analyzing their symmetry measures. Thesis of Otterloo [10] presents a thorough treatment of the problem of shape similarity and symmetry based on parametric contour representations. An interested reader can find there also a detailed discussion of the related literature.

The aim of this paper is to propose efficient algorithms for computation of reflection, rotation and skew symmetry measures for polygonal shapes. All these algorithms are based on the nice algorithm proposed in [1] for comparing polygonal shapes.

Definition 1. *A transformation $g : \mathbb{R}^2 \longrightarrow \mathbb{R}^2$ of the Euclidean plane into itself is called a* symmetry *of polygon $P \subset \mathbb{R}^2$ if $g(P) = P$ up to translation. In this case polygon P is called g-symmetric.*

---

* The authors were supported by the INTAS grant N 96-785

Denote by r_β the rotation around the origin over an angle β, and by ℓ_α the reflection with respect to the line L_α through the origin which makes an angle α with the positive x-axis.

Let g be a rotation around the origin over the angle $2\pi/m$ for some integer $m > 1$ and polygon P is g-symmetric. In this case we say that polygon P is *rotation symmetric of order m*.

Let g be a reflection ℓ_α and polygon P is g-symmetric. Then we say that polygon P is *reflection symmetric with respect to the line L_α*. Polygon P is called *reflection symmetric* if it is reflection symmetric with respect to some line L_α. This line is called the *axis of reflection symmetry*.

Polygon P is called *skew symmetric (skew rotation symmetric of order m)* if there exists an affine transformation g such that polygon $g(P)$ is reflection symmetric (respectively, rotation symmetric of order m).

In practical applications one is often interested in the quantitative evaluation of shape symmetry. Towards this goal Grünbaum [3] introduced the notion of *central symmetry measure*. We adopt this definition in the paper. Denote by $\mathcal{P}$ the set of polygons in $\mathbb{R}^2$.

Definition 2. *A real-valued function $\mu : \mathcal{P} \longrightarrow \mathbb{R}$ is called a measure of g-symmetry for polygons in $\mathbb{R}^2$ if the following conditions hold:*

1. $0 \le \mu(P) \le 1$ for every polygon $P \in \mathcal{P}$.
2. $\mu(P) = 1$ if and only if P is a g-symmetric polygon.
3. $\mu(P)$ is a continuous function of P (in a given topology).

Given a group of transformation H the measure μ is called H-invariant if $\mu(P) = \mu(h(P))$ for every $P \in \mathcal{P}$ and $h \in H$. It is supposed that $h(P) \in \mathcal{P}$ for $h \in H$.

There are known in the literature different approaches to introduce symmetry measures. In this paper we do it via dissimilarity metric specially defined on the set of shapes.

2 Comparing polygonal shapes

In this section we describe an idea of the algorithm developed in [1] for comparing polygonal shapes.

Plane regions bounded by one closed curve (in particular, polygons) may be represented up to translation and scaling using a turning function.

Let A be a shape bounded by a curve of unit length (if the length is not equal 1 then the shape is scaled). Let o be an initial point on the curve. Then every point $p(s)$ on the curve can be be parameterized by the curve length s, $0 \le s < 1$, from the initial point o to $p(s)$ while passing the curve in a counterclockwise direction. Denote by $\Theta_A(s)$ the angle between the x-axis and the normal to the curve at the point $p(s)$. The function $\Theta_A(s)$ determines A up to translation and is called a *turning function*. Evidently, $\Theta_A(1) = \Theta_A(0) + 2\pi$. Now consider function $\Theta_A(s)$ to be defined for all $s \in \mathbb{R}$ (this corresponds to passing the curve several times; negative s correspond to passing the curve in the reverse direction). In this case the equality $\Theta_A(s+1) = \Theta_A(s) + 2\pi$ is true. For convex regions $\Theta_A(s)$ is monotonic but for non-convex regions $\Theta_A(s)$ can take arbitrarily large positive and negative values for $s \in [0,1]$.

Further we assume that A is a polygon. In this case the turning function $\Theta_A(s)$ is piecewise constant. The example of a polygon and its turning function is presented at the Fig. 1 a), b).

The turning function is highly sensitive to noise and this is probably the main drawback of the representation (see Fig. 1 c), d)). Although the shapes shown at Fig. 1 a), c) look similar the difference between their turning functions is quite significant. Therefore preliminary smoothing of boundary might be useful when using this representation for comparing shapes.

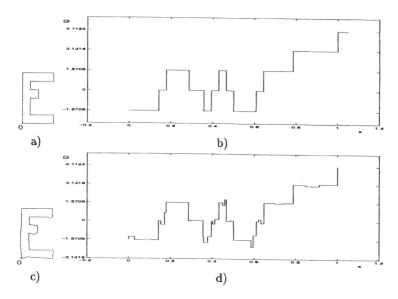

Fig. 1. Turning function representation. a) Original polygon P; b) Turning function of polygon P; c) Noisy version of polygon P; d) Turning function of the noisy polygon.

We can introduce a metric on the set of functions (defined on the segment $[0,1]$) by different ways. In [1] they use L_2-metric. L_2-distance between functions $\Theta_A(s)$ and $\Theta_B(s)$ equals by definition $(\int_0^1 (\Theta_A(s) - \Theta_B(s))^2 ds)^{\frac{1}{2}}$.

The turning function of the polygon depends on the choice of the initial point o belonging to the shape boundary. But dissimilarity metric should depend only on the shapes of polygons. Therefore scaling, translation and rotation invariant dissimilarity metric for shapes can be introduced as [1]:

$$d_2(A, B) = \left(\min_{\theta \in \mathbb{R}, t \in [0,1]} \int_0^1 (\Theta_A(s+t) - \Theta_B(s) + \theta)^2 ds \right)^{\frac{1}{2}}$$

$$= \left(\min_{\theta \in \mathbb{R}, t \in [0,1]} h(t, \theta) \right)^{\frac{1}{2}}, \tag{1}$$

where $h(t, \theta) = \int_0^1 (\Theta_A(s + t) - \Theta_B(s) + \theta)^2 ds$.

Note that if we are interested in shape similarity up to scaling and translation only (i.e., no invariance to rotation is assumed) then the minimization is performed for $\theta = 2k\pi$, $k \in \mathbb{Z}$. We denote further the scaling and translation invariant metric by $d_2'(A, B)$.

Denote $f(s) = \Theta_A(s)$, $g(s) = \Theta_B(s)$. The minimum on θ of function $h(t, \theta)$ is achieved at $\theta = \theta^*(t) = \int_0^1 (g(s) - f(s + t)) ds = \alpha - 2\pi t$, where $\alpha = \int_0^1 g(s) ds - \int_0^1 f(s) ds$ is a constant.

The function $h(t, \theta)$ for every fixed θ is piesewise linear on t and the minimum of it on t is achieved in one of the fracture points of this piecewise linear function, i.e. in one of t, where a breakpoint of $f(s + t)$ coincides with some breakpoint of $g(s)$. The number of such points is finite, all of them are determined by the turning functions of polygons A and B and do not depend on θ.

Sometimes invariance to affine transformations is required while comparing shapes. It seems that the minimization of the L_2-distance for the whole affine transformation group cannot be done efficiently. Therefore a preliminary normalization to affine transformations may be reasonable.

Given two shapes A and B it is possible to transform them to the canonical shapes $A^\bullet$ and $B^\bullet$ such that shapes A and B are equivalent with respect to G_+ (affine transformations group preserving an orientation) if and only if $A^\bullet$ and $B^\bullet$ are equivalent with respect to rotations [4, 5]. The definition of a canonical shape is based on the concept of ellipse of inertia [6, p.48-53] and any compact set could be easily transformed into the canonical form by some transformation from G_+.

3 Symmetry measures

Having a dissimilarity metric $d_2(A, B)$ one can introduce a shape similarity measure

$$\sigma(A, B) = \exp(-d_2(A, B)).$$

The similarity measure takes values from $[0, 1]$ and the value 1 is achieved iff the the shapes are similar with respect to the considered group of transformations.

Now we can introduce and compute efficiently reflection, rotation and skew symmetry measures of polygonal shapes.

Denote as earlier by r_β the rotation around the origin over an angle β, and by ℓ_α the reflection with respect to the line through the origin which makes an angle α with the positive x-axis. It is known that $\ell_\alpha r_\beta = \ell_{\alpha - \beta/2}$, $r_\beta \ell_\alpha = \ell_{\alpha + \beta/2}$.

Reflection symmetry. Denote by $\sigma'(A, B) = \exp(-d_2'(A, B))$ the similarity measure which is invariant to translations and scalings. Then reflection symmetry measure $\mu_1(P)$ of the polygon P can be introduced as follows:

$$\mu_1(P) = \max_{0 \le \alpha < \pi} \sigma'(P, l_\alpha(P)).$$

Denote by $l_0(P)$ the reflection of P with respect to x-axis. Then

$$\mu_1(P) = \max_{0 \le \beta < 2\pi} \sigma'(P, r_\beta l_0(P)).$$

If $f(s)$ is the turning function of P then $-f(-s)$ will be the turning function of $l_0(P)$. Therefore applying algorithm for dissimilarity metric computation for polygons P and $l_0(P)$ we can compute efficiently the reflection symmetry measure $\mu_1(P)$ and the angle $\beta^\star$ for which the minimum in the dissimilarity metric $d_2(P, l_0(P))$ is achieved. Then the angle $\alpha^\star$ of the μ_1-nearest reflection symmetry axis equals $\beta^\star/2$. One is usually interested to see the exact position of the axis in the shape, i.e. to find a point through which the axis passes. Suppose that the minimal value of the dissimilarity metric $d_2(P, l_0(P))$ is achieved for the critical point t. Then we can find two points s_1 and s_2 belonging to the contour of P with distances $t/2$ and $(t+1)/2$ from the initial point. These two points are equivalent in the sense that t and $t+1$ define the same shift of the turning function. If the polygon P is reflection symmetric then the both lines having angle $\alpha^\star$ with the x-axis and passing through points s_1 and s_2 coincide with the exact reflection symmetry axis. Otherwise they do not coincide in general. In this case it seems reasonable to choose the middle point of the segment $[s_1, s_2]$ as the point through which the μ_1-nearest reflection symmetry axis passes.

Given any $\epsilon > 0$ the approach allows also to obtain all axes corresponding to critical points for which the values $\sigma'(P, l_\alpha(P))$ differ from the maximal one no more than by ϵ.

Rotation symmetry. Rotation symmetry measure $\mu'_{2,m}$ of order m for polygon P can be introduced as follows

$$\mu'_{2,m}(P) = \sigma'(A, r_{2\pi/m}(P)).$$

This definition has the following drawback. For $m > 3$ the values $\sigma'(P, r_{2k\pi/m}(P))$, $k = 1, 2, \ldots, m-1$, are in general different although they characterize the same symmetry of P (for k and m being mutually prime). Therefore it is preferable to introduce the following rotation symmetry measure

$$\mu_{2,m} = \max_{k \text{ mutually prime with } m} \sigma'(P, r_{2k\pi/m}(P)).$$

This measure can be computed using the algorithm below.

Algorithm 1 Computation of the rotation symmetry measure $\mu_{2,m}$.
Let $f(s)$ be the turning function of polygon P.

- *Look over in the increasing order all t_i, $0 \le t_i < 2\pi$, such that $f(s+t_i)$ and $f(s)$ have a common breakpoint;*
- *For every t_i compute $h(t_i, 0) = \int_0^1 (f(s+t_i) - f(s))^2 ds;$*

- *Compute* $d_i = h(t_i, 2k\pi/m) = h(t_i, 0) - 4k\pi\theta^*(t_i)/m + (2k\pi/m)^2$, *where* $\theta^*(t_i) = 2\pi t_i$, *k is the nearest integer to* $m\theta^*(t_i)/2\pi$ *and k is mutually prime with m.*
  If d_i *is less than the current value of d (at the beginning d is set to be a large number) set* $d = d_i$. *Go to* t_{i+1}.

At the end of the algorithm $\exp(-\sqrt{d})$ *is the value of the rotation symmetry measure.*

The algorithm allows to compute rotation symmetry measures $\mu_{2,m}$ simultaneously for all m, $1 < m \leq M$, or maximum of these symmetry measures. This algorithm is also easily modified for the computation of the symmetry measure which can be introduces as $\mu''_{2,m} = \min_k,\ _{0 \leq k < m}\ \sigma'(P, r_{2k\pi/m}(P))$.

Affine symmetry. From the properties of the canonical form it follows that a figure possess a skew symmetry (skew rotational symmetry of order m) if and only if its canonical form is reflection symmetric (rotation symmetric of order m). Therefore measures of skew reflection symmetry (skew rotation symmetry of order m) of polygon P can be introduced as $\mu_1(P^\bullet)$ (respectively, $\mu_{2,m}(P^\bullet)$). Having found the turning function of canonical form $P^\bullet$ these measures can be computed efficiently.

4 Discussion

The developed algorithms were applied for computation of symmetry measures for a number of polygonal shapes given at Fig. 2.

Table 1 presents the values of reflection symmetry measures computed for original polygons and their canonical forms. It is clear that the latter ones define skew reflection symmetry measures. Since canonical forms could be even less reflection symmetrical than the original polygons themselves, the skew reflection symmetry measures computed using a normalization approach can be smaller than the reflection symmetry measures. This is in fact the drawback of the normalization approach. However, for shapes possessing exact or nearly exact skew symmetries there will be no contradictions between the values of reflection and skew symmetry measures. Fig. 2 show positions of μ_1-nearest reflection symmetry axes for original polygons. Note that rotation symmetric shapes of order $m \geq 3$ coincide with their canonical forms (up to uniform scaling).

Table 1. Reflection symmetry measures for polygons given at Fig. 2 and their canonical forms.

Polygons	P_1	P_2	P_3	P_4	P_5	P_6	P_7	P_8
$\mu_1(P)$	0.552	0.357	1	1	1	0.743	0.65	0.467
$\mu_1(P^\bullet)$	0.583	0.303	1	1	1	1	0.577	1

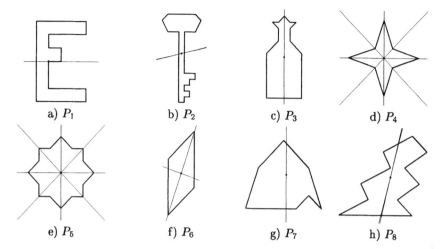

Fig. 2. Polygonal shapes analyzed in this paper and positions of μ_1-nearest reflection symmetry axes.

We tested also the sensitivity of reflection symmetry measure μ_1 and the position of μ_1-nearest reflection symmetry axis to noise. Additional polygon vertices were randomly chosen either along the whole perimeter of original polygons or some part of it.

The reflection symmetry measure values usually decrease with increasing the noise amplitude. This effect is stronger (as it was expected) for polygons with high value of symmetry measure. The increase of vertices number also results in decreasing symmetry measure values.

The values of rotation symmetry measure $\mu_{2,m}$ are shown in Table 2. This measure is more suitable for computation for small rotation orders m (in the table we give the values for $2 \le m \le 6$). For asymmetrical shapes and high rotation orders the shape is compared with itself rotated at the small angle. In this case the distance metric value is quite small and it is usually achieved for a critical point $t = 0$. As a result the rotation symmetry measure turns to 1 while increasing a rotation order and doesn't depend on the figure itself. Therefore it is preferable to use another rotation symmetry measure for high rotation orders, for example, the measure $\mu_{2,m}''$ already mentioned above.

References

1. E. M. Arkin, L. P. Chew, D. P. Huttenlocher, K. Kedem, and J. S. B. Mitchell. An efficiently computable metric for comparing polygonal shapes. *IEEE Transactions on Pattern Analysis and Machine Intelligence*, 13:209–216, 1991.
2. T. J. Cham and R. Cipolla. Symmetry detection through local skewed symmetries. *Image and Vision Computing*, 13:439–450, 1995.
3. B. Grünbaum. Measures of symmetry for convex sets. In *Proc. Sympos. Pure Math.*, volume 7, pages 233–270. Providence, USA, 1963.

Table 2. Measures of rotation symmetry for polygons $P_1 - P_8$ and their canonical forms.

Rotation order	2	3	4	5	6
$\mu_{2,m}(P_1)$	0.113	0.140	0.208	0.285	0.351
$\mu_{2,m}(P_1^\bullet)$	0.097	0.133	0.208	0.285	0.351
$\mu_{2,m}(P_2)$	0.321	0.222	0.208	0.285	0.351
$\mu_{2,m}(P_2^\bullet)$	0.246	0.189	0.208	0.285	0.351
$\mu_{2,m}(P_3)$	0.437	0.389	0.32	0.377	0.351
$\mu_{2,m}(P_3^\bullet)$	0.297	0.422	0.386	0.385	0.351
$\mu_{2,m}(P_4)$	1	0.592	1	0.73	0.592
$\mu_{2,m}(P_5)$	1	0.659	1	0.73	0.659
$\mu_{2,m}(P_6)$	1	0.382	0.38	0.534	0.382
$\mu_{2,m}(P_6^\bullet)$	1	0.592	1	0.73	0.592
$\mu_{2,m}(P_7)$	0.385	0.476	0.426	0.444	0.406
$\mu_{2,m}(P_7^\bullet)$	0.378	0.485	0.414	0.45	0.393
$\mu_{2,m}(P_8)$	0.300	0.272	0.253	0.286	0.351
$\mu_{2,m}(P_8^\bullet)$	0.271	0.262	0.256	0.287	0.351

4. H. J. A. M. Heijmans and A. Tuzikov. Similarity and symmetry measures for convex shapes using Minkowski addition. *IEEE Transactions on Pattern Analysis and Machine Intelligence*, 20(9):980–993, 1998.
5. J. Hong and X. Tan. Recognize the similarity between shapes under affine transformation. In *Proc. Second Int. Conf. Computer Vision*, pages 489–493. IEEE Comput. Soc. Press, 1988.
6. B. K. P. Horn. *Robot Vision*. MIT Press, Cambridge, 1986.
7. G. Marola. On the detection of the axes of symmetry of symmetric and almost symmetric planar images. *IEEE Transactions on Pattern Analysis and Machine Intelligence*, 11:104–108, 1989.
8. T. Masuda, K. Yamamoto, and H. Yamada. Detection of partial symmetry using correlation with rotated-reflected images. *Pattern Recognition*, 26:1245–1253, 1993.
9. W. H. Tsai and S. L. Chou. Detection of generalized principal axes in rotationally symmetric shapes. *Pattern Recognition*, 24:95–104, 1991.
10. P. J. van Otterloo. *A Contour-Oriented Approach to Digital Shape Analysis*. PhD thesis, Delft University of Technology, Delft, The Netherlands, 1988.
11. H. Weyl. *Symmetry*. Prinston University Press, Prinston, 1952.
12. H. Zabrodsky, S. Peleg, and D. Avnir. Symmetry as a continuous feature. *IEEE Transactions on Pattern Analysis and Machine Intelligence*, 17:1154–1166, 1995.

An Interpolative Scheme for Fractal Image Compression in the Wavelet Domain

M. Ghazel[2], E.R. Vrscay[1] and A.K. Khandani[2]

[1] Department of Applied Mathematics
University of Waterloo, Waterloo, ON, Canada N2L 3G1.
ervrscay@links.uwaterloo.ca http://links.uwaterloo.ca
[2] Department of Electrical and Computer Engineering, University of Waterloo
mghazel@inverness.math.uwaterloo.ca khandani@shannon.uwaterloo.ca

Abstract. Standard fractal image coding methods seek to find a contractive *fractal transform* operator T that best scales and copies subsets of a target image $\mathcal{I}$ (its *domain blocks*) onto smaller subsets (its *range blocks*). The fixed point of this operator is an approximation to the image $\mathcal{I}$ and can be generated by iteration of T. Although generally good fidelity is achieved at significant compression rates, the method can suffer from blockiness artifacts. This inspired the introduction of *fractal-wavelet transforms* which operate on the wavelet representations of functions: Wavelet coefficient subtrees are scaled and copied onto lower subtrees. We propose a simple adaptive and unrestricted fractal-wavelet scheme that adopts a dynamic partitioning of the wavelet decomposition tree, resulting in intermediate representations between the various dyadic levels. In this way, one may (i) generate a continuous and relatively smooth rate distortion curve and (ii) encode images at a pre-defined bit rate or representation tolerance error.

Keywords: Fractal image compression, fractal-wavelet compression, IFS.

1 Introduction

In this paper we introduce a systematic and simple interpolative scheme to perform fractal image compression in the wavelet domain using the method of *fractal-wavelet transforms* [6, 11, 13]. We emphasize the word *interpolative* since the method overcomes the usual limitations of working in the discrete hierarchy of wavelet subbands. Standard fractal image compression methods work in the spatial domain, mapping "domain" or "parent" pixel blocks onto smaller "range" or "child" pixel blocks [4, 5, 7, 9, 12]. In most applications the pixel blocks are dyadic squares, with typical block configurations being (i) 16×16 pixel domain blocks mapped onto 8×8 range blocks and (ii) 8×8 blocks onto 4×4 blocks. (Quadtree schemes can, of course, employ a range of dyadic domain-range configurations.)

Fractal-wavelet transforms, discovered independently by a number of workers ([3, 6, 8] to name only a few), were introduced in an effort to reduce the blockiness inherent in fractal image compression. Their action involves a scaling and copying of wavelet coefficient subtrees to lower subtrees, quite analogous to the action of fractal image coders in the spatial domain. The strength of the fractal-wavelet transform lies in its ability to combine the best of two worlds (i) *fractal transforms* with their inherent properties of scaling and local self-similarity and (ii) *wavelet transforms* based on multiresolution analysis and the discrete wavelet transform. In [13], a generalized 2D fractal-wavelet transform allowing independent scaling of the three fundamental wavelet subbands was introduced. As expected, the fidelity of image representations is increased but at the expense of compression rate. Various strategies for the quantization of the fractal coefficients have been explored in an attempt to maximize the compression capabilities.

In spite of their advantages, fractal-wavelet schemes can still be rather restrictive in that they adopt the standard three-subband wavelet tree decomposition for the selection of "parent" and "child" quadtrees. The result is a static representation that varies significantly from one resolution level to the next. However, it may be desirable to produce intermediate approximations for purposes of satisfying a particular rate-distortion relationship. One way to achieve such approximations is the method of adaptive quadtrees. In this paper, we propose and implement a simpler adaptive and unrestricted fractal-wavelet scheme that adopts a dynamic partitioning of the wavelet decomposition tree, resulting in intermediate representation between the various levels. Some of the benefits of such an adaptive fractal-wavelet scheme include the ability to (i) generate a continuous and relatively smooth rate distortion curve for fractal-wavelet schemes, and (ii) encode images at a pre-defined bit rate or representation tolerance error. Due to space limitation, we refer the reader to Ref. [13] for background reading on fractal transforms and their relationship to fractal-wavelet transforms - a preprint version may be downloaded from http://links.uwaterloo.ca.

2 Generalized 2D Fractal-Wavelet Transforms

2.1 Mathematical Setting

We consider the standard construction of orthonormal wavelet bases in $\mathcal{L}^2(\mathbf{R}^2)$ using suitable tensor products of subspaces V_i and W_j in $\mathcal{L}^2(\mathbf{R})$ as discussed in [2, 10]. Define the sequence of nested subspaces $\mathbf{V}_k \in \mathcal{L}^2(\mathbf{R}^2)$, $k \in \mathbf{Z}$, where $\mathbf{V}_k \subset \mathbf{V}_{k+1}$ so that $\cap_n \mathbf{V}_n = \{\mathbf{0}\}$ and $\lim_{n \to \infty} \cup \mathbf{V}_n = \mathcal{L}^2(\mathbf{R}^2)$. For each $\mathbf{V}_k$ define its orthogonal complement $\mathbf{W}_k$ so that $\mathbf{V}_{k+1} = \mathbf{V}_k \oplus \mathbf{W}_k$. For any $m \in \mathbf{Z}$ and $n > 0$,

$$\mathbf{V}_m \oplus \mathbf{W}_m \oplus \mathbf{W}_{m+1} \oplus \ldots \oplus \mathbf{W}_{m+n} = \mathbf{V}_{m+n+1}. \tag{1}$$

Then

$$\begin{aligned} \mathbf{V}_k &= V_k \otimes V_k, \\ \mathbf{W}_k &= (V_k \otimes W_k) \oplus (W_k \otimes V_k) \oplus (W_k \otimes W_k). \end{aligned} \tag{2}$$

Of particular interest is the case $m = 0$ and a subset of functions belonging to the subspaces $\mathbf{V}_k^0 \subset \mathbf{V}_k$ and $\mathbf{W}_k^0 \subset \mathbf{W}_k$, $k \geq 0$, defined as follows:

$$\begin{aligned} \mathbf{V}_{k+1}^0 &= \mathbf{V}_k^0 \oplus \mathbf{W}_k^0, \\ \mathbf{W}_k^0 &= \mathbf{W}_k^h \oplus \mathbf{W}_k^v \oplus \mathbf{W}_k^d, \quad k \geq 0, \end{aligned} \tag{3}$$

where

$$\begin{aligned} \mathbf{V}_k^0 &= \overline{\operatorname{span}\{\phi_{kij}(x,y) = \phi_{ki}(x)\phi_{kj}(y), \ 0 \leq i,j \leq 2^k - 1\}} \\ \mathbf{W}_k^h &= \overline{\operatorname{span}\{\psi_{kij}^h(x,y) = \phi_{ki}(x)\psi_{kj}(y), \ 0 \leq i,j \leq 2^k - 1\}} \\ \mathbf{W}_k^v &= \overline{\operatorname{span}\{\psi_{kij}^v(x,y) = \psi_{ki}(x)\phi_{kj}(y), \ 0 \leq i,j \leq 2^k - 1\}} \\ \mathbf{W}_k^d &= \overline{\operatorname{span}\{\psi_{kij}^d(x,y) = \psi_{ki}(x)\psi_{kj}(y), \ 0 \leq i,j \leq 2^k - 1\}}. \end{aligned} \tag{4}$$

The superscripts h, v and d stand for *horizontal*, *vertical* and *diagonal*, respectively [2]. We shall be concerned with functions having the following wavelet expansions:

$$f(x,y) = b_{000}\phi_{000}(x,y) + \sum_{k=0}^{\infty}\sum_{i,j=0}^{2^k-1}[a_{kij}^h\psi_{kij}^h(x,y) + a_{kij}^v\psi_{kij}^v(x,y) + a_{kij}^d\psi_{kij}^d(x,y)]. \quad (5)$$

The wavelet expansion coefficients are conveniently arranged in a standard matrix fashion [2, 10], the first three blocks of which are shown in Figure 1. The blocks $\mathbf{A}_k^h, \mathbf{A}_k^v, \mathbf{A}_k^d$, $k \geq 0$, each contain 2^{2k} coefficients $a_{kij}^h, a_{kij}^v, a_{kij}^d$, respectively. The three collections of blocks

$$\mathbf{A}^h = \bigcup_k^{\infty}\mathbf{A}_k^h, \quad \mathbf{A}^v = \bigcup_k^{\infty}\mathbf{A}_k^v, \quad \mathbf{A}^d = \bigcup_k^{\infty}\mathbf{A}_k^d, \quad (6)$$

comprise the fundamental *horizontal*, *vertical* and *diagonal* quadtrees of the coefficient tree.

Now consider any wavelet coefficient a_{kij}^λ, $\lambda \in \{h, v, d\}$ in this matrix and the unique (infinite) quadtree with this element as its root. We shall denote this (sub)quadtree as A_{kij}^λ. In the Haar case, for a fixed set of indices $\{k, i, j\}$ the three quadtrees A_{kij}^h, A_{kij}^v and A_{kij}^d correspond to the *same spatial portion* of the function or image. This feature was illustrated nicely by Davis [3].

2.2 Generalized 2D Fractal-Wavelet Transforms

The two-dimensional fractal-wavelet transforms involve mappings of "parent" quadtrees of wavelet expansions to lower "child" quadtrees. For simplicity in presentation and notation, we consider a particular case in which the roots of all parent quadtrees appear in a given block and the roots of all child quadtrees appear in another given block.

Select two integers, the parent and child levels, k_1^* and k_2^*, respectively, with $0 \leq k_1^* < k_2^*$. Then for each possible index $0 \leq i, j \leq 2^{k_2^*} - 1$ define the three sets of affine block transforms:

$$W_{ij}^\lambda : A_{k_1^*,i^\lambda(i,j),j^\lambda(i,j)}^\lambda \to A_{k_2^*,i,j}^\lambda, \quad A_{k_2^*,i,j}^\lambda = \alpha_{ij}^\lambda A_{k_1^*,i^\lambda(i,j),j^\lambda(i,j)}^\lambda, \quad (7)$$

where $\lambda \in \{h, v, d\}$. The procedure is illustrated in Figure 1. Notice how the child quadtrees at level k_2^* are replaced by scaled copies of parent quadtrees from level k_1^*.

These block transforms will comprise a unique fractal-wavelet (FW) operator M, also known as an Iterated Function System on Wavelets (IFSW) operator. The use of the indices i^h, j^h, etc. emphasizes that the parent quadtrees corresponding to a given set of child quadtrees $A_{k_2^*,i,j}^h, A_{k_2^*,i,j}^v$ and $A_{k_2^*,i,j}^d$ need *not* be the same. As well, the scaling coefficients $\alpha_{ij}^h, \alpha_{ij}^v$ and α_{ij}^d are *independent*. In the original fractal-wavelet transforms, i.e. [3, 8], common parents and common scaling factors are used for the various subbands, that is:

$$i^h(i,j) = i^v(i,j) = i^d(i,j)$$
$$j^h(i,j) = j^v(i,j) = j^d(i,j)$$
$$\alpha_{ij}^h = \alpha_{ij}^v = \alpha_{ij}^d \quad (8)$$

In other words, the h, v and d subbands are *not* treated independently.

The "fractal code" associated with an IFSW operator M consists of:

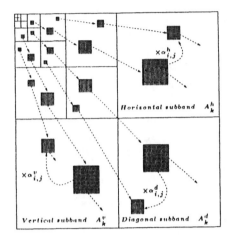

Fig. 1. The two-dimensional fractal-wavelet transform.

1. The parent-child index pair (k_1^*, k_2^*).
2. The wavelet coefficients in blocks $\mathbf{B}_0$ (i.e. b_{000}) and $\mathbf{A}_k^\lambda$, $\lambda \in \{h, v, d\}$ for $1 \leq k \leq k_2^* - 1$: $4^{k_2^*}$ coefficients.
3. The scaling factors α_{ij}^λ and parent block indices $(i^\lambda(i,j),\ j^\lambda(i,j))$ for elements a_{ij}^λ in each of the three blocks $A_{k_2^*}^\lambda$. Total number of parameters: (i) $3 \cdot 4^{k_2^*}$ scaling factors, (ii) $2 \cdot 3 \cdot 4^{k_2^*}$ indices.

It has been shown [13] that, under certain conditions, the fractal-wavelet transform M is contractive in an appropriate complete metric space (l_2-type square summable sequences) of wavelet coefficients. For the special transform given in Eq. (7), contractivity is guaranteed when

$$c_Q = 2^{k_2^* - k_1^*} \max_{\lambda, i, j} |\alpha_{ij}^\lambda| \ < \ 1, \tag{9}$$

where $\lambda \in \{h, v, d\}$ and $0 \leq i, j \leq 2^{k_2^*} - 1$.

From the Banach Fixed Point Theorem for contraction maps, the condition $c_Q < 1$ guarantees the existence of a unique fixed point of the operator M, that is, a unique wavelet coefficient tree, $\bar{c}$ such that $M\bar{c} = \bar{c}$. Moreover, the wavelet tree $\bar{c}$ may be generated by iteration of M.

3 Inverse Problem for Fractal-Wavelet Transforms

Fractal-based approximation methods are based upon the following idea. (We refer the reader to [13] for mathematical details.) Let (Y, d_Y) denote a complete metric space whose elements represent the "images" that we wish to approximate. Suppose that $y \in Y$ is a "target" image. Then one seeks to find (if possible!) a contraction mapping T on Y whose fixed point $\bar{y}$ approximates y as closely as possible. The fractal parameters that define T are then stored in the computer. Reconstruction of the approximation $\bar{y}$ is achieved by the iteration sequence $y_{n+1} = T y_n$ which, by Banach's Theorem, is guaranteed to converge to $\bar{y}$ for any "seed" image $y_0 \in Y$ (e.g., a blank screen, $y_0 = 0$).

Unfortunately, the determination of such optimal contractive operators T is quite complicated. A considerable simplification is afforded by the "Collage Theorem" [1], a simple corollary of Banach's Theorem. Given a target image $y \in Y$ we seek to find a contraction map T that makes the *collage distance* $d_Y(Ty, y)$ as small as possible. In standard fractal image compression, Y is an appropriate space of image functions. In this study, the space Y consists of wavelet coefficient quadtrees of a target image. Therefore the inverse problem for fractal-wavelet transforms becomes:

> Given a target image with wavelet coefficient matrix $\mathbf{A}$ composed of blocks A_k^λ, $0 \le k \le N$, with $\lambda \in \{h, v, d\}$ and typically $N = 8$ (512×512 image), find a fractal-wavelet operator M that maps the matrix $\mathbf{A}$ of wavelet coefficients as close as possible to itself, i.e. that makes the collage distance $\Delta = \|\mathbf{A} - M\mathbf{A}\|$ as small as possible.

For a given wavelet matrix $\mathbf{A}$ the collage distance Δ can be defined as

$$\Delta = \max_{\lambda, i, j} \|A_{k_2^*, i, j}^\lambda - \alpha_{i,j}^\lambda A_{k_1^*, i^\lambda(i,j), j^\lambda(i,j)}^\lambda\|, \tag{10}$$

where the maximum is taken over *all* child quadtrees i, j lying in *all* subbands $\lambda \in \{h, v, d\}$.

3.1 The Encoding Process

The fractal-wavelet encoding scheme for a fixed set of parent-child level values, (k_1^*, k_2^*), is illustrated in Figure 2. For each subband $\lambda \in \{h, v, d\}$ and each child quadtree $A_{k_2^*, i, j}^\lambda$ we find the parent quadtree $A_{k_1^*, i', j'}^\lambda$ for which the collage distance associated with that child, namely,

$$\Delta_{i,j,i',j'}^\lambda = \|A_{k_2^*, i, j}^\lambda - \alpha_{i,j,i',j'}^\lambda A_{k_1^*, i', j'}^\lambda\|, \tag{11}$$

is minimized. For each (i', j') pair, the determination of the optimal scaling coefficient α is a simple least-squares problem.

As in the case of normal fractal-based compression methods, there exists a variety of strategies for "optimal" parent block assignment, involving some kind of searching over feasible parent block indices (i', j'). The optimal strategy is to perform a full search of all possible parent blocks within a subband λ in level k_1^*. However, this is expensive from both a computational as well as coding point of view. As we show below, restrictive searches requiring much less computational time often yield good results with relatively small sacrifices in accuracy. Four schemes corresponding to various ways of choosing the parents and the fractal parameters are described in Table 1. These schemes will be examined below.

3.2 The Decoding Process

The decoding process of the fractal-wavelet transforms involves copying wavelet coefficient subtrees onto lower subtrees in a parallel manner. Note that all wavelet coefficients to level $k_2^* - 1$ are untouched. The copying is performed one level at a time, beginning with the computation of the wavelet coefficients at level k_2^* (from those at level k_1^*. The computation of each successive level $k_2^* + i$, $i = 0, 1, 2, \ldots$, represents an application of the FW operator M. For a 512×512 image, the procedure is stopped with the calculation of level 8. Extending the process to higher levels results in a zooming of the target image [13].

Scheme	α's	Parents	Description
I	$(\alpha^h, \alpha^v, \alpha^d)$	Independent	Generalized FW scheme with independent parent blocks, i.e. the three indices $(i^\lambda(i,j), j^\lambda(i,j))$, and independent coefficients α^λ, $\lambda \in \{h,v,d\}$.
II	$(\alpha^h, \alpha^v, \alpha^d)$	Common	Particular case of scheme I where the parent blocks $(i^\lambda(i,j), j^\lambda(i,j))$ are the same for the subbands $\lambda \in \{h,v,d\}$.
III	$(\alpha^h = \alpha^v, \alpha^d)$	Common	Particular case of Scheme II where $\alpha^h = \alpha^v$, but α^d is distinct.
IV	$(\alpha^h = \alpha^v = \alpha^d)$	Common	Quite restrictive scheme; the indices $(i^\lambda(i,j), j^\lambda(i,j))$ and the coefficients α^λ are the same for $\lambda \in \{h,v,d\}$.

Table 1. The four fractal-wavelet schemes implemented in this paper.

4 Adaptive Fractal-Wavelet Schemes

As illustrated in Table 2, a significant improvement in fidelity of image representation can be achieved by going from the level $(k_1^*, k_2^*) = (4,5)$ to the $(k_1^*, k_2^*) = (5,6)$ level. In some cases, this visible improvement is achieved at the expense of a drastic reduction in the compression ratio. Sometimes, however, an intermediate level approximation may be desired. In this section, we present an adaptive algorithm for achieving such an interme-

	$(k_1^*, k_2^*) = (4,5)$			$(k_1^*, k_2^*) = (5,6)$		
Scheme	RMSE	PSNR	C_R	RMSE	PSNR	C_R
I	11.73	26.75	47.15	6.44	31.95	10.30
II	11.64	26.81	75.26	7.30	30.87	18.35
III	12.38	26.28	89.89	7.75	30.35	21.86
IV	12.81	25.98	112.26	8.04	30.03	27.12

Table 2. Results for the representation of the test image of "Lenna" using the various schemes at $(k_1^*, k_2^*) = (4,5)$ and $(k_1^*, k_2^*) = (5,6)$ levels. RMSE denotes root mean squared error. PSNR expressed in dBs. C_R denotes the compression ratio.

diate representation. The proposed scheme is less restrictive than the schemes described above, in the sense that the parent and child blocks do not have to be restricted in size or location to the various level- subbands of the spatial orientation tree. Continuous and relatively smooth rate distortion curves are also generated for the adaptive versions of the various schemes described in the previous section. These curves are illustrated in Figure 3.

The adaptivity of the proposed schemes stems from two different sources. The first is concerned with the size and location of the child and parent blocks $A_{k_1}^\lambda$ and $A_{k_2}^\lambda$, $\lambda \in \{h,v,d\}$. However the second criteria of adaptivity is related to the search for the optimal parent quadtrees corresponding to the child quadtrees.

4.1 Adaptive Partitioning of Wavelet Coefficients

In the following method, the parent and child blocks are chosen from locations that move increasingly farther down the wavelet coefficient tree. As a result, their corresponding sizes increase. Both features are illustrated in Figure 2.

Adaptive partitioning algorithm:

1. **STEP 1:** Starting at the initial level $(l = l_0)$ $(k_{1,l}^*, k_{2,l}^*) = (3, 4)$:
   - The parent and child blocks $A_{k_{1,l}^*}^\lambda$ and $A_{k_{2,l}^*}^\lambda$ are of sizes 8×8 and 16×16, respectively.
2. **STEP 2:** Slide and expand the parent block $A_{k_{1,l}^*}^\lambda$ by one pixel, in the direction of $\lambda \in \{h, v, d\}$:
   - This preserves its square geometry, to obtain the block $A_{k_{1,l+1}^*}^\lambda$ that is now of size 9×9 pixels.
   - This results in sliding the corresponding child block $A_{k_{2,l+1}^*}^\lambda$ by 2 pixels in the direction of $\lambda \in \{h, v, d\}$.
3. **STEP 3:** Expand each child block $A_{k_{2,l+1}^*}^\lambda$, in the direction of $\lambda \in \{h, v, d\}$, by 2 pixels:
   - This results in new expanded child blocks $A_{k_{2,l+1}^*}^\lambda$ of size 18×18 pixels.
   - This also maintains the $4 : 1$ ratio between the size of a parent block and that of the child block.
4. **STEP 4:** This process is then extended to the higher levels to cover the entire wavelet decomposition tree.
   - The fractal-wavelet scheme of choice can then be applied using this new partition of parent and child quadtrees.
5. **STEP 5:** Check if the *stopping criterion* (prescribed bit rate or distortion) is achieved:
   - If so, **STOP.**
   - Otherwise: $l = l + 1$, **GOTO STEP 2.**

Figure 2 illustrates an arbitrary level of this adaptive partitioning of the wavelet decomposition tree. Note how the conventional borders of the standard wavelet tree partitioning are redesigned. Clearly, starting from the initial level $(k_{1,l_0}^*, k_{2,l_0}^*) = (3, 4)$, and repeating the process described in the above algorithm, we reach the $(k_{1,l_0+8}^*, k_{2,l_0+8}^*) = (4, 5)$ level after 8 iterations and the level $(k_{1,l_0+24}^*, k_{2,l_0+24}^*) = (5, 6)$ after 24 iterations. This results in generating continuous and relatively smooth rate distortion curves for the fractal-wavelet schemes. Figure 3, illustrates these curves for adaptive versions of schemes II and IV, described in the previous section.

Remarks

- It is obvious that the starting level $(k_{1,l_0}^*, k_{2,l_0}^*)$ can be anywhere in the wavelet decomposition tree.
- We can also slide and expand the child and parent blocks by larger factors so that we move more quickly from coarse approximations to more refined representations of the image.

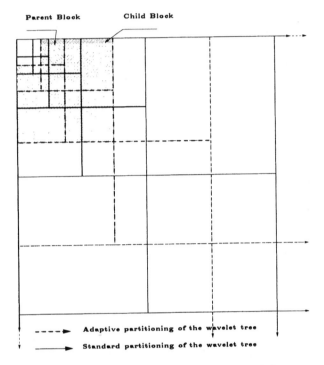

Fig. 2. Adaptive partitioning of the wavelet coefficients tree.

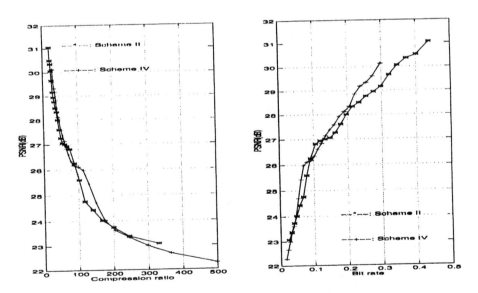

Fig. 3. Rate Distortion Curves for the Adaptive versions of Schemes II and IV. *left*: PSNR *vs.* Compression Ratio, *right*: PSNR *vs.* Bit-Rate (test image : "Lenna", using Daubechies-8 wavelets).

- When sliding the parent and child blocks by a relatively small number of pixels, there is no need to perform an exhaustive search for the optimal parent quadtree for each child quadtree. In fact, one can use the same parent block obtained in the previous $(l-1)^{th}$ level, for most of them, except the new child quadtrees or those quadtrees whose parents are no longer in the current parent blocks. This speeds up the scheme and introduces some element of progressiveness or embeddedness in the scheme.
- The scheme also works backwards. That is, one can start at a higher level, say $(k_{1,l_0}^*, k_{2,l_0}^*) = (5, 6)$, and go down to the level $(k_{1,l_f}^*, k_{2,l_f}^*) = (3, 4)$ by applying the inverse transformations to those described in the above algorithm.
- This algorithm allows us to perform fractal-wavelet image compression at a predetermined bit rate or PSNR. One can converge to the desired bit-rate by simply sliding the parent and child blocks, as described in the algorithm.

4.2 Adaptive Parent-Child Quadtree Matching

Another source of adaptivity in fractal-wavelet schemes lies in the search for the optimal parent-quadtree. As illustrated in Figure 4, restrictive search requiring much less computational time often yield comparable results at the expense of relatively small sacrifices in accuracy. Similar results were obtained when using the other schemes. This factor of adaptivity was in fact integrated in the process of developing fast and adaptive fractal-wavelet schemes.

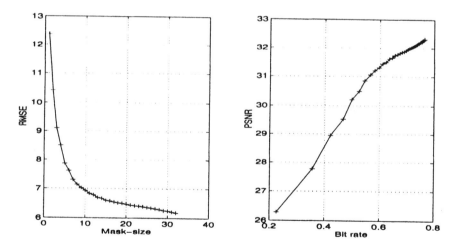

Fig. 4. (a) RMSE vs. mask-size, (b) PSNR vs. Bit-rate. using Scheme II at level $(k_1^*, k_2^*) = (5, 6)$ (test image: "Lenna", using Daubechies-8 wavelets).

5   Discussion and Conclusion

This paper represents a preliminary attempt to study and implement a truly adaptive fractal scheme in the wavelet domain. We have proposed and implemented adaptive and unrestricted fractal-wavelet algorithm that adopts a dynamic partitioning of the wavelet decomposition tree, resulting in intermediate representation between the various levels. Some of the benefits of such an adaptive fractal-wavelet scheme include (i) generating a continuous and relatively smooth rate distortion curve for fractal-wavelet schemes, and (ii) encoding images at a pre-defined bit rate or representation tolerance error. This scheme is characterized by its flexibility in terms of the partitioning of the wavelet decomposition tree. In that, an adaptive strategy for partioning the wavelet coefficients was proposed and implemented. The conventional borders of the standard wavelet decomposition tree are redesigned, resulting in a new partition that overlaps with the various subbands. The scheme is also reversible, as one can start at any refined approximation level and move up to a coarser approximation. More importantly, it is capable of achieving fractal-wavelet image compression at a predetermined bit rate or approximation error. There is also some element of progressiveness in this scheme in the sense that it re-uses the parent-child matching obtained in the previous iterations. It also does not require a fully decomposed wavelet tree of the image - in fact, only a tree decomposed partially to an appropriate level, is needed. This, of course, speeds up the decoding process.

6 Acknowledgements

This research was supported by an Interdisciplinary Research Grant from the University of Waterloo (ERV and AKK) as well as grants from NSERC - Natural Sciences and Engineering Council of Canada (ERV), and CITO - Communications and Information Technology Ontario (AKK). All support is gratefully appreciated and acknowledged.

References

1. M.F. Barnsley, *Fractals Everywhere*, Academic Press, NY (1988).
2. I. Daubechies, *Ten Lectures on Wavelets*, SIAM Press, Philadelphia (1992).
3. G.M. Davis, A wavelet-based analysis of fractal image compression, IEEE Trans. Image Proc. (1998).
4. Y. Fisher, Ed., *Fractal Image Compression, Theory and Application*, Springer, NY (1995).
5. Y. Fisher, Ed., *Fractal Image Encoding and Analysis*, Springer, Heidelberg (1998).
6. B. Forte and E.R. Vrscay, Inverse Problem Methods for Generalized Fractal Transforms, *Fractal Image Encoding and Analysis, ibid.*
7. A. Jacquin, "Image coding based on a fractal theory of iterated contractive image transformations," *IEEE Trans. Image Proc.*, 1, no. 1, pp. 18-30 (1992).
8. H. Krupnik, D. Malah, and E. Karnin. Fractal Representation of Images via the Discrete Wavelet Transform, *IEEE 18th Conv. EE*, Tel-Aviv, Israel, March 1995.
9. N. Lu, *Fractal Imaging*, Academic Press, NY (1997).
10. S.G. Mallat, A theory for multiresolution signal decomposition: The wavelet representation, *IEEE Trans. PAMI*, 11, No. 7, 674-693 (1989).
11. F. Mendivil and E.R. Vrscay, Correspondence between fractal-wavelet transforms and Iterated Function Systems, in *Fractals in Engineering, From Theory to Industrial Applications*, J. Lévy Véhel, E. Lutton and C. Tricot, Eds., Springer, NY (1997).
12. D.M. Monro and F. Dudbridge, Fractal block coding of images, Elect. Lett. **28** no. 11, pp. 1053-1055 (1992).
13. E.R. Vrscay, "A generalized class of fractal-wavelet transforms for image representation and compression," Can. J. Elect. Comp. Eng. **23**, Nos. 1,2, pp. 69-84 (1998).

On Convergence of Discrete and Selective Fractal Operators

Władysław Skarbek [*]

Multimedia Laboratory
Department of Electronics and Information Technology
Warsaw University of Technology
Email:W.Skarbek@ire.pw.edu.pl

Abstract. It is proved that any iterative sequence for a fractal operator which is contractive in l_∞ norm with contractivity c^*, after clamping and rounding to integer levels, is perceptually convergent at the threshold $\tau \geq 1/(1-c^*)$. Eventual contractivity is a sufficient condition for the convergence of iterations of a fractal operator. This paper shows that it is also necessary if the averaging operation in the definition of the fractal operator is replaced by the selection operation.

1 Introduction

By $\mathcal{I}$ we denote the class of all d dimensional images, i.e. the real valued functions of the form $f : D \to [v_{min}, v_{max}]$, which are defined on a visual surface $D \subset R^d$ with values in the grayscale interval $[v_{min}, v_{max}] \subset R$. Typically $d = 2$ and D is a finite discrete grid of points, so called pixels and integer values from $[v_{min}, v_{max}]$ are used as pixel grayscale values.

The fractal compression of such images is based on representing the given image f by a fractal operator $F : \mathcal{I} \to \mathcal{I}$ which is defined by local operators mapping image fragments g_S defined on source domains $S \subset D, S \in \Gamma$ onto image fragments g'_T defined on target domains $T \subset D, T \in \Pi$ (cf. [2, 5]). Actually for each T there exists exactly one $S(T)$ and g'_T is obtained from $g_{S(T)}$ by reducing operation R_T, contrasting the result by c_T and offsetting o_T. Since target domains are disjoint and they cover the domain D we have:

$$F(g) \doteq \sum_{T \in \Pi} [c_T R_T(g) + o_T 1_T] \tag{1}$$

where 1_T is the characteristic function of the domain T. The reducing operation effectively replaces a small block (typically 2×2) of pixels by one pixel. There are two popular ways of such replacement: make averaging of pixels in the block or select one of them. In the latter case the fractal operator is called *the selective* fractal operator.

---

[*] This work was supported in part by European ESPRIT INCO CRIT2 project.

The decoder receives for each $T : c_T, o_T$, the displacement T from $S(T)$ and optionally the permutation index p_T. It starts with arbitrary initial image $g_0 \in \mathcal{I}$ and iterates in sequence for $i > 0$:

$$F^{\circ i}(g_0) \doteq g_i = F(g_{i-1}) \tag{2}$$

In practice the process is terminated after about $i = 15$ iterations as usually no further improvement of image quality measure PSNR is observed.

The sufficient condition for the convergence g_i is for instance contractivity of F (cf. [2,4]) as the Banach fixed point theorem can be applied. In the infinity norm l_∞ the contractivity in l_∞ norm is equivalent with the condition:

$$c^* \doteq \max_T |c_T| < 1 \tag{3}$$

where c^* appears to be the Lipshitz coefficient for F.

The limit image $\tilde{f} \doteq \lim_{i \to \infty} F^{\circ i}(g_0)$ is the fixed point of F. By the collage theorem ([1]) $\tilde{f}$ is close to the original image f as the coder builds F to minimize the distance between f and $F(f)$.

2 Discrete fractal operator

Even when the image g has only integer values $v = g(p)$, $p \in D$, the image $g' = F(g)$ has already real values $v' = g'(p)$. Moreover, the result of contrasting and offsetting can give results outside the interval $[v_{min}, v_{max}]$. In order to save image storage, the clamping operation C and the rounding operation Q is usually applied to g' before storing pixel values. These postprocessing operations together with the original operator create *the discrete* fractal operator.

If $h' = C(h)$ and for the pixel $p \in D$ $v = h(p)$ then $v' = h'(p)$ is defined by:

$$v' \doteq \begin{cases} v_{min} & \text{if } v < v_{min}, \\ v_{max} & \text{if } v > v_{max}, \\ v & \text{otherwise.} \end{cases} \tag{4}$$

Now, if $h' = Q(h)$ and for the pixel $p \in D$ $v = h(p)$ then $v' = h'(p)$ is defined by:

$$v' \doteq \begin{cases} \lfloor v \rfloor & \text{if } v < \lfloor v \rfloor + 0.5 , \\ \lceil v \rceil & \text{otherwise.} \end{cases} \tag{5}$$

The composition of C and Q operations can be viewed as postprocessing for each local operator:

$$F_{QC}(g) \doteq \sum_{T \in \Pi} QC \left[c_T \cdot \mathbf{R}_T(g) + o_T \cdot 1_T \right] \tag{6}$$

While clamping the fractal operator to the grayscale range of values of the images preserves its contractivity, performing the rounding process to discrete levels spoils this property and the given iterative sequence is generally not convergent. In practice this lack of convergence is not observed by HVS (Human Visual System) on decoder's output. We explain this phenomenon by presenting a strict notion of the perceptual convergence at the given threshold.

The sequence of images $g_i \in \mathcal{I}$ is *perceptually convergent* at threshold τ if there exists i_0 such that for all $j, i > i_0$: $\|g_j - g_i\|_\infty \leq \tau$.

On the gray scale with 256 levels, the threshold $\tau = 4$ makes individual pixel values indistinguishable.

The notion of perceptual convergence was introduced by the author in [6] and applied in case of rounding process only. Here stronger notion of perceptual convergence is considered (but more suitable from the application point of view), the results are extended to the case of composition of clamping and rounding operations.

3 Convergence analysis of discrete sequence

The clamping operation C acts independently for every pixel. If two images have for the given pixel values v_1 and v_2 respectively, then after clamping their values v_1' and v_2' are not going apart. Namely:

$$|v_1' - v_2'| \leq |v_1 - v_2| .$$

In terms of the infinity norm, this local property can be written in a global way:

$$\|C(g) - C(h)\| \leq \|g - h\| \tag{7}$$

Therefore if F is contractive with contractivity coefficient α then the operator $F_C \doteq C \circ F$ is contractive too:

$$\|F_C(g) - F_C(h)\| = \|C(F(g)) - C(F(h))\| \leq \|F(g) - F(h)\| \leq \alpha\|g - h\| \tag{8}$$

The rounding operation Q acts independently for every pixel too. If two images have for the given pixel close values $v_1 = \lfloor v_1 \rfloor + 0.5(1 - \epsilon)$ and $v_2 = \lfloor v_1 \rfloor + 0.5(1 + \epsilon)$ respectively, then after rounding their values v_1' and v_2' are going apart. Namely:

$$|v_1' - v_2'| = 1 > \epsilon .$$

In terms of the infinity norm, there exist two close images g, h such that:

$$\|g - h\| = \epsilon \text{ and } \|Q(g) - Q(h)\| > \|g - h\| \tag{9}$$

Therefore the rounding operation Q can "damage" the contractivity of the operator F_C. Namely, if F is the contractive operator then after the clamping and rounding F_{QC} usually is not contractive. However, a modified Lipshitz condition is still true for F_{QC}:

Lemma 1.

Let c^* be the Lipshitz factor in the infinity norm of the affine fractal operator F. Then, for any $g, h \in \mathcal{I}$:

1.

$$\|F_{QC}(g) - F_{QC}(h)\|_\infty \leq c^*\|g - h\|_\infty + 1 \qquad (10)$$

2.

$$\|F_{QC}^{oi}(g) - F_{QC}^{oi}(h)\|_\infty \leq c^{*i}\|g - h\|_\infty + c^{*(i-1)} + \cdots + c^* + 1 \qquad (11)$$

Proof: Let $\epsilon_T(g) \doteq QC[c_T R_T(g) + o_T 1_T] - C[c_T R_T(g) + o_T 1_T]$ and analogously $\epsilon_T(h)$ be the rounding error on subimages. The rounding error on the given pixel is not greater than 0.5. Hence the infinity norm of error subimages is not greater than 0.5. We estimate the norm of operator values:

$$\|F_{QC}(g) - F_{QC}(h)\| =$$

$$= \max_{T \in \Pi} \|QC[c_T R_T(g) + o_T 1_T] - QC[c_T R_T(h) + o_T 1_T]\|$$

$$= \max_{T \in \Pi} \|C[c_T R_T(g) + o_T 1_T] + \epsilon_T(g) - C[c_T R_T(h) + o_t 1_T] - \epsilon_T(h)\|$$

$$\leq \max_{T \in \Pi} \|C[c_T R_T(g) + o_T 1_T] - C[c_T R_T(h) + o_T 1_T]\| +$$

$$max_{T \in \Pi} \|\epsilon_T(g) - \epsilon_T(h)\|$$

$$\leq \max_{T \in \Pi} \|c_T R_T(g - h)\| + \max_{T \in \Pi} \|\epsilon_T(g) - \epsilon_T(h)\|$$

$$\leq c^*\|g - h\|_\infty + 1$$

The second point of lemma we get by the applications of the inequality from the point one. □

For the contractive operator F, the Lipshitz-like condition for F_{QC} keeps the distance between images within a fixed bound. Namely, from the last lemma we get easily the following corollary:

Corollary 1.

If the affine fractal operator F has $c^* < 1$ then

$$\|g - h\|_\infty \leq \frac{1}{1 - c^*} \implies \|F_{QC}(g) - F_{QC}(h)\|_\infty \leq \frac{1}{1 - c^*} \qquad (12)$$

From the above corollary we see that trajectories starting within certain bound they are kept within the same bound. However, the matter of the perceptual convergence is opened.

Due to the finiteness of the space of possible discrete images with fixed domain D, any iterative sequence enters a cycle. In a rare case this cycle could consist of single element only. It appears that the infinity distance for any two images from any cycle is bounded by the threshold $\tau = 1/(1 - c^*)$:

Theorem 1.

Let g_0 be any initial image and g, h belong to a cycle entered by the iterative sequence $\{F_{QC}^{\circ i}(g_0)\}$. If $c^* < 1$ then

$$\|g - h\|_\infty \leq \frac{1}{1 - c^*} \tag{13}$$

Hence images from the given cycle are distant from each other not more than $1/(1 - c^*)$.

Proof: Let elements g, h be from the cycle of length K, have the form $g = F_{QC}^{\circ q}(g_0)$ and $h = F_{QC}^{\circ(q+d)}(g_0)$ for certain iterations $q, q + d$. Using the point 2 of lemma 1 and the periodicity of the iterative sequence, we get for any $t \geq 0$:

$$\|g - h\| = \|F_{QC}^{\circ q}(g_0) - F_{QC}^{\circ(q+d)}(g_0)\| = \|F_{QC}^{\circ(q+tK)}(g_0) - F_{QC}^{\circ(q+d+tK)}(g_0)\|$$

$$\leq (c^*)^{q+tK}\|g_0 - F_{QC}^{\circ d}(g_0)\| + (1 - (c^*)^{q+tK})/(1 - c^*) \overset{t \to \infty}{\longrightarrow} 1/(1 - c^*)$$

$\square$

Combining the corollary 1 and the theorem 1, we have immediately:

Corollary 2.

Let $\tau \geq 1/(1 - c^*)$ be a threshold. If $c^* < 1$ then for any initial image g_0, the iterative sequence $\{F_{QC}^{\circ i}(g_0)\}$ is perceptually convergent at the threshold τ. In practical circumstances we get perceptual convergence for discrete operator if $c^* \leq 0.75$ and $\tau = 4$.

4 Convergence of selective fractal operators

Using the classical contractive mapping theorem of Banach (Dugundi [3]) the fixed point $\tilde{f}$ is approximated by successive iterations $F^{\circ i}(x_0)$ of the contractive operator F (see Barnsley and Hurd [2], Fisher [4]). However, practical use of this scheme encountered several difficulties as for most existing encoding schemes the essential condition of contractivity for F cannot be guaranteed.

It was observed that if F is not contractive then a certain iteration $F^{\circ k}$ may be contractive. This situation is recognized as eventual contractivity [4].

In this paper we use the graph perspective to show that the eventual contractivity is a consequence of the finite image domain.

Related, preliminary results for this research were included in one of the sections of the author's paper [6].

4.1 Pixel influence graph

We say that a pixel p_i is *influenced* by a pixel p_j (denote $p_i \leftarrow p_j$) if there exists a local mapping in which new value v_i of the pixel p_i is computed using the current value v_j of the pixel p_j. For selective fractal operators each pixel p_i is *influenced* by exactly one pixel p_j.

The *pixel influence graph* $G_p = (V_p, E_p)$ can be defined using the relation $\leftarrow$. The set of vertices V_p is identified with integer labels of pixels, i.e. $V_p \doteq \{1, \ldots, n\}$, while the set of directed edges E_p is defined as follows:

$$E_p \doteq \{(j, i) \mid p_i \leftarrow p_j\} \ .$$

Note, that the input degree of any vertex in the pixel influence graph G_{ps} for the selective fractal operator equals one.

We can assign contrasts as labels of edges in G_p. Namely the edge (i, j) is labelled by $|c_k|$ if and only if $p_j \in T_k$. We can also speak about the total contrast of any path in the graph G_p as the product of all labels on this path.

We say that the graph G_p is *stable* if and only if for all its cycles the total contrast is less than one.

4.2 Pixel graph stability versus operator convergence

In case of selective fractal operator the pixel value v_i is computed by affine scalar equation:

$$v_i(t + 1) = c_i v_j(t) + o_i \tag{14}$$

From the fact that the image domain is finite and from the equation 14 we conclude:

Lemma 2.

1. *The graph G_{ps} is a union of connected components which have a simple structure: a cycle with optionally joined directed trees.*
2. *If for certain pixel on a cycle its values obtained by iterating the selective fractal operator are convergent then values of all other pixels located in the same connected component are convergent too.*
3. *If one of contrasting coefficients, assigned to an edge located in a cycle, is equal to zero then all pixels in the same connected component have values convergent.*
4. *If for certain pixel on a cycle its values obtained by iterating the selective fractal operator are not converging then values of all other pixels located in the same connected component are not converging too provided that all contrasting coefficients are not equal zero.*

By the last lemma we see that for the given cycle $\gamma = (i_1, \ldots, i_r)$ of length r it is enough to choose one vertex, suppose i_1 and analyze only convergence of this pixel values.

Theorem 2.

For the given pixel i located on a cycle γ in graph G_{ps} the discrete time sequence of its values $v_i(t)$ is convergent if and only if the total contrast of this cycle $c(\gamma) < 1$. Therefore $v_i(t)$ is convergent for all i from the same connected component of G_{ps} if and only if the total contrast $c(\gamma) < 1$ for the cycle γ included in this component.

Proof: Without loss of generality we consider a simple case of a cycle $\gamma = (1, 2, 3, 4)$. Then we can find explicit formulas for values $v_i(t)$:

$$v_2(t + 1) = c_2 v_1(t) + o_2;$$
$$v_3(t + 2) = c_3 c_2 v_1(t) + c_3 o_2 + o_3;$$
$$v_4(t + 3) = c_4 c_3 c_2 v_1(t) + c_4 c_3 o_2 + c_4 o_3 + o_4;$$
$$v_1(t + 4) = c_1 c_4 c_3 c_2 v_1(t) + c_1 c_4 c_3 o_2 + c_1 c_4 o_3 + c_1 o_4 + o_1;$$

Hence $v_1(t + 4)$ is of the form $a v_1(t) + b$, where $a = c_1 c_2 c_3 c_4$. This type of sequence is convergent if and only if $|a| < 1$. The limit does not depend on the initial value $v_1(0)$. Therefore four subsequences $v_1(t + 4)$, $v_1(t + 5)$, $v_1(t + 6)$, $v_1(t + 7)$ have the same limit.

In our case $|a|$ is the total contrast of the cycle γ, what concludes the proof of the first statement of the theorem.

The second statement follows from the fact that values in other pixels of the same connected component are delayed affine combinations of the value $v_1(t)$ for the chosen pixel from the cycle. □

The last theorem gives us a straight way to make a link to eventual contractivity for the selective fractal operators.

Namely the selective fractal operator restricted to a connected component with the cycle of length r is eventually contractive at the power r. If we take k as the least common multiple for all cycle lengths r we obtain contractivity for $F^{\circ k}$. This leads us to the following conclusions:

Corollary 3.

1. *Selective fractal operator F is convergent if and only if its pixel influence graph G_{ps} is stable.*
2. *Pixel influence graph of selective fractal operator F is stable if and only if F is eventually contractive.*
3. *Selective fractal operator F is convergent if and only if F is eventually contractive.*

5   Conclusions

Two important in practice subclasses of fractal operator class are considered: the discrete (quantised and clamped) operators and selective operators. The convergence conditions for them are presented.

Clamping the fractal operator to the given grayscale range of values of the images preserves its contractivity, but the rounding process to discrete levels removes this property and the given iterative sequence is generally not convergent. However, in practice this lack of convergence is not observed by HVS (Human Visual System) on decoder's output. This phenomenon is explained here by introducing a strict notion of the perceptual convergence at the given threshold. It is proved that any iterative sequence for a fractal operator which is contractive in l_∞ norm with contractivity $c^* < 1$, after clamping and rounding to integer levels, is perceptually convergent at the threshold $\tau \geq 1/(1 - c^*)$.

The graph stability condition in pixel influence graph appears to be sufficient and necessary for convergence of selective fractal operators.

Eventual contractivity for selective fractal operators is not only sufficient but also necessary condition for the fractal operator convergence.

References

1. M. F. Barnsley, *Fractals everywhere* Addison Wesley, 1988.
2. M. F. Barnsley and L. P. Hurd, *Fractal image compression*, AK Peters. Ltd, Wellesley, MA, 1993.
3. J. Dugundi and A. Granas, *Fixed point theory*, Polish Scientific Publishers, Warszawa, 1982.
4. Y. Fisher ed., *Fractal image compression, Theory and Application*, Springer Verlag, New York, 1995.
5. A. E. Jacquin, "Image coding based on a fractal theory of iterated contractive image transformations", *IEEE Trans. on Image Processing*, vol. 1, no. 1, pp. 18-30, 1992.
6. W. Skarbek, On convergence of affine fractal operators, *Image Processing and Communications*, 1(1), 1995.

Optimized Fast Algorithms for the Quaternionic Fourier Transform[*]

Michael Felsberg and Gerald Sommer

Christian-Albrechts-University of Kiel
Institute of Computer Science and Applied Mathematics
Cognitive Systems
Preußerstraße 1–9, 24105 Kiel, Germany
Tel: +49 431 560433, Fax: +49 431 560481
{mfe,gs}@ks.informatik.uni-kiel.de

Abstract. In this article, we deal with fast algorithms for the quaternionic Fourier transform (QFT). Our aim is to give a guideline for choosing algorithms in practical cases. Hence, we are not only interested in the theoretic complexity but in the real execution time of the implementation of an algorithm. This includes floating point multiplications, additions, index computations and the memory accesses. We mainly consider two cases: the QFT of a real signal and the QFT of a quaternionic signal. For both cases it follows that the row-column method yields very fast algorithms. Additionally, these algorithms are easy to implement since one can fall back on standard algorithms for the fast Fourier transform and the fast Hartley transform. The latter is the optimal choice for real signals since there is no redundancy in the transform. We take advantage of the fact that each complete transform can be converted into another complete transform. In the case of the complex Fourier transform, the Hartley transform, and the QFT, the conversions are of low complexity. Hence, the QFT of a real signal is optimally calculated using the Hartley transform.

1 Introduction

In image processing, the complex and the quaternionic Fourier transform are important tools. Hence, it is advantageous to have fast algorithms for these transforms. In the past, many algorithms for the complex Fourier transform have been proposed [1, 9, 4]. Most of them use the radix-n principle. By new algebraic embeddings, the theoretic complexities of the algorithms have been decreased. Mostly, the considerations only deal with the number of floating point multiplications.

In this paper, we mainly present fast algorithms for the quaternionic Fourier transform [6, 2]. We specialize in radix-2 algorithms since we think that higher radices yield unnecessary extensions of the spatial domain (the domain must be a power of n). Furthermore, we consider both, the additive and the multiplicative complexity since there is no difference in the execution time nowadays. Our aim is to equip the reader with the necessary basics on choosing the right algorithm in a certain case.

We underline our results by experiments on a real computer. Many algorithms and combinations have been tested, though we have not implemented every algorithm in the same environment.

---

[*] This work was supported by the DFG (So-320/2-1).

2 Basics

In this section we define the *discrete quaternionic Fourier transform* (DQFT), which is based on the continuous transform proposed independently by Ell and Bülow [6, 2]. Furthermore, we state the standard 1-D *fast Fourier transform* (FFT1) and the 1-D *fast Hartley transform* (FHT1). We suppose that the reader is familiar with the algebra of quaternions $\mathbb{H}$, the Fourier transform and the Hartley transform.

2.1 The Quaternionic Fourier Transform

Let $\{1, i, j, k\}$ denote the basis of the algebra of quaternions $\mathbb{H}$, where we have $i^2 = j^2 = k^2 = -1$ and $ij = -ji = k$. The coefficient-selection operators are defined by $\mathrm{Re}(q) + i\mathrm{Im}(q) + j\mathrm{Jm}(q) + k\mathrm{Km}(q) = q$ for $q \in \mathbb{H}$. The *discrete quaternionic Fourier transform* (QFT) is defined by sampling the QFT

$$F_{u,v}^q = \sum_{m=0}^{M-1} \sum_{n=0}^{N-1} e^{-i2\pi um/M} f_{m,n} e^{-j2\pi vn/N} \ , \tag{1}$$

according to [3].

The *inverse transform* is obtained by changing the signs in the exponential terms and by multiplying an additional normalizing factor $(MN)^{-1}$. Except for the different imaginary units, the QFT is calculated in the same way as the complex Fourier transform.

The quaternionic spectrum of a real signal is *quaternionic Hermite symmetric*, i.e. we have

$$F^q(u, -v) = \alpha(F^q(u, v)) \qquad \text{and} \qquad F^q(-u, v) = \beta(F^q(u, v)) \ , \tag{2}$$

where $\alpha(q) = -iqi$ and $\beta(q) = -jqj$ are two *non-trivial involutions* of the quaternion algebra. Hence, the QFT of a real signal consists of 75% redundant data.

2.2 The Fast Fourier Transform and the Fast Hartley Transform

In this paper, we use the *discrete Fourier transform* (DFT) without a normalizing factor. Applying the *decimation of time* method, we obtain the formula for the FFT

$$F_u = F_u^e + e^{-i2\pi u2^{-k}} F_u^o \ , \tag{3}$$

where F, F^e, and F^o denote the spectra of f, f^e, and f^o, respectively, and $f_m^e = f_{2m}$ and $f_m^o = f_{2m+1}$.

The Hartley transform [1] is a real valued integral transform. The only formal difference to the complex (inverse) Fourier transform is that the sine-function in the kernel is not multiplied by the imaginary unit.

The *discrete Hartley transform* (DHT) is defined by

$$H_u = \sum_{m=0}^{M-1} f_m \mathrm{cas}(2\pi mu/M) = \sum_{m=0}^{M-1} f_m(\cos(2\pi mu/M) + \sin(2\pi mu/M)) \ . \tag{4}$$

The shift-theorem of the DHT reads [1]

$$\sum_{m=0}^{M} f_{m-a} \mathrm{cas}(2\pi mu/M) = H_u \cos(2\pi au/M) + H_{M-u} \sin(2\pi au/M) \ . \quad (5)$$

Consequently, the decimation of time method yields a formula, which is different from (3):

$$H_u = H_u^e + H_u^o \cos(2\pi u 2^{-k}) + H_u^o \sin(2\pi u 2^{-k}) \ , \quad (6)$$

where H, H^e, and H^o denote the Hartley transforms of f, f^e, and f^o, respectively, and $f_m^e = f_{2m}$ and $f_m^o = f_{2m+1}$.

3 Fast Algorithms for 2-D Signals

In this section, we describe the *row-column method*, which yields a fast algorithm for an n-D transform using 1-D transforms. Another approach for developing fast algorithms is the 2-D *decimation method* which divides the signal into four parts [8]. Finally, the mappings between the Hartley transform, the QFT and the complex spectrum of a 2-D signal are given.

3.1 The Row-Column Method

Consider a 2-D integral transform whose kernel is separable. Such a transform can be calculated by applying 1-D transforms on each coordinate. Without loss of generality, we firstly apply the transform with respect to the x-coordinate and secondly wrt. the y-coordinate.

Since the kernels of the complex 2-D Fourier transform and the QFT are separable, we can state the *row-column algorithm* of FFT1s (RC-FFT1).

1. The 2-D Fourier transform is separable. Therefore, the complex Fourier transform of a 2-D signal $f_{m,n}$ can be calculated by applying the 1-D FFT to each column of $f_{u,n}$. Thereby, the signal $f_{u,n}$ is obtained by the application of the 1-D FFT to each row of $f_{m,n}$. Hence, the spectrum can be calculated by $M + N$ 1-D transforms.
2. The QFT is separable, too.

$$F_{u,v}^q = \sum_{n=0}^{N-1} \left(\underbrace{\sum_{m=0}^{M-1} e^{-i2\pi um/M} (\mathrm{Re}(f_{m,n}) + i\mathrm{Im}(f_{m,n}))}_{f_{u,n}^{ri}} \right.$$

$$\left. + \underbrace{\sum_{m=0}^{M-1} e^{-i2\pi um/M} (\mathrm{Jm}(f_{m,n}) + i\mathrm{Km}(f_{m,n})) j}_{f_{u,n}^{jk}} \right) e^{-j2\pi vn/N}$$

$$= \sum_{n=0}^{N-1} \underbrace{(\mathrm{Re}(f_{u,n}^{ri}) + \mathrm{Re}(f_{u,n}^{jk})j) e^{-j2\pi vn/N}}_{\hat{f}_{u,n}^{R}} + i \sum_{n=0}^{N-1} \underbrace{(\mathrm{Im}(f_{u,n}^{ri}) + \mathrm{Im}(f_{u,n}^{jk})j) e^{-j2\pi vn/N}}_{\hat{f}_{u,n}^{I}} \ . \quad (7)$$

Therefore, the QFT of a 2-D signal $f_{m,n}$ can be calculated by applying the 1-D FFT to each column of both $\hat{f}_{u,n}^{R}$ and $\hat{f}_{u,n}^{I}$. Hence, the QFT can be calculated by $2M + 2N$ 1-D FFTs.

Note 1. The Hartley kernel is *not* separable, since

$$\operatorname{cas}(2\pi(um + vn)) \neq \operatorname{cas}(2\pi um)\operatorname{cas}(2\pi vn) \ . \tag{8}$$

Nevertheless, the application of the row-column method to FHT1s (RC-FHT1) yields an interesting transform, since the DQFT of a real signal can be synthesized from the RC-FHT1 (see Sect. 3.3). The only difference between the 2-D DHT and the result of the RC-FHT1 is the sign of the component, which is odd wrt. both coordinates.

3.2 The 2-D Decimation Method

Fast algorithms for the 2-D transforms can be developed by dividing the spatial domain with respect to *both* coordinates:

$$f^{ee}_{m,n} = f_{2m,2n} \quad f^{oe}_{m,n} = f_{2m+1,2n} \quad f^{eo}_{m,n} = f_{2m,2n+1} \quad f^{oo}_{m,n} = f_{2m+1,2n+1} \ . \tag{9}$$

Note that the signal must be of quadratic shape $2^k \times 2^k = N \times N$. The (quaternionic) Fourier transforms of f^{ee}, f^{oe}, f^{eo}, and f^{oo} are denoted by F^{ee}, F^{oe}, F^{eo}, and F^{oo}, respectively. The corresponding Hartley transforms are denoted by H^{ee}, H^{oe}, H^{eo}, and H^{oo}. Using this notation, we can state the following equations:

$$F_{u,v} = F^{ee}_{u,v} + F^{oe}_{u,v}e^{-i2\pi u/N} + F^{eo}_{u,v}e^{-i2\pi v/N} + F^{oo}_{u,v}e^{-i2\pi(u+v)/N} \tag{10}$$

$$F^q_{u,v} = F^{ee}_{u,v} + e^{-i2\pi u/N}F^{oe}_{u,v} + F^{eo}_{u,v}e^{-j2\pi v/N} + e^{-i2\pi u/N}F^{oo}_{u,v}e^{-i2\pi v/N} \tag{11}$$

$$H_{u,v} = H^{ee}_{u,v} + H^{eo}_{u,v}\cos(2\pi v/N) + H^{eo}_{u,\bar{v}}\sin(2\pi v/N) + H^{oe}_{u,v}\cos(2\pi u/N) \tag{12}$$
$$+H^{oe}_{\bar{u},v}\sin(2\pi u/N) + H^{oo}_{u,v}\cos(2\pi(u+v)/N) + H^{oo}_{\bar{u},\bar{v}}\sin(2\pi(u+v)/N)$$

where $N = 2^k \geq 2$, $\bar{u} = N - u$ and $\bar{v} = N - v$.

Note 2. We cannot apply the n-D decimation method to the *Clifford Fourier transform* (CFT), since this method requires a commutative algebra if $n > 2$. The CFT is the generalization of the QFT for higher dimensions [2]. This problem concerning the decimation can be solved by embedding the transform into a different, commutative algebra [7].

3.3 The Mappings between the Spectra

Since we can convert each complete transform into each other, we have six mappings which describe the relation between the DHT, the DQFT and the DFT of a real signal. We define four operators, which yield the even and odd part of a 2-D signal wrt. the x- and the y-coordinate

$$\begin{aligned}
\text{Ee}(f(x,y)) &= \tfrac{1}{4}(f(x,y) + f(-x,y) + f(x,-y) + f(-x,-y)) \\
\text{Eo}(f(x,y)) &= \tfrac{1}{4}(f(x,y) + f(-x,y) - f(x,-y) - f(-x,-y)) \\
\text{Oe}(f(x,y)) &= \tfrac{1}{4}(f(x,y) - f(-x,y) + f(x,-y) - f(-x,-y)) \\
\text{Oo}(f(x,y)) &= \tfrac{1}{4}(f(x,y) - f(-x,y) - f(x,-y) + f(-x,-y)) \ .
\end{aligned} \tag{13}$$

Using this operators, we obtain the following Table 1.

Note that the row-column method applied to 1-D FHTs yields a transform which differs from the 2-D DHT wrt. $Oo(H)$ only. Let H denote the transform which is calculated by the RC-FHT1. Then, $Oo(H) = -Oo(H)$. Hence, H is a complete representation and we can obtain the DQFT by applying the RC-FHT1 algorithm

transform	relation to DHT
$F =$	$Ee(H) - iEo(H) - iOe(H) + Oo(H)$
$F^q =$	$Ee(H) - jEo(H) - iOe(H) - kOo(H)$
	relation to DFT
$H =$	$Ee(F) + iEo(F) + iOe(F) + Oo(F)$
$F^q =$	$Ee(F) + kEo(F) + Oe(F) - kOo(F)$
	relation to DQFT
$H =$	$Ee(F^q) + jEo(F^q) + iOe(F^q) + kOo(F^q)$
$F =$	$Ee(F^q) - kEo(F^q) + Oe(F^q) + kOo(F^q)$

Table 1: Relations between the transforms

$$F^q = Ee(H) - jEo(H) - iOe(H) + kOo(H) \ . \tag{14}$$

4 Complexities

In this section, we consider the complexities of the presented algorithms, in order to decide which one is the fastest. Since the DFT and the DQFT of a real signal contain *redundancy*, we present one approach which reduces the redundancy in order to speed up the algorithm. Besides the theoretic complexities, we consider the execution time of real implementations.

4.1 Optimizations

First of all, we have one effect which can be used to speed up *any* of the presented transforms. We know that a phase-shift by π yields a change of sign. Due to this fact, we can calculate the values at the frequencies u and $u + N/2$ using the same addends, e.g.

$$\begin{pmatrix} F_u \\ F_{u+N/2} \end{pmatrix} = \begin{pmatrix} 1 & 1 \\ 1 & -1 \end{pmatrix} \begin{pmatrix} F_u^e \\ e^{-i2\pi u/N} F_u^o \end{pmatrix} \ . \tag{15}$$

Consequently, the number of multiplications for each n-D fast transform is divided by 2^n. Furthermore, we decrease the number of additions for the FHT since we have

$$\begin{pmatrix} H_u \\ H_{u+N/2} \end{pmatrix} = \begin{pmatrix} 1 & 1 \\ 1 & -1 \end{pmatrix} \begin{pmatrix} H_u^e \\ H_u^o \cos(2\pi u/N) + H_{N-u}^o \sin(2\pi u/N) \end{pmatrix} \tag{16}$$

in the 1-D case.

Due to Hermite symmetry, we have a redundancy of 50% for the DFT and of 75% for the DQFT. We can use the Hermite symmetry directly by copying the data and changing the signs in the imaginary parts. This method yields a lower complexity of arithmetic operations, but increases the number of memory accesses. Therefore, it is advantageous to use an implicit method, called *overlapping* [4].

The idea of the latter is to create a new, complex signal $\bar{f} = f^e + if^o$ of the length $N/2$. Since the DFT is linear, the spectra F^e and F^o can be extracted from the spectrum $\bar{F}$ of $\bar{f}$ ($\bar{F} = F^e + iF^o$). Hence, the complexity is divided by an asymptotic factor of two. For the FQFT, we can apply the same approach, i.e. we have $\bar{f} = f^{ee} + if^{oe} + jf^{eo} + kf^{oo}$. This increases the asymptotic complexity by four.

4.2 Evaluation of the Complexities

Since modern processors evaluate floating point multiplications as fast as additions [11, 5, 12], we do not only consider the number of floating point multiplications, but also the number of additions and the total number of floating point operations (flops). Since we can easily convert one spectrum to another, it is not crucial for the estimation of the complexity, if the calculated spectrum is real, complex or quaternionic. Nevertheless, our aim is to develop the fastest way to calculate the DQFT.

In the Table 2, the flops of the presented 1-D fast algorithms can be found. Every algorithm uses the effect of the π-phase, and the FFT1 algorithm for real signals uses overlapping. For the sake of clarity we suppose that overlapping halves the complexity. Note that this is *not* true for finite signal lengths. The complexity is reduced by a factor *less* than two in this case.

transform	multiplications	additions	flops
FHT1	$N \log_2 N$	$\frac{3}{2} N \log_2 N$	$\frac{5}{2} N \log_2 N$
FFT1 ($\mathbb{R}$)	$N \log_2 N$	$\frac{3}{2} N \log_2 N$	$\frac{5}{2} N \log_2 N$
FFT1 ($\mathbb{C}$)	$2N \log_2 N$	$3N \log_2 N$	$5N \log_2 N$

Table 2: Complexities of the presented 1-D algorithms

In the Table 3, the flops of the presented 2-D decimation algorithms and the row-column algorithms can be found. Every algorithm uses the effect of the π-phase, where the matrix of combinations of the signs can be separated [9]:

$$\begin{pmatrix} 1 & 1 & 1 & 1 \\ 1 & -1 & 1 & -1 \\ 1 & 1 & -1 & -1 \\ 1 & -1 & -1 & 1 \end{pmatrix} = \begin{pmatrix} 1 & 0 & 1 & 0 \\ 0 & 1 & 0 & 1 \\ 1 & 0 & -1 & 0 \\ 0 & 1 & 0 & -1 \end{pmatrix} \begin{pmatrix} 1 & 1 & 0 & 0 \\ 1 & -1 & 0 & 0 \\ 0 & 0 & 1 & 1 \\ 0 & 0 & 1 & -1 \end{pmatrix} . \tag{17}$$

The separation decreases the number of additions needed for the four quadrants by two thirds. Furthermore, the FQFT algorithm for real signals uses overlapping. For the sake of clarity we provide that overlapping quarters the complexity. Again, this is *not* true for finite signal sizes.

transform	multiplications	additions	flops
RC-FHT1	$2N^2 \log_2 N$	$3N^2 \log_2 N$	$5N^2 \log_2 N$
RC-FFT1 ($\mathbb{R}$)	$3N^2 \log_2 N$	$\frac{9}{2} N^2 \log_2 N$	$\frac{15}{2} N^2 \log_2 N$
RC-FFT1 ($\mathbb{H}$)	$8N^2 \log_2 N$	$12N^2 \log_2 N$	$20N^2 \log_2 N$
FHT2	$\frac{3}{2} N^2 \log_2 N$	$\frac{11}{4} N^2 \log_2 N$	$\frac{17}{4} N^2 \log_2 N$
FFT2 ($\mathbb{H}$)	$6N^2 \log_2 N$	$11N^2 \log_2 N$	$17N^2 \log_2 N$
FQFT ($\mathbb{R}$)	$2N^2 \log_2 N$	$2N^2 \log_2 N$	$4N^2 \log_2 N$
FQFT ($\mathbb{H}$)	$8N^2 \log_2 N$	$8N^2 \log_2 N$	$16N^2 \log_2 N$

Table 3: Complexities of the presented 2-D algorithms

Note that the RC-FFT1 (IR) algorithm uses three overlapping FFT1s since in (7) $f^{jk} = 0$ and both f^R and f^I are real valued. The FFT2 (IR) is omitted since overlapping cannot be applied in this case, unless the DQFT is used [4].

4.3 Comparison of the Algorithms

According to Table 2, the fastest algorithm for calculating the 1-D Fourier transform of a real signal is the FHT1 or the FFT1 with overlapping. Since the overlapping yields a high linear complexity, we prefer the FHT1.

In a real implementation, the time for the memory access cannot be neglected. Therefore, the FHT1 and the FFT1 with overlapping are not twice as fast as the FFT1 without overlapping. Our implementation of the FHT1 takes less than two thirds of the execution time of the FFT1 from the numerical recipes [10], but it takes more than one half of the time (e.g. 5.0 s versus 8.9 s).

For complex signals, the FFT1 is the fastest algorithm. The execution of two FHT1s takes more execution time than one FFT1.

The quaternionic Fourier transform of a 2-D signal can be calculated in many ways. According to Table 3, we have the following ranking of algorithms for real signals: FQFT, FHT2, RC-FHT1, RC-FFT1. Note that for the execution time we have another ranking. The overlapping in the FQFT algorithm yields a high quadratic complexity. Hence, for realistic sizes the Hartley transforms have lower complexities. In real implementations, the RC-FHT1 algorithm takes the least execution time, since the memory access of the FHT2 is more complicated. Depending on the implementation, we have the execution time ranking RC-FHT1, FHT2, RC-FFT1, FQFT (Fig. 1).

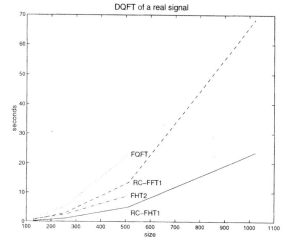

Fig. 1: Execution time for the estimation of the DQFT on a SPARCstation 10

For quaternionic signals, we have the following ranking according to Table 3: FQFT, FFT2, RC-FFT1. Though the FFT2 should be faster than the RC-FFT1 (as well for the evaluation of the DQFT as for the DFT), most implementations of the FFT2 are RC-FFT1 algorithms [10]. Depending on the implementation, we have the execution time ranking RC-FFT1, FFT2[1], FQFT[2].

For both cases of the DQFT (real and quaternionic signal), the row-column method has the advantage that both coordinates are independently extended to a power of two. For the 2-D decimation algorithms, the signal must in addition be quadratic.

---

[1] We provide that the FFT2 is implemented by FFT1s.

[2] The memory access is distributed over the whole data, which yields a slower execution.

5 Conclusion

We have presented different algorithms for evaluating the DQFT of real and quaternionic signals. The comparison of these algorithms, their complexities and their execution times yield some characterizing properties:

- We have algorithms which are based on well known transforms (e.g. on the FFT1) and we have some which are new (FQFT).
- Some algorithms are modular (row-column) and some are monolithic.
- The implementation can yield a simple and short code (FFT1) or a complicated and extensive code (FQFT with overlapping).
- The algebraic representation can be simple (FHT1) or complicated (FQFT).

In our experiments we ascertained that those algorithms which can be implemented most easily and which are modular have the shortest execution time though their theoretic complexity is not the lowest. Obviously, the implementations are faster, if they use simple algebraic representations. On the other hand, the FHT1 (6) has more addends than the FFT1 (3). Hence, the FFT1 is the optimal transform for signals which are not real valued.

Since we can construct a very fast algorithm using the standard FFT1, there is only little programming effort to obtain an implementation which evaluates the DQFT. For real signals, the FHT1 should be used instead of the FFT1, which yields a shorter execution time and less programming effort than an implementation of overlapping.

References

[1] R. N. Bracewell. *The Fourier transform and its applications.* McGraw Hill, 1986.
[2] T. Bülow and G. Sommer. Algebraically Extended Representation of Multi-Dimensional Signals. In *Proceedings of the 10th Scandinavian Conference on Image Analysis*, pages 559–566, 1997.
[3] T. Bülow and G. Sommer. Multi-Dimensional Signal Processing Using an Algebraically Extended Signal Representation. In G. Sommer and J.J. Koenderink, editors, *Int'l Workshop on Algebraic Frames for the Perception-Action Cycle, AF-PAC'97, Kiel*, volume 1315 of *LNCS*, pages 148–163. Springer, 1997.
[4] V. M. Chernov. Discrete orthogonal transforms with data representation in composition algebras. In *Proceedings of the 9th Scandinavian Conference on Image Analysis*, pages 357–364, 1995.
[5] Digital Equipment Corporation. Digital Semiconductor Alpha 21164PC Microprocessor Data Sheet[3], 1997.
[6] T. A. Ell. *Hypercomplex Spectral Transformations.* PhD thesis, University of Minnesota, 1992.
[7] M. Felsberg. Signal Processing Using Frequency Domain Methods in Clifford Algebra[4]. Master's thesis, Christian-Albrechts-University of Kiel, 1998.
[8] M. Felsberg et al. Fast Algorithms of Hypercomplex Fourier Transforms. In G. Sommer, editor, *Geometric Computing with Clifford Algebra*, Springer Series in Information Sciences. Springer, Berlin, 1999. to appear.
[9] B. Jähne. *Digitale Bildverarbeitung.* Springer, Berlin, 1997.
[10] W. Press et al. *Numerical Recipes in C.* Cambridge University Press, 1994.
[11] Silicon Graphics, Inc. MIPS RISC Technology R10000 Microprocessor Technical Brief[5], 1998.
[12] Sun Microsystems, Inc. UltraSPARC-II Data Sheet[6], 1998.

---

[3] http://ftp.digital.com/pub/DECinfo/semiconductor/literature/164pcds.pdf
[4] http://www.ks.informatik.uni-kiel.de/~mfe/research.html
[5] http://www.sgi.com/processors/r10k/tech_info/Tech_Brief.html
[6] http://www.sun.com/microelectronics/datasheets/stp1031/index2.html

Image Block Coding Based on New Algorithms of Shortlength DCT with Minimal Multiplicative Complexity

Marina A. Chichyeva, Vladimir M. Chernov

Image Processing System Institute of RAS
151, Molodogvardejskaya st., 443001, Samara, Russia
Fax: +7 (8462) 322763
e-mail: mchi@smr.ru, vche@smr.ru

Abstract. A quality of block coding algorithms based on the DCT with use of $N{\times}N$-blocks ($8{\le}N{<}16$) is researched in the paper. A time of the image compression does not increase at increasing of block size. It is achieved due to decrease of computational complexity by means of a new approach to synthesis of the short-length DCT algorithms. This approach is connected to interpretation of the DCT calculation as operations in associated algebraic structures.

1   Introduction

A discrete cosine transform (DCT) [1]:

$$\tilde{x}(m)=\lambda_m \sum_{n=0}^{N-1}x(n)\cos\left(\frac{\pi\left(n+\frac{1}{2}\right)m}{N}\right), \ (m=0,...,N\text{-}1) \tag{1}$$

is widely used in image processing, in particular, the DCT of length $N=8$ is a base of a lot of the modern coding standards such as JPEG, MPEG, ITU-T [1, 2] (λ_m are normalizing coefficients). However a method of block coding based on the DCT (and so the mentioned standards) does not completely take into account features of a concrete image. Usually, a creation of adaptive algorithms of a DCT coding [3] leads to increase of computational complexity as a result of DCT size increasing. It is connected to the fact that the fast algorithms (FA) of DCT at $N>8$ are much less investigated than the DCT of length 8, which accounts for a large number of publications [4, 5].

The most of the known FAs DCT synthesized for arbitrary lengths have comparatively low *asymptotic* arithmetic complexity, but they do not take into account the specific of "short" lengths: comparatively small number of different values of basic functions, high requirements to *structural* complexity, which is the dominant characteristic of speed at small length of transform, etc.

Acknowledgements. This work was performed with financial support from the Russian Foundation of Fundamental Investigations (Grant No 97-01-009000)

For example, the method of reduction of the DCT to the discrete Fourier transform (DFT) of a real double-length sequence proposed in [1] requires 34 operations of real multiplication and 56 operations of real addition [6] at $N=8$ and using of split-radix algorithm of a real DFT of length 16. The specific algorithm of work [4] requires only 12 multiplications and 29 additions.

This work is aimed at the research of quality of block coding algorithms with use of $N \times N$-blocks ($8 \leq N < 16$). The decrease of a computational complexity of the FA DCT (and, hence, the time of the image compression/restoration) is achieved with the help of a new approach to the synthesis of such algorithms. The principles of the approach were specified in [7, 8, 9]. It was connected to an interpretation of the calculation of the DCT as operations in the associated algebraic structures (some finite-dimensional algebras). In this work the authors obtain and use the algorithms with the minimal arithmetic complexity following these principles.

2 Algebraic principles of synthesis of fast algorithms of short-length DCT

The proposed method of synthesis of the fast algorithms of the discrete cosine transform is based on the following ideas.

1. The DCT matrix has *a block structure*. The multiplication of such a matrix by an input vector is reduced to multiplications of matrixes of the smaller sizes with specific properties of "symmetry" by vectors from subspaces of signal space.
2. The multiplication of these submatrices by vectors of the corresponding subspaces is equivalent to a multiplication of elements in some *finite-dimensional algebras*.
3. In the most of considered cases (but not in all) these algebras are group algebras of cyclical groups or their quotient algebras. The multiplication of elements in such algebras is equivalent to a multiplication in *polynomial rings* (or a cyclical convolution).

3 Finite-dimensional algebras associated with FA DCT

We consider some d-dimensional algebras $\mathbb{A}$ over $\mathbb{R}$ with a basis $\{1, e_1, ..., e_{d-1}\}$.

The typical element $a \in \mathbb{A}$ can be represented in the form $a = \alpha_0 + \sum_{j=1}^{d-1} \alpha_j e_j$.

An addition of elements is performed component-wise and a multiplication is induced by rules of the basic elements multiplication.

In this paper the following finite-dimensional algebras are considered.

1. The two-dimensional algebra $\mathbb{A}_1^{(2)}$ with the basis $\{1, e_1\}$ and rules of the multiplication of the basic elements: $1^2 = 1$, $e_1^2 = -1$.

(The algebra $\mathbb{C}$ of complexes).

2. The two-dimensional algebra $A_2^{(2)}$ with the basis $\{1, e_1\}$ and rules of the multiplication of the basic elements: $1^2 = 1$, $e_1^2 = 1$.

(The algebra of "double" numbers is isomorphic to direct sum $\mathbb{R} \oplus \mathbb{R}$ [10])

3. The three-dimensional algebra $A^{(3)}$ with the basis $\{1, e_1, e_2\}$ and rules of the multiplication of the basic elements: $e_1^2 = e_2$, $e_2^2 = -e_1$, $e_1 e_2 = e_2 e_1 = -1$.

4. The four-dimensional algebra $A_1^{(4)}$ with the basis $\{1, e_1, e_2, e_3\}$ and rules of the multiplication of the basic elements:

$$e_1^2 = -e_2, \ e_2^2 = -1, \ e_3^2 = e_2, \ e_1 e_2 = e_2 e_1 = e_3, \ e_2 e_3 = e_3 e_2 = -e_1, \ e_1 e_3 = e_3 e_1 = 1.$$

5. The four-dimensional algebra $A_2^{(4)}$ with the basis $\{1, e_1, e_2, e_3\}$ and rules of the multiplication of the basic elements:

$$e_1^2 = e_2, \ e_2^2 = 1, \ e_3^2 = e_2, \ e_1 e_2 = e_2 e_1 = e_3, \ e_2 e_3 = e_3 e_2 = e_1, \ e_1 e_3 = e_3 e_1 = 1.$$

6. The four-dimensional algebra $A_3^{(4)}$ with the basis $\{1, e_1, e_2, e_3\}$ and rules of the multiplication of the basic elements:

$$e_1^2 = -1, \ e_2^2 = 1, \ e_3^2 = -1, \ e_1 e_2 = e_2 e_1 = -e_3, \ e_2 e_3 = e_3 e_2 = -e_1, \ e_1 e_3 = e_3 e_1 = e_2.$$

Proposition 1. a) The multiplication of elements $(\alpha_0 + \alpha_1 e_1)$, $(\beta_0 + \beta_1 e_1) \in A_1^{(2)}$ is equivalent to the multiplication of the polynomials $(\alpha_0 + \alpha_1 t)(\beta_0 + \beta_1 t)(\mod(t^2 + 1))$ and requires 3 real multiplications and 3 real additions according to [11].

b) The multiplication of elements $(\alpha_0 + \alpha_1 e_1)$, $(\beta_0 + \beta_1 e_1) \in A_2^{(2)}$ is equivalent to the multiplication of the polynomials $(\alpha_0 + \alpha_1 t)(\beta_0 + \beta_1 t)(\mod(t^2 - 1))$ and requires 2 real multiplications and 4 real additions according to [11].

c) The multiplication of elements $(\alpha_0 + \alpha_1 e_1 + \alpha_2 e_2)$, $(\beta_0 + \beta_1 e_1 + \beta_2 e_2) \in A^{(3)}$ is equivalent to the multiplication of the polynomials

$$(\alpha_0 + \alpha_1 t + \alpha_2 t^2)(\beta_0 + \beta_1 t + \beta_2 t^2)(\mod(t^3 + 1))$$

and requires 4 real multiplications and 14 real additions according to [11].

d) The multiplication of elements $(\alpha_0 + \alpha_1 e_1 + \alpha_2 e_2 + \alpha_3 e_3)$, $(\beta_0 + \beta_1 e_1 + \beta_2 e_2 + \beta_3 e_3) \in A_1^{(4)}$ is equivalent to the multiplication of the polynomials

$$(\alpha_0 + \alpha_1 t - \alpha_2 t^2 - \alpha_3 t^3)(\beta_0 + \beta_1 t - \beta_2 t^2 - \beta_3 t^3)(\mod(t^4 + 1))$$

and requires 9 real multiplications and 15 real additions according to [11].

e) The multiplication of elements $(\alpha_0 + \alpha_1 e_1 + \alpha_2 e_2 + \alpha_3 e_3)$, $(\beta_0 + \beta_1 e_1 + \beta_2 e_2 + \beta_3 e_3) \in A_2^{(4)}$ is equivalent to the multiplication of the polynomials

$$(\alpha_0 + \alpha_1 t + \alpha_2 t^2 + \alpha_3 t^3)(\beta_0 + \beta_1 t + \beta_2 t^2 + \beta_3 t^3)(\mod(t^4 - 1))$$

and requires 5 real multiplications and 15 real additions according to [11].

Proposition 2. The algebra $A_3^{(4)}$ is isomorphic to the direct sum $\mathbb{C} \oplus \mathbb{C}$:

$$C \oplus C = \left\{ (z_1, z_2) : z_1 = a_1 + b_1 i_1, z_2 = a_2 + b_2 i_2, i_1^2 = i_2^2 = -1, a_1, a_2, b_1, b_2 \in \mathbb{R} \right\}$$

Corollary 1. A multiplication of a constant element $a = \alpha_0 + \alpha_1 e_1 + \alpha_2 e_2 + \alpha_3 e_3$ by a vector $b = \beta_0 + \beta_1 e_1 + \beta_2 e_2 + \beta_3 e_3$ of the algebra $\mathbb{A}_3^{(4)}$ requires 6 real multiplications and 10 real additions.

4   The base example: DCT of length 8

In the non-normalized matrix form the DCT (1) takes the form $\mathbf{X} = \mathbf{Fx}$, where $\mathbf{X}^t = (X(0), \ldots, X(7))$, $\mathbf{x}^t = (x(0), \ldots, x(7))$, $\mathbf{F}$ is the DCT matrix, t is the transposition sign. After rearranging components of the input and output vectors the matrix representation of the DCT can be reduced to the form $\mathbf{Y} = \mathbf{Ty}$, where

$$\mathbf{T} = \left(\begin{array}{cccc|cccc}
a & c & -d & -b & -a & -c & d & b \\
c & d & -b & a & -c & -d & b & -a \\
d & b & a & c & -d & -b & -a & -c \\
b & -a & c & d & -b & a & -c & -d \\
\hline
f & -l & -f & l & f & -l & -f & l \\
l & f & -l & -f & l & f & -l & -f \\
\hline
g & -g & g & -g & g & -g & g & -g \\
1 & 1 & 1 & 1 & 1 & 1 & 1 & 1
\end{array}\right),$$

$$a = \cos\left(\frac{\pi}{16}\right),\ b = \cos\left(\frac{3\pi}{16}\right),\ c = \cos\left(\frac{5\pi}{16}\right),\ d = \cos\left(\frac{7\pi}{16}\right),\ l = \cos\left(\frac{3\pi}{8}\right),\ f = \cos\left(\frac{\pi}{8}\right),\ g = \cos\left(\frac{\pi}{4}\right).$$

The forming of the auxiliary variables from the components of a vector $\mathbf{y}$:

$$z(0) = y(0) - y(4),\ z(1) = y(1) - y(5),\ z(2) = y(2) - y(6),\ z(3) = y(3) - y(7),$$
$$z(4) = (y(0) + y(4)) - (y(2) + y(6)),\ z(5) = (y(1) + y(5)) - (y(3) + y(7)),$$
$$z(6) = [(y(0) + y(4)) + (y(2) + y(6))] - [(y(1) + y(5)) + (y(3) + y(7))],$$
$$z(7) = [(y(0) + y(4)) + (y(2) + y(6))] + [(y(1) + y(5)) + (y(3) + y(7))].$$

requires 14 operations of real addition. Then the implementation of the cosine transform is reduced to the following matrix calculations:

$$\begin{pmatrix} Y(0) \\ Y(1) \\ Y(2) \\ Y(3) \end{pmatrix} = \begin{pmatrix} a & c & -d & -b \\ c & d & -b & a \\ d & b & a & c \\ b & -a & c & d \end{pmatrix} \begin{pmatrix} z(0) \\ z(1) \\ z(2) \\ z(3) \end{pmatrix},\quad \begin{pmatrix} Y(4) \\ Y(5) \end{pmatrix} = \begin{pmatrix} f & -l \\ l & f \end{pmatrix} \begin{pmatrix} z(4) \\ z(5) \end{pmatrix},\quad \begin{array}{l} Y(6) = g z(6), \\ Y(7) = z(7). \end{array}$$

Proposition 3. a) The calculation of the first matrix product is equivalent to the calculation of the product of elements $s, p \in \mathbb{A}_1^{(4)}$:

$$sp = (c + ae_1 + be_2 + de_3)(z(0) + z(1)e_1 + z(2)e_2 + z(3)e_3).$$

According to Proposition 1(d) it requires 9 operations of real multiplication and 15 operations of real addition.

b) The calculation of the second matrix product is equivalent to the calculation of the product of elements $q, r \in A_1^{(2)}$: $qr = \left(f + le_1\right)\left(z(4) + z(5)e_1\right)$.

According to Proposition 1(a) it requires 3 operations of real multiplication and 3 operations of real addition.

c) The other calculations require one operation of real multiplication.

The total complexity of the DCT algorithm of the length 8 including the forming of the auxiliary variables $z_0, ..., z_7$ is equal to $9+3+1=13$ operations of multiplication and $14+15+3=32$ operations of addition.

Remark. The structure of the considered algorithm does not depend on particular values of the parameters $a, b, ..., g$. Let $g' = l' = c' = 1$, $f' = f/_l$, $a' = a/_c$, $d' = d/_c$, $b' = b/_c$.

Then the part of multiplications becomes trivial. Multiplications by g, e, c are united with the normalization of the DCT-spectrum components (with multiplications by λ_m) in (1). Thus the proposed DCT algorithm of length 8 requires $2+8=10$ operations of multiplication and 32 operations of addition.

5   DCT algorithms for $N = 9, 10, 12, 15$

After permutation of rows and columns the DCT matrices of lengths 9 and 10 take the form

$$
T_9 = \begin{pmatrix}
a & c & d & -d & -c & -a & b & -b & 0 \\
c & -d & a & -a & d & -c & -b & b & 0 \\
d & a & -c & c & -a & -d & -b & b & 0 \\
e & -f & -g & -g & -f & e & h & h & -1 \\
g & -e & f & f & -e & g & -h & -h & 1 \\
f & g & -e & -e & g & f & -h & -h & 1 \\
b & -b & -b & b & b & -b & 0 & 0 & 0 \\
h & h & h & h & h & h & -1 & -1 & -1 \\
1 & 1 & 1 & 1 & 1 & 1 & 1 & 1 & 1
\end{pmatrix},
$$

Proposition 4. A calculation of the matrix products

$$
\begin{pmatrix} a & c & d \\ c & -d & a \\ d & a & -c \end{pmatrix}\begin{pmatrix} z(0) \\ z(1) \\ z(2) \end{pmatrix}, \quad \begin{pmatrix} e & -f & -g \\ g & -e & f \\ f & g & -e \end{pmatrix}\begin{pmatrix} z(5) \\ z(6) \\ z(7) \end{pmatrix},
$$

is equivalent to the calculation of the products of elements of the algebra $A^{(3)}$:

$$
\left(-ae_2 - ce_1 + d\right)\left(z(0)e_1 + z(1)e_2 + z(2)\right) \text{ and } \left(fe_2 - ge_1 - e\right)\left(z(6)e_2 - z(5)e_1 + z(7)\right)
$$

respectively and according to Proposition 1(c) requires 4 real multiplications and 14 real additions for each product.

Corollary 2. The DCT of length 9 requires 8 multiplications and 44 additions or less than one multiplication and about five additions per pixel.

Analogous propositions can be proved for $N=10, 12, 15$.

Proposition 5. A multiplication of the DCT matrix for $N=10$ by an input vector is equivalent to

a) a multiplication of a variable element of the algebra $A_2^{(4)}$ by a constant element of the same algebra;

b) a multiplication of a variable element of the algebra $A_1^{(2)}$ by a constant element of the same algebra;

c) a multiplication of a variable element of the algebra $A_2^{(2)}$ by a constant element of the same algebra;

d) additional multiplications of constants by variables and auxiliary additions.

Corollary 3. The DCT of length 10 requires 9 multiplications and 43 additions.

Proposition 6. A multiplication of the DCT matrix for $N=12$ by an input vector is equivalent to

a) a multiplication of a variable element of the algebra $A_3^{(4)}$ by a constant element of the same algebra;

b) a multiplication of a variable element of the algebra $A_1^{(2)}$ by a constant element of the same algebra;

c) a multiplication of a variable element of the algebra $A_2^{(2)}$ by a constant element of the same algebra;

d) additional multiplications of constants by variables and auxiliary additions.

Corollary 4. The DCT of length 12 requires 13 multiplications and 55 additions.

Proposition 7. A multiplication of the DCT matrix for $N=15$ by an input vector is equivalent to

a) two multiplications of variable elements of the algebra $A_2^{(4)}$ by constant elements of the same algebra;

b) a multiplication of a variable element of the algebra $A_1^{(2)}$ by a constant element of the same algebra;

c) a multiplication of a variable element of the algebra $A_2^{(2)}$ by a constant element of the same algebra;

d) additional multiplications of constants by variables and auxiliary additions.

Corollary 5. The DCT of length 15 requires 24 multiplications and 83 additions.

The algorithm of Ref. [12] synthesized for arbitrary lengths was used as a base for the comparative analysis of computational complexity. At $N = 2^k$ the estimates of a computational complexity of this algorithm coincide with the estimates of the complexity of the best DCT algorithms from the known ones [4, 5].

Table contains the number of operations for the implementation of the DCT by the proposed and known methods.

N	Proposed algorithms		The algorithm of Ref. [12]	
	*	+	*	+
8	10	32	12	29
9	8	44	11	44
10	9	43	15	36
12	13	51	20	43
15	21	82	35	89

6   Discussion of experimental results

In the publications [13] the block coding method using the discrete cosine transform is described. An input image is divided into square blocks of $N{\times}N$ pixels with the cosine spectrum to be calculated, and selecting and coding of the resulting transforms to be performed for each block. A set of selected and coded transforms forms the contents of compressed data.

At decoding the values of transforms are reconstructed, the inverse DCT is performed for each block, and a reconstructed image is formed.

Below the mean square error introduced in data by the procedure of compression/restoration is termed a compression error.

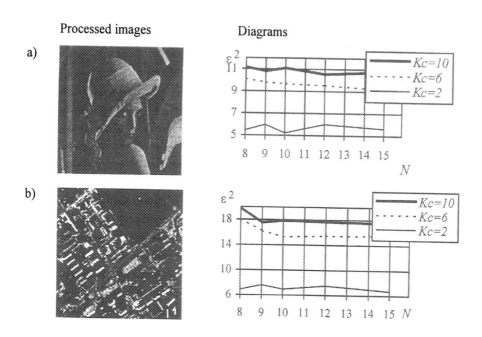

Fig. 1. The dependencies of the compression errors versus block size for image:
a) "Portrait"; b) "Airspace view".

Fig. 1 depicts the compression errors ε^2 with pregiven coefficients of compression Kc for various N. An analysis of shown dependencies allows us to conclude that there is in some sense an optimal block size for one or another class of images. The proposed DCT algorithms guarantee that a speed of image processing is closed on a speed of image processing by the best of known algorithms of the DCT for $N=8$. The research of the time of image processing by block DCT is shown that synthesized algorithms guarantees the time close to the best of known DCT algorithms of length 8.

In the present work a dispersion method selecting of transforms was applied. The use of selecting method analogous to JPEG standard instead of dispersion method could additionally increase the speed of processing. Unfortunately, nobody considers such standards for $N\neq8$. The authors consider a development of such standards as perspective direction of further investigations.

References

1. Rao, K. R., Yip, P.: Discrete Cosine Transform. Academic Press, San Diego, 1990.
2. Wallace, G. K.: The JPEG still picture compression standard. //Communications of the ACM, Vol. 34, No 4, pp. 31-44, 1991.
3. Krupiczka, A.: Interblock variance as a segmentation criterion in image coding. //Mashine Graphics and Vision, Vol. 5, Nos 1/2, pp. 229-235, 1996
4. Chan-Wan, Y.-H., Siu, C.: On the realization of discrete cosine transform using the distributed arithmetic. //IEEE Transactions on Circuits and Systems - I: Fundamental Theory and Applications, Vol. 39, No 9, pp. 705-712, 1992
5. Hou, H. S., Tretter, D. K.: Interesting properties of the discrete cosine transform. //J. Visual Commun. and Image Represent., Vol. 3, No 1, pp. 73-83, 1992
6. Duhamel, P.: Implementation of split-radix FFT algorithms for complex, real, and real-symmetric data. //IEEE Trans. Accoust., Speech, and Signal Process., 1984, Vol.32, No 4, pp.750-761.
7. Feig, E., Ben-Or, M.: On algebras related to the discrete cosine transform. //Linear Algebra and Its Applications, Vol. 266, pp. 81-106, 1997
8. Bazensky, G., Tasche, M.: Fast polynomial multiplication and convolutions related to the discrete cosine transform. //Linear Algebra and Its Applications, Vol. 252, Issue 1-3, pp. 1-25, 1997
9. Chichyeva, M. A., Chernov, V. M.: "One-step" short-length DCT algorithms with data representation in the direct sum of associative algebras. //Proceedings CAIP'97, Springer, LNCS 1296, pp.590-596, 1997
10. Jacobson, N.: Structure and representation of Jordan algebras. Providense, R.I., 1968
11. Nussbaumer, H. J.: Fast Fourier transform and convolution algorithms. Springer Verlag, 1971.
12. Chan, S.-C., Ho, K.-L.: Fast algorithms for computing the discrete cosine transform. //IEEE Trans. on Circuits and Systems, Vol. 39, No 3, pp.185-190, 1992
13. Pratt, W. K.: Digital Image Processing, Vol.2. A Wiley-Interscience Publication, New York, 1978.

Object Recognition with Representations Based on Sparsified Gabor Wavelets Used as Local Line Detectors

Norbert Krüger

Institut für Informatik und praktischer Mathematik
Christian-Albrechts-Universität zu Kiel
Preußerstrasse 1-9, 24105 Kiel
Germany
nkr@ks.informatik.uni-kiel.de

Abstract. We introduce an object recognition system (called ORAS-SYLL) in which objects are represented as a sparse and spatially organized set of local (bent) line segments. The line segments correspond to binarized Gabor wavelets or banana wavelets, which are bent and stretched Gabor wavelets. These features can be metrically organized, the metric enables an efficient learning of object representations. Learning can be performed autonomously by utilizing motor–controlled feedback. The learned representation are used for fast and efficient localization and discrimination of objects in complex scenes.

ORASSYLL has been heavily influenced by an older and well known vision system [4, 9], and has also been influenced by Biederman's comments to this older system [1]. A comparison of ORASSYLL and the older system, including some remarks about the specific role of Gabor wavelets within ORASSYLL, is given at the end of the paper.

1 Introduction

In this paper we describe a novel object recognition system called ORAS-SYLL (**O**bject **R**ecognition with **A**utonomously learned and **S**parse **SY**mbolic representations based on **L**ocal **L**ine detectors). In ORASSYLL representations of object classes can be learned autonomously. The learned representations are used for a fast and efficient localization and identification of objects in complicated scenes.

We facilitate and guide learning by carefully selected a priori knowledge. Important constraints are the restriction of features to localized (bent) line segments (PF1), their metric organization (PF2), their hierarchical processing (PF3) and the sparse representations of objects by these features (see figure 1). Other contraints, discussed in detail in [3], are concerned with the division of the feature space in independent subspaces (PL1: Independence), its temporal organization (PL2: Correspondence) and statistical criteria for the evaluation of significant features for an object class (Invariance Maximization (PE1) and Redundancy Reduction (PE2)).

A crucial aspect of our work and an important difference to an older system [4, 9] is the specific application of Gabor wavelets as expressed in the constraints

PF1, PF2 and PF4. PF1 states that we apply Gabor wavelets as local line detectors, i.e., we assign a symbolic meaning to them. This meaning also enables the imbedding of Gabor wavelets into a metrically organized space (PF2) as well as for a sparse object representation (PF4).

Our representation of a certain view of an object class comprises only important features, learned from different examples (see figure 2 left). In section 2 we formalize PF1 by assigning a local line segment to Gabor wavelets or banana wavelets respectively (see figure 1a,b). In addition to the parameters frequency and orientation banana wavelets possess the properties curvature and elongation. The space of Gabor or banana wavelet reponses is very large. An object can be represented as a configuration of a few of these features, therefore it can be coded sparsely (PF4) (see figure 1c). The feature space can be understood as a metric space (PF2), its metric representing the similarity of features. This metric is essential for feature extraction and the learning algorithm (section 3). The banana wavelet responses can be derived from Gabor wavelet responses by hierarchical processing (PF3) to gain speed and reduce memory requirements. The sparse representation combined with the hierarchical feature processing allows a fast and effective locating (section 4).

In order to avoid the necessity of manual intervention for the generation of ground truth we equip the system with a mechanism which can produce controlled training data by moving an object with a robot arm and following the object by fixating the robot hand. The robot produces training data on which a certain view of an object is shown with varying background and illumination but with corresponding landmarks having the same pixel position in the image (see figure 2 left). We apply the learning algorithm to this data to extract an object representation (see figure 2 left,v). Another way to avoid manual intervention is one–shot learning (see figure 2 left), which already allows for the extraction of representations successfully applicable to difficult discrimination tasks.

ORASSYLL has been heavily influenced by an older and well known vision system [4, 9], and has also been influenced by Biederman's comments to this older system [1]. A comparison of ORASSYLL and the older system, including some remarks about the specific role of Gabor wavelets within ORASSYLL, is given in section 6.

2   The Feature Space

The principle PF1 gives us a significant reduction of the search space. Instead of allowing, e.g., all linear filters as possible features, we restrict ourself to a small

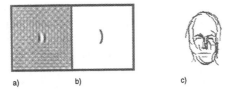

a)   b)   c)

Fig. 1. a: Arbitrary wavelet. b: Corresponding path. c: Visualization of a representation of an object class. Gabor or Banana wavelets with lower frequencies are represented by line segments with larger width.

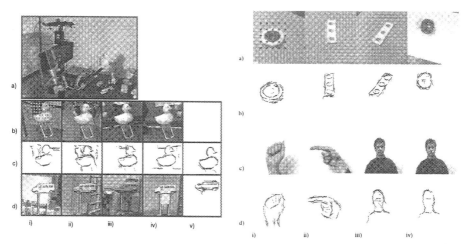

Fig. 2. Left: Autonomous learning: a) The robot arm with the camera. **b)** The "retinal" images produced by following the robot arm holding a toy–duck. **c,i–iv)** Significant Features per Instance extracted in an rectangular region (shown in b,i). **c,v)** Learned representation. **d)** Training data and learned representation for a toy car. **Right: One–shot learning:** Row a) and c) show the objects to be learned in front of homogeneous backgound. Row b) and d) show the extracted representations. For all objects a rectangular grid was roughly positioned on the object as in the first image a,i).

subset. Considering the risk of a wrong feature selection it is necessary to give good reasons for our decision. We argue that nearly any 2D–view of an object can be composed of localized curved lines. Furthermore, the fact that humans can easiliy handle line drawings of objects strengthens our assumption PF1.

Gabor and Banana Wavelets: A banana wavelet B^b is a complex–valued function, parameterized by a vector $\mathbf{b}$ of four variables $\mathbf{b} = (f, \alpha, c, s)$ expressing the attributes frequency (f), orientation (α), curvature (c) and size (s). It can be understood as a product of a rotated (and curved) complex wave function F^b and a stretched two–dimensional Gaussian G^b rotated (and bent) according to F^b (see figure 1a).

Our basic feature is the magnitude of the filter response extracted by a convolution with an image. A Gabor or banana wavelet B^b causes a strong response at pixel position $\mathbf{x}$ when the local structure of the image at that pixel position is similar to B^b (see [3]).

The Feature Space: The six–dimensional space of vectors $\mathbf{c} = (\mathbf{x}, \mathbf{b})$ is called the *feature space* with $\mathbf{c}$ representing the banana wavelet B^b with its center at pixel position $\mathbf{x}$ in an image. In [3] we define a metric $d(\mathbf{c}_1, \mathbf{c}_2)$. Two coordinates $\mathbf{c}_1, \mathbf{c}_2$ are expected to have a small distance d when their corresponding kernels are similar, i.e., they represent similar features (PF2).

Approximation of Banana Wavelets by Gabor Wavelets: To reduce computational requirements for the extraction of the large feature space we have defined an algorithm to approximate banana wavelets from Gabor wavelets and banana wavelet responses from Gabor wavelet responses (see [3]). By this hierarchical processing (PF3) we achieve a speed up to a factor 5.

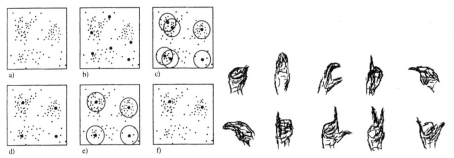

Fig. 3. Left: Clustering: a) Distribution of the significant features per instance extracted at a certain landmark. b) Codebook initialization. c) Codebook vectors after learning. d) Substituting sets of codebook vectors with small distance by their center of gravity. e) Counting the number of elements within a certain radius. f) Deleting codebook vectors representing insignificant features. **Right: Learned Representations of 10 different hand postures** The manually provided ground truth consists of 6 pictures per hand posture with a grid consisting of approximately 40 landmarks is placed.

3   Learning

Extracting Significant Features Per Instance: Our aim is to extract the local structure in an image I in terms of (curved) line segments expressed by Gabor or banana wavelets. We define a *significant feature per instance* of an object by two qualities. Firstly it has to cause a strong response, secondly it has to represent a maximum within a local area of the feature space. Figure 2 right b,c) and left c,i–iv) show the significant features per instance for some objects (each wavelet is described by a curve with same orientation, curvature and size).
One–shot learning: By positioning a rectangular grid on a roughly segmented object (see figure 2 left a,i) in front of homogeneous background and extracting significant features per instance as described above suitable representations of objects can already be extracted. These representations are successfully applied to difficult discrimination tasks.
Clustering: After extracting the significant features per instance in different pictures we apply an algorithm to extract invariant local features for a *class of objects*. Here the task is the selection of the *relevant features* for the object class from the noisy features extracted from our training examples (see figure 2 left c,i–iv). We assume the correspondence problem to be solved, i.e., we assume the position of certain landmarks of an object to be known on pictures of different examples of these objects. In some of our simulations we determined corresponding landmarks manually, for the rest we replaced this manual intervention by motor controlled feedback (see section 5).

In a nutshell the learning algorithm works as follows (illustrated for two dimensions in figure 3 left): a–c) For each landmark we express the significant features per instance of all training examples by six dimensional codebook vector $(\mathbf{x}, \mathbf{b})$, representing the pixel position and the parameter frequency, orientation, curvature and elongation. We optimize the codebook vectors by the LBG vector quantization algorithm [5]. d) Codebook vectors with small distances are substituted by their center of gravity (PE2: reduction of redundancy). e,f) A significant feature for an object class is defined as a codebook vector expressing many data points. That means the feature corresponding to the code book vector or a similar feature (according to our metric d) often occurs in our training set, i.e., has high invariance (PE1). We end up with a graph with its nodes labeled with banana wavelets representing the learned significant features (see figure 2 left dv, ev). The edges of the graph labeled with metric relations of the landmarks.

4 Matching

To use our learned representation for location and classification of objects we define a similarity function between a graph labeled with the learned banana wavelets and a certain position in the image. A *total similarity* averages *local similarities*. The local similarity expresses the system's confidence whether a pixel in the image represents a certain line segment. The graph is adapted in position and scale by optimizing the total similarity. The graph with the highest similarity determines the size and position of the objects within the image.

In a nutshell the local similarity is defined as follows (for details see [3]): The magnitude of the filter responses depends significantly on the strength of edges in the image. However, here we are only interested in the presence and not in the strength of edges. Thus, in a second step a function $N(\ )$ normalizes the real valued filter responses into the interval $[0,1]$. The value $N(\mathbf{c})$ represents the likelihood of the presence or absence of a local line segment corresponding to $\mathbf{c} = (\mathbf{x}, \mathbf{b})$. This normalization is based on the "Above Average Criterion":

AAC a line segment corresponding to the banana wavelet $\mathbf{c}$ is present if the corresponding banana wavelet response is distinctly above the average response.

For a learned feature and pixel position in the image we simply check whether the corresponding banana response is high or low, i.e., we look at the normalized wavelet response $N(\mathbf{c})$. The total similarity (which is optimized during matching) is simply the average over all these local confidences. Because of the sparseness (PF4) of our representation only a few of these checks have to be made, therefore the matching is fast. Because we make use only of the important features, the matching is efficient.

5 Simulations

Learning of Representation: Firstly we apply the learning algorithm to data consisting of manually provided landmarks. Our training sets consist of a set of approximately 60 examples of an object viewed in a certain pose. As objects we use cans, faces, and hand postures. Corresponding landmarks are defined manually on the different representatives of a class of objects for the learned representation for hand postures.

To avoid the manual generation of ground truth we can either apply one–shot learning (see section 3) or make use of motor controlled feedback: By moving an object with a robot arm and following the object by keeping fixation relative to the robot hand using its known 3D position, we produce training data in which a certain view of an object is shown with varying background and illumination

Repres.		Trafo		Performance	
nb. reps	rep	approx	sec.	sec. match	Recog.
1) 10	standard	no approx	17.0	9.5	93 %
2) 10	one instance	approx	4.9	12.4	80 %
3) 10	bunch graph		0.9	18.0	93 %
4) 10	standard	no approx	17.0	9.5	90 %
5) 10	standard	approx	4.9	9.5	80 %
6) 10	bunch graph		0.9	18.0	65 %

Table 1. Matching results for hand posture recognition (for interpretation see text).

but with corresponding landmarks in the same pixel position within the image (see fig 2 left b,d). Then we can apply our learning algorithm with a rectangular grid roughly positioned on the object (see figure 2 left b,i). For the generation of ground truth for frontal faces we recorded a sequence of pictures in which a person is sitting fixed on a chair. Illumination and background is changed as for cans. To extract representations for different scales we apply the learning algorithm to the very same pictures of the different sequences scaled accordingly. **Matching:** For the problem of face finding in complex scenes with large size variation a significant improvement in terms of performance and speed compared to the older system [4, 9] could be achieved. In [6] face recognition with binarized banana wavelets was performed on a very large data set (more than 700 pictures) with size variation of faces between 40 and 60 pixel, inhomogeneous background and uncontrolled illumination. For this set performance was 95%.

Our test sets of hand postures contain images of 10 different hand postures (figure 3 (right) shows the learned representations) in front of homogeneous background with controlled illumination (row 1–3, 240 images) and with a second set containing images with inhomogeneous background and varying illumination (row 4–6, 200 images). Matching with ten representations (one for each hand posture) takes 9.5 seconds and recognition rate for the first set was 93% (first row). The simulations corresponding to the second row were performed with representations extracted by one-shot learning. The performance is still remarkably high (80%). The performance with the bunch graph approach as described in [8] is given in the third row. Results for the test set with uncontrolled background and illumination is shown in row 4–6. For the first test set performance within the bunch graph approach [9, 8] is comparable to ORASSYLL. For the second and more difficult, set performance of ORASSYLL is significantly better.

6 Comparison with the Jet-Based System

ORASSYLL has been heavily influenced by an older and well known vision system [4, 9, 8] in the following called jet–based system, and has also been influenced by Biederman's comments to this older jet–based system [1]. The system [4, 9] was very successfully applied to face recognition. High correlation between the system's and human's face recognition performance has been shown [1]. However, Biederman and his associates [1] also have shown that the system [4, 9] has only low correlation to human object recognition.

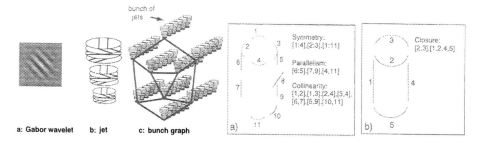

a: Gabor wavelet   **b: jet**   **c: bunch graph**

Fig. 4. Left: Representation of objects within the jet–based system: a: a Gabor wavelet (real part). **b:** a jet calculated as a set of Gabor wavelets (the discs symbolize the different frequencies and directions of **k**). **c:** a bunch graph. **Right: Gestalt principles as low order relations of local line segments: a:** Sparse representation of a can with local curved lines and lists of second order Gestalt relations between. **b:** Smoothing and grouping of the representation in a).

As models for objects the older system also employs labeled graphs. The edges of graphs are labeled with distance vectors between node positions. Nodes are labeled with jets [4] or bunches of jets [9] respectively. In a bunch of jets each jet is derived from the image of a different example of the view of an object. A is bunch thus covering a variety of forms a single landmark may take. This structure is called *bunch graph* [9].

Jets are derived from a set of linear filter operations in the form of convolutions of the image $I(\mathbf{x})$ with a set of Gabor wavelets, of different wavelength and orientation (see figure 4 left a). A jet is formed by the set of complex values rendered by all wavelets centered at a given position of the image (see figure 4 left b). Due to the spatial extent of the wavelets, jets describe a local area around their position. A bunch $\mathcal{B}$ of jets taken at the same landmark (that is, at corresponding positions) of different examples of a certain view of an object class forms a generalized representation of this landmark (see figure 4 left c).

Jet components a_j (the index j standing for length and orientation of the components' wave vectors) are the magnitude a_j of Gabor wavelet reponses. The similarity between two jets $\mathcal{J}$ and $\mathcal{J}'$ is defined as the normalized scalar product of the two jets:

$$S(\mathcal{J}, \mathcal{J}') = \frac{1}{\sqrt{\sum_j a_j^2 \sum_j a_j'^2}} \cdot \sum_j a_j a_j' \tag{1}$$

Conceptional Differences of Object Representations in the Jet–based System and ORASSYLL: The object representation on ORASSYLL shows four conceptional (D1–D4) differences to the representation based on jets and bunches of jets.

- **(D1)** The distinction curvature vs. straightness can be explicitly used as a feature.
- **(D2)** A restriction to a specific set of binary features of high complexity (more precisely, local (curved) line segments) is imposed for object representation.

– **(D3)** The object representations are sparse.
– **(D4)** A metric is utilized as an additional structure of the feature space.

In the next subsection we will discuss how these differences influence the recognition process.

6.1 Comparison with Jets and Bunches of Jets: Two Arguments in Favor of ORASSYLL for the Task "Object Recognition"

In this section two advantages of the object representation within ORASSYLL are discussed.

First Argument: In a Jet Significant and Insignificant Features are Lumped together whereas in ORASSYLL Features are Stored Separately. Therefore Matching and Extraction of Significant Features is Facilitated: In a jet, significant and insignificant features are lumped together. Even when a single Gabor wavelet response gives information about the occurrence of a local line with a certain orientation, a jet always represents the whole local image patch. The jet similarity (1) reflects the relative strengths of a complete set of Gabor wavelet responses at the actual pixel position, and therefore reflects the fit to a whole local region. For example, a local area of an object may have an edge with a certain orientation resulting in a strong response of the corresponding Gabor wavelet. The occurrence of an edge with different orientation in the background causes a strong response for the Gabor wavelet with different orientation. Because the denominator in equation (1) increases by the "background–response", the relative strength of Gabor wavelet responses, and therefore the similarity (1) changes. In [3] it has been demonstrated that — because the relative strength of Gabor wavelet responses varies significantly with changes of background and illumination — the similarity (1) is more sensitive to these sources of noise compared to the similarity function applied within ORASSYLL.

In ORASSYLL two different distance functions for learning and matching are used: Firstly, the metric defines a distance between features (D4). For learning, features at a close distance are grouped together within one cluster but features at a large distance are treated as *separate*. The metric of the feature space reflects the difference of properties of features such as difference in space, curvature or orientation. This allows to distinguish between significant features and insignificant features (e.g., corresponding to the background) and to keep only the significant features within the object representation (see figure 2 left). Secondly, the similarity of a binarized banana wavelet to a local image patch (based on the Above Average Criterion) indicates the presence or absence of the learned feature fairly independent of background and illumination (as shown in [3]) and allows for a comparison of only the learned and significant features to the image.

Second Argument: Coding and Detection of Important Relations such as Collinearity, Parallelism and Symmetry is more Difficult with Jets than with Binarized Banana Wavelets: An important issue within ORASSYLL is the definition of a local criterion for the presence of a local (curved) line. Maybe there does not even exist such a completely satisfying *local* criterion and the presence of a local line segment depends on the *context*. Compared to the older system, invariance to changes in illumination and background is significantly higher, but still, variation occurs as it has been demonstrated in [3].

In [2] within the framework of ORASSYLL, it could been shown that collinearity and parallelism can be detected and mathematically characterized in natural images with binarized Gabor responses. Without binarization, i.e., without a transformation to a feature of higher complexity corresponding to a localized line (D2), these two Gestalt principles are barely detectable in the statistics of natural images.

A sparse representation of objects (D3) allows for the description and detection of Gestalt principles as low–order statistics of feature relations. (see figure 4 right) We suggest that this coding also facilitates the reliable recognition of a local line segment by integrating contextual information because interactions between features such as inhibition and reinforcement can be defined according to Gestalt principles within the feature space. Coding of Gestalt principles with jets would presuppose higher order statistics. Therefore learning and integration of Gestalt principles becomes more difficult within the jet approach.

Acknowledgment: I would like to thank Christoph von der Malsburg, Gabriele Peters, Laurenz Wiskott and Michael Pötzsch for fruitful discussion.

References

1. I. Biederman and P. Kalocsai. Neurocomputational bases of object and face recognition. *Philosophical Transactions of the Royal Society: Biological Sciences*, 352:1203–1219, 1997.

2. N. Krüger. Collinearity and parallism are statistically significant second order relations of complex cell responses. *Neural Processing Letters*, 8(2), 1998.

3. N. Krüger. *Visual Learning with a priori Constraints (Phd Thesis)*. Shaker Verlag, Germany, 1998.

4. M. Lades, J.C. Vorbrüggen, J. Buhmann, J. Lange, C. von der Malsburg, R.P. Würtz, and W. Konen. Distortion invariant object recognition in the dynamik link architecture. *IEEE Transactions on Computers*, 42(3):300–311, 1992.

5. Y. Linde, A. Buzo, and R.M. Gray. An algorithm for vector quantizer design. *IEEE Transactions on communication*, vol. COM-28:84–95, 1980.

6. H.S. Loos, B. Fritzke, and C. von der Malsburg. Positionsvorhersage von bewegten objekten in groformatigen bildsequenzen. *Proceedings in Artificial Intelligence: Dynamische Perzeption*, pages 31–38, 1998.

7. Michael Pötzsch, Thomas Maurer, Laurenz Wiskott, and Christoph von der Malsburg. Reconstruction from graphs labeled with responses of gabor filters. In C. v.d. Malsburg, W. v. Seelen, J.C. Vorbrüggen, and B. Sendhoff, editors, *Proceedings of the ICANN 1996*, Springer Verlag, Berlin, Heidelberg, New York, Bochum, July 1996.

8. J. Triesch and C. von der Malsburg. Robust classification of hand postures against complex background. *Proceedings of the Second International Workshop on Automatic Face- and Gesture recognition, Vermont*, pages 170–175, 1996.

9. L. Wiskott, J.M. Fellous, N. Krüger, and C. von der Malsburg. Face recognition by elastic bunch graph matching. *IEEE Transactions on Pattern Analysis and Machine Intelligence*, pages 775–780, 1997.

Effective Implementation of Linear Discriminant Analysis for Face Recognition and Verification

Yongping Li, Josef Kittler, Jiri Matas

Centre for Vision Speech and Signal Processing,
University of Surrey, Guildford, Surrey GU2 5XH, England
{Y.Li, J.Kittler, G.Matas}@ee.surrey.ac.uk

Abstract. The algorithmic techniques for the implementation of the Linear Discriminant Analysis (LDA) play an important role when the LDA is applied to the high dimensional pattern recognition problem such as face recognition or verification. The LDA implementation in the context of face recognition and verification is investigated in this paper. Three main algorithmic techniques: matrix transformation, the Cholesky factorisation and QR algorithm, the Kronecker canonical form and QZ algorithm are proposed and tested on four publicly available face databases (M2VTS, YALE, XM2FDB, HARVARD)[1]. Extensive experimental results support the conclusion that the implementation based on the Kronecker canonical form and the QZ algorithm accomplishes the best performance in all experiments.

1   Introduction

The linear discriminant analysis (LDA) approach to feature extraction is well known [5]. A detailed description of the LDA for pattern recognition can be found in [2]. Theoretically, LDA-based features should exhibit classification performance superior to that achievable with the features computed using Principal Components Analysis (PCA). However, in the context of face recognition or verification the LDA method has only occasionly been reported to outperform the PCA approach [7]. Better performance of the LDA method as compared with the PCA approach was reported in [1][3][4]. Notably, no details regarding the implementation of the LDA algorithm were presented in these papers.

So far little attention has been paid to the implementation of LDA. In this work we test the hypothesis that poor performance of LDA in face recognition experiments (as an example of high-dimensional problems with a small training set) can be at least partially explained by incorrect selection of the numerical method for solving the associated eigenvalue problem.

---

[1] See   *http://www.tele.ucl.ac.be/M2VTS*   for   M2VTS   database;   see *http://cvc.yale.edu/projects/yalefaces/yalefaces.html* for YALE face database; see *http://www.ee.surrey.ac.uk/Research/VSSP/xm2fdb* for XM2FDB database; see *ftp://ftp.hrl.harvard.edu/pub/faces* for HARVARD face database.

The work presented here investigates the efficacy of various algorithms for face recognition and verification based on Linear Discriminant Analysis. An implementation which employs the Kronecker canonical form and the QZ algorithm is recommended because of its stability. This robust algorithm achieves the best performance in all experiments. The experiments were undertaken on four publicly available face databases: the M2VTS database, the XM2FDB database, the Yale face database and the Harvard face database.

The paper is organised as follows. In Section 2 we briefly describe the theoretical framework of face recognition approaches based on the eigenvalue analysis of matrices of second order statistical moments. Computational considerations and algorithmic techniques for the implementation of LDA are presented in Section 3. A description of the experiments including the face databases, experimental protocol, algorithms involved and the experimental results obtained are given in Section 4. Finally a summary and some conclusions are presented in Section 5.

2 Linear Discriminant Analysis

For a set of vectors $x_i, i = 1, \ldots, M$, $x_i \in R^D$ belonging to one of c classes $\{C_1, C_2, \ldots, C_c\}$, the between-class scatter matrix S_B and within-class scatter matrix S_W are defined as

$$S_B = \sum_{k=1}^{c} (\mu_k - \mu)(\mu_k - \mu)^T \tag{1}$$

$$S_W = \sum_{i=1}^{c} \sum_{x_k \in C_i} (x_k - \mu_i)(x_k - \mu_i)^T \tag{2}$$

where μ is the grand mean and μ_i is the mean of class C_i.

The objective of LDA is to find a transformation matrix W_{opt} maximising the ratio of determinants $\frac{|W^T S_B W|}{|W^T S_W W|}$. W_{opt} is known to be the solution of the following eigensystem problem ([2]):

$$S_B W - S_W W \Lambda = 0 \tag{3}$$

Premultiplying both sides by S_W^{-1}, (3) becomes:

$$(S_W^{-1} S_B) W = W \Lambda \tag{4}$$

where Λ is a diagonal matrix whose elements are the eigenvalues of Equation (3). In the context of face recognition, the column vectors w_i $(i = 1, \ldots, c - 1)$ of matrix W are referred to as fisherfaces [1].

Dimensionality reduction. In high dimensional problems (e.g. in the case where x_i are images and D is $\approx 10^5$) S_W is almost always singular, since the number of training samples M is much smaller than D. Therefore a dimensionality reduction must be applied before solving the eigenproblem (3). Commonly, the dimensionality reduction is achieved by Principal Component Analysis [11][1]; the first $(M - c)$ eigenprojections are used to represent vectors x_i. This also allows S_W and S_B to be calculable in computer with a normal memory size. The optimal linear feature extractor W_{opt} is then defined as:

$$W_{opt} = W_{fld} * W pca \qquad (5)$$

where W_{pca} is the PCA projection matrix and W_{fld} is the optimal projection obtained by maximising

$$W_{fld} = arg \max_W \frac{|W^T W_{pca}^T S_W W_{pca} W|}{|W^T W_{pca}^T S_B W_{pca} W|} \qquad (6)$$

3 Algorithms for the $S_B - \lambda S_w$ (Pencil) Eigenproblem

Though both S_B and S_W are real symmetric matrices, the product matrix S_P where $S_P = S_W^{-1} S_B$ needs not be symmetric [6]. Therefore directly solving the eigenproblem of single matrix (S_P) will lead to unstable eigenvalues and eigenvectors.

Several algorithmic techniques can be used to solve the eigenproblem of $S_B w - \lambda S_W w$. The set of matrices in the form of $S_B - \lambda S_W$ is referred to as a linear *pencil*. As both S_B and S_W are covariance matrices, they are either positive definite or semi-positive definite real symmetric matrices.

The matrix transformation technique. A matrix transformation technique described in [2] is used to convert the pencil to a transformed real symmetric matrix. In order to obtain this transformation, the eigenvalues and eigenvectors of S_W are firstly computed. They are denoted as Λ_w and V_w and satisfy

$$S_W V_w - V_w \Lambda_w = 0 \qquad (7)$$

If S_W is a symmetric positive definite matrix, there exists a matrix B such that

$$B^T S_W B = I \qquad (8)$$

Comparing (7) and (8), we have

$$B = V_w \Lambda_w^{-1/2} \qquad (9)$$

Let us define a tranformation matrix $S_B' = B^T S_B B$. Then the eigensystem problem becomes

$$S_B' V - V \tilde{\Lambda} = 0 \qquad (10)$$

As the rank of S_B' is at most $c - 1$, it will have only d $(d \leq c - 1)$ non-zero eigenvalues. This means that all the relevant information is compressed into d eigenvectors associated with the non-zero eigenvalues of S_B'. Denoting the system of these eigenvectors by V', i.e.,

$$V' = \{v_1, v_2, ..., v_d\} \qquad (11)$$

the optimal feature extractor W_{opt} is given as

$$W_{opt} = BV' = V_w \Lambda_w^{-1/2} V' \tag{12}$$

Cholesky factorisation and QR algorithm. When the dimensions of S_W and S_B are reduced from D to $n(n \leq (M - c))$, S_W becomes positive definite and the *pencil* turns out to be a *symmetric definite pencil*. The problem of (3) can then be converted to (13) below using the congruence transformation.

$$(X^T S_B X)W - (X^T S_W X)W\Lambda = 0 \tag{13}$$

where $S_W = S_W^T \in IR^{n \times n}$, $S_B = S_B^T \in IR^{n \times n}$ and the matrix X satisfies

$$X^T S_W X = I, \quad X^T S_B X = diag(\lambda_1, \ldots, \lambda_n) \tag{14}$$

The steps to compute W_{opt} are defined as follows

- The Cholesky factorisation $S_W = LL^T$ is computed first using the method given in [8]. Formula (4) now becomes

$$L^{-T}L^{-1}S_B W = W\Lambda \tag{15}$$

Multiplying both sides by L^T, we get

$$L^{-1}S_B W = L^T W\Lambda \tag{16}$$

which can be rewritten as

$$(L^{-1}S_B L^{-T})(L^T W) = (L^T W)\Lambda$$

or

$$PY - Y\Lambda = 0 \tag{17}$$

where $P = L^{-1}S_B L^{-T}$ and $Y = L^T W$
- Then the symmetric QR algorithm can be applied to compute the Schur decomposition

$$Q^T PQ = diag(\lambda_1, \ldots, \lambda_n) \tag{18}$$

- Finally, the eigenvectors are calculated by

$$W_{opt} = L^{-T}Q \tag{19}$$

Kronecker canonical form and QZ algorithm. The more general situation is that both S_W and S_B are near singular, thus neither the matrix transformation nor the Cholesky factorisation can be applied. In such situations the QZ algorithm must be employed.

The main idea of the QZ algorithm introduced in [9] is to transform matrices S_W and S_B simultaneously to triangular matrices $\tilde{B}$ and $\tilde{A}$ that satisfy

$$\tilde{B} = QS_W Z, \quad \tilde{A} = QS_B Z \tag{20}$$

where matrices Q and Z are derived as a product of the Gauss transformation. Hence the eigenproblem of $S_B W = S_W W \Lambda$ is equivalent to

$$\tilde{A}W' = \tilde{B}W'\Lambda \tag{21}$$

If the diagonal elements of matrix $\tilde{B}$ are non zero, i.e., $\tilde{b}_{ii} \neq 0$, then the eigenvalues and eigenvectors of the original pencil are obtained by

$$\lambda_i = \tilde{a}_{ii}/\tilde{b}_{ii}, \qquad W_{opt} = ZW' \tag{22}$$

The behaviour of the QZ algorithm on pencils that are not only *regular* but also nearly *singular* is analysed in [15]. The results reported in that paper strongly support that when the pencil is singular or near singular the QZ algorithm should be preceded by an algorithm which extracts the singular part of the pencil. This situation is likely to arise when LDA is involved in face identification. If the pencil is converted into the *Kronecker canonical form*(see [13]), the general QZ algorithm always works well. This is also proven by our experimental results shown in Section 4.

4 Experiments

4.1 Face Verification Experiments

The comparative performance of the algorithms presented in Section 3 was tested as part of a face verification experiment performed on data from three publicly available face databases. All implementations of LDA were tested using the same protocol.

Experimental ensembles

Three experimental ensembles were selected from the three different face databases. They are all registered using an approach based on the eyes positions.

- The EPFL ensemble: 37 persons, 4 images per person selected from four shots of the M2VTS database.
- The YALE ensemble: 15 subjects, 11 images per subject containing all images in the YALE face database.
- The SURREY ensemble: 100 individuals, 4 sessions, one image per session selected from the XM2FDB database

Experimental protocol

The experimental protocol was firstly designed for the performance evaluation of various methods carried on the M2VTS database. It combines the 'leave-one-out' strategy and the rotation scheme [2]. For a general ensemble of c persons, s sessions(shots), each person in turn is labelled as an *imposter*, whilst the $(c-1)$ others are considered as *clients*. The training set consists of $(s-1)$ shots of $(c-1)$ clients. The remaining one shot is used as the test set. Each client tries to access under his or her own identity (ID) and the imposter tries to access under the ID of $(c-1)$ clients. After all rotations, the number of client and imposter tests is

$c \times s \times (c-1)$. This procedure leads to 5328 tests for EPFL ensemble, 2310 tests for YALE ensemble and 39600 tests for SURREY ensemble.

Results of the face verification experiments

The following implementation methods are applied in the experiments for comparison.

- Eigenface approach based on the Principal Components Analysis abbreviated as "*PCA*".
- LDA method implemented using the Cholesky factorisation and the QR algorithm abbreviated as "*QR*".
- LDA method implemented using the Kronecker canonical form and the QZ algorithm abbreviated as "*QZ*".
- LDA method implemented using matrix transformation techniques abbreviated as "*MT*".

The eigenface (PCA) method used all of the available eigenfaces to make the results as good as possible, so did all the 'LDA' algorithms. Experimental results are presented in terms of the receiver operating characteristics curve(ROC). All ROC curves are displayed in Figures 1, 2, 3. The equal error rate(EER) for all experiments are given in Table 1. We can find from these results that the "QZ" algorithm achieves the best performance on every ensemble.

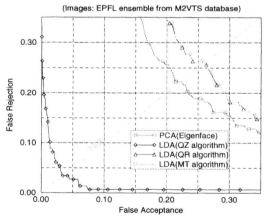

Fig. 1. Performances on the EPFL ensemble

	Results of algorithms			
Ensemble	"PCA"	"MT"	"QR"	"QZ"
EPFL	22.3	26.2	26.2	3.1
YALE	33.3	18.2	6.5	2.3
SURREY	9.3	19.6	19.6	2.9

Table 1. Equal error rates of verification experiments using the eigenprojection method (PCA), Cholesky factorisation (QR), Kronecker canonical form (QZ) and the Matrix transformation technique (MT).

Results analysis and improvement consideration

From Table 1, we find not only that there are big differences between the results when LDA is implemented using various algorithms but also that the eigenface

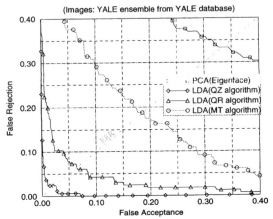

Fig. 2. Performances on the YALE ensemble

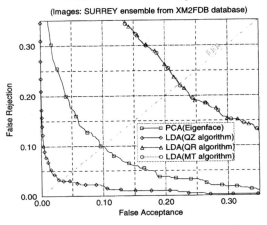

Fig. 3. Performances on the SURREY ensemble

method in some cases achieves better results than the "QR" and "MT" algorithms. Similar results were also obtained by other researchers. One example can be found in [7] where the results of the eigenface method were better than those yielded by the MDF(most discriminant feature) method[2]. This contrasts with the theory which favours LDA.

We were surprised that the LDA method was seldom reported to exhibit better results than the eigenface approach. Eventually, we developed a robust implementation ("QZ") of LDA which always outperforms the eigenface method. While comparing the immediate results, we found that an important processing step was omitted in both "QR" and "MT" algorithms.

Checking the background of these algorithmic techniques, some pre-requisites play an important role in the application. For example, the Cholesky factorisation requires the matrix to be "well-conditioned enough"; the Jacobi method

---

[2] The MDF method is a different name of the LDA approach.

which finds the eigensolution for a real symmetric matrix performs the eigenvector normalisation at every iteration with a roundoff thresholding precision. If the matrix is quite well-conditioned, the orthonormalisation of the eigenvectors will be satisfactory for both the "QR" algorithm and the "MT" algorithm. The QZ algorithm is designed to deal directly with the pencil problem(in the form of $Ax = \lambda Bx$) and its performance is unaffected by singularity or near-singularity of A, B or $A - \lambda B$.

However, except when S_W is really singular, both "QR" and "MT" algorithms can achieve nearly the same results as "QZ" algorithm by applying a supplementary normalisation of the final eigenvectors.

4.2 Face Recognition Experiments

As a comparison, the face recognition experiments described in [1] were repeated using the LDA approach implemented with "QZ" algorithm. Images from both the Harvard database and the Yale database were centred and cropped to emulate the data shown in [1]. Results are given in Tables 2, 3.

Error Rate(%) of face recognition experiments on Harvard Database							
Method	Reduced	Extrapolating from subset 1			Interpolating between subsets 1, 5		
	Space	Subset 1	Subset 2	Subset 3	Subset 2	Subset 3	Subset 4
fisherface[4]	4	0.0	0.0	4.6	0.0	0.0	1.2
LDA ("QZ")	4	0.0	0.0	3.1	0.0	0.0	0.0

Table 2. Extrapolation and interpolation experimental results of variation in lighting

"Leave-one-out" of Yale Database			
Method	Reduced	Error Rate(%)	
	Space	Close Crop	Full Face
fisherface[4]	14[5]	7.3	0.6
LDA ("QZ")	14	3.6	0.6

Table 3. Experimental results of variation in facial expression and lighting

The results show that the LDA approach implemented using the "QZ" algorithm achieves better results in all three face recognition experiments compared with the best results reported in [1].

5 Conclusions

In this paper, a detailed study of three numerical algorithms for linear discriminant analysis(LDA) in the context of face verification has been investigated. A

---

[4] No description for the fisherface method implementation was given in [1].

[5] There are only 15 subjects(11 images/subject) in the Yale database; the maximum reduced space is 14.

robust algorithm has been proposed and tested. It achieves a very good performance when face images are registered using a semi-automatic face image registration method based on eyes positions.

The LDA is a powerful classifier, it is robust to lighting conditions, facial expressions and small pose changes. The experiments performed show that the selection of an appropriate implementation of LDA influences significantly the verification performance. The "QZ" algorithm always outperforms other algorithms regardless of the face ensemble.

Acknowledgements

The research work has been carried out within the framework of the European ACTS-M2VTS project. Acknowledgement is also due to the centre for Computational Vision & Control at Yale University and Harvard Robotics Laboratory for the permission to use their face databases.

References

1. P. Belhumeur et al, "Eigenfaces vs. Fisherfaces: Recognition Using Class Specific Linear Projection," in *Proc. of ECCV'96*, pp 45–58, Cambridge, UK, 1996.
2. Pierre A. Devijver and Josef Kittler. "Pattern Recognition: A Statistical Approach," Prentice-Hall, Englewood Cliffs, N.J., 1982.
3. K. Etemad and R.Chellappa, "Face Recognition using Discriminant Eigenvectors," In *Proc. of ICASSP'96*, pp. 2148-2151, 1996.
4. K. Etemad and R.Chellappa, "Discriminant Analysis for Recognition of Human Face Images," In *Proc. of AVBPA'97*, pp. 127-142, 1997.
5. R.Fisher, "The use of multiple measures in taxonomic problems," *Ann. Eugenics*, vol.7, pp. 179-188, 1936.
6. G.H. Golub, and C.F.Van Loan, *Matrix computations*, The Johns Hopkins University Press, Baltimore and London, 1989.
7. C.Liu and H.Wechsler, "Face Recognition Using Evolutionary Pursuit," in *Proc. of ECCV'98*, Vol. II, pp. 596-612, June 1998.
8. R.S. Martin et al, "Symmetric decomposition of a positive define matrix," *Numer. Math.*, Vol. 7, pp. 362-383, 1965.
9. C.B.Molei et al., "An algorithm for generalized matrix eigenvalue problem $Ax = \lambda Bx$," *SIAM J.Numer. Anal.*, Vol. 10, pp. 99-130, 1973.
10. William H. Press, et al., *Numerical recipes in C: the art of scientific computing*, Cambridge Press, 1989.
11. L. Sirovich et al, "Low-dimensional procedure for the characterization on human faces," *J. Opt. Soc. Am. A*, vol. 4, no. 3, pp 519-524, 1987.
12. M.Turk and A.Pentland, "Eigenface for Recognition," *Journal of Cognitive Neuroscience*, Vol. 3, no. 1, pp. 70-86, 1991.
13. P. Van Dooren, "The computation of Kronecker's form of a singular pencil," *Linear Algebra and Appl.*, Vol. 27, pp. 103-140, 1979.
14. J. Wilkinson et al., "Handbook for Automatic Computation Vol II," *Linear Algebra*, Springer-Verlag, Berlin Heidelberg, New York, 1971.
15. J. H. Wilkinson, "Kronecker's Canonical Form and the QZ Algorithm," *Linear Algebra and Appl.*, vol. 28, pp. 285-303, 1979.

A Scale-Space Approach to Face Recognition from Profiles

Zdravko Lipoščak, and Sven Lončărić

University of Zagreb, Faculty of Electical Engineering and Computing
Department of Electronic Systems and Information Processing
Unska 3, Zagreb, Croatia
sven.loncaric@fer.hr

Abstract. A method for face recognition using profile images based on the scale-space filtering is presented in this paper. A grey-level image of profile is thresholded to produce a binary, black and white image, the black corresponding to face region. A pre-processing step then extracts the outline curve of the front portion of the silhouette that bounds the face image. From this curve, a set of twelve fiducial marks is automatically identified using scale-space filtering with varying the scale parameter. A set of twenty-one feature characteristics is derived from these fiducial marks. After normalising the feature characteristics using two selected fiducial marks, the Euclidean distance measure is used for measuring the similarity of the feature vectors derived from the outline profiles. Experiments were performed on a total of 150 profiles of thirty persons. Experimental results are presented and discussed. Finally, recognition rates and conclusions are given.

1. Introduction

Since the early 1990's Face Recognition Technology (FRT) become an active research area. A general statement of the problem of face recognition can be formulated as follows: Given still or video images of a scene, identify one or more persons in the scene using a stored database of faces [2].

The solution of the problem involves segmentation of faces from cluttered scenes, extraction of features from face region, identification, and matching.

Face recognition problems and techniques can be separated in two large groups: dynamic (video) and static (no video) matching. Dynamic matching is used when a video sequence is available. The video images tend to be of low quality, the background is very cluttered and often is more than one face present in the picture. However, since a video sequence is available, one could use motion as a strong cue for segmenting faces of moving persons.

Static matching uses images with typically reasonably controlled illumination, background, resolution, and distance between camera (or 3D scanner) and the person. Some of the images that arise in this group can be acquired from a video camera.

Machine recognition of faces has several applications, ranging from static matching of controlled photographs as in mug shots matching and credit card verification to surveillance video images. Mug shots matching is the most common application in static matching group. Typically, in mug shots photographs one frontal and one or more side views of a person's face are taken. Profile images provide a detailed structure of the face that is not seen in frontal images [7]. Face recognition from profiles concentrates on locating points of interest, called fiducial points. Recognition involves the determination of relationships among these fiducial points.

In this work we try to develop simple and fast method for detecting these fiducial points. For that purpose we treat the outline of a profile like a function and we use scale-space filtering [1], [3] for detection of extrema in that function and its first few derivatives. The profile is first expanded by convolution with Gaussian masks over a continuum of sizes. From this "scale-space" image we determine scale parameters which are used for detection of specific fiducial points. A set of twenty-one feature characteristics is derived from these fiducial marks. After normalising the feature characteristics using two selected fiducial marks, Euclidean distance measure was used for measuring the similarity of the feature vectors derived from the outline profiles [8]. Results of the proposed profile matching method in the presence of rotation, translation and size variance of profile faces are included in this paper.

2. Scale-Space Filtering

Scale-space filtering [1] is a method that describes signals qualitatively, in terms of extrema in the signal or its derivatives, in a manner that deals effectively with the problem of scale-precisely localising large-scale events, and effectively managing the ambiguity of descriptions at multiple scales, without introducing arbitrary thresholds or free parameters. The extrema in signal and its first few derivatives provide a useful general-purpose qualitative description for many kinds of signals.

Descriptions that depend on scale can be computed in many ways. As a primitive scale-parameterisation, the Gaussian convolution is attractive for a number of its properties, amounting to "well-behavedness": the Gaussian is symmetric and strictly decreasing about the mean, and therefore the weighting assigned to signal values decreases smoothly with distance. The Gaussian convolution behaves well near the limits of the scale parameter, σ, approaching the un-smoothed signal for small σ, and approaching the signal's mean for large σ. The Gaussian is also readily differentiated and integrated.

The Gaussian convolution of signal $f(x)$ depends booth on x, the signal's independent variable, and on σ, the gaussian's standard deviation. The convolution is given by

$$F(x,\sigma) = f(x) * g(x,\sigma) = \int_{-\infty}^{\infty} f(u) \frac{1}{\sigma\sqrt{2\pi}} e^{-\frac{(x-u)^2}{2\sigma^2}} du \tag{1}$$

where "*" denotes convolution with respect to x. This function defines a surface on the (x, σ)-plane, where each profile of constant σ is a Gaussian-smoothed version of f(x), the amount of smoothing increasing with σ. (x, σ)-plane is called scale space, and the function, F, defined in (1), the scale-space image of *f*.

At any value of σ the extrema in the *n*th derivative of the smoothed signal are given by the zero-crossings in the (n+1)th derivative. Although the scale-space filtering methods apply to zeros in any derivative, Witkin [1] restricted his attention to those in the second. These are extrema of slope, i.e. inflection points. In terms of the scale-space image, the inflections at *all* values of σ are the points that satisfy

$$F_{xx} = 0, \; F_{xxx} \neq 0, \tag{2}$$

using subscript notation to indicate partial differentiation.

The contours of $F_{xx} = 0$ mark the appearance and motion of inflection points in the smoothed signal, and provide the raw material for a qualitative description over all scales, in terms of inflection points. Witkin applies two simplifying assumptions to these contours: (1) the *identity* assumption, that extrema observed at different scales, but lying on a common zero-contour in scale space, arise from a single underlying event, and (2) the *localisation* assumption, that the true location of an event giving rise to a zero-contour is the contour's x location as σ → 0.

Referring to Fig. 1, notice that the zero contours form arches, closed above, but open below. The localisation assumption is motivated by the observation that linear smoothing has two effects: qualitative simplification- the removal of fine-scale features- and spatial distortion- dislocation, broadening and flattening of the features that survive. The latter undesirable effect may be overcome, by tracking coarse extrema to their fine scale locations. Thus, a coarse scale may be used to *identify* extrema, and a fine scale, to *localise* them. Each zero-contour therefore reduces to an (x, σ) pair, specifying its fine-scale location on the x-axis, and the coarsest scale at which the contour appears.

While coarse-to-fine tracking solves the problem of localising large-scale events, it does not solve the multi-scale integration problem. Witkin in [1] reduced the scale-space image to a simple tree, concisely but completely describing the qualitative structure of the signal over all scales of observation. In general, each undistinguished interval, observed in scale-space, is bounded on each side by the zero contours that define it, bounded above by the singular point at which it merges into an enclosing interval, and bounded below by the singular point at which it divides into sub-intervals.

3. Profile Face Analysis Using Scale-Space Filtering

The input in our face processing and analysis system is a gray-scale image of a scene containing the side view of a human face. The goal of our approach is to find a simple and fast method for recognition of profile faces. For that purpose we are using scale-space filtering and scale-space images of different faces for detecting the possible unique one or more parameters σ for all faces (see Fig.1.). The Gaussian convolution of a signal f(x), which represents profile line, with this σ parameter may be used to identify and localise extrema in f(x). These extrema represent interest points like a nose peak, nose bottom, mouth point, chin point etc. (see Fig.2.a).

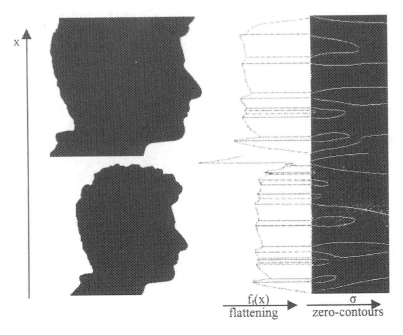

Fig. 1. Scale-space filtering of different profile face images (profile face images from University of Bern profile database[1])

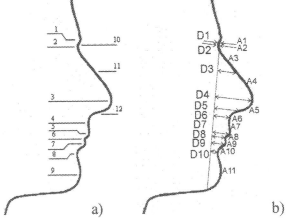

Fig. 2. a) The twelve fiducial points of interest for profile face recognition, b) Feature vector has twenty-one component; ten distances D_1- D_{10} (normalised with $/(D_4 + D_5)$) and eleven profile arcs A_1- A_{11}(normalised with $/(A_5 + A_6)$)

In first step the image was tresholded to produce a binary, black and white image, the black corresponding to the face region. A pre-processing step then extracts the outline curve of the front of the silhouette. Front silhouette line bounds the hair, face

---

and clothes of the person- this whole line we call the *profile line*. This profile line is converted from 2D picture into 1D signal *f(x)* which we call the *profile vector*; where *x* is a row index and *f(x)* is a column index of a pixel inside a profile line. For any row *x* without profile line we set *f(x) = 0*.

In the next step the profile vector f(x) is flattened using a Gaussian convolution $F(x, \sigma_L)$ of a profile with large parameter σ_L;

$$f_f(x) = f(x) - F(x, \sigma_L) \qquad (3)$$

This step transforms all fiducial points in extrema and ensures better rotation invariance of the method (see Fig.1.).

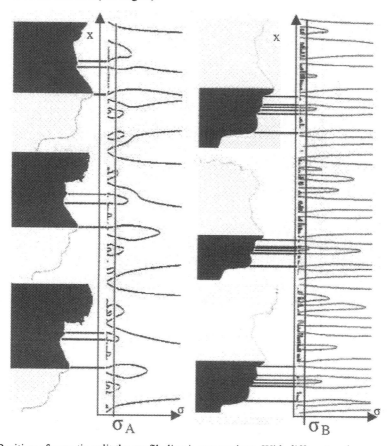

Fig. 3. Position of nose tip split the profile line in two regions. With different scale parameters σ_A and σ_B we detect all fiducial points.

The one-dimensional profile vector is expanded into a two-dimensional scale-space image. Comparing scale-space images of different profile faces we detect two unique parameters σ that gives extrema in desired fiducial points in all profile faces. After finding the fiducial points we can use the methods from similar papers [3],[4],[5] for describing profile faces using these points. From twelve fiducial points (see Fig.2.a; extrema points 1 to 9 and inflection points 10 to 12), we derive a simple set of twenty-one features. After normalising characteristics using two selected fiducial marks, Euclidean distance measure was used for measuring the similarity of the feature vectors derived from the outline profiles. A ranking of most similar faces is obtained by ordering the Euclidean distances.

We use two scale-space parameters σ - one for the profile region above the nose and one for the profile region below the nose. Two parameters are required because of different scales of facial features above and below of the nose tip.

Position of the nose tip splits the profile line in two regions; A – region above the nose and B – region below the nose. For this two regions we use different parameters σ_A and σ_B for detection of other points of interest (see Fig.3.). In region A and in region B of different profile images we can find unique parameters σ_A and σ_B for computing Gaussian convolution of flatten profile line and detection of global extrema that belong to the fiducial points 1- 2 and 4 – 9. For the inflection points 10 – 12 we can use the same σ_A and zero-contours of the inflection points.

Nose is an *event* that persists across large changes in scale. Next short algorithm is used for nose tip detection;
1. Extract the profile line $f(x)$ from input graylevel picture.
2. Smooth the profile line using Gaussian convolution with small parameter σ_s (eq. 1.);
   $$F(x, \sigma_s) = f(x) * g(x, \sigma_s)$$
3. Smooth the profile line using Gaussian convolution with large parameter σ_L;
   $$F(x, \sigma_L) = f(x) * g(x, \sigma_L)$$
4. Flatten the profile line; $f_f(x, \sigma_s, \sigma_L) = F(x, \sigma_s) - F(x, \sigma_L)$
5. Compute Gaussian convolution with σ_N; $F_N(x, \sigma_N, \sigma_s, \sigma_L) = f_f(x, \sigma_s, \sigma_L) * g(x, \sigma_N)$
6. Find extrema in $F_N(x, \sigma_N, \sigma_s, \sigma_L)$.
7. Using simple rule $((\alpha_1 > 0)\ \&\ (\alpha_2 < 0)\ \&\ (d_1 > d_2))$ for detection tip of nose we are find position of nose in all 150 pictures using only three unique parameters σ_s, σ_L and σ_N (see Fig.4.);

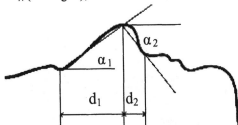

Fig. 4. Nose tip detection

The extreme in a smoothed and flattened profile that satisfy this rule belongs to the contour $F_x = 0$ of nose tip. Tracking the extreme by this contour we find position of nose tip. Coarse-to-fine tracking step is necessary because of large σ_L and σ_N.

The above algorithm with different values of σ_s, σ_L and σ_N is used for localising others fiducial points. Fiducial points are first two extrema above the nose tip and first six extrema below the nose tip and for this extrema we don't use coarse-to-fine tracking. All fiducial points are shown in Fig. 2a. From these points we derive the feature vector for each profile face in database. Feature vector has twenty-one component; ten distances D_1- D_{10} (normalised with $/(D_4 + D_5)$) and eleven profile arcs A_1- A_{11} (normalised with $/(A_5 + A_6)$) (see Fig. 2.b). D_1- D_{10} are distances between the *profile axis* and fiducial points. Profile axis is straight line, which pass through fiducial points one and nine.

4. Results and Discussion

Using the two face regions A and B with appropriate parameters σ we obtained the vector corresponding to each input profile face image. The dimensions of the profile vector depend on the number of fiducial points. In works based on different fiducial point extraction procedures [4], [5] authors use different numbers of fiducial points. From these fiducial marks they derived the sets of features. In [6] are defined 17 fiducial points which appears to be the best combination for face recognition. Most methods use the minimum Euclidean distance between the unknown and the reference feature vector to determine the correct identification of a profile, and some use thresholding windows for population reduction during the search for the reference feature vector. In our work we use twelve fiducial points; this number is detrmined by the nature of the method. Additional points can possible improve recognition rate and can be obtained by changing the scale parameters σ_s, σ_L and σ_N.

Table 1. Scale parameters for detecting fiducial points

Point	σ_s	σ_L	σ_N
nose tip (point 3)	6	17	28
points 1, 2 and 10, 11, 12	1	12	12.3
points 4, 5, 6, 7, 8 and 9	1	8	9

For training and testing set we used a profile face images Database University of Bern which contains profile views of the 30 people. For each person they took the five graylevel images with variations of the head position, the size and the contrast (1, 2, 3 big profiles with high contrast, looks like binarized, and 4, 5 small profiles with normal gray levels). Pictures are with controlled uniform background and without background clutter. Size of these images is 342 x 512 pixels.

Table 2. Feature vectors distance matrix (for the three persons)

Euclidean distances		Person 1			Person 2			Person 3		
		a	b	c	a	b	c	a	b	c
Person 1	a	0	0.16	0.53	0.56	0.61	0.62	1.08	0.99	0.98
	b	0.16	0	0.48	0.63	0.68	0.66	1.05	0.96	0.94
	c	0.53	0.48	0	0.84	0.95	0.85	0.66	0.56	0.54
Person 2	a	0.56	0.63	0.84	0	0.17	0.25	1.29	1.24	1.25
	b	0.61	0.68	0.95	0.17	0	0.29	1.41	1.37	1.37
	c	0.62	0.67	0.85	0.25	0.29	0	1.33	1.27	1.28
Person 3	a	1.08	1.05	0.66	1.29	1.41	1.33	0	0.17	0.25
	b	0.99	0.96	0.56	1.24	1.37	1.27	0.17	0	0.17
	c	0.98	0.93	0.54	1.25	1.37	1.27	0.25	0.17	0

Euclidean distances between feature vectors of nine profile pictures of three persons are shown in Table 2. We can see grouping of distances amount the three vectors that belongs to the same profile face.

In recognition experiment we use four profile images for each of 30 people in training set. The fifth profile face image for each of 30 person we used in testing set. In the first step we extract ten most similar profiles using only components $D_1 - D_{10}$. In the second step we used ten profiles from the first step and find the profile face with smallest Euclidean distance using component $A_1 - A_{11}$. The algorithm is successively recognised 27 out of 30 persons. Profile faces from three people were recognised incorrectly. Obtained recognition rate is 90%.

5. Conclusion

In this paper, a new approach for face recognition from profile images is proposed. The method is based on scale-space analysis of the profile, followed by the extraction of fiducial points on the profile. Scale-space analysis gives the scale parameters for the family of profile faces needed for detection fiducial points. Using only nine constant scale parameters (see Table 1.) we find fiducial points in all 150 profile images. From these fiducial points a set of feature vectors is created. A ranking of most similar faces was obtained by ordering the Euclidean distances. Method is simple and fast and has shown promising results.

References

1. A. P. Witkin: Scale-space filtering, Proc. 8th. Int. Joint Conf. AI 1983, pp 1019-1022.
2. R. Chellappa, C. L. Wilson, and S. Sirohey: Human and Machine Recognition of Faces: A Survey, Proc of the IEEE, Vol. 83, No. 5, May 1995. pp. 705-740.
3. S. Loncaric: A Survey of Shape Analysis Techniques, Patt. Rec. Vol. 31, No. 8, 1998, pp. 983- 1001
4. L. Harmon and W. Hunt, "Automatic recognition of human face profiles," Computer Graphic and Image Process., vol.6, pp. 135-156, 1977.
5. C.Wu and J. Huang, "Human face profile recognition by computer," Patt. Recog, vol.23, pp. 255-259, 1990.
6. L. Harmon, M. Khan, R. Lasch, and P. Ramig, "Machine identification of human faces," Patt. Recog., vol.13, pp.97-110, 1981.
7. Pigeon, S.: Multimodal Face Recognition and Identification. UCL. PhD Thesis. (1999)
8. Parker, J. R.: Algorithms for Image Processing and Computer Vision. Wiley Comp. Pub. (1998)

Pattern Recognition Combining De-noising and Linear Discriminant Analysis within a Real World Application

Volker Roth[1], Volker Steinhage[1], Stefan Schröder[2],
Armin B. Cremers[1], and Dieter Wittmann[2]

[1] Institut für Informatik, Römerstrasse 164, D-53117 Bonn, Germany
{roth,steinhag,abc}@cs.uni-bonn.de
[2] Institut für landwirtschaftl. Zoologie und Bienenkunde, Melbweg 42,
D-53127 Bonn, Germany
ULT404@ibm.rhrz.uni-bonn.de

Abstract. Computer aided systems based on image analysis have become popular in zoological systematics in the recent years. For insects in particular, the difficult taxonomy and the lack of experts greatly hampers studies on conservation and ecology. This problem was emphasized at the UN Conference of Environment, Rio 1992, leading to a directive to intensify efforts to develop automated identification systems for pollinating insects. We have developed a system for the automated identification of bee species which employs image analysis to classify bee forewings. Using the knowledge of a zoological expert to create learning sets of images together with labels indicating the species membership, we have formulated this problem in the framework of *supervised learning*. While the image analysis process is documented in [5], we describe in this paper a new model for classification that consists of a combination of Linear Discriminant Analysis with a de-noising technique based on a nonlinear generalization of principal component analysis, called Kernel PCA. This model combines the property of visualization provided by Linear Discriminant Analysis with powerful feature extraction and leads to significantly improved classification performance.

1   Introduction

Within the project *Automated Classification of Bee Species* (Hymenoptera: Apoidea) *by Morphological Analysis of Digital Images* we developed an automated identification system for bees based on image analysis of their forewings. Using diffuse background illumination, the wings of the bees show a clear venation within a transparent skin, see fig. 1. They are clipped under a microscope slide and a digitized image is taken with a CCD camera with connected frame grabber. The following image analysis procedure consists of extracting the whole venation (see fig. 2) and representing the wing by a data vector whose components are morphometric features of the veins. With labelled learning sets of some 50 specimens per species, being delivered by a zoological expert, the problem can be interpreted in the framework of *supervised learning*. A learning machine is used to find a functional dependency between the labels which indicate the class membership and the numerical attributes extracted from the images of the wings.

Our recent work has shown that our approach consisting of Linear Discriminant Analysis for classification allows the identification of many bee species with high performance, see [5] and [2]. In this paper we present an advanced model for classification that has shown great advantage for the case of bee species living in *social communities* which constituted some problems for our previous system.

Fig. 1. A wing image.   **Fig. 2.** Extracted venation.

1.1 Classification Based on Fishers Linear Discriminant

For the purpose of classification, it is useful to subdivide the approximately 800 bee species living in Middle Europe into 2 different groups, namely solitary and social bees. The former can bee seen as independent individuals whereas the latter live in large social communities with distinct castes. This division is important since it permits us to employ two models of inner species variability. Solitary bees are equal individuals and within one species they should vary only due to random genetic differences which can be assumed to be normally distributed. On the other hand, the inner species distribution of social bees depends heavily on their specific caste system which leads to more complex models and an increased scatter in the modeling data.

Our first attempt to classify the bees was to employ classical linear discriminant analysis (known as Fishers Linear Discriminant or LDA) [1]. LDA is a statistical method that projects the data vectors into a lower dimensional space in such way that the ratio of *between group scatter* S_B and *within group scatter* S_W is maximized. Denoting K the number of classes, LDA formally consists of an eigenvalue decomposition of $S_W^{-1} S_B$ leading to the so called *canonical variates*. These can be ordered by decreasing eigenvalue size such that the first variates contain the major part of the classification information. As a consequence, this procedure allows low dimensional representations and therefore a *visualization* of the data. Such visualization is very important in an interdisciplinary project since it makes an intuitive interpretation of classification results possible for the zoological user and therefore helps to increase the acceptance of our computer aided classification system within the zoological community.

Classification based on LDA indeed showed very good results for solitary bees, indicating that the Gaussian assumptions of the class conditional distributions were justified. Using the same model for the classification of social bees, however, led to much worse classification rates. This indicates that in this case the model assumptions do not hold and that one should use a more sophisticated classification approach and/or extract 'better' features from the data using appropriate preprocessing methods.

The first point addresses some theoretical shortcomings of the classical LDA approach. LDA is a simple one prototype classifier, the class prototypes are considered to be the mean vectors of the class conditional distributions. The problem is, that such kind of prototypes may be appropriate for *representation* of the classes but not necessarily for *separation*. Moreover, LDA requires the estimation of a covariance matrix and the mean vectors from the given sample, which may be a very hard estimation task in situations where the input dimensionality is high and only small samples sizes are available. A classification approach which overcomes these shortcomings is the so called *Support Vector Machine* [6]. The 'prototypes' used by the SVM classifier are constructed to be optimal for *separation*. They consist of those input patterns that are lying closest to the separating hyperplane and are called *support vectors*. SVMs implement the *Structural Risk Minimization Principle* which allows high generalization performance in high dimensional spaces if only small sample sizes are available. In cases of 'hard' classification problems that afford highly nonlinear decision functions, the SVMs have shown great success [6].

However, in special classification tasks, where the model assumptions made by the LDA approach are not violated strongly and where a rather 'simple' decision boundary may be sufficient, the advantages of the SVM classifier become less and their missing visualization ability may favour a LDA-like model in spite of theoretical shortcomings. For the classification of bees (which we consider as such case) we will show that the combination of suitable data preprocessing and LDA leads to a classifier that still has the convenient property of data visualization and even outperforms the SVM approach. Such data preprocessing employs an unsupervised feature extraction phase within the previously described supervised learning framework.

1.2   Combination of Kernel PCA and LDA

In [3] a method of unsupervised preprocessing using a nonlinear generalization of principal component analysis, called *Kernel PCA*, as feature extractor for an SVM is proposed. The straightforward combination of Kernel PCA with the LDA approach leads to the problem that LDA fails to extract suitable class boundaries due to difficulties involved with parameter estimation in a higher-dimensional space. Therefore we used a reconstruction technique described in [4] for calculating pre-images of the extracted features in input space. This application of Kernel PCA can be interpreted as *de-noising* of the input patterns resulting in decreased inner class scatter leading to better classification results.

Following this idea we interpret our bee data as if it was generated from one or more point sources per species and disturbed by additive noise due to genetical mutations and differences in castes and populations. We propose a classification model which consists of de-noising based on Kernel PCA as a preprocessing step for the LDA classifier. This combination was able to reduce misclassification rates by a factor of 4 compared to pure LDA. We consider this model to be suitable for the special type of classification problems characterized by the following features. Firstly, the class conditional distributions are 'Gaussian-like' due to a random process (e.g. random genetical mutations) which may be disturbed by substructures (e.g. induced by distinct castes and populations). Secondly, there exists a low-dimensional representation of each pattern, i.e. a small set of highly informative features that can be extracted from the data.

2 Kernel PCA as Nonlinear Generalization of PCA

PCA is a classical technique for finding an approximation of data sets in a lower dimensional subspace. The new coordinates are given by projections onto the eigenvectors of the sample covariance matrix and are called *principal components*. PCA therefore is readily performed by diagonalizing the sample covariance matrix $C = \frac{1}{l} \sum_{i=1}^{l} x_i x_i^T$ (assuming that the data are centred). This corresponds to solving the eigenvalue equation $\lambda v = C v$. The main idea of *Kernel PCA* [3] is to apply standard PCA not in the space of observations but in a *feature space* F that is related to the former by a nonlinear map

$$\Phi : \mathbb{R}^N \to \mathbb{F}, \qquad x \to \Phi(x). \tag{1}$$

Assuming that the mapped data are centred in $\mathbb{F}$, one needs to find eigenvalues $\lambda > 0$ and eigenvectors $V \in \mathbb{F} \backslash \{0\}$ satisfying

$$\lambda V = \overline{C} V \qquad \text{with} \quad \overline{C} = \frac{1}{l} \sum_{j=1}^{l} \Phi(x_j) \Phi(x_j)^T. \tag{2}$$

Defining an $l \times l$ matrix K by $K_{ij} := (\Phi(x_i) \cdot \Phi(x_j))$, the eigenvalue equation translates into: $l \lambda \alpha = K \alpha$. For the purpose of extracting principle components one has to compute projections onto the eigenvectors V^m in $\mathbb{F}$. Let x be a test point with image $\Phi(x)$ in $\mathbb{F}$, then the projection onto the m-th component is given by

$$\beta_m := (V^m \cdot \Phi(x)) = \sum_{i=1}^{l} \alpha_i^m (\Phi(x_i) \cdot \Phi(x)). \tag{3}$$

It is important that computing the projections β_m can be done *without* performing the map $\Phi(x)$ explicitly: only dot products $(\Phi(x_i) \cdot \Phi(x))$ are needed. Computation of dot products in feature space can be done efficiently by using *kernel functions* k (for details see [3]). For some choices of k there exists a map Φ into some feature space $\mathbb{F}$ such that k acts as a dot product in $\mathbb{F}$.

2.1 Reconstruction and De-noising

Reconstruction of an image of a vector x under the map Φ in feature space F from the first n principal components can be done by applying a projection operator

$$P_n \Phi(x) = \sum_{m=1}^{n} \beta_m V^m. \tag{4}$$

If we think about *data de-noising*, however, we are interested in reconstructions in input space rather than in feature space. That means, we are looking for a vector $z \in \mathbb{R}^N$ satisfying $\Phi(z) = P_n \Phi(x)$. But depending on the kernel functions employed, such a z may not exist [4]. However, reasonable pre-images can be approximated by minimizing the squared distance $\rho(z)$ between the Φ-image of a vector z and the reconstructed pattern in $\mathbb{F}$:

$$\rho(z) = \|\Phi(z) - P_n\Phi(x)\|^2. \tag{5}$$

Substituting (4) in (5) and dropping terms independent of z leads to an equation which can again be expressed solely in terms of the kernel function:

$$\rho(z) = k(z, z) - 2\sum_{m=1}^{n} \beta_m \sum_{i=1}^{l} \alpha_i^m k(z, x_i). \tag{6}$$

For kernels satisfying $k(x, x) = \text{const}$, $\forall x$ an optimal z can be determined by an iterative update scheme as follows:

$$z_{t+1} = \frac{\sum_{i=1}^{l} \gamma_i \exp(-\|z_t - x_i\|^2/c) x_i}{\sum_{i=1}^{l} \gamma_i \exp(-\|z_t - x_i\|^2/c)} \tag{7}$$

with $\gamma_i = \sum_{m=1}^{n} \beta_m \alpha_i^m$. A popular kernel type which satisfies ($k(x, x) = \text{const}$) are e.g. the Radial Basis Function (RBF) kernels of the form $k(x, y) = e^{-\|x-y\|^2/c}$.

2.2 A Two Dimensional Model for Bee Species

To give insight into point movements under Kernel PCA, we used 2 dimensional artificial datasets that allow visualization. We computed the distance between the Φ-image of an input pattern and its own reconstruction after applying kernel PCA:

$$\tilde{\rho}(z) = \|\Phi(z) - P_n\Phi(z)\|^2 = \|\Phi(z)\|^2 - 2\left(\Phi(z) \cdot P_n\Phi(z)\right) + \|P_n\Phi(z)\|^2 \tag{8}$$

The idea is that a point in input space whose Φ-image is a good approximation of its own reconstruction should be a fixed point of the whole procedure. Under the assumption, that one can reach always the same approximation quality, i.e. the same minimal value of (5), (8) can be seen as a measure of movement. Similar to (6) there is an expression of (8) in terms of the kernel function:

$$\tilde{\rho}(z) = k(z, z) - 2\sum_{i=1}^{l} \tilde{\gamma}_i k(z, x_i) + \sum_{i,j=1}^{l} \tilde{\gamma}_i \tilde{\gamma}_j k(x_i, x_j) \tag{9}$$

with $\tilde{\gamma}_i = \sum_{k=1}^{n} \tilde{\beta}_k \alpha_i^k$ and $\tilde{\beta}_k = (V^k \cdot \Phi(z)) = \sum_{i=1}^{l} \alpha_i^k k(z, x_i)$.

We used a toy example with artificial data coming from 3 different point sources with additive Gaussian noise (we repeated the experiment in [4]). This toy example can be seen as a simple model of bee specimens belonging to 3 different species with normally distributed inner species scatter. Reconstruction based on the first 2 principal components could effectively reduce the noise as can be seen in Fig. 3. All the data points are moved towards their original point sources and the 'inner species' scatter is reduced completely (middle). The right plot shows contour lines of the 'movement' of data points calculated according to (8). White indicates areas in input space with the largest amount of movement, while black indicates fixed points. The white circles indicating the position of the calculated pre-images within this contour plot coincide with the fixed points of the whole procedure. Obviously, the first 2 principal components contain all the structural information and the higher components only pick up the noise.

In our experiments with real world data from the bee project, we could show that Kernel PCA leads to similar de-noising effects even in the case of high-dimensional data with non-Gaussian distributed classes (see part 4).

Fig. 3. De-noising of artificial data generated from 3 point sources with Gaussian noise (left), point movements (middle) and contour lines of 'minimal movement' (right).

3 Using Kernel PCA as Feature Extractor for Classification

In [3] an interpretation of feature extraction using Kernel PCA as a two layer network is given. (See the left part of Fig. 4.)

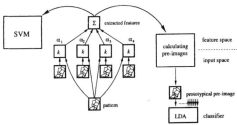

Fig. 4. Feature extraction for the SVM (left) and calculation of pre-images combined with the LDA classifier (right). The patterns show extracted venations of wings.

In the first layer, an input vector is compared with the sample via the kernel function k. In the second layer the outputs are linearly combined with weights α_i. The authors propose a combination of this form of Kernel PCA with the SVM to form a multi-layer SVM by using the extracted features as inputs for the classifier. This classifier reaches the best performance if the number of extracted features is very large, indicating that Kernel PCA is able to provide features in a high-dimensional space that are convenient for classification based on a SVM.

However, LDA would have severe problems in such high-dimensional spaces. Moreover it is not clear why the new features should improve the parameter estimation anyway. Therefore, if one wishes data visualization, simply substituting the SVM by a LDA classifier does not seem to be promising. However we have shown experimentally that picking up the information contained in the extracted features and calculating pre-images of the mapped data in input space forms a preprocessing step useful for LDA. (See the right part of Fig. 4.) Compared to the original patterns, the pre-images are a more 'prototypical' version of the former leading to a *class inherent de-noising* of the input data.

4 Classification of Bees

In this section we illustrate classification of bees using kernel PCA as a preprocessor for LDA employing following example. Within the genus *Bombus* (bumble bees) there is a complex of 3 very similar species which is known to be a very difficult identification problem for taxonomical experts. For instance, it is almost impossible to correctly classify the female specimens of the worker casts of those 3 species.

We applied the LDA approach on 35-dimensional datasets representing the wings of the bees. The learning set consisted of 70 specimens per class, testing for the generalization ability was done using a leave-one-out cross-validation procedure. The learning set was not separable, there were 4 misclassifications. Figure 5 shows the projected data after applying LDA. Since we investigated 3 classes, the projections lie in a (3-1)-dimensional subspace that can be visualized without any loss of information. The classification rule in this space assigns an observation to the nearest class center. The leave-one-out procedure yields 12 misclassifications (5.7% misclassification rate). After de-noising with kernel PCA there is only 1 misclassified bee in the learning set and the cross-validation error drops down to 3 specimens (a misclassification rate of 1.4%). As one can see in Fig. 6, the inner class scatter is significantly reduced compared to Fig. 5. This results both in better separation of the learning set and in an increased generalization ability by a factor of 4.

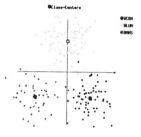 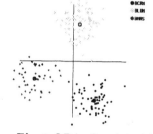

Fig. 5. Standard LDA.   **Fig. 6.** LDA after de-noising.

The question of what type of kernel function used in Kernel PCA will lead to *suitable* features is still unsolved. On the other hand, experiments with different types of kernel functions used for *classification* with SVMs have shown that the results are rather independent of the *type* of kernel [6]. If one thinks of Kernel PCA only as a preprocessor for classification, this property may carry over to the unsupervised situation and gives rise to the use of RBF kernels for which fast update schemes exist (7). The remaining question then is what parameter c should be used. In [4] a heuristic choice for c is given by the dimension of the the input vectors multiplied by twice the mean variance in every dimension. Following this idea we rescaled our data to have unit variance in every dimension by the transformation $x_{ij} \leftarrow \frac{x_{ij} - m_j}{s_j}$, with s_j being the sample variance in the j-th dimension and m_j being the sample means respectively. Our experiments show that this normalization is suitable for the use of RBF kernels and moreover provides a good estimation of the optimal parameter c. In all experiments with bee data we obtain the best results when c is between one and two times the dimension of the input vectors: $c \sim 1.5 \cdot \dim(\boldsymbol{x})$. Moreover, this type of normalization allows good predictions of RBF kernels for the case of *classification* with SVMs. In this case we obtain the best result (a leave-one-out rate of 2.8 %) with exactly the same RBF kernel.

5 Discussion

Within the project *Automated Classification of Bee Species by Morphological Analysis of Digital Images* we have studied models for classification and feature extraction. Starting with the classical LDA approach, we could show that for many solitary bee species the model assumptions made by LDA hold rather well. This especially favours the LDA classifier because of the ease with which data may be visualized. However, for the case of social bees, this model showed worse performance and we therefore studied the advanced methods of *Support Vector Machines* for classification and *Kernel PCA* for feature extraction. From the theoretical point of view, the SVM classifier overcomes many theoretical shortcomings of LDA. Using the SVM for the classification of social bees, however, we discovered that this classifier outperforms the LDA only to some extent Moreover it does not allow data visualization. We therefore focused on feature extraction using Kernel PCA and propose a combination of de-noising and LDA for our classification problem. This leads to a classifier that inherits the convenient properties of LDA and even outperforms the SVM.

We consider the classification of social bees to be typical for a whole class of classification problems for which the model assumptions made by LDA are not violated too strongly and nonlinear classifiers allow only a slight improvement. To prove the applicability of this model to similar problems will be subject of future work.

Acknowledgement. This work was supported by Deutsche Forschungsgemeinschaft, DFG. Thanks to T. Arbuckle and M. Held for discussions and comments.

References

1. DUDA R.O., HART P.E. (1973) *Pattern Classification and Scene Analysis*, Wiley & Sons
2. ROTH V., SCHRÖDER S., CREMERS A.B., DRESCHER W., STEINHAGE V., WITTMANN D. (1999) *Computergestützte Klassifikation von Wildbienen mit Methoden der Bildanalyse*, Wissenschaftliche Berichte FZKA 6252, ISSN 0947-8620
3. SCHÖLKOPF B., SMOLA A., MÜLLER K.R. (1998) *Nonlinear Component Analysis as a Kernel Eigenvalue Problem*, In: *Neural Computation*, Vol. 10, Issue 5, pp. 1299-1398, MIT Press
4. SCHÖLKOPF B., MIKA S., SMOLA A., RÄTSCH G., MÜLLER K.R. (1998) *Kernel PCA Pattern Reconstruction via Approximate Pre-Images*, In: Niklasson L., Boden M. and Ziemke T. (eds.), *Proceedings of the 8th International Conference on Artificial Neural Networks*, Springer, Perspectives in Neural Computing.
5. STEINHAGE V., KASTENHOLZ B. SCHRÖDER S., DRESCHER W (1997), *A Hierarchical Approach to Classify Solitary Bees Based on Image Analysis*, In: Mustererkennung 1997, 19. DAGM-Symposium, Braunschweig, Informatik aktuell, Springer
6. VAPNIK V. (1998) *Statistical Learning Theory*, Wiley & Sons

Invariant Reference Points Methodology and Applications

Krystian Ignasiak*, Władysław Skarbek, Miloud Ghuwar

Multimedia Laboratory, Department of Electronics and Information Technology
Warsaw University of Technology
tel:(48 22)6605315, fax: (48 22)8255248, email:Skarbek@ire.pw.edu.pl

Abstract. A methodology for pattern recognition based on design of invariant reference points is described. Within this framework many classical pattern classifiers can be described (e.g. the k-NN distance classifier) and few new classifiers are defined (e.g. LPCAS classifier). LPCAS is applied for handwritten digit recognition reaching performance of 99.4% of recognition rate on NIST database.

1 Introduction

Pattern recognition is an area in artificial intelligence relating to synthesis and analysis of procedures for classifying objects on the basis of their physical measurements (for instance visual images). In case of images, the object recognition process consists of three main stages: (1) object localization in the image (*segmentation stage*); (2) feature vector extraction (*measurement stage*); (3) object classifying (*decision stage*).

This work concerns a new classifying methodology, called here the *IRP method (Invariant Reference Points)*.

The IRP method as a methodology which is capable to generate a number of classifiers, gives no restrictions on segmentation methods and feature extraction methods. It is based on a construction for the given class a pair of type $(x_i, F_i) =$ (the reference point from the space of measurements, the operator in the space of measurements) such that x_i is an invariant point of F_i, i.e. $F_i(x_i) = x_i$.

In any classification problem, we deal with a distinguished object class Ω, which is subdivided into nonintersecting subclasses Ω_i, $i = 1, \ldots, n$:

$$\Omega = \Omega_1 \cup \cdots \cup \Omega_n, \quad \Omega_i \cap \Omega_j = \emptyset \quad i, j = 1, \ldots, n, \quad i \neq j.$$

For each object $\omega \in \Omega$ we can measure its N–dimensional feature vector $x = X(\omega)$. The classifier using a particular vector $X(\omega) \in R^N$ elaborates a decision $\delta(x) = i$ of the membership for the object ω from the class Ω_i :

$$\delta : R^N \to \{1, \ldots, n\}.$$

If $\omega \in \Omega_i$, but $\delta(X(\omega)) \neq i$, then the decision is wrong. An optimal decision procedure δ gives the minimum for the recognition error.

---

* This work was supported in part by European ESPRIT INCO CRIT2 project for two first authors

2 Invariant reference points

Let us assume that the learning set $X^{(i)} \subset R^N$, $i = 1, \ldots, n$, consists of measurement vectors obtained for objects $\omega \in \Omega_i$.

The IRP method is based on a construction of an *IRP collection* $\mathcal{Z}_i$, $i = 1, \ldots, n$ which is built using information included in the learning set X_i. $\mathcal{Z}_i$ consists of certain number k_i of pairs $(x_j^{(i)}, F_j^{(i)})$, where $x_j^{(i)}$ is the j-th reference point in the collection $\mathcal{Z}_i$, while $F_j^{(i)}$ is the operator in measurement space ($F_j^{(i)} : R^N \to R^N$) (we call it also as *eigenoperator*) such that $x_j^{(i)}$ is its invariant point:

$$F_j^{(i)}(x_j^{(i)}) = x_j^{(i)}, \qquad \mathcal{Z}_i = \left(k_i, (x_1^{(i)}, F_1^{(i)}), \ldots, (x_{k_i}^{(i)}, F_{k_i}^{(i)}) \right)$$

where $j = 1, \ldots, k_i$, $i = 1, \ldots, n$.

Let $\mathcal{R}_i = \{x_1^{(i)}, \ldots, x_{k_i}^{(i)}\}$ be the set of reference points for the class Ω_i and $\mathcal{F}_i = \{F_1^{(i)}, \ldots, F_{k_i}^{(i)}\}$ be the set of eigenoperators for the class Ω_i.

Roughly, the IRP collection $\mathcal{Z}_i$ approximates the learning set X_i with reference points. Eigenoperators in proximity of their invariant points have a low dynamics. This property is used in the classifier. We define here three general [6, 7] schemes for constructing reference sets $\mathcal{R}_i$:

1. **Centroid method:** we define only one reference point in the class ($k_i = 1$) which is defined as the centroid of the learning set X_i :

$$x_1^{(i)} \doteq \frac{1}{|X_i|} \sum_{y \in X_i} y, \qquad \mathcal{R}_i = \{x_1^{(i)}\};$$

2. **Clustering method:** the algorithm finds a set of reference points $\mathcal{R}_i = \{x_1^{(i)}, \ldots, x_{k_i}^{(i)}\}$ optimizing a certain cost function, e.g.:

$$\text{cost}(\mathcal{R}_i) = \alpha \text{MSE}(\mathcal{R}_i, X_i) + \beta k_i \tag{1}$$

where α, β are weights,

$$\text{MSE}(\mathcal{R}) \doteq \frac{1}{|X_i|} \sum_{y \in X_i} (y - x_{j(y)}^{(i)})^2, \qquad j(y) = \arg \min_{1 \le j \le k_i} \|y - x_j^{(i)}\|.$$

Let us notice that the cost function 1 makes possible adequate choice of the number of reference points;

3. **Learning set technique:** here the set of reference points is equal to the learning set, i.e. $\mathcal{R}_i = X^{(i)}$. This technique has a special meaning when the class is represented by several measurement vectors (for instance in face recognition we usually have only few face poses of the given person).

3 Design of eigenoperators

The design of the eigenoperator $F_j^{(i)}$ such that $F_j^{(i)} = x_j^{(i)}$ is generally hard problem if the only criterion is the recognition error for the resulting classifier. We present here three universal design techniques:

1. **Constant operator technique:** for each reference point $x_j^{(i)}$, the operator $F_j^{(i)}$ is defined as follows: $F_j^{(i)}(x) \doteq x_j^{(i)}$ for each $x \in R^N$. It will be shown how this technique combined with a clustering method for reference point design, leads to k–NN method.

2. **Technique of projection onto principal subspace [8]:** let $X_j^{(i)} \doteq \{y \in X^{(i)} : i(y) = j\}$ be the set of learning vectors which are closer to $x_j^{(i)}$ than to any other reference point in $\mathcal{R}_i$. If the reference points are found by the centroid algorithm then $x_j^{(i)}$ is the centroid of the set $X_j^{(i)}$.

Assuming an adequate cardinality of this set we can implement the principal component analysis (PCA – [2, 5]) with K components. PCA gives K principal vectors which are placed into columns of the matrix $W_j^{(i)}$. Then the eigenoperator can be defined by the following formula:

$$F_j^{(i)}(x) \doteq x_j^{(i)} + W_j^{(i)} W_j^{(i)}{}^T (x - x_j^{(i)}) \quad \text{for any } x \in R^N.$$

It is obvious that $F_j^{(i)}$ is invariant in $x_j^{(i)}$, i.e. $F_j^{(i)}(x_j^{(i)}) = x_j^{(i)}$.

We should emphasize that the above design has a practical sense for large learning sets.

If PCA analysis is performed locally in each cluster of data then we call this process Local Principal Component (LPCA) Analysis [10].

3. **Fractal operator technique:** the algorithm performs an extensive search of local mappings of fragments of the vector $x = x_j^{(i)}$ of cardinality pk, $p > 1$ (so called source domains) to fragments of cardinality k (so called target domains) [4].

Let us assume that k divides N, $(N = kl)$, and let $\Pi = \{T_1, \ldots, T_l\}$ be the set of target domains, i.e. a partition of index set $\{1, \ldots, N\}$ into l subsets of cardinality k. Let $\gamma = \{S_1, \ldots, S_l\}$ be the set of the source domains which are best matched to corresponding target domains, $|S_i| = pk$. Matching here denotes an optimal local affine mapping $L_a^{(x)}$ which reduces the fragment $x|_{S_a}$ to $x|_{T_a}$, where $\cdot|_{T_a}$ denotes a restriction to the domain T_a. Then the fractal operator, i.e. the eigenoperator for the reference point $x = x_j^{(i)}$ is defined as follows:

$$F_j^{(i)}(y) = \sum_{a=1}^{l} L_a^{(x)}(y).$$

4 Class selection techniques

In the IRP method, for the classification of a measurement vector x we consider a sequence of distance values $D_i(x) = (\|x - F_1^{(i)}(x)\|, \ldots, \|x - F_{k_i}^{(i)}(x)\|)$, $i = 1, \ldots, n$. The classifier uses a selection technique S_a which for the given x on the basis of $D_i(x)$ returns a class id, x probably belongs to.

Let $D(x) = (D_1(x), \ldots, D_{k_i}(x))$. There exists many possibilities to aggregate information about distances, and next the class selection. Few of them are defined below:

1. **Minimum distortion technique:** $S_1 \doteq \arg \min_{1 \le i \le n} \min(D_i(x))$. The class for which the minimum distortion $\|x - F_j^{(i)}(x)\|$ is achieved;

2. **Minimum average distortion technique:** $S_2 \doteq \arg \min_{1 \le i \le n} \text{avg}(D_i(x))$,

where $\text{avg}(D_i)$ is the average value of the sequence D_i. Here the average distortion introduced by eigenoperators in the given class decides about the choice of the given class.

3. **Basis functions technique:**

$$S_3 \doteq \left| 0.5 + \sum_{i=1}^{n} \sum_{j=1}^{k_i} c_{ij} \phi_{ij}(x) \right|,$$

where $\phi_{ij}(x) \doteq e^{-\|x - F_j^{(i)}(x)\|}$, $x \in R^N$ is a basis function concentrated in the point $x_j^{(i)}$. The coefficients c_{ij} can be found using recursive least square method for the learning data sequence $X = \bigcup_{i=1}^{n} X^{(i)}$.

4. **k least distortions technique k–NZ:** Let $I_k(x) \doteq (i_1, \ldots, i_k)$ be the sequence of class ids for which we have found k least distortions from the sequence $\|x - F_j^{(i)}(x)\|$. Then $S_4 \doteq \arg\max_{1 \le j \le n} |\{a : i_a = j\}|$, where i_a is a-th coordinate of the sequence $I_k(x)$. In this approach we choose the class id which occurs in k least distortions most frequently.

5 The IRP method definition

Let $X^{(i)} \subset R^N$ be the set of learning sequence for the class Ω_i, $i = 1, \ldots, n$. Suppose that for these learning sets we can build IRP ensembles $Z_i = (k_i; (x_1^{(i)}, F_1^{(i)}),$
$\ldots, (x_{k_i}^{(i)}, F_{k_i}^{(i)}))$.

Let $x \in R^N$ be a testing measurement vector. Then the classification procedure in the IRP method is of the form:

1. Compute sequences $D_i(x) = (\|x - F_i^{(j)}(x)\|)$, $1 \le j \le k_i$, $i = 1, \ldots, n$;
2. Let $D(x) = (D_1(x), \ldots, D_n(x))$;
3. Assign x to the class with id $\delta(x) \doteq S_a(D(x))$.

The above algorithm is of generic type. A specific algorithm is obtained by specifying: (1) a method for constructing the IRP ensemble Z_i, $i = 1, \ldots, n$; (2) a norm $\|x - y\|$; (3) a selection technique S_a.

6 Specifying classifiers by IRP method

In this section, we show several classifiers specified as special cases of the IRP method.

1. **The minimum distance from reference points method [6, 7, 3]:** Let us assume that in the i-th class there are reference points $R_i = \{x_1^{(i)}, \ldots, x_{k_i}^{(i)}\}$. Building an IRP ensemble we choose constant operator technique, i.e. we choose $F_j^{(i)}(x) = x_j^{(i)}$ for each $x \in R^N$. Let us notice that the distortion introduced by the operator $F_j^{(i)}$ equals to the distance of x to the reference point $x_j^{(i)}$:

$$\|x - F_j^{(i)}(x)\| = \|x - x_j^{(i)}\|$$

Therefore by choosing the selection technique S_1 which is based on the minimum of the distortions we get a classifier which chooses the class to which the closest reference point belongs:

$$\delta(x) \doteq S_1(D(x)) = \arg \min_{1 \leq i \leq n} \min_{1 \leq j \leq k_i} \|x - x_j^{(i)}\|;$$

2. **k nearest neighbors method (k–NN) [3]:** Similarly to minimum distance method, we take constant operators for the IRP ensemble too. As the selection technique we choose the technique S_4 of k least distortions. Then the sequence $I_k(x) = (i_1, \ldots, i_k)$ is the sequence of k least distances to reference points. The technique S_4 chooses the most frequent class in the sequence $I_k(x)$, i.e. the class which is the most frequent class within k nearest neighbors of the testing vector x :

$$\delta(x) \doteq S_4(D(x)) = \arg \max_{1 \leq j \leq n} |\{a : i_a = j\}|.$$

This is a decision function of the very popular k-NN classifier which appeared to be very effective in many applications (e.g.: crops estimates based on satellite LANDSAT pictures);

3. **Radial basis functions classifier [1, 9]:** Suppose that we have only single reference point $x_1^{(i)}$ for the class Ω_i. Choosing the constant eigenoperators $F_1^{(i)}$ and the selection technique S_3, we get the following decision function:

$$\delta(x) \doteq S_3(D(x)) = \left\lfloor 0.5 + \sum_{i=1}^{n} c_{i1} e^{-\|x - x_1^{(i)}\|} \right\rfloor.$$

Denoting $w_i = c_{i1}$, $x^{(i)} = x_1^{(i)}$ let us consider the function

$$g(x) \doteq \sum_{i=1}^{n} w_i e^{-\|x - x^{(i)}\|}.$$

The function $g(x)$ is a linear combination of Gaussian basis functions. There are well known techniques of searching parameters $x^{(i)}$ and w_i such that $g(x)$ estimates the given function f which is known only by specifying of learning pairs (y, i), i.e. measurement vectors y for learning objects and class ids i, $(i = 1, \ldots, n)$, the object belongs to. The obtained classifier is the known radial basis functions neural network.

4. **Local Principal Component for Subspace method (LPCAS) [11]:** In this method both reference points and eigenoperators are obtained in mutually joined processes of clustering and principal component searching. Both processes have neural nature.

Actually we have joined modified LBG vector quantization with modified Oja-RLS algorithm. The modification of LBG is in replacing from the second epoch, of the distance from the reference point by the distance from the current subspace anchored in this reference point (see [11]).

As the selection technique for LPCAS, we choose the minimum distortion technique S_1.

7 Application of LPCAS to handwritten digit recognition

We have applied LPCAS method for handwritten digits stored in NIST database (see sample input data in Fig. 1(a)) which were collected from zip codes handwritten on envelopes. The database includes about 200 thousands of pictures for already segmented digits. More extended comparison with other methods is shown in this paper than in previous author's paper [11].

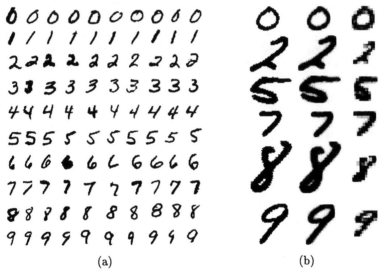

(a)   (b)

Fig. 1. (a) Sample data from NIST data base. (b) Feature vector extraction (from left to right): original images, gray versions, reduced to 10 by 10 pixels grid.

Vector data extraction from image data. Each picture in NIST database has the binary form with the fixed resolution of 128 rows and 128 columns. However, the digit itself occupies a small part of the picture. For convenience of file handling on PC for experiments about 14 thousands pictures were chosen randomly from the original database. Each experimental session of our recognition program begins from a random choice of about seven thousands pictures for local model learning and from the remaining set, the same number of random testing pictures. The results of recognition at such circumstances appear to be stable, i.e. they change a little from one experimental session to another.

The vector data extraction is implemented in two steps (cf. Figure 1(b)). Firstly, the conversion from binary image to gray scale image is performed by counting the percentage of black pixels in the neighborhood of the current pixel. Secondly, the background is cropped out and the remaining foreground part is resized to the resolution $n \times n$ where typically $n = 10 - 20$. The resizing operation is actually a decimation filter implemented by simple averaging of those pixel values which enter into the given superpixel.

Results for distance to class centroids. Our program can be specialized to a distance classifier by setting the subspace dimension to zero and the number of clusters in each class to one. The results of classification for a random sample of about seven thousands testing examples is shown in the Table 1. The entry n_{ij} of i-th row and j-th column contains the number of data samples which belong to the class i and are recognized as digits from the class j. We obtain for this classical method about 86% of recognition rate what is rather a poor result.

digit	0	1	2	3	4	5	6	7	8	9	all
%	92.98	81.92	86.66	89.96	87.75	82.78	91.98	83.69	76.74	83.40	85.79

Table 1. Results for the distance method: $N = 20 \times 20$, $K = 0$, $L = 1$.

Results for distance to class subspaces. In these experiments we test the subspace method which is obtained in our program by setting the number of clusters in each class to one. The dimension K of the subspace for each class is set to the same value. From the Table 2, we see that increasing K from one to 16, results in increase of the global recognition rate. However, further increase of K up to maximum possible value N results in worse performance.

	0	1	2	3	4	5	6	7	8	9	all
1	95.1	94.5	91.2	93.6	87.6	90.8	95.7	92.3	83.0	85.0	90.87
2	96.8	96.8	90.8	93.7	91.9	94.2	96.7	94.2	86.7	92.3	93.41
4	98.5	100.0	90.9	95.1	92.3	96.7	97.7	95.5	89.9	94.6	95.13
8	98.5	99.9	95.9	97.0	96.8	97.1	99.0	97.3	92.6	97.8	97.19
16	99.2	99.4	97.4	96.7	97.7	97.6	99.4	99.2	95.6	98.3	98.08

Table 2. Results for the subspace method: $N = 20 \times 20$, $K = 1, 2, 4, 8, 16$, $L = 1$.

Results for distance to class local subspaces. The LPCAS method was tested for different dimensionality K of local subspaces and different number L of clusters in classes. Results are presented in the Table 3. Similarly to the subspace method further increase of K deteriorates algorithm's performance too. The increase of L to values more than six makes results less confident as the number of samples per cluster decreases to the level at which learning of the local model by the neural approach is not possible.

In order to tune the classifier we have decided to set number of clusters in classes individually. This increases the recognition rate up to 99.36%, see Table 4.

	0	1	2	3	4	5	6	7	8	9	all
1	97.5	99.4	95.0	95.1	95.1	97.1	97.0	96.5	87.4	94.6	95.47
2	98.6	99.6	94.7	96.2	95.2	98.5	98.9	97.0	89.2	96.7	96.46
4	99.0	99.6	96.6	96.7	96.5	97.6	98.9	98.4	93.8	97.5	97.46
8	100.0	99.7	98.1	97.4	98.9	99.7	99.4	99.2	96.3	98.5	98.71
16	99.9	99.6	98.5	98.3	99.0	99.1	99.7	98.7	96.9	98.7	98.84

Table 3. Recognition rates for LPCAS: $N = 20 \times 20$, $K = 1, 2, 4, 8, 16$, $L = 4$.

	0	1	2	3	4	5	6	7	8	9	all
L	4	4	4	4	4	4	4	4	4	4	
%	99.9	99.6	98.5	98.3	99.0	99.1	99.7	98.7	96.9	98.7	98.84
L	4	4	5	5	5	5	4	5	5	5	
%	99.6	99.7	99.1	99.1	98.9	99.2	99.9	99.1	98.2	98.7	99.14
L	4	4	5	5	**6**	5	4	5	**6**	**6**	
%	99.7	99.7	99.1	99.1	99.4	99.2	99.9	99.1	99.2	99.2	**99.36**

Table 4. Tuning the classifier: setting the number of local subspaces in classes individually , $N = 20 \times 20, K = 16$.

8 Conclusions

New methodology for pattern recognition was elaborated. It is based on design of invariant reference points. The methodology is capable to define new classifiers such as fractal operator classifying system. It is shown that many prominent classical recognition schemes, for instance the k-NN distance classifier, are special cases of the IRP method.

Local Principal Components Analysis for Subspace method is also expressed in this new formalism. LPCAS applied for pattern recognition problem, appears to be a better tool than classical subspace method. The reason is in ability of discrimination for existing subclasses in the given pattern class. For handwritten numerals the technique reaches the recognition rate of about 99.4%.

References

1. C. Bishop (1995) *Neural Networks for Pattern Recognition*, Clarendon Press, Oxford.
2. K.I. Diamantaras, S.Y. Kung (1996) *Principal Component Neural Networks*, John Wiley & Sons, New York.
3. R.O. Duda, P.E. Hart (1973) *Pattern Classification and Scene Analysis*, Wiley, New York.
4. Y. Fisher, ed. (1995) *Fractal Image Compression – Theory and Application*, Springer Verlag.
5. H. Hotelling (1933) *Analysis of a complex of statistical variables into principal components*, Journal of Educational Psychology 24, 417-441.
6. T. Kohonen (1995) *Self-Organizing Maps*, Springer, Berlin.
7. Y. Linde, A. Buzo, R.M. Gray (1980) An algorithm for vector quantizer design, *IEEE Trans. Comm.*, COM-**28** 1980 28-45.
8. E. Oja (1983) *Subspace methods of pattern recognition*, Research Studies Press, England.
9. B.D. Ripley (1996) *Pattern Recognition and Neural Networks*, Cambridge University Press, Cambridge.
10. Skarbek, W.: Local Principal Components Analysis for Transform Coding. 1996 Int. Symposium on Nonlinear Theory and its Applications, NOLTA'96 Proceedings, Research Society NTA, IEICE, Japan, Oct. 1996 381-384
11. Skarbek, W., Ghuwar, M., Ignasiak, K.: Local Subspace Method for Pattern Recognition. CAIP'97, Lecture Notes 1296, 1997.

A New Framework of Invariant Fitting

Klaus Voss and Herbert Suesse

Friedrich-Schiller-University Jena, Department of Computer Science
Ernst-Abbe-Platz 1-4
D-07743 Jena, Germany
{nkv,nbs}@uni-jena.de
http://pandora.inf.uni-jena.de

Abstract. This paper is an extension of the already published paper Voss/Suesse [11]. In that paper we have developed a new region-based fitting method using the method of normalization. There we have demonstrated the zero-parametric fitting of *lines, triangles, parallelograms, circles* and *ellipses*. In the present paper we discuss this normalization idea for fitting of closed regions using circular segments, elliptical segments and rectangles. As features we use the area-based low order moments. We show that we have to solve only one-dimensional optimization problems in these cases

1 Introduction

There are many ellipse fitting methods available in the literature, see for example [4]. The fitting methods known from the literature face two problems. Firstly, a statistically unbiased estimate of the parameters is nessecary for an accurate reconstruction of objects, see for example [5,6]. Secondly, for the recognition of objects using ellipses, we need an invariant ellipse fitting method, see for example I.Weiss in [13]. Much of the work in this area is contained in the papers of Bookstein [1] and Forsyth et.al. But they are all concerned with fitting techniques for a discrete set of points suppose to be on an ellipse. In contrast, our proposed method is a region-based fitting method for a closed region surrounded by an elliptic arc and a line segment. As features we use the area-based moments of the given closed region. There are many advantages of region-based fitting but solving the resulting multidimensional optimization problems is in most cases numerically expensive. Our approach reduces the numerical effort drastically.

2 A New Invariant Procedure for Elliptical Segments

Besides the fitting of ellipses, in practice also the fitting of elliptical segments is important. In the following we give a short review of our fitting method, see [11]. Let O be a given object, let $P(\theta)$ be a given class of *primitives* , and let $\mathbf{T}$ be a class of transformations. The primitives $P(\theta)$ are described by m parameters $\theta_1, ..., \theta_m$, and the investigated transformation $t \in \mathbf{T}$ is characterized by

n parameters $\xi_1, ..., \xi_n$, where the tuples $(\xi_1, ..., \xi_n)$ are elements of the transformation space **T**. Furthermore, we choose a typical *representative* of the class of primitives (e.g. the square as a representative of the class of all parallelograms). Now we derive features of the object, for example, the feature tuple $\mathbf{f}(O) = (M_{00}, M_{10}, M_{01}, ...)$ of the area-moments M_{kl} of the object O.

The well-known feature-based fitting procedure solves the problem

$$Minimize \ \|\mathbf{f}(O) - \mathbf{f}(P(\theta))\|^2 \tag{1}$$

by a search in the m-dimensional space Θ of all primitives $p \in P(\theta)$. If we use the idea of simultaneously normalizing the *object* and the *primitive*, then we get the following fitting problem with respect to the canonical frames:

$$Minimize \ \|\mathbf{f}(O') - \mathbf{f}(P'(\theta^*))\|^2 \ . \tag{2}$$

The dimension of the fitting problem (2) is zero for all examples discussed in our paper [11], since the normalization procedure reduces the dimension of the "free-parameter-space" to zero.

As features of the objects and the primitives we use the area moments:

$$M_{pq} = \iint\limits_{object} x^p y^q dx dy \ .$$

The here investigated primitives $P(\theta)$ are elliptical segments that can in general be described by $m = 7$ parameters. As a typical prototype $P(\theta)$ of all elliptical segments, we choose a standard circular segment, see Fig. 1, since each elliptical segment can be affinely transformed into a circular segment. The transformations **T** are elements of the affine transformation group with $n = 6$ parameters. Thus the space θ^* of the free parameters has the dimension $l = m - n = 1$. This implies that we have only a one-dimensional fitting problem to solve.

In the following we describe the *normalization procedure*. We choose the special *standard method* , see [9, 11]. Using this standard method with respect to the central moments m_{kl}, we have to carry out the following procedure step by step:

- Normalize the translation by the centroid, i.e., $m_{10} = m_{01} = 0$.
- Normalize the x-shearing by $m_{11} = 0$.
- Normalize the anisotrope scaling by $m_{20} = m_{02} = 1$.
- Normalize the rotation by $m_{30} + m_{12} = 0$, additionally $m_{12} + m_{03} \geq 0$.

Firstly, we have to do this for the object by a numerical procedure. Secondly, we have to do this analytically for all equivalence classes of elliptical segments. We describe this normalization in the following.

An elliptical area segment A ist the intersection of an elliptical region E and a half space H. Thus, the affine mapping $E \rightarrow E' = C$ to a circular region C and the mapping $H \rightarrow H'$ of the half space by the same affine transformation yield

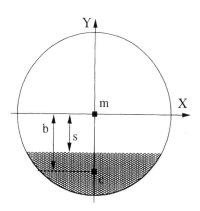

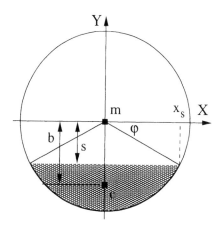

Fig. 1. Circular Segment, distance $b \geq 0$, coordinate s with $-1 \leq s \leq +1$

Fig. 2. Sector and triangle, $-\frac{\pi}{2} \leq \varphi \leq +\frac{\pi}{2}$

a circular segment as their intersection, $C \cap H'$. Hence each elliptical segment can be affinely mapped to any circular segment. All circular segments can be arranged as a one-parametric set, see Fig. 1, and it is only necessary to investigate circular segments. A circular segment $C(s)$ can be interpreted as a prototype of the equivalence class of all elliptical segments that can be affinely mapped to $C(s)$. The essential parameter of a circular segment is the parameter s (the radius of the circle is $r = 1$), and we assume that the moments M_{pq} of this segment are given. Thus, we have to calculate the moments $M_{pq}(s)$ of a circular segment, displayed in Fig.1. To do this, we decompose the circular segment $C(s)$ into a circular sector $C_s(\varphi)$ and a triangle $C_t(\varphi)$, see Fig. 2. Note that $M_{pq}(C(s)) = M_{pq}(C_s(\varphi)) + M_{pq}(C_t(\varphi))$. The zeroth to fourth order moments of the circular sector and the triangle with $\varphi = \arcsin s, x_s = \sqrt{1 - s^2}$ you can find in [12].

Now, we have to carry out the normalizing steps. First, we calculate the centroid $\mathbf{c} = (x_c, y_c) = (0, -b)$ of the circular segment, and then we shift the centroid into the origin of the coordinate system. Then, the new central moments are

$$m'_{pq} = \sum_{i=0}^{q} \binom{q}{i} b^i M_{p,q-i} , \ b \geq 0 .$$

The new location of the circular segment is displayed in Fig. 3. Here, the two first conditions of the affine invariant mapping are fulfilled, namely $m'_{10} = 0$ and $m'_{01} = 0$. In the next step, we calculate the new second order moments. Because of the symmetry of the problem, we have already $m'_{11} = 0$. Finally, we perform a simple anisotrope scaling $x'' = \gamma \cdot x'$ and $y'' = \delta \cdot y'$ for which we use the normalization

$$m''_{20} = 1 , \ m''_{02} = 1 .$$

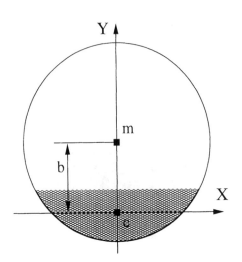

Fig. 3. Shifted Circular Segment

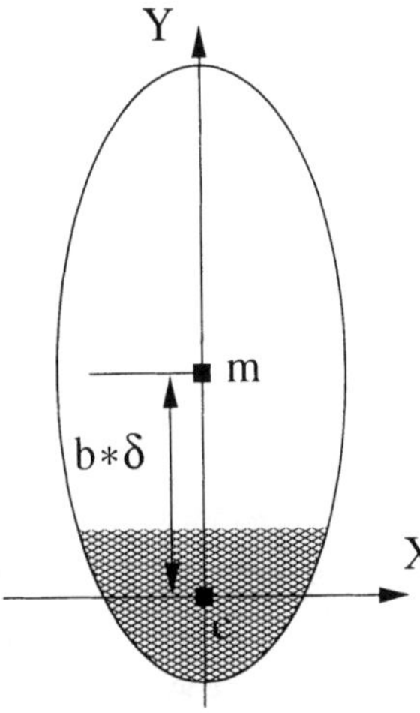

Fig. 4. Canonical Frame of a Circular Segment

These equations have the solutions

$$\gamma = \sqrt[8]{\frac{m'_{02}}{m'^3_{20}}}, \quad \delta = \sqrt[8]{\frac{m'_{20}}{m'^3_{02}}}.$$

Using this solutions, we have to transform the moments by $m''_{pq} = \gamma^{p+1}\delta^{q+1}m'_{pq}$. Note that these two remaining constraints for an affine mapping $m''_{30} + m''_{12} = 0$ and $m''_{21} + m_{03} > 0$ are fulfilled automatically. This follows from the fact that all moments m_{kl} vanish for odd numbers k when the circular segment is located in a symmetrical orientation. During these algorithmic steps, the centroid is always unchanged in the origin. However, the middle point **m** of the original circle is first shifted to $(0, b)$ and then to $(0, b \cdot \delta)$. The diameter of the original unit circle is scaled in x-direction by the factor γ, and in y-direction by the factor δ. So we find the equation

$$\frac{x''^2}{\gamma^2} + \frac{(y'' - b \cdot \delta)^2}{\delta^2} = 1 \tag{3}$$

for the ellipse of the canonical frame, see Fig. 4. Now it is possible *analytically* to determine the higher order moments of the canonical frame (Fig. 4), which depend on parameter s:

$$m''_{pq}(s) = \int\int x''^p y''^q dx'' dy''.$$

The particularly chosen prototype $P(\theta)$ is the circular segment displayed in Fig. 1. The expression $P'(\theta^*)$ is the canonical frame of the circular segment, displayed in Fig. 4. The features $\mathbf{f}(P'(\theta^*)$ are the moments $m''_{pq}(s)$ of the canonical frame. The moments $m''_{pq}(s)$ can be calculated analytically. The fitting problem is now reduced to finding the solution of the optimization problem (2). This optimization problem depends on one parameter s only. To solve the optimization problem (2), we use only some moments up to the order 4 by

$$Minimize \left[(m''_{00}(s) - m''_{00}(O'))^2 + \sum_{3 \leq p+q \leq 4} (m''_{pq}(s) - m''_{pq}(O'))^2 \right]. \quad (4)$$

The result of the optimization process (4) is an optimal parameter, s_{opt}.

How can we find the equation of the fitted ellipse of the object? The equation of the ellipse of the canonical frame is described by equation (3), where we choose the parameters b, γ, δ depending on s_{opt}. Thus we can describe this ellipse by the equation $\mathbf{x}''^T \mathbf{E}'' \mathbf{x}'' = 0$, where

$$\mathbf{E}'' = \begin{pmatrix} \frac{1}{\gamma^2} & 0 & 0 \\ 0 & \frac{1}{\delta^2} & -\frac{b}{\delta} \\ 0 & -\frac{b}{\delta} & b^2 - 1 \end{pmatrix}, \quad \mathbf{x}'' = \begin{pmatrix} x'' \\ y'' \\ 1 \end{pmatrix}.$$

Now, let $\mathbf{A}$ be an affine transformation derived by the standard method that transforms the given object O to its canonical frame O'. That is,

$$\mathbf{A} = \begin{pmatrix} a_{11} & a_{12} & a_{10} \\ a_{21} & a_{22} & a_{20} \\ 0 & 0 & 1 \end{pmatrix}, \quad \mathbf{x} = \begin{pmatrix} x \\ y \\ 1 \end{pmatrix}.$$

Then we get the equation of the fitted ellipse by $\mathbf{x}^T \mathbf{E} \mathbf{x} = 0$ with $\mathbf{E} = \mathbf{A}^T \mathbf{E}'' \mathbf{A}$. Additionally, we get the two endpoints of the line segment by the inverse transformation $\mathbf{A}^{-1}$ of the two endpoints $(-x_{s_{opt}}, s_{opt}), (x_{s_{opt}}, s_{opt})$ in the optimal canonical frame, see Fig. 5.

3　A New Invariant Fitting Procedure for Circular Segments

Using the fitting procedure of the last section we can also fit objects by circular segments. We have only to modify the transformation group and the normalization procedure. We choose the similarity transformation group that is described by four parameters, two for the translation, one for the rotation, and one for an isotrope scaling.

– Firstly, we normalize the translation by the centroid, i.e. $m'_{10} = m'_{01} = 0$.

– Secondly, we normalize the rotation by the constraint $m''_{11} = 0$, and we get the well-known expression for the angle φ:

$$\tan 2\varphi = \frac{2m'_{11}}{m'_{20} - m'_{02}}$$

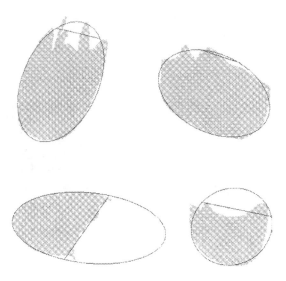

Fig. 5. Examples of the fitted elliptical segments

We obtain four possible solutions for the angle φ:

$$\varphi_1 = \tfrac{1}{2} \arctan \frac{2m'_{11}}{m'_{20} - m'_{02}}$$
$$\varphi_2 = \varphi_1 + \tfrac{\pi}{2}$$
$$\varphi_3 = \varphi_1 + \pi$$
$$\varphi_4 = \varphi_1 + \tfrac{3\pi}{2} .$$

We have to choose an angle φ such that the correspondence to the normalized circular segment is satisfied. Therefore, the additional constraints

1. $m''_{20} \geq m''_{02}$ implies that the major axis of the inertial ellipse is the x-axis,
2. $m''_{21} \geq 0$ implies the correspondence that the "line segment" is above the x-axis.

– Thirdly, we have to normalize an isotrope scaling by the constraint $m'''_{00} = 1$.

We have to carry out this normalization procedure for the object and the circular segment (Fig. 1). For the circular segment we can use the procedure displayed in Figs. 1, 2 and 3. The difference between Fig. 4 and a new Figure is only an isotrope scaling with $m'''_{00} = 1$.

4 A New Invariant Fitting Procedure for Rectangles

Following the last section we choose the similarity transformation group that is described by four parameters, two for the translation, one for the rotation, and one for an isotrope scaling.

– Firstly, we normalize the translation by the centroid, i.e. $m'_{10} = m'_{01} = 0$.

– Secondly, we normalize the rotation by the constraint $m''_{11} = 0$, and we get again for the angle φ:

$$\tan 2\varphi = \frac{2m'_{11}}{m'_{20} - m_{02}}$$

We obtain four possible solutions for the angle φ:

$$\begin{aligned}
\varphi_1 &= \tfrac{1}{2} \arctan \tfrac{2m'_{11}}{m'_{20} - m'_{02}} \\
\varphi_2 &= \varphi_1 + \tfrac{\pi}{2} \\
\varphi_3 &= \varphi_1 + \pi \\
\varphi_4 &= \varphi_1 + \tfrac{3\pi}{2} \ .
\end{aligned}$$

We have to choose any of the four solutions. Only if we have an object that is closely a square then additionally we have the solutions $\varphi_i^* = \varphi_i + \tfrac{\pi}{4}$.

– Thirdly, we have to normalize an isotrope scaling by the constraint $m'''_{00} = 1$.

We have to carry out this normalization procedure for the object and the rectangle. Chosing a rectangle with the middle point in the origin and the sides parallel to the axes, and the area of the rectangle is $m_{00} = 1$ then the normalized constraints are satisfied. The ratio of both sides $\frac{a}{b}$ is unknown. We estimate this ratio by an one-dimensional optimization process using the higher order moments not used for the normalization. Additionally, if an angle φ_i^* is a solution then the optimization process is also to do for this angle. Last not least we chose the angle with the best result. In Fig. 6, some examples are displayed.

5   A New Invariant Fitting Procedure for Super Ellipses

This fitting procedure also depends on one optimization parameter, for a detailled description, see [12].

References

1. F.Bookstein, "Fitting Conic Sections to Scattered Data," *Computer Graphics and Image Processing*, vol. 9, pp. 56-71, 1979
2. A.W.Fitzgibbon, M.Pilu, and R.B.Fisher, "Direct Least Squares Fitting of Ellipses," *Proceedings 13th ICPR, Wien 1996* pp. 253-257
3. D. Forsyth et.al., "Invariant Descriptors for 3D-Object Recognition and Pose," *IEEE Trans. PAMI*, vol. 13, pp. 971-991, 1991
4. R.M. Haralick, and L.G.Shapiro, "Computer and Robot Vision,", Vol. I, Addison-Wesley, 1993
5. K.Kanatani, "Statistical Bias of Conic Fitting and Renormalization," *IEEE Trans. PAMI* , vol. 16, pp. 320-326, 1994

274

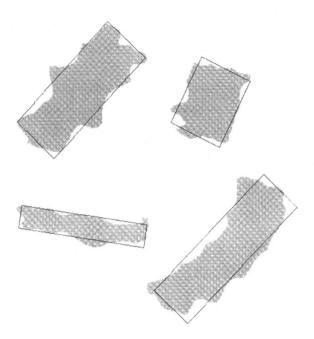

Fig. 6. Examples of the fitted rectangles

6. Y.Kanazawa, and K. Kanatani, "Optimal Conic Fitting and Reliability Evaluation," *IEICE Trans.Inf.Syst.*, vol. 9, pp. 1323-1328, 1996

7. D.Kapur, and J.L.Mundy, "Fitting Affine Invariant Conics to Curves," In *Mundy J.L. and Zisserman A.: Geometric Invariance in Computer Vision, MIT Press 1992* pp.252-266, 1992

8. P.Rosin, "A Note on Least Square Fitting of Ellipses," *Pattern Recognition Letters* , vol. 14, pp. 799-808, 1993

9. I.Rothe, H.Suesse, and K.Voss, "The Method of Normalization to Determine Invariants," *IEEE Trans. PAMI* , vol. 18, pp. 366-375, 1996

10. F.Solina, and R.Bajcsy, "Recovery of parametric models from range images: The case for superquadrics with global deformations," *IEEE Trans. PAMI* , vol. 12, pp. 131-147, 1990

11. K. Voss., and H.Suesse, "Invariant Fitting of Planar Objects by Primitives," *IEEE Trans. PAMI*, vol. 19, pp. 80-83, 1997

12. K. Voss., and H. Suesse, "A New One-Parametric Fitting Method for Planar Objects," *accepted for IEEE Trans. PAMI*, 1999

13. I.Weiss, "Geometric Invariants and Object Recognition," *IJCV* , vol. 10, pp. 207-231, 1992

3D Reconstruction of Volume Defects from Few X-Ray Images

C. Lehr, C.-E. Liedtke

Institut für Theoretische Nachrichtentechnik und Informationsverarbeitung,
Division: Automatic Image Interpretation, Prof. Dr.-Ing. C.-E. Liedtke,
Universität Hannover, Appelstr. 9A, 30167 Hannover, Germany
Phone: +49-511-7625328, Fax: +49-511-7625333, Email: lehr@tnt.uni-hannover.de

Abstract. In nondestructive testing for quality control of industrial objects the standard X-ray analysis produces a 2D projection of the 3D objects. Defects can be detected but cannot be localized in 3D position, size and shape. Tomographic testing equipment turns frequently out to be too costly and time consuming for many applications. Here a new approach for 3D reconstruction is suggested using standard X-ray equipment without costly positioning equipment. The new approach requires only a small number of X-ray views from different directions in order to reduce the image acquisition time.

The geometric and photometric imaging properties of the system are calibrated using different calibration patterns. The parameters of a CAHV camera model are obtained for each view permitting the exact registration of the acquired images. The efficiency of the 3D reconstruction algorithm has been increased by limiting the reconstruction to regions of interest around the defects. This requires an automated segmentation. The 3D reconstruction of the defects is performed with an iterative procedure. Regularization of the reconstruction problem is achieved on the basis of the maximum entropy principle. The reliability and robustness of the method has been tested on simulated and real data.

Keywords: X-ray, tomography, calibration

1 Motivation

Non destructive testing constitutes a major part of quality control in industrial production. The standard X-ray analysis produces a 2D projection of the 3D objects. Defects can be detected but cannot be localized in 3D position, size, and shape. For this purpose tomographic testing equipment has to be used which is described at several places in the literature [1],[2]. The use of tomographic testing methods within industrial quality control is limited by several facts. Traditional tomographic systems are too complex and expensive for several applications. They require usually sophisticated detectors and very accurate and expensive mechanical devices for positioning. The inspection time which is caused by the analysis of a large number of views taken from around the object is often considered to be too long [3],[4]. Frequently it is the aspect ratio of the test objects which prevents the radiographic analysis from all sides for the volume reconstruction as in the case of platelike objects. For these reasons there exists a demand for tomographic systems, which operate on the basis of simple radiographs and which permit the volume reconstruction from a small number of views. The first aspect would tend to reduce the costs, the second to reduce the time for the analysis. In connection with the reduction of the number of views as compared to standard tomographic equipment additional problems become apparent. The reconstruction problem becomes strongly underdetermined and may produce inconsistent results due to calibration errors, noise and other effects and the conventional tomographic algorithms like filtered backprojection, transformation methods, etc are not suitable anymore. Additional information like prior information about the geometry or material properties of the object under investigation is needed in order to arrive at a unique solution.

A special application is the reconstruction of casting defects, i.e. the reconstruction of an inclusion like gas within an otherwise homogeneous object. In this case the image reconstruction problem can be reduced to the estimation of binary object properties, i.e. object material vs. inclusion. Different approaches have been suggested in the literature [5],[6] where some of them require the use of parametrized functions in order to describe the objects under investigation. Because of the large variety of possible casting defects these model based approaches do not seem to be suitable or seem at least to limit the scale of possible applications. Other approaches are based on modelling the image as a binary Markov random field [7],[3]. The reconstruction is achieved by minimizing an error function dependent on the measured projections and on a a-priori choice of image modelling assumptions. The main difficulty of this type of approach is to achieve a reasonable fast and reliable minimization considering the very large number of unknown voxels.

In this paper a two-stage procedure for the 3D reconstruction of volume defects from a few radiographs is presented, which exploits the prior knowledge of the material parameters. No a-priori assumptions about the shape of the object under investigation have to be made. An overview about the processing steps is depicted in Fig. 1. After a calibration of the projective properties of the imaging device and a registration of the different views the complexity of the reconstruction problem is reduced in a first step by limiting the reconstruction to regions of interest around the defects. The defect areas are segmented and the spatial extent of the defects in beam direction is estimated from the measured data using calibration curves. In a second step the defects are reconstructed with an iterative procedure. Regularization of the reconstruction problem is achieved on the basis of the maximum entropy principle in connection with an iterative procedure for the binarization of the reconstruction results. In Chapter 2 the camera model and the calibration procedure is explained. Chapter 3 describes the segmentation process and Chapter 4 the procedure for the 3D reconstruction. Finally in Chapter 5 results are presented based on simulated and real data.

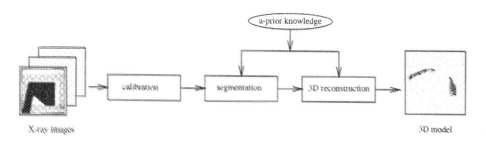

Fig. 1. System overview

2 Camera Parameters

The components of the X-ray system are depicted in Fig. 2. A microfocus X-ray tube serves as X-ray source. For imaging an image intensifier and a CCD-camera are used. The focus of the radiation-source is in the magnitude of μm and serves effectively as point source. The rays which traverse the test object are attenuated according to its geometric and material properties. The mapping process of the rays can be described by perspective projection. The imaging system itself exhibits a number of (primary radial) geometric distortions. This is partly due to the spheric shaped screen of the image intensifier, partly caused by magnetic fields which influence the electronic beam and other reasons. The radial distortions of the optical system of the CCD-camera appear to be much smaller and can be neglected.

The imaging properties of the system are mathematically described by the camera model. The parameters of the camera model are estimated in an calibration procedure using a calibration pattern. Since for tomographic reconstructions several camera views of the same object are required, the images have to be registered. For this purpose test object and calibration pattern need to remain in a fixed position relative to each other during the image acquisition step. The calibration procedure is applied to each viewing position in order to obtain precise position parameters for each individual view. The nonlinear geometric distortions are considered in a second step following the perspective projection.

For modelling the perspective projection the CAHV camera model from Yakimovski and Cunningham [8] has been employed using the following nomenclature:

C: position of the radiation source in space
H_0: unit vector pointing in the horizontal direction of the imaging target
V_0: unit vector pointing in the vertical direction of the imaging target
A: unit vector describing the optical axis
h_x and h_y: image point where the optical axis traverses the imaging target
s_x, s_y: pixel size
f: distance of radiation source from imaging plane.

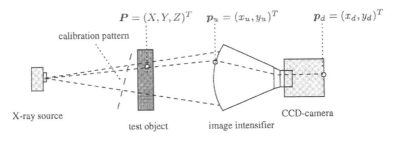

Fig. 2. Components of the X-ray system

Following the principles of geometric projection a point P in space is projected onto a point p_u in the image plane disregarding geometrical distortions following Eq. 1.

$$p_u = \begin{pmatrix} x_u \\ y_u \end{pmatrix} = \frac{1}{A^T \cdot (P - C)} \cdot \begin{pmatrix} (f/s_x \cdot H_0 + h_x \cdot A)^T \cdot (P - C) \\ (f/s_y \cdot V_0 + h_y \cdot A)^T \cdot (P - C) \end{pmatrix} \quad (1)$$

The line of projection from the radiation source to an image point p_u is calculated as

$$S(p_u) = C + s \cdot S_0(p_u) \quad (2)$$

where $S_0(p_u)$ represents the line of sight. The intensity of the beam is attenuated according to the geometric extensions along the line of sight depending on the material properties and the particular spectrum of the radiation source as described in Eq. 3.

$$i(p_u) = \int_{W_{min}}^{W_{max}} \left(i_0(p_u) \cdot \exp \left(- \int_0^{s_{BV}} \mu\left(S(p_u)\right) ds \right) \right) dW \quad (3)$$

In Eq. 3 $i_0(p_u)$ refers to the un-attenuated intensity in the image point p_u and W to a particular wavelength of the radiation source.

Geometrical distortions are corrected according to Eq. 4 using two-dimensional polynomial functions. The parameters of these two-dimensional functions have been estimated during a separate calibration process employing a planar calibration pattern with equidistant calibration marks. The coefficients of the polynomials have been obtained during an optimization procedure, where the mean squared distance between the estimated positions of the calibration marks and their prior known correct positions has been minimized [9]. From the calibration the pixel size s_x and s_y are obtained as well.

The calibration remains valid as long as the geometric relation between radiation source and the imaging device does not change.

$$\boldsymbol{p_d} = \begin{pmatrix} x_d \\ y_d \end{pmatrix} = \begin{pmatrix} f_1(x_u, y_u) \\ f_2(x_u, y_u) \end{pmatrix} \tag{4}$$

$$f_1(), f_2() : polynomial\, functions$$

The estimation of the parameters of the projection model is done for each individual position in the sequence of camera views which are acquired for one tomographic reconstruction. The setup is indicated in Fig. 3. The test object under investigation is recorded together with a calibration pattern. The calibration pattern consists of a planar plate with circular calibration marks arranged on its periphery. The presence of the calibration pattern does not affect the analysis of the investigated object. It reduces only the image area which can be utilized. The calibration procedure does not require the use of all calibration marks but only those which are detected with a high degree of reliability.

The parameter estimation uses a modified version of the method described by Tsai [10] for the calibration of CCD-cameras. It works iteratively starting with a reduced set of parameters and increases the number of parameters step by step. Since the camera parameters influence each other small errors in the localization of the calibration marks may result in significant errors of the derived camera parameters. The reliability can be improved by increasing the number of measurements. This can be achieved by using the measurements from several views for the estimation of those parameters which remain constant in all views, the so called inner camera parameters f, h_x and h_y.

3 Image Segmentation

The efficiency of the tomographic reconstruction could be increased considerably by reducing the reconstruction problem to the area which contains the defects. This requires an automated segmentation of the images which will be used for the reconstruction. The steps of the image segmentation procedure are shown in Fig. 4.

In a first step the measured intensity values $i(\boldsymbol{p_u})$ are converted into estimates for the material thickness $g_b(\boldsymbol{p_u})$ using a calibration function G.

$$g_b(\boldsymbol{p_u}) = G\left(i(\boldsymbol{p_u})\right) \tag{5}$$

This is referred to in Fig. 4 by the term "linearization". G depends on the intensity and the spectrum of the radiation source and the material of the object under investigation. In order to consider different radiation effects like backscattering the calibration function is obtained from experiments using calibration objects of different thickness.

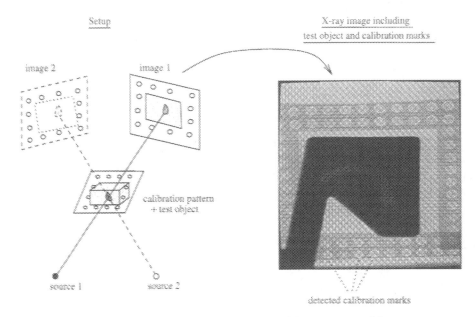

Fig. 3. Estimation of the parameters of the camera model

For the estimation of the spatial extent of the defect, called defect thickness $g_d(p_u)$, an estimate for the material thickness without the defect has to be estimated from the values of $g_b(p_u)$ outside the defect area. This is called a background model $g_h(p_u)$. Assuming locally smooth surfaces of the objects under investigation a two dimensional polynom of 3rd order according to Eq. 6 has been chosen for background interpolation.

$$g_h(x,y) = h_0 + h_1 \cdot x + h_2 \cdot y + h_3 \cdot x^2 + h_4 \cdot xy + ... + h_{14} \cdot x^2 y^3 + h_{15} \cdot x^3 y^3$$

$$(6)$$

The segmentation starts by using some arbitrary automated defect detection method or a semiautomated interactive method incorporating a manual marking of regions of interest around the defect regions. This first estimate needs not to be accurate. From this first estimate the defect-free thickness $g_h(p_u)$ is estimated. The defect thickness $g_d(p_u)$ is calculated from the difference between $g_h(p_u)$ and $g_b(p_u)$. Using $g_d(p_u)$ the defect region can be localized more accurately. In further iterations the improved defect localization leads to an improved background estimation resulting in a further improvement of the defect segmentation. From the final result a good estimate of the defect extent in all three dimensions can be obtained.

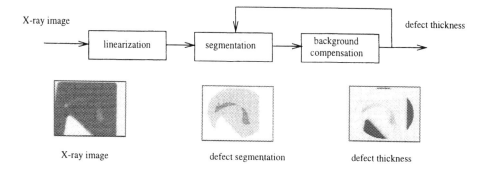

Fig. 4. Image segmentation: The X-ray image serves as input image. The defect segmentation constitutes a binary mask, indicating the precise location of the defect within the 2D image. The defect thickness represents a map indicating the estimated defect thickness for each point.

4 3D Reconstruction

The 3D reconstruction of the defect areas is carried out in an relaxation loop. The 3D volume which encloses the defect area is described by the discrete 3D voxel vector μ. μ describes the material properties of the object under investigation. The values are normalized $0 \leq \mu(i) \leq 1$. Since the material is assumed to be homogeneous except for the defects, binary values are expected for $\mu(i)$: $\mu(d) = 1$ for the voxels belonging to the defect area and $\mu(o) = 0$ for the voxels belonging to the non-defect area. The relation between the measured and preprocessed data in the area of interest p and the binary material properties μ are given by the matrix equation (7) where $\underline{A}$ describes the spatial relations, i.e. projection of the 3D volume space onto the different 2D radiographic views which have been recorded. The matrix value a_{ij} corresponds to the influence of the i-th voxel on the j-th projection.

$$p = \underline{A} * \mu \tag{7}$$

When only a few radiographs are used as had been proposed for the presented method, Eq. 7 turns out to be underdetermined. Inconsistent results may turn up due to errors in preprocessing, calibration errors, from both, the densiometric and the geometric calibration as well as due to different noise sources. The standard approach is to formulate additional constraints for the reconstruction process. These additional constraints are added to an energy function which has to be minimized and they have to be weighted by a regularization factor as is indicated in Eq. 8.

$$E = ||\underline{A} * \mu - p||^2 + \beta * E_R(\mu) \tag{8}$$

As additional constraint the negative entropy $E_R(\mu)$ of the voxel vector has been chosen. Minimizing $E_R(\mu)$ is equivalent to maximizing the entropy. The minimization of the energy function E in Eq. 8 is achived using a modification of the MART algorithm for entropy maximization. Using the MART method the choice of the explicite regularizaton factor β in Eq. 8 is replaced by the choice of the relaxation parameter λ in Eq. 9.2. The modified MART algorithm, which has been used is the following:

1. Start with a strictly positive vector μ^k, $k = 0$
2. Calculate for all pixels j the actual projections $\tilde{p}_j^k$ and the new voxel values

$$\tilde{p}_j^k = \sum_i a_{ij} * \mu_i^k \tag{9.1}$$

$$\mu_i^{k+1} = \mu_i^k * \left(\frac{p_j}{\tilde{p}_j^k}\right)^{\lambda * a_{ij}} \tag{9.2}$$

3. Update all voxels according

$$\mu_i^{k+1} = \begin{cases} \mu_i^{k+1} & if \ \mu_i^{k+1} \leq 1 \\ 1 & if \ \mu_i^{k+1} > 1 \end{cases} \tag{9.3}$$

4. Update all voxel i every K.th iteration

$$\mu_i^{k+1} = median(\mu_i^{k+1}) \tag{9.4}$$

5. k=k+1
6. Repeat steps 2 to 6 until convergence

The first steps contain the basic MART algorithm. Depending on the noise properties of the data an adequate relaxation factor λ is selected automatically in order to guarantee a stable convergence of the procedure. Eq. 9.3 imposes the material constraints on the reconstruction. Finally, the repeated median filtering of the intermediate results leads to a clear improvement of the convergence behaviour. The noise properties of the measured data used for the choice of an adequate relaxation factor λ are estimated during the preprocessing procedure.

The minimization of the energy function E in Eq. 8 results in continuous material values μ where low values indicate the tendency toward an non-defect property and high values the tendency towards a defect property. After the minimum of the E has been obtained using the modified MART algorithm thresholding is applied to μ in order to update the defect area. Only the voxels with the largest material values are identified as belonging to the defect area. With the updated vector a new optimization cycle is started until the binary solution has been reached. In each iteration the threshold is automatically adapted to the current μ. The idea of this binarization procedure is to assign only a few voxels to the defect area in each iteration in order to ensure a gradual adjustment of the solution to the binary constraint.

5 Results

The performance of the suggested reconstruction method had at first been tested on simulated data. The complex test object of Fig. 5 has been chosen. It consists of a ball positioned within a convex hull. From this test object two sets of simulated X-ray views have been generated. Each set contains five different views within a range of 90°. The two sets differ in that one set has been augmented with noise in order to test the robustness of the reconstruction method. The surface of the simulated object, one sample of a simulated view from each set and the surface of the reconstructed 3D object are shown in Fig. 5. As can be seen the reconstructions almost match the shape of the original object except for some slight differences at the lower part of the convex hull in case of the noise deteriorated data.

Real data have been processed with known and unknown geometric properties. In both cases the X-ray images have first been rectified for geometrical distortions and the camera parameters have been estimated. Geometric distortions of sometimes more than 11 pixels could be reduced to less than 0.3 pixel.

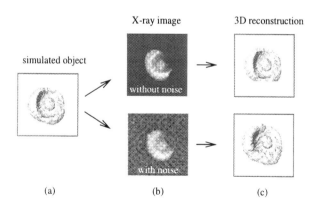

Fig. 5. Reconstruction of a simulated object: (a) shape of the test object, (b) simulated X-ray views with and without added noise, (c) shape of reconstructed object

A test object with known geometric properties consists of a plate with drilling holes of different depth and diameter. Five views have been acquired within a range of 60°. All reconstruction errors appear in the border region on the surface. The deviation from the real shape is in the average less than one voxel. The maximum deviation amounted to 2 voxels. The test object and its 3D reconstruction are depicted in Fig. 6 on the left side.

Fig. 6 (left side) shows the reconstruction of casting defects with geometric properties which have not be known before. This and the other tests demon-

284

strate very well that it is possible to reconstruct the 3D shape of objects from a few X-ray images taken from different views with standard X-ray systems and without special positioning equipment.

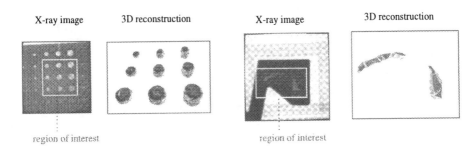

X-ray image   3D reconstruction   X-ray image   3D reconstruction

region of interest   region of interest

Fig. 6. Reconstruction of drilling holes and casting defects

References

1. Herman,G.T.: Image Reconstruction from Projections. The fundamentals of Computerized Tomography. Academic Press, 1980.
2. Kak, A. C.; Slaney, M.: Principles of Computerized Tomographic Imaging. New York, IEEE Press, 1988.
3. Robert, N.; Peyrin, F.; Yaffe, M.J.: Blodd vessel reconstruction from a limited number of cone-beam projections: Application to cerebral blood vessel projections and to an excise animal heart. Annual SPIE Conference on Machine Vision Applications in Industrial Inspection, San Jose, Cal., 1995, SPIE proc. series 2423.
4. Stegemann, D.: Zerströungsfreie Prüfverfahren: Radiographie und Radioskopie. B.G. Teubner, Stutgart, 1995.
5. Djafari, A.M.: Shape Reconstruction in X-Ray Tomography. Image Reconstruction and Restoration 2, San Diego, California, 1997, SPIE Vol. 3170.
6. Milanfar, P.; Karl, W.C.; Willesky, A.S.: Reconstructing Binary Polygonal Objects from Projections: A Statistical View. Graphical Models and Image Processing, Vol 56, 5, 1994.
7. Retraint, F.; Peyrin, F.; Dinten, J.M.: Three-Dimensional Regularized Binary Image Reconstruction From Two-Dimensional Projections Using a Randomized ICM Algorithmus. Int. J. of Imag. Systems Technology, pp. 135-146, Vol 9, 1998.
8. Yakimovski, Y.; Cunningham, R.: A System for Extraction Three-Dimensional Measurement from a Stereo Pair of TV Cameras. Intern. Journal on Computer Graphics and Image Processing.Vol 7, S. 195-210, 1978.
9. Lehr, C.; Feiste, K.; Stegeman, D.; Liedtke, C.-E.: Three dimensional defect analysis using stereoradioscopy based on camera modelling. 7.ECNDT Conference, Copenhagen, 1998.
10. Tsai, R.Y.: A versatile camera calibration technique for high-accuracy 3D machine vision meterology using off-the-shell tv cameras and lenses. IEEE J.Robotics Automation, Vol. RA-3(4), August 1988, S. 323-344.

Geodesic Path Based Interpolation Using Level Sets Propagation

Bruno Migeon, Fabien Boissé, Philippe Deforge, Pierre Marché

Laboratoire Vision et Robotique
63, avenue de Lattre de Tassigny
18020 Bourges Cedex, France
{Bruno.Migeon, Philippe.Deforge, Pierre.Marche}@bourges.univ-orleans.fr

Abstract. The main task of medical 3D reconstruction is to build accurately an isotropic volume consisting of cubic voxels from 2D serial cross sections. However, the spacing between the slices is typically much greater than the size of a pixel within the slice. Therefore, interpolation between the slices is of great importance. This paper presents a new kind of contour interpolation method in order to overcome the drawbacks of the existing methods. It assumes being applied after a segmentation step extracting the contours, and after a contour association step in order to avoid pathological cases. Moreover, it is based on the geodesic metric instead of Euclidean metric, and allows the interpolation of complex contours with concavities.

1 Introduction

In many medical applications, a 3D object must be reconstructed from serial cross-sections, either to aid in the comprehension of the object's structure for the diagnostic or to facilitate its automatic manipulation and analysis. Major applications can be found in the study of the structural and morphological characteristics of organs in biomedical sciences [1], [2], [3], radiation treatment planning, surgical planning [4], [5], etc...

The main task of medical 3D reconstruction is to build accurately an isotropic volume consisting of cubic voxels from 2D serial cross sections. However, the spacing between the slices is typically much greater than the size of a pixel within the slice. Therefore, interpolation between the slices is of great importance.

Two typical and widely used interpolation methods are *Dynamic Elastic Interpolation* [6], [7] (which is a interpolation method of contours) and *Shape-Based Interpolation* [8], [9] (which is a interpolation method of regions). The complicated implementation and the computing effort involved prevent *Dynamic Elastic Interpolation* from extensive practical applications. Moreover, it fails in case of complex shapes with severe concavities. *Shape-Based Interpolation* is a practical interpolation method for its easy implementation. However, this method fails to interpolate the slices when there is no overlapping area between the two objects and when the difference between the two objects is too important. Other methods have come to light since these two well known methods [10], [11] but they also have drawbacks and are not so commonly used.

This paper presents a new kind of contour interpolation method in order to overcome the drawbacks of the existing methods. It is based on the geodesic metric instead of Euclidean metric so that interpolation of complex contours is possible. Moreover, it assumes being applied after a segmentation step extracting the contours, and after a contour association step [12] allowing the building of a 3D scene as a set of different entities of contours. Thus, the presented contour interpolation method has to be used in a 3D reconstruction process for each portion of entity, without any pathological case, thanks to the previous application of a contour association method.

2 Contour Interpolation Based on Geodesic Paths Using Level Sets Propagation

2.1 Principle

Formally stated, the contour interpolation problem between two consecutive slices for a portion of an entity of an object amounts to :

- given a source contour $\mathcal{C}_S$ at $z = z_S$ and a target contour $\mathcal{C}_T$ at $z = z_T$,
- compute new contours $\mathcal{C}_\alpha$ for $0 \le \alpha \le 1$ corresponding to $z_S \le z \le z_T$.

Considering a source contour and a target contour, the principle of the proposed method is firstly to determine the XOR regions after overlapping. Then, each point of the two contour portions delimiting the XOR regions are linearly matched. Finally, from each pair of matched points, an interpolated point is computed and corresponds to a point along the geodesic path reaching these two original points. The set of all interpolated points consists of an interpolated contour.

2.2 Centroid Registration Transform

The aim of the centroid registration is to take into account the object translation for slice to slice in order to compensate for large shifts in object position. Firstly, the centroid C_S and C_T of each region $\mathcal{R}_S$ and $\mathcal{R}_T$ are computed. Then, $\mathcal{R}_S$ and $\mathcal{R}_T$ are transformed by translation into $\tilde{\mathcal{R}}_S$ and $\tilde{\mathcal{R}}_T$ respectively in order to superimpose the centroids C_S and C_T on a unique point O of the image.

2.3 Matching

After the centroid registration transform, the two regions overlap and then XOR regions are determined. They correspond to the regions belonging only to $\tilde{\mathcal{R}}_S$ or $\tilde{\mathcal{R}}_T$.

$\tilde{C}_S$ and $\tilde{C}_T$ being the contours delimiting two regions $\tilde{R}_S$ or $\tilde{R}_T$, each XOR region X_i is delimited conjointly by a part of $\tilde{C}_S$ and a part of $\tilde{C}_T$. For each X_i region, pixels belonging to $\tilde{C}_S$ are matched linearly with pixels belonging to $\tilde{C}_T$. Let P_{Si} and P_{Ti} be respectively the part of $\tilde{C}_S$ and $\tilde{C}_T$ for the X_i region, where N_{Si} and N_{Ti} are respectively the number of pixels of P_{Si} and P_{Ti}. Then, the pixel of P_{Si} having the index n_{Si} is matched with the pixel of P_{Ti} having the index n_{Ti} computed by :

$$n_{Ti} = \frac{N_{Ti} - 1}{N_{Si} - 1} n_{Si} + \frac{N_{Si} - N_{Ti}}{N_{Si} - 1} . \tag{1}$$

2.4 Geodesic Paths Using Level Sets Progagation

For each pair of matched points, a geodesic path of deformation is computed. A geodesic path between two points in a region is the shortest path remaining in the region. A graph technique using Dijkstra's algorithm [13] is possible, but it is extremely time consuming and does not give satisfactory results. So, a level sets propagation technique is used here for the geodesic paths computation [14].

Let us consider a pair of matched points P_{Sj} and P_{Tk}. A first distance map $f_{ij}(x,y)$ is computed from the initial point P_{Sj} :

$$\forall \ P(x,y) \ : \ \begin{cases} \text{if } P \in X_i & \text{then } f_{ij}(x,y) = d_{Xi}(P_{Sj}, P) \\ \text{if } P \notin X_i & \text{then } f_{ij}(x,y) \equiv \infty \\ \text{if } P = P_{Sj} & \text{then } f_{ij}(x,y) = 0 \end{cases} . \tag{2}$$

where $d_{Xi}(P_{Sj}, P)$ denotes the geodesic distance between P_{Sj} and P inside X_i.

A second distance map $g_{ik}(x,y)$ is computed from the arrival point P_{Tk} in the same way.

It can be noticed that :

$$f_{ij}(P_{Sj}) = 0 \quad \text{and} \quad f_{ij}(P_{Tk}) = d_{Xi}(P_{Sj}, P_{Tk}) . \tag{3}$$

$$g_{ik}(P_{Tk}) = 0 \quad \text{and} \quad g_{ik}(P_{Sj}) = d_{Xi}(P_{Tk}, P_{Sj}) . \tag{4}$$

Thus, for a point Q belonging to the geodesic path reaching P_{Sj} to P_{Tk} :

$$f_{ij}(Q) + g_{ik}(Q) = d_{Xi}(P_{Sj}, Q) + d_{Xi}(P_{Tk}, Q) = d_{Xi}(P_{Sj}, P_{Tk}) \ . \tag{5}$$

Building a third distance map h as :

$$h_{ijk}(x, y) = f_{ij}(x, y) + g_{ik}(x, y) \ . \tag{6}$$

the geodesic path G_{ijk} between P_{Sj} and P_{Tk} is given by [14] :

$$G_{ijk} = \{(x, y) / h_{ijk}(x, y) = g_m\} \quad \text{with} \quad g_m = \min_{Xi} h_{ijk}(x, y) \ . \tag{7}$$

In practice, the desired minimal geodesic path is achieved by applying a contour finder on h_{ijk} to find the level set $g_m + \varepsilon$, for some very small ε, and then by applying a simple thinning algorithm that operates on the interior of the minimal level set [14].

2.5 Interpolation Process

For a given α, the contour $\tilde{\mathcal{C}}_\alpha$ to be computed from $\tilde{\mathcal{C}}_S$ and $\tilde{\mathcal{C}}_T$ is composed of a set of portions of contours $\tilde{\mathcal{C}}_{\alpha i}$ in each X_i region. $\tilde{\mathcal{C}}_{\alpha i}$ consists of the set of points taken linearly along all the geodesic paths G_{ijk} .

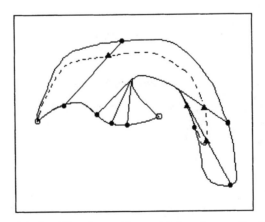

Fig. 1. Some geodesic paths in a Xi region and interpolated contour portion for a given α.

Figure 1 illustrates the principle with an example. It shows a region X_i delimited by two portions of contours $\mathcal{P}_{Si}$ and $\mathcal{P}_{Ti}$. Four pairs of matched points (black points) are represented with their geodesic path. The interpolated contour obtained for a given α is drawn, passing through the selected points on each path (black triangles).

2.6 Inverse Centroid Registration Transform

After the interpolation process, an interpolated contour is obtained for each α. Then, each real interpolated contour $\mathcal{C}_\alpha$ is computed by taking into account the inverse centroid registration transform which consists in computing the right position of the interpolated contour.

2.7 Interpolation in Case of Branching

The proposed contour interpolation is used in a 3D reconstruction context based on a 3D representation built with a contour association method. After the interpolation process, N_k contours are computed for the slice at $z = z_k$. These N_k contours $\mathcal{C}_{kl}$ are associated to N_k regions $\mathcal{R}_{kl}$. But, a merging of different regions is possible due to branching. In that case, the merged regions have a common contour.

Finally, after the contour interpolation of the 3D scene, all the right contours of a slice are obtained by filling in all the computed contours to obtain regions, and by extracting the contours of these regions as described in [6].

3 Results

This part presents the capability of the presented method.

Figure 2 shows the two contours to be interpolated superimposed on the same image (a simple circle and a complex contour) with a severe translation between them. Figure 3a represents the two superimposed contours after the centroid registration transform, dedicated to the compensation of this large shift in position. Figure 3b points out the XOR regions. There are two XOR regions in this example : a small one and a very extended one. Figure 3c shows some geodesic paths. Figure 4 represents superimposed results of the interpolation process, with several computed contours.

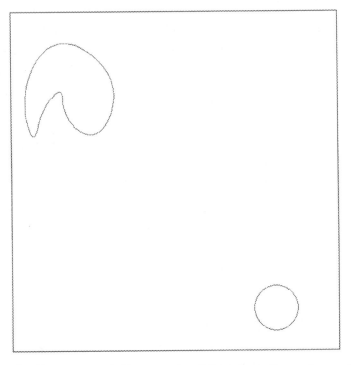

Fig. 2. Example with two contours with a severe translation between them and a complex shape

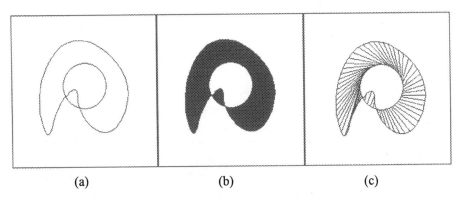

| (a) | (b) | (c) |

Fig. 3. Intermediary steps of the process. a) : overlapped contours after centroid registration transform, b) : determination of XOR regions, c) : geodesic paths

As illustrated by this example, the results given by the presented method are very interesting. It works even in the case of non overlapping areas between the two original contours; and in the case of complex shapes with severe concavities, thanks to using the geodesic metric. On the other hand, it is very time-consuming. Indeed, for different examples tested, the proposed method takes often about 2 minutes while *Shape-based interpolation* method requires 15 seconds on a PC 133 MHz. This time consumption is due to the geodesic path computation.

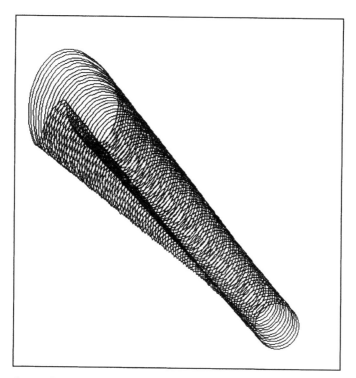

Fig. 4. Superimposed results after the interpolation process relative to the example of figure 2.

If the presented method gives quite interesting results, unfortunately it fails when the two original contours do not intersect after centroid registration. Moreover, interpolation results may depend on the centroid registration transform but this has yet to be studied.

4. Conclusion

A new kind of contour interpolation has been presented in this paper. Assuming being applied on each entity portion of a 3D scene in a 3D reconstruction process, it has still to solve a simple problem of interpolating new contours between a source contour and a target contour.

Thanks to using the geodesic metric to compute the deformation paths between each pair of matched points, it allows interpolation of complex shapes. It shows the geodesic metric is interesting to avoid pathological cases encountered with classical methods, usually due to matching errors. This method gives quite good results, but its main drawback is the time consumption due to the long computation of all the geodesic paths, even using level sets propagation. Further work will be carried out to reduce the time consumption.

References

1. Macagno, E.R., Levinthal, C., Sobel, I.: Three dimensional computer reconstruction of neurons and neuronal assemblies. Ann. Rev. Biophys. Bioeng. 8 (1979) 323-351
2. Burton, P.R.: Computer assisted three-dimensional reconstruction of the ultrastructure of the frog olfactory axon. Norelco Rep. 32 (1985) 1-10
3. Migeon, B., Marché, P.: in vitro 3D reconstruction of long bones using B-scan image processing. Medical & Biological Engineering & Computing. 35 (1997) 369-372
4. Bloch, P., Udupa, K.: Application of computerized tomography to radiation therapy and surgical planning. In Proc. of IEEE. 71 (1983) 351-372
5. Cutting, C., Grayson, D.D.S., Bookstein, F., Fellingham, L., McCarthy, J.G.: Computer-aided planning and evaluation of facial and orthognathic surgery. Comput. Plastic Surg. 13 (1986) 449-462
6. Lin, W.C., Liang, C.C., Chen, C.T.: Dynamic elastic interpolation for 3D medical image reconstruction from serial cross sections. IEEE Trans Medical Imaging. 7 (1988) 225-232
7. Chen, S.Y., Lin, W.C., Liang, C.C., Chen, C.T.: Improvement on dynamic elastic interpolation technique for reconstructing 3D objects from serial cross-sections. IEEE Trans. Med. Imaging. 9 (1990) 71-83
8. Raya, S.P., Udupa, J.K.: Shape-based interpolation of multidimensional objects. IEEE Trans. Medical Imaging. 9 (1990) 32-42
9. Herman, G.T., Zheng, J.S., Bucholtz, C.A.: Shape-based interpolation. IEEE CG&A. 12 (1992) 69-79
10. Migeon, B., Vieyres, P., Marché, P.: Interpolation of star-shaped contours for the creation of lists of voxels. Application to 3D visualisation of long bones. Int. J. CADCAM Comput. Graph. 9 (1994) 579-587
11. Dufrenois, F., Durin, H., Reboul, S., Dubus, J.P.: Contour interpolation using snakes. Application to 3D reconstruction from tomograph images. Innov. Tech. Biol. Med. 16, (1995)
12. Migeon, B., Deforge, P., Marché, P.: A contour association method for the 3D reconstruction of long bones obtained with the URTURIP Technique. In Proc. of IEEE EMBS (1998)
13. Cormen, T.; Leiserson, C.; Rivest; R.: Introduction to Algorithms. Mc Graw Hill. (1990)
14. Kimmel, R., Amir, A., Burckstein.: Finding shortest paths on surfaces using level sets propagation. IEEE Trans. on PAMI. 17 (1995) 635-640

Free-Form Surface Description in Multiple Scales: Extension to Incomplete Surfaces

Nasser Khalili, Farzin Mokhtarian and Peter Yuen

Centre for Vision, Speech, and Signal Processing
Department of Electronic and Electrical Engineering
University of Surrey, Guildford, England GU2 5XH, UK

Tel: +44 1483 876035,
Fax: +44-1483-259554,
Email: N.Khalili@ee.surrey.ac.uk
Web: *http://www.ee.surrey.ac.uk/Research/VSSP/demos/css3d/index.html*

Abstract

A novel technique for multi-scale smoothing of a free-form 3-D surface is presented. Diffusion of the surface is achieved through convolutions of local parametrisations of the surface with a 2-D Gaussian filter. Our method for local parametrisation makes use of semigeodesic coordinates as a natural and efficient way of sampling the local surface shape. The smoothing eliminates the surface noise together with high curvature regions such as sharp edges, therefore, sharp corners become rounded as the object is smoothed iteratively. During smoothing some surfaces can become very thin locally. Application of decimation followed by refinement removes very small/ thin triangles and segments those surfaces into parts which are then smoothed separately. Furthermore, surfaces with holes and surfaces that are not simply connected do not pose any problems. Our method is also more efficient than those techniques since 2-D rather than 3-D convolutions are employed. It is also argued that the proposed technique is preferable to *volumetric smoothing* or *level set methods* since it is applicable to incomplete surface data which occurs during occlusion. Our technique was applied to closed as well as open 3-D surfaces and the results are presented here.

1 Introduction

This paper introduces a new technique for multi-scale shape description of free-form 3-D surfaces represented by polygonal or triangular meshes. Although there are several methods available to model a surface, triangular meshes are the simplest and most effective form of polygons to cover a free-form surface. The common types of polygonal meshes include the triangular mesh and the four sided spline patches [5]. Triangular meshes have been utilised in our work. The multi-scale technique proposed here can be considered a generalisation of earlier multi-scale representation theories proposed for 2-D contours [12] and space curves [9].

In our approach, diffusion of the surface is achieved through convolutions of local parametrisations of the surface with a 2-D Gaussian filter. *Semigeodesic coordinates* [4] are utilised as a natural and efficient way of locally parametrising surface shape. The most important advantage of our method is that unlike other diffusion techniques such as volumetric diffusion [7] or level set methods [16], it has *local support* and is therefore applicable to partial data corresponding to surface-segments. This property makes it suitable for object recognition applications in presence of occlusions. The organisation of this paper is as follows. Section 2 gives a brief overview of previous work on 3-D object representations including the disadvantage(s) of each method. Section 3 covers implementation issues encountered when adapting semigeodesic coordinates coordinates to 3-D triangular meshes. Section 4 presents diffusion results and discussion. Section 5 contains the concluding remarks.

2   Literature Survey

This section presents a survey of previous work in representation of 3-D surfaces. Comprehensive surveys of 3-D object recognition systems are presented by Besl and Jain [1], *Generalised cones* or *cylinders* [19] approximate a 3-D object using globally parametrised mathematical models, but they are not applicable to detailed free-form objects. A form of 3-D surface smoothing has been carried out in [23] but this method has drawbacks since it is based on weighted averaging using neighbouring vertices and is therefore dependent on the underlying triangulation. In *volumetric diffusion* [7] or *level set methods* [16], an object is treated as a filled area or volume. The major shortcoming of these approaches is lack of local support. In other words, the entire object data must be available. This problem makes them unsuitable for object recognition in presence of occlusion.

3   Semigeodesic Parametrisation

Free-form 3-D surfaces are complex hence, no global coordinate system exists on these surfaces which could yield a natural parametrisation of that surface. Studies of local properties of 3-D surfaces are carried out in differential geometry using local coordinate systems called *curvilinear coordinates* or *Gaussian coordinates* [4]. Each system of curvilinear coordinates is introduced on a patch of a regular surface referred to as a *simple sheet*. A simple sheet of a surface is obtained from a rectangle by stretching, squeezing, and bending but without tearing or gluing together. Given a parametric representation

$\mathbf{r} = \mathbf{r}(u,v)$ on a local patch, the values of the parameters u and v determine the position of each point on that patch. Construction and implementation of semigeodesic coordinates in our technique is described in [10].

3.1   Construction of Semigeodesic Coordinates

A geodesic line is defined as a line, which represents the shortest distance between two given points on a 3-D surface. Semigeodesic coordinates are constructed at each vertex of the mesh which becomes the local origin. The following procedure is employed:

1. Construct a geodesic from the origin in an arbitrary direction such as the direction of one of the incident edges.

2. Construct the other half of that geodesic by extending it through the origin in the reverse direction.

3. Parametrise that geodesic by the arclength parameter at regular intervals to obtain a sequence of sample points.

4. At each sample point on the first geodesic, construct a perpendicular geodesic and extend it in both directions.

5. Parametrise each of the geodesics constructed in the previous step by the arclength parameter at regular intervals.

Due to the displacement of vertices which occurs as a result of smoothing, very small and/or very thin triangles can be generated during smoothing. These odd triangles can cause computational problems and are therefore removed or merged with neighbouring triangles using known existing algorithms for mesh decimation and refinement [6]. Detection of these triangles is based on the length of the shortest side or the smallest angle. When the smallest side or the smallest angle of a triangle is less than a small threshold, that triangle is removed by merging it with neighbouring triangles. Decimation and refinement are applied after each iteration to simplify the mesh. As a result, the number of triangles gradually decreases during smoothing. It is also possible for a surface to become very thin locally as a result of smoothing. When this happens, smoothing can not continue without segmentation of the surface into parts. Such a segmentation also occurs as a result of mesh decimation and refinement since the thinned area of the surface always consists of very small and thin triangles. Smoothing can then continue after segmentation with each part of the object smoothed independently.

3.2   Semigeodesic Coordinates on Open Surfaces

The algorithm described above should be modified to make it also applicable to open surfaces. The algorithm for smoothing an open surface is defined in the following way:

- Grid construction and smoothing at internal vertices is carried out as on closed surfaces. Any geodesic line that reaches the boundary will stop. The last sample point at or near the boundary will be duplicated until the grid is filled. Likewise, if some geodesic lines can not be constructed, the last geodesic line near the boundary will be duplicated until the grid is filled.

- If the vertex V of triangle T resides on the boundary, measure the angle α between the two edges of T that are incident on V. Choose the first geodesic line as the bisector of α. Only half of the first geodesic line is constructed because the other half falls outside the surface boundary.

- At the same vertex, construct another geodesic line perpendicular to the first one.

- One of those geodesic lines might soon intersect the boundary, so compare the lengths of those lines and choose the longer one. This allows the maximum size grid to be constructed.

- Construct the second family of geodesic lines as perpendicular to the longer geodesic line determined above.

- As before, any geodesic line that reaches the boundary will stop, and the last sample point at or near the boundary will be duplicated until the grid is filled.

3.3 Gaussian Smoothing of a 3-D Surface

The procedures outlined above can be followed to construct semigeodesic coordinates at every point of a 3-D surface S. In case of semigeodesic coordinates, local parametrisation yields at each point P:

$$\mathbf{r}(u, v) = (x(u, v),\ y(u, v),\ z(u, v)).$$

The new location of point P is given by:

$$\mathbf{R}(u, v, \sigma) = (\mathcal{X}(u, v, \sigma),\ \mathcal{Y}(u, v, \sigma),\ \mathcal{Z}(u, v, \sigma))$$

where

$$\mathcal{X}(u, v, \sigma) = x(u, v)\ \otimes\ G(u, v, \sigma)$$
$$\mathcal{Y}(u, v, \sigma) = y(u, v)\ \otimes\ G(u, v, \sigma)$$
$$\mathcal{Z}(u, v, \sigma) = z(u, v)\ \otimes\ G(u, v, \sigma)$$

and

$$G(u, v, \sigma) = \frac{1}{2\pi\sigma^2}\, e^{-\frac{(u^2+v^2)}{2\sigma^2}}$$

$\otimes$ denotes convolution. This process is repeated at each point of S and the new point positions after filtering define the smoothed surface. Since the coordinates constructed are valid locally, the Gaussian filters have $\sigma = 1$. In order to achieve multi-scale descriptions of a 3-D surface $\mathcal{S}$, the smoothed surface is then considered as the input to the next stage of smoothing. This procedure is then iterated many times to obtain multi-scale descriptions of $\mathcal{S}$.

4   Results and Discussion

The smoothing routines were implemented entirely in C++ and complete triangulated models of 3-D objects used for our experiments are constructed at our canter [5]. In order to experiment with our techniques, 3-D objects with different numbers of triangles were used. Triangular meshes have been utilised in our work. The first test object was a dinosaur with 2996 triangles and 1500 vertices as shown in Figure 1. The object becomes smoother gradually and the legs, tail and ears are removed after 10 iterations. The second test object was a cow with 3348 triangles and 1676 vertices as shown in Figure 2. The surface noise is eliminated iteratively with the object becoming smoother gradually where after 12 iterations the legs, ears and tail are removed. Figure 3 shows the third test object which was a telephone handset with 11124 triangles and 5564 vertices. Notice that the surface noise is eliminated iteratively with the object becoming smoother gradually and after 15 iterations the object becomes very thin in the middle. Decimation and refinement then removes the thin handset and segments the object into two parts. Smoothing then continues for each part as shown in Figure 3.

These examples show that our technique is effective in eliminating surface noise as well as removing surface detail. The result is gradual simplification of object shape. Animation of surface diffusion can be observed at the web site: *http://www.ee.surrey.ac.uk/Research/VSSP/demos/css3d/index.html* Our smoothing technique was also applied to a number of open/incomplete surfaces. Figure 4 shows the results obtained on a part of the telephone handset shown in Figure 4. This object also has a triangle removed in order to generate an internal hole. Figure 5 shows smoothing results obtained on a partial rabbit. The object is smoothed iteratively and the ears disappear as well.

5   Conclusions

A novel technique for multi-scale smoothing of a free-form triangulated 3-D surface was presented. This was achieved by convolving local parametrisations of the surface with 2-D Gaussian filters iteratively. Our method for local parametrisation made use of semigeodesic coordinates as natural and efficient ways of sampling the local surface shape. The smoothing eliminated the surface noise and small surface detail gradually, and resulted in gradual

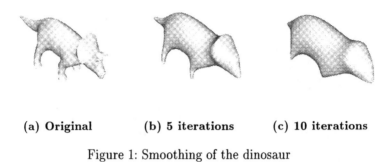

(a) **Original**   (b) **5 iterations**   (c) **10 iterations**

Figure 1: Smoothing of the dinosaur

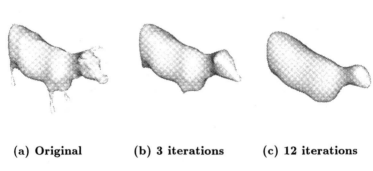

(a) **Original**   (b) **3 iterations**   (c) **12 iterations**

Figure 2: Smoothing of the cow

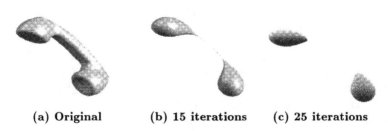

(a) **Original**   (b) **15 iterations**   (c) **25 iterations**

Figure 3: Diffusion of the telephone handset

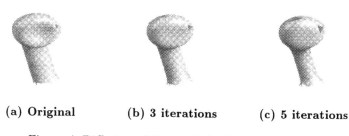

(a) Original　　　　**(b) 3 iterations**　　　　**(c) 5 iterations**

Figure 4: Diffusion of the partial telephone handset

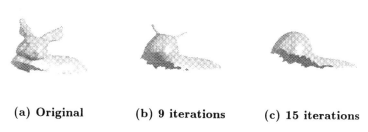

(a) Original　　　　**(b) 9 iterations**　　　　**(c) 15 iterations**

Figure 5: Smoothing of the rabbit

simplification of object shape. The method was independent of the underlying triangulation. During smoothing some surfaces can become very thin locally. Application of decimation followed by refinement removes very small/thin triangles and segments those surfaces into parts which are then smoothed separately. Our approach is preferable to *volumetric smoothing* or *level set methods* since it is applicable to incomplete surface data which occurs during occlusion. Finally, surfaces with holes and surfaces that are not simply connected do not pose any problems. Our approach is preferable to *volumetric smoothing* or *level set methods* since it is applicable to incomplete surface data which occurs during occlusion.

References

[1] P J Besl and R C Jain. Three dimentional object recognition. *ACM Computing Surveys*, 17:75–145, 1985.

[2] T W Chen and W C Lin. A neural network approach to csg-based 3-d object recognition. *IEEE Trans. on Pattern Analysis and Machine Intelligence*, 16(7):719–726, 1994.

[3] R T Chin and C R Dyer. Model-based recognition in robot vision. In *ACM Computing Surveys*, volume 18, pages 67–108, 1986.

[4] A Goetz. *Introduction to differential geometry.* Addison-Wesley, Reading, MA, 1970.

[5] A Hilton, A J Stoddart, J Illingworth, and T Windeatt. Marching triangles: Range image fusion for complex object modelling. In *Proc IEEE International Conference on Image Processing*, pages 381–384, Lausanne, Switzerland, 1996.

[6] H Hoppe. Progressive meshes. In *Proc SIGGRAPH*, pages 99–106, 1996.

[7] J J Koenderink. *Solid shape*. MIT Press, Cambridge, MA, 1990.

[8] A K Mackworth and F Mokhtarian. Scale-based description of planar curves. In *Proc Canadian Society for Computational Studies of Intelligence*, pages 114–119, London, Ontario, 1984.

[9] F Mokhtarian. A theory of multi-scale, torsion-based shape representation for space curves. *Computer Vision and Image Understanding*, 68(1):1–17, 1997.

[10] F Mokhtarian, N Khalili, and P Yuen. Multi-scale 3-d free-form surface smoothing. In *Proc British Machine Vision Conference*, pages 730–739, 1998.

[11] F Mokhtarian and A K Mackworth. Scale-based description and recognition of planar curves and two-dimensional shapes. *IEEE Trans Pattern Analysis and Machine Intelligence*, 8(1):34–43, 1986.

[12] F Mokhtarian and A K Mackworth. A theory of multi-scale, curvature-based shape representation for planar curves. *IEEE Trans Pattern Analysis and Machine Intelligence*, 14(8):789–805, 1992.

[13] M Pilu and R Fisher. Recognition of geons by parametric deformable contour models. In *Proc European Conference on Computer Vision*, pages 71–82, Cambridge, UK, 1996.

[14] H Samet. The design and analysis of spatial data structures. *Addison-Wesley*, 1990.

[15] M Seibert and A M Waxman. Adaptive 3-d object recognition from multiple views. In *IEEE Trans Pattern Analysis and Machine Intelligence*, volume 14, pages 107–124, 1992.

[16] J A Sethian. *Level set methods*. Cambridge University Press, 1996.

[17] S S Sinha and R Jain. Range image analysis. In *Handbook of Pattern Recognition and Image Processing: Computer Vision (T Y Young, ed.)*, volume 2, pages 185–237, 1994.

[18] F Solina and R Bajcsy. recovery of parametric models from range images: Thee case for superquadrics with global deformations. *IEEE Trans. on Pattern Analysis and Machine intelligence*, 12:131–147, 1990.

[19] B I Soroka and R K Bajcsy. Generalized cylinders from serial sections. In *Proc IJCPR*, 1976.

[20] A J Stoddart and M Baker. Reconstruction of smooth surfaces with arbitrary topology adaptive splines. In *Proc ECCV*, 1998.

[21] P Suetens, P Fua, and A J Hanson. Computational strategies for object recognition. *ACM Computing Surveys*, 24(2):5–61, 1992.

[22] G Taubin. Curve and surface smoothing without shrinkage. In *Proc ICCV*, pages 852–857, 1995.

[23] G Taubin. Optimal surface smoothing as filter design. In *Proc ECCV*, 1996.

Complex Analysis for Reconstruction from Controlled Motion*

R. Andrew Hicks   David Pettey   Kostas Daniilidis   Ruzena Bajcsy
GRASP Laboratory,
Department of Computer and Information Science
University of Pennsylvania

{rah, djpettey, kostas}@grip.cis.upenn.edu, bajcsy@central.cis.upenn.edu

Abstract

We address the problem of control-based recovery of robot pose and the environmental lay-out. Panoramic sensors provide us with a 1D projection of characteristic features of a 2D operation map. Trajectories of these projections contain information about the position of a priori unknown landmarks in the environment. We introduce here the notion of spatiotemporal signatures of projection trajectories. These signatures are global measures, characterized by considerably higher robustness with respect to noise and outliers than the commonly applied point correspondence. By modeling the 2D motion plane as the complex plane we show that by means of complex analysis the reconstruction problem is reduced to a system of two quadratic - or even linear in some cases - equations in two variables. The algorithm is tested in simulations and real experiments.

1 Introduction

Suppose that two points in the plane are given, z_1 and z_2, and a sensor is available that can measure the angle between z_1 and z_2 from any position in the plane at which it is placed. If the sensor undergoes a circular motion (figure (1)) then we may record the angles in a list and then plot each angle against its index in the list. For example, in the right of figure (1) we see such a graph for two points.

The primary motivation for this paper is to investigate the extent to which the data displayed on the right in figure (1) characterizes the two unknown points. In other words, we want to solve the inverse problem: given this data, how can the scene be reconstructed ? In the world, of course, there is no natural Cartesian frame, so what we really solve for is the magnitudes of z_1 and z_2, and the angle between them with respect to the center of the circular motion. Throughout this paper the scale will always be determined by taking the radius of the circle to be our unit.

* This work was supported by the following grants: NSF: IIS97-11380, NSF CISE: CDS97-03220-001, NSF Training: GER93-55018, ARO MURI/DARPA ONR: DAAH04-96-1-0007.

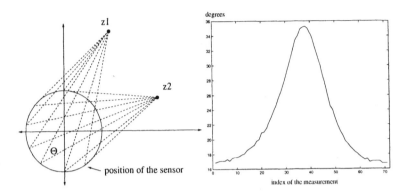

Figure1. The problem: find the values of z_1 and z_2 by moving on a circle and measuring the angle Θ during the motion. On the left we see the actual motion of the sensor, while on the right we see a plot of the measured angles between z_1 and z_2 vs. the index of the measurement in the list of all measurements. This data is taken from a real experiment.

1.1 Signatures

In many vision problems one must obtain geometric information about the environment and to do this a great deal of data is recorded. This data can be thought of as a long list of numbers, i.e. we measure a vector $\mathbf{c} = (c_1, c_2, ..., c_n)$. Then one seeks to solve a system of equations $F_j(\mathbf{c}, \mathbf{z}) = 0$, $j = 1, ..., m$, where the unknown is $\mathbf{z} = (z_1, z_2, ..., z_k)$, and $z_i \in \mathbf{R}$ or $\mathbf{C}$. Generally k, m and n are large integers and the system is non-linear and over-constrained. This is not a desirable situation for a number of reasons, the most obvious being that a "solution" may be hard to come by.

A more compact way to write the above system is as a single vector equation $\mathbf{F}(\mathbf{c}, \mathbf{z}) = \mathbf{0}$. For our purposes, if the problem can be solved for $k = 2$, then certainly the more general problem can solved by repeated application of the same method (see section 1.3 below). Therefore, rather than consider equations of this last type, we instead consider an equation of the form $\mathbf{G}(\mathbf{f}(\mathbf{c}), z_1, z_2) = 0$ where $\mathbf{f}$ is a smooth function mapping the space that $\mathbf{c}$ lives in to $\mathbf{R}$ or $\mathbf{C}$. Here we will take $\mathbf{f}$ to be the composition of the averaging operator, $\mathbf{A}$, with various analytic functions that act pointwise on $\mathbf{c}$. We say that $\mathbf{f}(\mathbf{c})$ is a *signature* of z_1, z_2. The fact that $\mathbf{f}$ is smooth makes the system robust to errors in measurement, while taking advantage of global information that may be embedded in the measured data. A signature is a quantity that can be measured experimentally, with the goal being to find the geometry of the configuration (z_1, z_2) from this value.

1.2 Structure from Motion

The problem of obtaining the geometry of a scene by moving a camera within it is a well-known one, and has many different forms. If the pose of the camera is known at all times, then reconstructing the geometry of the scene is easier than if the position is an unknown. The latter problem is known as the *structure from motion problem.* In this problem, the variables representing the poses of the camera are unknown, so one needs to take enough snapshots of the environment in order to solve for these variables plus the variables that represent the geometric structure of the environment. Because measurements from the camera have errors, one usually takes many "extra" snapshots in order to over-constrain the problem.

In this paper we will consider only the 2D version of the structure from motion problem, and take a very different approach than that described above. We investigate certain signatures associated with moving a camera on a circle and show how they can be used to reconstruct the positions of an unknown points.

1.3 Problem Statement

The general form of our problem is the following: Suppose that $z_1, ..., z_n$ are complex numbers representing n landmarks in the plane, and a sensor is available with the ability to measure the angles between the landmarks with respect to any given position of the sensor, w (i.e. it can measure the angle $\angle z_i w z_j, i, j = 1, ..., n$). Then for m positions $w_1, ..., w_m$ of the sensor[1], can one determine $z_1, ..., z_n$ and $w_1, ..., w_m$ given only the measurements $\angle z_i w_j z_k, i, k = 1, ..., n, j = 1, ..., m$?

One approach to the above is to write down all of the trigonometric equations associated with the configuration. Since the angles are known, the resulting polynomial system has many more equations than variables. Thus the system is over-constrained, which may be helpful since in a real experiment the angular measurements will contain noise. Then one may try to find a solution in the "least squares sense" by minimizing the sum of the squares of these equations.

In the problem we address here we will make the assumption that the points w_i all lie on a circle i.e. $|w_i| = 1, i = 1, ..., m$, but we will not assume where on the circle the w_i are. Additionally, we will assume that the w_i are densely distributed in the circle in an approximately uniform fashion. Also, notice that if the problem can be solved for two landmarks then it can be solved for more than two landmarks by applying the same method to pairs of the landmarks.

1.4 Panoramic Sensors

Recently, many researchers in the robotics and vision community have begun to investigate the use of curved mirrors to obtain panoramic and omni-directional views. Typically, such systems consist of a camera pointing upward at a convex mirror, as in figure (2). How to interpret and make use of the information obtained by such sensors, e.g. how it can be used to control robots, is not immediately clear. It does seem likely that panoramic systems may be able to

---

[1] In our model of the panoramic sensor, the pose of the sensor means only its position, and not orientation. For real applications it is possible to define and compute and orientation with our method if it is needed.

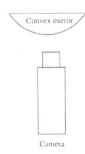 

Figure2. On the left we see one way to create a panoramic sensor: point a camera upwards at a curved mirror. On the right we see an image from such a panoramic sensor, with the camera visible in the center.

handle some problems that are difficult or impossible to address with a standard camera. To address our problem experimentally, we employed such a panoramic sensor, using a spherical mirror. A natural feature to consider for extraction from an image taken on such a system is any edge that is vertical in the real world, because these edges appear as radial lines in the image (see figure (2)). In the center of this image the lens of the camera is clearly visible. To measure the angles between the vertical edges we extend them to the center of the lens and compute the angle at their intersection.

1.5 Contributions

Our reconstruction method has a number of features that other methods do not.

(1) Rather than constructing one or more equations for each measurement, we first process the measured data to produce two numbers (signatures), which are used to construct two equations in the positions of the unknown landmarks. **These equations are quadratic in general and linear if a single reference point is known.**

(2) Due to the simplicity of the equations, optimization methods are not needed, and when the solutions are found they are known to be true solutions, as opposed to possible spurious solutions produced by numerical optimization methods.

(3) Due to the stable means of processing the data (integration), our method handles noise well. In addition, if one desired to carry out a detailed error analysis, it would be possible since the equations are so simple. This is not the case with a method that produces a very large number of nonlinear equations.

1.6 Related Work

Due to space limitations we briefly summarize here the related work. The precursor to this work is [3], where a similar problem was considered, but explicit solutions were not possible. Regarding the reconstruction aspect of our work, there is work in the eighties - early nineties on reconstruction using vertical edges and known motion (Kriegman [6], Kak [5]). There is work in structure from motion from circular trajectories (Shariat and Price [10], Sawhney [9]). This work, however, uses a set of equations with the constraint that the projection centers are on a circle.

Recently, a number of approaches to navigation and reconstruction using omnidirectional systems have been proposed (Nayar [7], Svoboda [12], Onoe [8], Srinivasan [2], Yagi [13], Medioni [11]). The work by Yagi and Medioni is very similar but uses an already known environmental map. The most relevant to this paper is the work on omnidirectional multibaseline stereo done by Kang and Szeliski [4], but this work uses conventional cameras and measurement equations.

2 Calculating Signatures

2.1 Angles and Complex Logarithms

Definition A function is **holomorphic** on an open subset $U \subset \mathbb{C}$ if it is complex differentiable at every point of U.

For any complex number $z = a + bi$, we define the exponential function by $\exp(z) = e^a(\cos(b) + i\sin(b))$. The exponential function is a holomorphic function which assumes all complex values except 0. If U is an open connected subset of $\mathbb{C}$, then a holomorphic function, ℓ, is called a **logarithm function** on U if for all $z \in U$, $\exp(\ell(z)) = z$.

We use $\mathbb{R}^-$ to denote the set of non-positive real numbers and $\mathbb{C}^-$ to denote the "slit" complex plane, $\mathbb{C} - \mathbb{R}^-$. If $z \in \mathbb{C}^-$, then there are unique real numbers θ and r with $r > 0$ and $-\pi < \theta < \pi$, such that $z = re^{i\theta}$. Given a number, z, written in this form, the **principle branch of the logarithm**, Log, is defined by

$$\text{Log}(z) = \ln(r) + i\theta, \tag{1}$$

where ln is the real natural logarithm. It is easy to check that the principle branch of the logarithm is a logarithm in the sense defined above. The set $\mathbb{R}^-$ is called the **branch cut** of the Log function.

It is possible to define logarithm functions other than Log using the above method. For example, we could choose $0 < \theta < 2\pi$ and take the branch cut to be the set of non-negative real numbers. Thus we can always construct a logarithm by choosing a ray emanating from the origin to slit the plane along an open interval of length 2π from which the angle is chosen. From this point on, we will use "log" to denote a fixed one of these branches of the logarithm, but we won't specify which one unless it is necessary.

Next, observe that $\theta = \Im(\log(z))$ where $\Im(z)$ means the imaginary part of z. This allows us to define the (signed) angle between $z_1, z_2 \in \mathbb{C}^-$ by

$$\angle(z_1, z_2) = \Im(\log(z_1) - \log(z_2)).^2 \tag{2}$$

2.2 Averaging Functions of Θ

We denote the unit circle in the complex plane as $S^1 = \{e^{it}|t \in [0, 2\pi]\}$. Let z_1 and z_2 be complex numbers. Then we define the angle between z_1 and z_2 with respect to a third point z as

$$\Theta(z) = \Im(\log(z_1 - z) - \log(z_2 - z)). \tag{3}$$

We emphasize here that our definition of Θ depends on the choice of logarithm and on z_1 and z_2. The domain of Θ is the complex plane minus two parallel slits, one emanating from z_1 and the other from z_2. On these slits, Θ is, of course, not defined.

The average angle between z_1 and z_2 with respect to a point moving on the unit circle can then be represented by the integral of Θ. Our first result is

Theorem 1. *If for a given z_1, z_2 and a choice of* log, Θ *is defined on all of S^1, then*

$$\frac{1}{2\pi} \oint_{S^1} \Theta dt = \angle(z_1, z_2), \tag{4}$$

where the same log function used to define Θ is also used to define the $\angle$ function.

Proof Viewed as a function of (x, y), Θ is a harmonic function (i.e. it is a solution to Laplace's equation $\nabla^2 \Phi = 0$) since it is the imaginary part of a holomophic function. It is a well known fact that if a harmonic function is defined on a disk then the average value of that function on the boundary of the disk is equal to the value of the function at the center of the disk. By hypothesis, the branch cuts of Θ do not cross the unit disk, so Θ is harmonic there and hence its average value on the boundary of disk (S^1) is equal to the value of Θ at the origin, i.e. $\frac{1}{2\pi} \oint_{S^1} \Theta dt = \Im(\log(z_1) - \log(z_2)) = \angle(z_1, z_2)$. $\Diamond$

If, for example, we chose to use the principle logarithm function, Log, then the theorem applies to any z_1 and z_2 as long as they do not lie to the left of S^1.

The above theorem tells us that integrating the angular data that can be measured tells us something about the geometry of z_1 and z_2. Given this, it seems likely that integrating a function of Θ could yield more information about z_1 and z_2.(From an experimental point of view, what this means is that the angular data is gathered in the form of a list of angles, and then this list is transformed by the above function and then averaged.) In particular, we would like to determine the magnitudes of z_1 and z_2. Using the identity $\Im(z) = \frac{z - \bar{z}}{2i}$ and the fact that $\bar{z} = \frac{1}{z}$ for $z \in S^1$, we have that for $z \in S^1$

$$\Theta(z) = \frac{1}{2i}(\log(z_1 - z) - \log(z_2 - z) - \log(\bar{z_1} - \frac{1}{z}) + \log(\bar{z_2} - \frac{1}{z})). \tag{5}$$

It is important to keep in mind that the right hand side of (5) is an expression for Θ restricted to S^1, and **does not agree with** Θ **off of** S^1. In particular, the right hand side of (5) defines a holomorphic function, whereas Θ is harmonic, but not holomophic, since it is real valued and non-constant.

Given (5), a natural function to integrate would appear to be $e^{2i\Theta}$. Thus, the quantity we want to compute is (for z_1 and z_2 fixed)

$$\frac{1}{2\pi} \oint_{S^1} e^{2i\Theta} dt. \tag{6}$$

This integral can be written as a complex contour integral by using the parameterization $z = e^{it}$, so that $\frac{dz}{dt} = ie^{it} = iz$. Therefore $dt = \frac{dz}{iz}$ and so (6) is equal to

$$\frac{1}{2\pi} \oint_{S^1} \frac{(z_1 - z)(\bar{z}_2 - \frac{1}{z})\, dz}{(\bar{z}_1 - \frac{1}{z})(z_2 - z)\, iz}. \tag{7}$$

The function $g(z) = \frac{(z_1-z)(\bar{z}_2-\frac{1}{z})}{(\bar{z}_1-\frac{1}{z})(z_2-z)} \frac{1}{iz}$ has singularities at $0, \frac{1}{\bar{z}_1}$ and z_2, all of which are simple poles. Therefore the residues at these poles can easily be computed. For example $Res(g,0) = \lim_{z \to 0} zg(z) = -i\frac{z_1}{z_2}$. Our next theorem follows from the calculation of these three residues, and the residue theorem(see [1]), which we now state for completeness.

The Residue Theorem *Let C be a positively oriented simple closed contour within and on which a function f is analytic except at a finite number of singular points $w_1, w_2, ..., w_n$ interior to C. If $R_1, ..., R_n$ denote the residues of f at those respective points then $\int_C f(z)dz = 2\pi i(R_1 + \cdots + R_n)$.*

Thus the value of (7) depends upon whether or not z_1 and z_2 lie inside or outside of S^1. (z_1 and z_2 may not lie on S^1 for this method to work, and we will not consider this case.) Therefore we have

Theorem 2. *The value of the integral (6) is*

(a) $\overline{\left(\frac{z_2}{z_1}\right)}$ *if* $|z_1| > 1, |z_2| < 1$,

(b) $\frac{-z_1\,\bar{z}_1 + z_2\,\bar{z}_1 - \bar{z}_2\,z_2 + z_1\,z_2\,\overline{z_2\,z_1}}{z_2\,(-1+z_2\,\bar{z}_1)\,\bar{z}_1}$ *if* $|z_1| > 1, |z_2| > 1$,

(c) $\frac{z_1\,\bar{z}_1 - 1 - z_1\,\bar{z}_2 + \bar{z}_2\,z_2}{-1+z_2\,\bar{z}_1}$ *if* $|z_1| < 1, |z_2| < 1$,

(d) $\frac{z_1}{z_2}$, *if* $|z_1| < 1, |z_2| > 1$.

To see how the above theorem can be applied to the problem stated in the introduction, suppose for example that z_2 is known and let A (for average) be the experimentally measured value of the integral (6). Then the above theorem tell us that if $|z_1| > 1$ and $|z_2| < 1$, then $z_1 = z_2 \cdot \frac{1}{A}$, i.e. we have a **linear solution** to the reconstruction problem.

Next we demonstrate how to solve the more general problem when z_1 and z_2 are both unknown. Suppose that $z_1 = r_1 e^{i\Theta_1}$ and $z_2 = r_1 e^{i\Theta_2}$, where $r_1, r_2 > 1$, i.e. we are in case (b) of theorem 2. Let A be the value of (6) and let ψ be the angle between z_1 and z_2 with respect to the origin, chosen with the orientation such that $z_2\bar{z}_1 = r_1 r_2 e^{i\psi}$. Theorem 2 part (a) may be rewritten as

$$\frac{-r_1^2 + r_1 r_2 e^{i\psi} - r_2^2 + r_1^2 r_2^2}{r_1 r_2 e^{i\psi} + r_1^2 r_2^2 e^{2i\psi}} = A. \tag{8}$$

where ψ may be considered as a known quantity because of theorem 1. This single equation between complex quantities can be converted to two real equations in r_1 and r_2. The easiest way to do this is first make the substitutions $u = r_1 r_2$ and $v = r_1^2 + r_2^2$ and rationalize. Then (8) becomes $-v + u e^{i\psi} + u^2 = (u e^{i\psi} + u^2 e^{2i\psi})A$. Taking the real and imaginary parts of the above equation gives linear equations for u and v. Once u and v are known, we must determine r_1 and r_2 from the equations $u = r_1 r_2$ and $v = r_1^2 + r_2^2$. These last two equations can clearly be converted to a single quadratic equation in r_1^2, allowing for the solution of r_1 and r_2.

3   Experimental Results

The main question to be addressed is the effect of noise on the above method when both z_1 and z_2 are unknown. The primarily source of error is in the data taken from the panoramic sensor, so we need to know the nature of the noise that we expect from using this device to measure angles. From experiment we have found the error in angular measurement between vertical edges to have a standard deviation, σ, of about .05 degrees at a range of one meter. To be conservative, in our simulations we varied σ from 0 to .2 degrees.

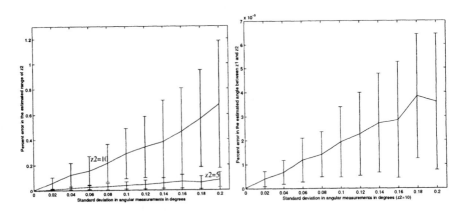

Figure3. Above on the left appears a plot of percent error in estimating the range of z_2 and versus the standard deviation in the noise (in degrees) of the measured angles, and on the right a similar plot for the estimate in the angle between z_1 and z_2. In both plots $z_1 = 5e^{\frac{i\pi}{3}}$. Note the extremely high accuracy of the estimate of the angle between z_1 and z_2

Another source of error is the computation of the averages of Θ and $e^{2i\Theta}$. We are approximating an integral with an average, and because we do not require any knowledge of where on the circle the sensor is, the average is only a good approximation if the density of the measurements is approximately uniform. Thus in the simulation we generated points uniformly spaced on the circle and added noise to the angular position of each point. We always took this noise to be zero mean Gaussian noise with standard deviation one-quarter of 2π divided by the number of sample points. This value of course depends on the experimental device being modeled.

There are many parameters that can be varied in simulation, such as the magnitude of the points being reconstructed, the number of data points used, the numerical integration scheme and so on. Therefore we first fixed the two known points to be $z_1 = 5e^{\frac{i\pi}{3}}$ and $z_2 = 5$, fixed the number of measurements taken on the circle to be 5000, and considered the effect of zero mean Gaussian noise with σ ranging from 0 to .2 degrees. Given all of the above parameters 100 trials were performed and in each trial the percent error in the magnitude of z_2 was computed and an average over all trials of this error was computed.

To demonstrate the effect of range we then repeated this entire simulation with $z_2 = 10$. Each choice of z_2 yields a plot of percent error versus the standard deviation in the noise, which both appear in figure (3).

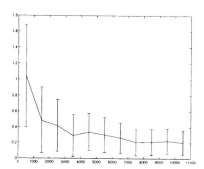

Figure4. A plot of the percent error in the estimated value of the magnitude of z_2 versus the number of sample points used on the circle. Here we fixed $z_1 = 5e^{\frac{i\pi}{3}}$ and $z_2 = 10$. The error bars represent the standard deviation in one hundred trials.

Figure 4 illustrates the percent error in reconstruction versus the number of sample points used, i.e. the number of angular pairs measured. In this case we fixed $\sigma = .1$ and let $z_1 = 5e^{\frac{i\pi}{3}}$ and $z_2 = 10$.

Using a panoramic sensor (consisting of a Sony XC-77 camera and a spherical mirror) on a turntable, we performed an experiment to reconstruct the positions of two vertical edges. The edges were the vertical edges of white paper on a black background placed at 2.8 (r_1) and 2.7 (r_2) units from the origin (which is the center of the turntable) with an angle of 22 degrees between them. An discussed above, vertical edges in the world appear as radial in the spherical image (see figure (2)). Recall that the unit of measure is the distance from the center of the turntable to the optical axis of the camera, which was 18 cm (The roughness of the measurements is due to the fact that the true values for the distances were found by hand using a ruler.). We then gathered 71 data points by rotating the turntable by hand. The estimates obtained were 2.7 units for r_1 and 2.7 units for r_2.

4 Conclusion

In this paper we have considered the use of controlled motion for the problem of recovering robot pose and the lay-out of the environment. To do this we have introduced certain spaciotemporal signatures that give rise to quadratic or linear equations for the solution of the problem. This is achieved by the application of the residue calculus from the theory of complex variables. The method we propose is robust to noise and outliers and takes into account global information that alternative methods do not.

References

1. C. Berenstein and R. Gay. *Complex variables : an introduction.* Springer-Verlag: New York, 1991.
2. J.S. Chahl and M.V. Srinivasan. Range estimation with a panoramic sensor. *J. Optical Soc. Amer. A*, 14:2144–2152, 1997.
3. R. A. Hicks, D. Pettey, K. Daniilidis, and R. Bajcsy. Global signatures for robot control and reconstruction. To appear *Proc. Workshop on Robust Vision for Vision-based control 1998 IEEE Conference on Robotics and Automation*, 1999.
4. S. Kang and R. Szeliski. 3-d scene data recovery using omnidirectional multibaseline stereo. *International Journal of Computer Vision*, 25:167–183., 1997.
5. A. Kosaka and A.C. Kak. Fast vision-guided mobile robot navigation using model-based reasoning and prediciton of uncertainties. *Computer Vision Image Understanding*, 56:271–329, 1992.
6. D.J. Kriegman, E. Triendl, and T.O. Binford. Stereo vision and navigation in buildings for mobile robots. *Trans. on Robotics and Automation*, 5:792–804, 1989.
7. S. Nayar. Catadioptric omnidirectional camera. In *Proc. Computer Vision Pattern Recognition*, pages 482–488, 1997.
8. Y. Onoe, H. Yokoya, and K. Yamazawa. Visual surveillance and monitoring system using an omnidrectional system. In *Proc. Int. Conf. on Pattern Recognition*, 1998.
9. H.S. Sawhney, J. Oliensis, and A.R. Hanson. Description and reconstruction from image trajectories of rotational motion. In *Int. Conf. computer Vision*, pages 494–498, 1990.
10. H. Shariat and K.E. Price. Motion estimation with more then two frames. *IEEE Trans. Patt. Anal. Mach. Intell.*, 12:417–434, 1990.
11. F. Stein and G. Medioni. Map-based localization using the panoramic horizon. *Trans. on Robotics and Automation*, 11:892–896, 1995.
12. T. Svoboda, T. Padjla, and V. Hlavac. Epipolar geometry for panoramic cameras. In *Proc. European Conference on Computer Vision*, 1998.
13. Y. Yagi, S. Nishizawa, and S. Tsuji. Map-based navigation for a mobile robot with omnidirectional image senso. *Trans. on Robotics and Automation*, 11:634–648, 1995.

Estimating Consistent Motion from Three Views: An Alternative to Trifocal Analysis

Stefan Trautwein, Matthias Mühlich, Dirk Feiden and Rudolf Mester

J. W. Goethe-Universität Frankfurt, Germany
trautwei@stud.uni-frankfurt.de,
(muehlich|feiden|mester)@iap.uni-frankfurt.de
Tel: (+49) 69 798-22387   Fax: (+49) 69 798-28510

Abstract. The main goal of this paper is to introduce methods for three-view motion analysis that do not need threefold correspondences in the image planes as the well-known trifocal tensor methods do. With this characteristic, the proposed method is a practically very advantageous approach for (ego-)motion analysis and structure from motion. The proposed method starts with three two-view parameter estimates generated by Hartley/Mühlich-equilibrated TLS solutions, enforces geometrical consistency and iteratively optimizes the distances from the set of epipolar lines.

1   Introduction

In this paper we will discuss several iterative and non-iterative methods for the estimation of the motion parameters resp. the relative poses of three calibrated cameras, and we will examine which methods are recommendable for applications. We will show that point matches over three images are not necessarily needed for exploiting the information that is hidden in three images. We consider the situation where only correspondences between every two images out of the triplet exist, i.e. $\mathcal{I}1 \leftrightarrow \mathcal{I}2$, $\mathcal{I}2 \leftrightarrow \mathcal{I}3$ and $\mathcal{I}1 \leftrightarrow \mathcal{I}3$ where $\mathcal{I}i$ stands for image no. i (see fig. 3 c). In this situation, the calculation of the trifocal tensor is not possible.

The main idea is to enforce geometrical consistency of the motions, i.e. the three motions have to fit together (see section 1.2). We will present two different approaches to the solution of this problem

Approach 1: Separate solutions of three two-image-problems with subsequent non-iterative enforcement of consistency (see section 3)

Approach 2: enforcement of consistency constraints during the iterative solution of the three-view-problem (section 4)

Approach 2 is closely related to a method proposed in [7] for a stereo camera. As all approach 2 methods are iterative methods, they need initial values that have to be generated with a two-view method. Therefore the two approaches only differ in the way in which the three-view consistency is enforced. For approach 1 methods, the three essential matrices (E-matrices) are the only input whereas the approach 2 methods use the point correspondences for a nonlinear optimization. The first approach can be done without any iterative optimization and is therefore much faster than the second approach which gives better results, though. The motion analysis is done in two steps (see figs. 1 and 2): in step 1,

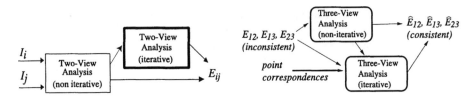

Fig. 1. Two-view estimation of E-matrices (applied 3 times)

Fig. 2. Enforcing consistency between the 3 E-matrices

three inconsistent E-matrices are calculated; these are the initial values for the enforcement of consistency in step 2.

1.1 Types of Matches

Most authors dealing with multi-view-analysis assume the existence of completely concatenated correspondences (c.c.c.), i.e. if there is an image point given in $\mathcal{I}_1$, the corresponding points in every other image are known. We will denote this matching type in our three-view-problem by "threefold-match" (see fig. 3 a). However, the assumption of c.c.c. is critical when we consider the block matching method [3] which gives covariance matrices for the estimated points. A block match from $\mathcal{I}_1$ to $\mathcal{I}_2$ and then from $\mathcal{I}_2$ to $\mathcal{I}_3$ will not yield the same

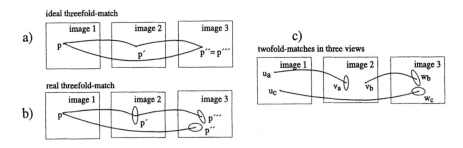

Fig. 3. Threefold and twofold-matches

point and covariance ellipse in $\mathcal{I}_3$ as a block match directly from $\mathcal{I}_1$ to $\mathcal{I}_3$ (fig. 3 b). To avoid these complications we will only consider twofold-matches (fig. 3 c) without excluding that e.g. points u_a and u_c could be identical.

1.2 Consistency Constraints

We use the following notations:
CS_i: Camera coordinate system no. i; u, v, w: coordinates of image points in $\mathcal{I}_1, \mathcal{I}_2$ and $\mathcal{I}_3$; $\mathbf{R}_{ij}, t_{ij}$: rotation and translation between CS_i and CS_j.

Given a point with coordinates x in CS_1, the following relation holds for its coordinates x' in CS_2: $x = \mathbf{R}_{12}x' + t_{12}$. It is important to notice that here the first index of a translation vector indicates the CS in which its coordinates are given, i.e. t_{12} is given in CS_1. If we explicitly transform a vector into another coordinate system, we indicate its new coordinate system by a superscript, e.g. t_{23} given in CS_1 is denoted by $t_{23}^{(1)}$.

The three motion entities $\mathbf{R}_{12}, t_{12}$, $\mathbf{R}_{13}, t_{13}$ and $\mathbf{R}_{23}, t_{23}$ are obviously not independent of each other. The motion from CS_1 to CS_2 followed by the motion from CS_2 to CS_3 must be the same as the motion from CS_1 to CS_3. This leads to the *consistency constraints*:

$$\mathbf{R}_{12}\mathbf{R}_{23} = \mathbf{R}_{13} \quad \text{rotational consistency} \tag{1}$$

$$t_{23}^{(1)} = t_{13} - t_{12} \quad \text{translational consistency} \tag{2}$$

In the second equation, t_{23} has been transformed into CS_1 so that the three translation vectors are given in the same coordinate system. The vector t_{23} is transformed from CS_2 to CS_1 by multiplication with $\mathbf{R}_{12}$ so that $t_{23}^{(1)} = \mathbf{R}_{12}t_{23}$. The translational consistency constraint, however, has to be weakened because we can determine translation vectors only up to a scaling factor

$$t_{12} = \alpha t_{13} + \beta t_{23}^{(1)} \quad \text{with} \quad \alpha, \beta \neq 0. \tag{3}$$

This means that the three translation vectors must be linearly dependent, i.e. they have to lie in a plane.

2 Two-View and Three-View Motion Analysis

There are two approaches for estimating rotation $\mathbf{R}$ and translation t (combined in the essential matrix $\mathbf{E}$) between two views of the same static scene. A family of so called 'linear' methods[1] is derived from the well-known 8-point algorithm.

Hartley [2, 1] showed that a normalization of image point coordinates before the calculation of the E-matrix or trifocal tensor increases the quality of the estimates. The coordinates x of a given image point are transformed by a matrix $\mathbf{S}$ so that the new coordinates $x' = \mathbf{S}x$ fulfill the conditions $\sum_i x_i' = 0$ and $\sum_i x_i' x_i'^T = N\mathbf{I}$. After the computation of the trifocal tensor $\mathbf{T}$ or the essential matrix $\mathbf{E}$, this normalization transformation has to be reversed to get the real motion parameters. This heuristically introduced transformation has been justified and improved in two papers by Mühlich and Mester [5, 4].

---

[1] These methods are not really linear as they are based on the calculation of eigenvectors which is certainly no linear operation.

Another two-view method involves the non-linear minimization of the sum of squared distances between image points and their corresponding epipolar lines. The squared euclidean distance between a point with the coordinates p and a line with the parameters[2] g is given by

$$d^2(p,\, g) = \frac{(g_1 p_1 + g_2 p_2 + g_3)^2}{g_1^2 + g_2^2} \, .$$ (4)

In order to estimate the essential matrix $\mathbf{E}_{12}$ between $\mathcal{I}1$ and $\mathcal{I}2$, the objective function is thus given by $F = \sum_{i=1}^{n_a} d^2(v_i, \mathbf{E}_{12}^T u_i)$. In the three-image-problem, the objective function is

$$F = \sum_{i=1}^{n_a} d^2(v_i, \mathbf{E}_{12}^T u_i) + \sum_{i=1}^{n_b} d^2(w_i, \mathbf{E}_{23}^T v_i) + \sum_{i=1}^{n_c} d^2(w_i, \mathbf{E}_{13}^T u_i) \, ,$$

where n_a, n_b and n_c denote the numbers of point correspondences in the image pairs. If the coordinate errors for each point correspondence are described by the same isotropic covariance matrix, the minimization of this choice of F is equivalent to a maximum likelihood estimation of the E-matrices.

3   Approach A1: Enforcing Motion Consistency

In this section we will present a very simple yet useful method to make use of the consistency constraints in three views. Let $\mathbf{E}_{12}$, $\mathbf{E}_{23}$ and $\mathbf{E}_{13}$ be three estimated essential matrices describing relative orientation between three different views (obtained by any arbitrary two-view method). Then these three matrices will usually be *not consistent*. Obviously, exploiting the consistency constraints allows an improvement of the estimation. The following question rises: Given these matrices, how can some *minimal change* (in a certain kind of error metrics) be added to enforce consistency in a second estimation step?

First of all, we determine rotation $\mathbf{R}$ and translation t from our essential matrices and get $\mathbf{R}_{12}$, $\mathbf{R}_{23}$, $\mathbf{R}_{13}$ and t_{12}, t_{23}, t_{13}. Then, we have to estimate the quantities $\hat{\mathbf{R}}_{12}$, $\hat{\mathbf{R}}_{23}$, $\hat{\mathbf{R}}_{13}$ and $\hat{t}_{12}$, $\hat{t}_{23}$, $\hat{t}_{13}$ that fulfill the consistency constraints (1) and (3).

3.1   Enforcing Rotational Consistency

We can rewrite our rotational consistency constraint as

$$\hat{\mathbf{R}}_{ij} \hat{\mathbf{R}}_{jk} \hat{\mathbf{R}}_{ki} \overset{!}{=} \mathbf{I}$$ (5)

with (ijk) being any permutation of 1, 2 and 3, i.e. (123), (231), (321), (132), (321), (213). If we solve (5) for *one single* choice of (ijk), it will hold for *all other* choices as well because (5) is invariant against cyclic permutations of (ijk) and we get the anticyclic permutations by transposing (5).

---

[2] a line with parameters g is defined as the set of all points x fulfilling x fulfill $g^T x = 0$.

If $\mathbf{R}_{ij}$ and $\hat{\mathbf{R}}_{ij}$ are rotation matrices, then there must be a rotation matrix $\mathbf{K}_{ij}$ that solves the equation

$$\hat{\mathbf{R}}_{ij} = \mathbf{K}_{ij}\mathbf{R}_{ij} . \tag{6}$$

These $\mathbf{K}_{ij}$ are the three correction terms and their rotation angles are to be minimized. For reasons of symmetry we can set

$$\mathbf{K}_{ij} = f(\mathbf{R}_{ij}\mathbf{R}_{jk}\mathbf{R}_{ki}) = f(\mathbf{F}) \tag{7}$$

with f being a certain function of its argument. This function shall be determined later on. The argument $\mathbf{F} = \mathbf{R}_{ij}\mathbf{R}_{jk}\mathbf{R}_{ki}$ has a certain geometrical interpretation: it is the rotation error of the three rotations and should be equal to the identity matrix for error-free measurements.

We can express $\mathbf{K}_{jk}$ and $\mathbf{K}_{ki}$ as

$$\mathbf{K}_{jk} = f(\mathbf{R}_{jk}\mathbf{R}_{ki}\mathbf{R}_{ij}) = f(\mathbf{R}_{ji}\mathbf{F}\mathbf{R}_{ij}) \tag{8}$$
$$\mathbf{K}_{ki} = f(\mathbf{R}_{ki}\mathbf{R}_{ij}\mathbf{R}_{jk}) = f(\mathbf{R}_{kj}\mathbf{R}_{ji}\mathbf{F}\mathbf{R}_{ij}\mathbf{R}_{jk}) . \tag{9}$$

The matrix $\mathbf{R}_{ji}\mathbf{F}\mathbf{R}_{ij}$ describes the same rotation as $\mathbf{F}$, transformed in the j-th coordinate system. If we demand that the function f must not depend on the coordinate system, we get

$$\mathbf{K}_{jk} = f(\mathbf{R}_{ji}\mathbf{F}\mathbf{R}_{ij}) = \mathbf{R}_{ji}\,f(\mathbf{F})\,\mathbf{R}_{ij} \tag{10}$$
$$\mathbf{K}_{ki} = f(\mathbf{R}_{kj}\mathbf{R}_{ji}\mathbf{F}\mathbf{R}_{ij}\mathbf{R}_{jk}) = \mathbf{R}_{kj}\mathbf{R}_{ji}\,f(\mathbf{F})\,\mathbf{R}_{ij}\mathbf{R}_{jk} . \tag{11}$$

Now we combine (7), (10) and (11) with (5):

$$f(\mathbf{F})\mathbf{R}_{ij}\mathbf{R}_{ji}f(\mathbf{F})\mathbf{R}_{ij}\mathbf{R}_{jk}\mathbf{R}_{kj}\mathbf{R}_{ji}f(\mathbf{F})\mathbf{R}_{ij}\mathbf{R}_{jk}\mathbf{R}_{ki} \overset{!}{=} \mathbf{I} \tag{12}$$
$$\Leftrightarrow \qquad f(\mathbf{F})f(\mathbf{F})f(\mathbf{F})\mathbf{F} \overset{!}{=} \mathbf{I} \tag{13}$$

We are looking for a minimal rotation that has to be applied three times after another to yield $\mathbf{F}^{-1}$. Obviously the desired rotation has the same rotation axis as $\mathbf{F}$ and $-\frac{1}{3}$ of its rotation angle. This matrix can easily be determined. Then we can determine $\hat{\mathbf{R}}_{ij}$ with eqn. (6), (7), (10) and (11).

3.2   Enforcing Translation Consistency

First of all we have to express all three translation vectors in the same coordinate system. The first coordinate system is most suitable for this purpose as t_{12} and t_{13} are already expressed in this system. The vector t_{23}, however, has to be transformed to $t_{23}^{(1)} = \hat{\mathbf{R}}_{12}t_{23}$.[3]

---

[3] The true rotation matrix $\mathbf{R}_{12}^0$ is unknown. In consequence, we transform t_{23} with a potentially wrong rotation matrix and, thus, increase the errors in this translation vector. To minimize this effect, it is useful to transform with the consistent matrix $\hat{\mathbf{R}}_{12}$ instead of $\mathbf{R}_{12}$.

Our requirement that all three translation vectors must lie in the same plane is equivalent to the statement that they are linearly dependent. This means that the matrix $\mathbf{M}$ defined as

$$\mathbf{M} = (\boldsymbol{t}_{12}, \boldsymbol{t}_{23}^{(1)}, \boldsymbol{t}_{13}) \tag{14}$$

must be singular. The Eckard-Young-Mirsky-theorem (see e.g. [5]) solves this problem using the singular value decomposition (SVD).[4]

If we assume that the error distribution in all vectors is equal and if we neglect the effect of the transformation of $\boldsymbol{t}_{23}^{(1)}$, all three vectors should be scaled to the same length, e.g. to unit vectors.

The columns of the rank-deficient matrix $\hat{\mathbf{M}}$ will be our consistent estimates for translation vectors and we will denote them as $\hat{\boldsymbol{t}}_{ij}$. The only thing left to do is to transform $\boldsymbol{t}_{23}^{(1)}$ back into $\mathcal{CS}_2$: $\hat{\boldsymbol{t}}_{23} = \hat{\mathbf{R}}_{12}^T \hat{\boldsymbol{t}}_{23}^{(1)}$. As a result, we now have three consistent rotation matrices $\hat{\mathbf{R}}_{ij}$ and three consistent translation vectors $\hat{\boldsymbol{t}}_{ij}$ that are better estimates of the underlying motion. This result is obtained with very little computational effort.

4 Approach A2: Simultaneous Estimation of Three Consistent E-Matrices

An alternative to the enforcement of consistency in a second estimation step (approach A1) is the direct estimation of three consistent E-matrices (approach A2). All methods described here will use nonlinear optimization and, thus, be slower than approach 1. On the contrary, we can expect to get better estimates.

Given three images of a static scene with correspondences between every pair of two images, our task is to estimate three consistent motions.

Our first idea was the following: We took the elements of the 3 E-matrices as a point in $\mathbb{R}^{27}$ and tried to minimize the distances of the measured points to their epipolar lines under the consistency constraints. These consistency constraints are functions in the elements of the E-matrices. We got the following constraints: 9 equations for rotational consistency (eq. (1)), 3 equations for translational consistency (eq. (2)) and 9 equations for the so called Demazure constraints (we used the reduced representation from [6] with only three independent equations) to be sure that the E-matrices are valid E-matrices, i.e. they can be decomposed into rotation and translation. These constraints are not independent of each other and they are rather complicated (linear combinations of products of up to 6 matrix elements).

This optimization approach did not produce useful results, i.e. the optimization routine did not converge which might have the following reasons: there are too many parameters and there are too many and too complicated constraints.

---

[4] This theorem decomposes $\mathbf{M}$ in two matrices $\mathbf{M} = \hat{\mathbf{M}} + \mathbf{D}$ with rank $(\hat{\mathbf{M}}) = 2$ and $\|\mathbf{D}\|_F^2 \rightarrow \min$. All matrix entries in the estimated error matrix $\mathbf{D}$ are given the same weight; therefore we obtain a maximum likelihood estimate if all entries are perturbed with independent identically distributed (i.i.d.) gaussian noise with the same standard deviation σ.

4.1 Minimal Parameterization of the Motion Parameters

To cope with the problem of too many and/or too complicated constraints, we will try to avoid constraints at all. This means that we are looking for a representation of our problem where the number of degrees of freedom (d.o.f.) and number of parameters are equal.

The motions $C1 \leftrightarrow C2$ (between camera 1 and 2) and $C2 \leftrightarrow C3$ both have 3 rotational and 2 translational d.o.f. (only the translation *direction* can be determined, not its *length*). The 11th d.o.f. is the relative scaling factor $\|t_{12}\| / \|t_{23}\|$. Once this scaling factor is known, the motion $C1 \leftrightarrow C3$ is completely determined, or in other words, this factor is the only new information contained in $C1 \leftrightarrow C3$.

The rotations are parameterized using the Rodrigues matrix [7]:

$$\mathbf{R} = \frac{1}{k} \begin{pmatrix} 1 + (a^2 - b^2 - c^2)/4 & -c + ab/2 & b + ac/2 \\ c + ab/2 & 1 + (-a^2 + b^2 - c^2)/4 & -a + bc/2 \\ -b + ac/2 & a + bc/2 & 1 + (-a^2 - b^2 + c^2)/4 \end{pmatrix}$$

with $k = 1 + (a^2 + b^2 + c^2)/4$ and the translation directions are parameterized by two angles so that $t = (\sin \alpha \sin \beta, \cos \alpha \sin \beta, \cos \beta)$ which implies $\|t\| = 1$. These parameterizations allow us to solve an *unconstrained* optimization problem.

After achieving good results for the three-view case, we also implemented a minimal parameterization method for the two-view motion analysis.

5 Experiments and Results

In summary, we have three methods for two-view analysis (Hartley, Hartley/Mühlich and minimal parameterization), which will be abbreviated by **HA**, **HM** and **MP**. The last method is an iterative method and needs a starting value generated by one of the first two methods. For the three view analysis we used approach **A1** or **A2**.

We conducted and compared the following experiments:

1. **HA** (no enforcement of consistency)
2. **HM** (no enforcement of consistency)
3. **MP2** (min. param. in two views with **HM** as starting value; no enforcement of consistency)
4. **HMk** (consistency approach **A1** with **HM** as starting value)
5. **MP2k** (consistency approach **A1** with **MP2** as starting value)
6. **MP3** (consistency approach **A2** with **HM** as starting value)

For comparison, we also generated a set of three completely concatenated point correspondences and calculated the trifocal tensor (**TFT**) using the method proposed in [2].

5.1 Comparison of Estimation Errors

In order to compare the quality of the estimates resulting from the different methods, we defined the following error measures:

translational error: sum of angular errors of the translation vectors

rotational error: sum of angular errors of the quaternions (4-dimensional unit vectors describing rotations). The angle between these two unit vectors describes both errors in rotation axis and angle in one value which is very advantageous. This error is more meaningful than the error in axis and angle separately (the error in the rotation angle is only useful if the axis is estimated without any error).

These errors are plotted as functions of the image noise in figs. 4 - 7.

5.2 Experimental Setup

We conducted experiments with two different setups. In any case there is a cloud of 3D-points which have randomized positions within a cube whose location is given by $0 \leq x \leq 10$, $0 \leq y \leq 10$, $20 \leq z \leq 30$. The first camera is the unity camera with camera matrix $\mathbf{M}_1 = (\mathbf{I}|\mathbf{0})$. The two other cameras are displaced in the following way: the rotation axes of $\mathbf{R}_{12}$ and $\mathbf{R}_{13}$ are $(1, 0.2, 0)^T$ resp. $(1, 0.1, 0)^T$ in both setups. The translation vectors t_{12} and t_{13} and the angles W2 und W3 of $\mathbf{R}_{12}$ and $\mathbf{R}_{13}$ summarized in the following table

Setup no.	t_{12}	t_{13}	W2	W3
1	$(1, 15, 13)^T$	$(5, 20, 25)^T$	45°	90°
2	$(0, 0, -2)^T$	$(0, 0, -4)^T$	1°	2°

In setup 2, the three cameras are placed on a line,[5] i.e. it is impossible to calculate the scaling factor $\lambda = \|t_{12}\| / \|t_{13}\|$ from the three translation directions. As a starting value for the optimization algorithm, we inserted the exact value $\lambda = 2$. To prevent miscomputations, we restricted λ to a value $\lambda < 4$. In setup 1, such a restriction was not necessary.

A two-view-problem always yields four (completely) different combinations of motion parameters. In our simulations we assumed that the true motion is approximately known so that we could single out the correct solution.

5.3 Results

The estimation errors both for translational and for rotational error are plotted as functions of the image noise. Each estimation error represents the mean value of 140 experimental evaluations. The rotations for the trifocal tensor method have not been calculated.

Looking at the figures 4 - 7 we come to the following general conclusions: as already shown in [5], the method **HM** increases the quality of the estimate,

---

[5] this is a common situation for a camera mounted in a car.

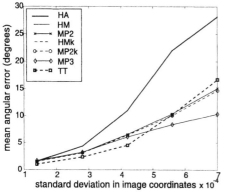 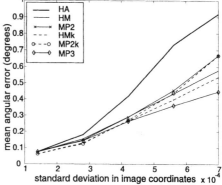

Fig. 4. setup 1: translational error. **HA** and **HM** lie on top of each other.

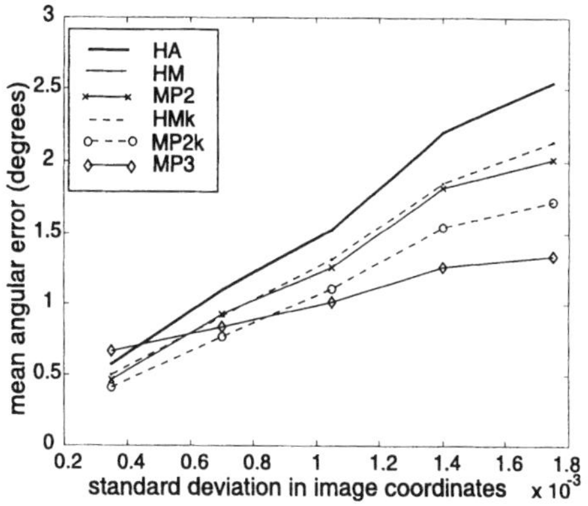

Fig. 5. setup 1: quaternion error. **HA** and **HM** lie on top of each other.

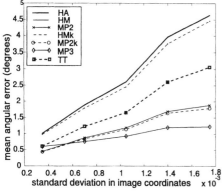

Fig. 6. setup 2: translational error. **HM** and **MP2** as well as **HMk** and **MP2k** lie on top of each other.

Fig. 7. setup 2: quaternion error

especially if the motion is parallel to the optical axis of the camera (setup 2). Comparing the graphs of **HM** and **MP2** in setup 1, we can see that a direct estimation of the motion parameters instead of estimating the fundamental matrix improves motion estimation considerably. This is not very surprising since, when estimating the fundamental matrix, we do not use the information that the camera is calibrated. For the same reason the trifocal tensor does not always yield the best results (see setup 1).

Comparing the graphs of **HM** and **HMk** or **MP2** and **MP2k** (especially the quaternion errors) we can see that the subsequent enforcement of consistency decreases the error both for rotations and for translations. The improvement is considerable because it is obtained with very little computational effort (i.e. without nonlinear optimization). There is a greater decrease in quaternion error than in translation error which may be explained by the fact that the enforcement of translation consistency depends on consistent rotations. In setup 2, a decrease in translational error is hardly visible.

The method **MP3** usually performs better than **MP2k** because we enforce the consistency constraints during the optimization of the motion parameters.

The trifocal tensor method can be compared with the other methods only up to a certain degree. It needs ideal threefold matches (which we don't assume to have in our methods). These matches surely contain more information than the twofold matches we need. Therefore, the trifocal tensor can produce better estimates. Nevertheless, our methods are often better because they are optimized for calibrated cameras.

We also have conducted experiments where all three cameras have taken images of the same 3D points, but there was not much difference to the experiments described above because the methods except for **TFT** do not assume the existence of completely concatenated correspondences and therefore they cannot exploit this additional information.

6 Summary

We have shown that the enforcement of motion consistency in a three-view-problem improves the quality of the estimated motion parameters considerably. As a result of our experiments we can propose the following motion estimation methods for situations where only twofold-matches are given in three views:

- If a method is sought that fits for a real-time application, our choice is method **HMk** for it does not involve nonlinear optimization and is therefore very fast.
- If primary interest is in good estimates without too much concern about computational time, we recommend **MP3**.

References

1. Richard I. Hartley. In Defense of the 8-Point-Algorithm. *IEEE Trans. on Pattern Analysis and Machine Intelligence*, 19(6):580–593, 1997.
2. Richard I. Hartley. Lines and Points in Three Views and the Trifocal Tensor. *International Journal of Computer Vision*, 22(2):125–140, 1997.
3. R. Mester and M. Hötter. Robust Displacement Vector Estimation including a Statistical Error Analysis. *Proc. 5th Intern. Conf. on Image Processing and its Applications, Edinburgh, pp. 168-172*, 1995.
4. M. Mühlich and R. Mester. Ein verallgemeinerter Total Least Squares-Ansatz zur Schätzung der Epipolargeometrie (in German). *in Mustererkennung 1998 (Proc. DAGM'98), Springer*, pages 349–356, 1998.
5. M. Mühlich and R. Mester. The Role of Total Least Squares in Motion Analysis. *in Proc. ECCV'98, Freiburg*, pages 305–321, 1998.
6. Bill Triggs. A New Approach to Geometric Fitting. *Available from http://www.inrialpes.fr/movi/people/Triggs*, 1997.
7. Zhengyou Zhang, Quang-Tuan Luong, and Olivier Faugeras. Motion of an Uncalibrated Stereo Rig: Self-Calibration and Metric Reconstruction. *INRIA Research Report 2079*, 1993.

Using Rigid Constraints to Analyse Motion Parameters from Two Sets of 3D Corresponding Point Pattern

Yonghuai Liu and Marcos A Rodrigues

AI and Pattern Recognition Research Group
Department of Computer Science, The University of Hull, Hull, HU6 7RX, UK

Abstract. Given a rigid body motion expressed by a set of 2D or 3D correspondences, the distance between feature points and angular information can be used as rigid constraints to calibrate motion parameters. In this paper, we first present a novel geometrical analysis of properties of reflected correspondence vectors. The analysis provides explicit expressions for distance between feature points and angle measurement synthesised into a single coordinate frame providing the closed form solutions to all motion parameters of interest. A novel calibration algorithm is proposed and compared with an algorithm based on the least squares method. The algorithm demonstrates the importance of the geometrical properties of reflected correspondence vectors to motion parameter estimation and that it is generally more accurate than algorithms based on the least squares method.

1 Introduction

Given that feature point correspondences between two different coordinate frames have been established, motion estimation is concerned with the determination of the corresponding parameters that accurately describe the object's motion in 3D space [1], [2], [6], [7], [10]. Motion estimation techniques find potential applications in many areas such as, for instance, navigation planning, analysis of the structure of 3D objects, and recognition of objects [4]. Many methods to calibrate 3D motion parameters have been put forward such as techniques based on least squares method and singular value decomposition [1], [7], [10], iterative algorithm [6], total least squares method [2]. When used to real world applications however, such methods present a number of problems, such as lack of efficient algorithms, high sensitivity to noise, and multiplicity of solutions [5].

A limitation of some of the above methods is that they do not make full use of distance and angle measurement as rigid constraints to calibrate motion parameters. This is mainly because no mathematical formalisation for distances and angle information in the context of motion estimation has been proposed. In this paper, we describe a novel geometrical analysis of reflected correspondence vectors. Based on such geometrical properties, we propose a novel geometrical algorithm for the calibration of 3D motion parameters. Experimental validation shows that the algorithm works well in the presence of noise with a performance that is superior to algorithms based on the least squares method.

2 Analysis of Reflected Correspondence Vectors

A rigid body motion can be expressed by the following relationship: $\mathbf{p}' = \mathbf{R}(\mathbf{p} - \mathbf{t})$, where $\mathbf{R}$ and $\mathbf{t}$ are the motion parameters of interest and represent, respectively, the rotation matrix and the translation vector of the rigid body motion, and $(\mathbf{p}, \mathbf{p}')$ is a correspondence between two different coordinate frames. The reflected correspondence ($\mathbf{RC}$) of any correspondence pair $(\mathbf{p}, \mathbf{p}'$ is defined as: $\mathbf{RC} = (\mathbf{p}, \mathbf{p}'') = (\mathbf{p}, -\mathbf{p}')$, leading to the definition of the correspondence vector $\mathbf{CV} = \mathbf{p} - \mathbf{p}'$, and reflected correspondence vector $\mathbf{RCV} = \mathbf{p} - \mathbf{p}''$.

A geometrical analysis of $\mathbf{RCV}$s in 2D is described as follows, assuming that the rotation angle of the motion is in the interval $(0, \pi)$. Given two sets of 2D correspondence data $(\mathbf{p_i}, \mathbf{p_i'})$ $(i = 1, 2, \cdots, n)$ determine the set of reflected correspondences $(\mathbf{p_i}, \mathbf{p_i''})$. Synthesize all such correspondences and the translation vector $\mathbf{t}$ into a single coordinate frame. Drawing the perpendicular bisectors of $\mathbf{RCV_i}$, it is found that they all intercept at a fixed point that we call the complementary pole $\mathbf{p_0}$ of a planar displacement. It is verified that the perpendicular bisector of the translation vector $\mathbf{t}$ also intercepts at the pole $\mathbf{p_0}$. The including angle between the lines passing through any reflected correspondence $(\mathbf{p_i}, \mathbf{p_i''})$ and the complementary pole $\mathbf{p_0}$ is equal to the supplement of the rotation angle θ of the transformation. Formalising the above description, we have the following property as depicted in Figure 1.

Property 1. There is one and only one non-zero point $\mathbf{p_0}$ in 2D which is uniquely determined by the rigid body's motion equidistant to any reflected correspondence $(\mathbf{p}, \mathbf{p}'')$ subject to the same rigid body motion and the including angle between $\mathbf{p} - \mathbf{p_0}$ and $\mathbf{p}'' - \mathbf{p_0}$ is equal to the supplement of the rotation angle θ of the motion.

The analysis of such 2D geometry reveals interesting and useful properties about the pole $\mathbf{p_0}$ for: (*i.*) the calibration of rigid body's motion parameters, (*ii.*) the analysis of the relationships between given $\mathbf{RCV}$s, and (*iii.*) limiting the number of solutions for the motion. Expressions to evaluate the pole $\mathbf{p_0}$, the rotation angle θ, the rotation matrix $\mathbf{R}$, and the translation vector $\mathbf{t}$ are described as follows. Given two non-parallel reflected correspondence vectors $\mathbf{RCV_1}$ and $\mathbf{RCV_2}$ their perpendicular bisectors intercept at the pole $\mathbf{p_0}$ which can be estimated by:

$$\mathbf{p_0} = \begin{pmatrix} \mathbf{RCV_1^T} \\ \mathbf{RCV_2^T} \end{pmatrix}^{-1} \begin{pmatrix} \frac{\mathbf{p_1^T p_1} - \mathbf{p_1''^T p_1''}}{2} \\ \frac{\mathbf{p_2^T p_2} - \mathbf{p_2''^T p_2''}}{2} \end{pmatrix} \tag{1}$$

In order to uniquely determine the pole $\mathbf{p_0}$, at least two non-parallel $\mathbf{RCV}$s are needed. The rotation angle θ can be estimated by:

$$\cos(\pi - \theta) = \frac{(\mathbf{p_i} - \mathbf{p_0})^{\mathbf{T}}(\mathbf{p_i''} - \mathbf{p_0})}{(\mathbf{p_i} - \mathbf{p_0})^{\mathbf{T}}(\mathbf{p_i} - \mathbf{p_0})} \tag{2}$$

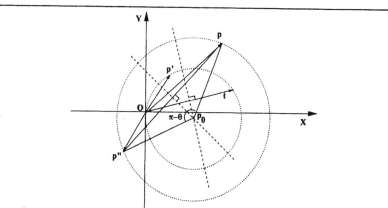

Fig. 1. A representation of the 2D geometry as defined by Property 1. After a rigid body's motion, points p and p″ lie on the same circle centred at the complementary pole $\mathbf{p_0}$, the perpendicular bisector of the translation vector t passes through $\mathbf{p_0}$, and the including angle between $\mathbf{p} - \mathbf{p_0}$ and $\mathbf{p}'' - \mathbf{p_0}$ is equal to the supplement of the rotation angle θ of the motion.

As a result, the rotation matrix $\mathbf{R}$ of the transformation is uniquely determined by:

$$\mathbf{R} = \begin{pmatrix} \cos\theta & \sin\theta \\ -\sin\theta & \cos\theta \end{pmatrix} \qquad (3)$$

leading to the estimation of the translation vector t as:

$$\mathbf{t} = (\mathbf{I} + \mathbf{R}^T)\mathbf{p_0} \qquad (4)$$

This completes the analysis in 2D. A step by step 3D geometrical analysis is described as follows. Given a set of correspondences $(\mathbf{p_i}, \mathbf{p_i'})$ $(i = 1, 2, \cdots, n)$ find the perpendicular bisector planes of their corresponding $\mathbf{RCV_i}$. It is verified that all planes intercept at a common point which we call the pole $\mathbf{p_0}$ of a spatial displacement. Then, find the vector difference $\mathbf{CV_i} - \mathbf{CV_j}$ $(i \neq j)$. The perpendicular bisector planes of such vectors are found to intercept at a fixed line. We call this fixed line the rotation axis $\mathbf{h}$. Projecting all reflected correspondences $(\mathbf{p_i}, \mathbf{p_i''})$ and the pole $\mathbf{p_0}$ on a plane perpendicular to the rotation axis $\mathbf{h}$ yields a set of projected RC $(\underline{\mathbf{p_i}}, \underline{\mathbf{p_i''}})$ and the projected pole $\underline{\mathbf{p_0}}$. Looking at this plane from a 2D perspective, we found that the including angle measured at the projected pole $\underline{\mathbf{p_0}}$ between projected RC $(\underline{\mathbf{p_i}}, \underline{\mathbf{p_i''}})$ is equal to the supplement of the rotation angle θ of the transformation. Formalising the above description, we have the following property as depicted in Figure 2:

Property 2. There is one and only one non-zero fixed pole $\mathbf{p_0}$ in 3D uniquely determined by the rigid body motion which is equidistant to all reflected correspondences $(\mathbf{p}, \mathbf{p}'')$ and one and only one non-zero fixed axis $\mathbf{h}$ where the including angle between projected reflected correspondences $(\underline{\mathbf{p}}, \underline{\mathbf{p}''})$ and projected pole $\underline{\mathbf{p_0}}$ on a plane perpendicular to $\mathbf{h}$ is equal to the supplement of the rotation angle θ of the motion.

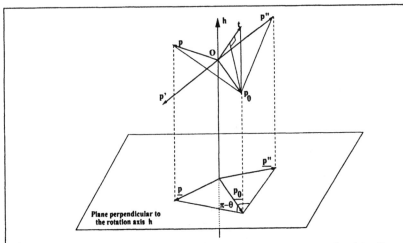

Fig. 2. 3D geometry of reflected correspondence vectors as defined by Property 2. After a rigid body motion, each pair $(\mathbf{p}, \mathbf{p}'')$ lie on the surface of concentric spheres centred at $\mathbf{p_0}$, the vertical bisector plane of the translation vector $\mathbf{t}$ passes through $\mathbf{p_0}$, and the including angle between the lines passing through the projected reflected correspondence $(\underline{\mathbf{p}}, \underline{\mathbf{p}''})$ and the projected pole $\underline{\mathbf{p_0}}$ on the plane perpendicular to the rotation axis $\mathbf{h}$ is equal to the supplement of the rotation angle θ of the motion.

The analysis of such interesting geometric properties leads to expressions to estimate the pole $\mathbf{p_0}$ in 3D, the rotation axis $\mathbf{h}$, and the transformation parameters rotation angle, rotation matrix, and translation vector $(\theta, \mathbf{R}, \mathbf{t})$ described as follows. Given three non-parallel $\mathbf{RCV_i}$ $(i = 1, 2, 3)$, the pole $\mathbf{p_0}$ can be uniquely determined as:

$$\begin{pmatrix} \mathbf{RCV}_1^T \\ \mathbf{RCV}_2^T \\ \mathbf{RCV}_3^T \end{pmatrix} \mathbf{p_0} = \begin{pmatrix} (\mathbf{p}_1^T \mathbf{p}_1 - \mathbf{p}_1''^T \mathbf{p}_1'')/2 \\ (\mathbf{p}_2^T \mathbf{p}_2 - \mathbf{p}_2''^T \mathbf{p}_2'')/2 \\ (\mathbf{p}_3^T \mathbf{p}_3 - \mathbf{p}_3''^T \mathbf{p}_3'')/2 \end{pmatrix} \tag{5}$$

If these RCVs are parallel, then their perpendicular bisector planes will coincide and the pole cannot be uniquely determined. Thus, in order to uniquely determine the pole $\mathbf{p_0}$ of a spatial displacement at least three non-parallel $\mathbf{RCV}$s are needed. Given three non-parallel correspondence vectors $\mathbf{CV_i}$ $(i = 1, 2, 3)$, the rotation axis $\mathbf{h} = (h_1, h_2, h_3)^T$ can be uniquely determined as:

$$\mathbf{h} = \frac{(\mathbf{CV_2} - \mathbf{CV_1}) \times (\mathbf{CV_3} - \mathbf{CV_1})}{\|(\mathbf{CV_2} - \mathbf{CV_1}) \times (\mathbf{CV_3} - \mathbf{CV_1})\|} \tag{6}$$

If the correspondence vectors $\mathbf{CV_i}$ $(i = 1, 2, 3)$ are parallel, then the rotation axis $\mathbf{h}$ can be uniquely determined as $\mathbf{h} = \mathbf{0}$. Once a valid rotation axis is known, all correspondences $(\mathbf{p_i}, \mathbf{p_i'})$ and the pole $\mathbf{p_0}$ can be projected on a plane perpendicular to $\mathbf{h}$:

$$\underline{\mathbf{p_i}} = (\mathbf{I} - \mathbf{h}\mathbf{h}^T)\mathbf{p_i}, \quad \underline{\mathbf{p_i'}} = (\mathbf{I} - \mathbf{h}\mathbf{h}^T)\mathbf{p_i'}, \quad \underline{\mathbf{p_0}} = (\mathbf{I} - \mathbf{h}\mathbf{h}^T)\mathbf{p_0} \tag{7}$$

where $(\underline{\mathbf{p}_i}, \underline{\mathbf{p}_i'})$ are 2D correspondences from which the 2D reflected correspondences $(\underline{\mathbf{p}_i}, \underline{\mathbf{p}_i''})$ can be determined. Thus, the rotation angle θ can be estimated as:

$$\cos(\pi - \theta) = \frac{(\mathbf{p_i} - \mathbf{p_0})^T(\mathbf{p_i''} - \mathbf{p_0})}{(\mathbf{p_i} - \mathbf{p_0})^T(\mathbf{p_i} - \mathbf{p_0})} \tag{8}$$

The rotation matrix $\mathbf{R}$ is estimated by the Rodrigues formula as:

$$\mathbf{R} = \mathbf{I} - \mathbf{H}\sin\theta + (1 - \cos\theta)\mathbf{H}^2 \tag{9}$$

where

$$\mathbf{H} = \begin{pmatrix} 0 & -h_3 & h_2 \\ h_3 & 0 & -h_1 \\ -h_2 & h_1 & 0 \end{pmatrix} \tag{10}$$

As a result, the translation vector $\mathbf{t}$ can be estimated as:

$$\mathbf{t} = (\mathbf{I} + \mathbf{R}^T)\mathbf{p_0} \tag{11}$$

This completes the analysis in 3D. Similarly to the case in 2D, such explicit expressions making full use of feature vectors and angular information are useful to estimation and analysis of solutions to rigid body motion parameters as described by the algorithm in the next Section.

3 Description of the Algorithm

In [2], the total least squares method (TLS) was proposed. Making use of the TLS method and based on the analysis of geometrical properties of **RCV**s described in the previous section, we propose the following algorithm called Geometrical Algorithm with Median Filter (GA-MF) to calibrate the motion parameters. Given that correspondences $(\mathbf{p_i}, \mathbf{p_i'})$ $(i = 1, 2, \cdots, n)$ where $n \geq 3$ between different coordinate frames have been established, the steps in the algorithm are:

1. Construct the reflected correspondences $(\mathbf{p_i}, \mathbf{p_i''})$ and their **RCV**$_i$.
2. Use the TLS method to calibrate the pole $\hat{\mathbf{p}}_0$ in 3D from Equation 5:

$$\begin{pmatrix} \mathbf{RCV}_1^T \\ \mathbf{RCV}_2^T \\ \vdots \\ \mathbf{RCV}_n^T \end{pmatrix} \hat{\mathbf{p}}_0 = \begin{pmatrix} (\mathbf{p}_1^T\mathbf{p}_1 - \mathbf{p}_1''^T\mathbf{p}_1'')/2 \\ (\mathbf{p}_2^T\mathbf{p}_2 - \mathbf{p}_2''^T\mathbf{p}_2'')/2 \\ \vdots \\ (\mathbf{p}_n^T\mathbf{p}_n - \mathbf{p}_n''^T\mathbf{p}_n'')/2 \end{pmatrix}$$

3. Construct the correspondence vectors: $\mathbf{CV}_i = \mathbf{p_i} - \mathbf{p_i'}$ and $\overline{\mathbf{CV}} = \bar{\mathbf{p}} - \bar{\mathbf{p}}'$ where $\bar{\mathbf{p}}$ and $\bar{\mathbf{p}}'$ are the centroids of the point sets $\mathbf{p_i}$ and $\mathbf{p_i'}$.
4. Use Equation 6 to estimate a set of rotation axes $\mathbf{h_i}$ corresponding to $\mathbf{CV}_i$, $\mathbf{CV}_{i+1}$, and $\overline{\mathbf{CV}}$.
5. Median filter $\mathbf{h_i}$ to obtain the finally calibrated rotation axis $\hat{\mathbf{h}} = \hat{\mathbf{h}}/||\hat{\mathbf{h}}||$

6. Use Equation 7 to project all correspondences $(\mathbf{p_i}, \mathbf{p'_i})$ and the pole $\hat{\mathbf{p}}_0$ on the plane perpendicular to the rotation axis $\hat{\mathbf{h}}$ yielding their 2D projected correspondences $(\underline{\mathbf{p_i}}, \underline{\mathbf{p'_i}})$ and 2D projected pole $\underline{\mathbf{p_0}}$.

7. From $(\underline{\mathbf{p_i}}, \underline{\mathbf{p'_i}})$, determine the 2D reflected correspondences $(\underline{\mathbf{p_i}}, \underline{\mathbf{p''_i}})$.

8. Use Equation 8 to estimate a set of the cosines of the including angle a_i between the lines passing through $(\underline{\mathbf{p_i}}, \underline{\mathbf{p''_i}})$ and the 2D pole $\underline{\mathbf{p_0}}$.

9. Calibrate $\hat{\theta}$ as: $\hat{\theta} = \pi - \arccos\{average\ of\ band\ pass\ filtered\ a_i\}$

10. Use Equation 9 to estimate the rotation matrix $\hat{\mathbf{R}}$.

11. Use Equation 11 to estimate the translation vector $\hat{\mathbf{t}}$.

4   Experimental Results

The GA-MF calibration algorithm has been implemented and, for comparison purposes, we have also implemented a well known calibration procedure based on the least squares method (LSM). The experiments consisted of using synthetic data described as follows. 30 points were randomly selected in a cube $[10, 30] \times [10, 30] \times [10, 30]$. The selected 3D points were then subject to a constant translation of $\mathbf{t} = (8, 8, 8)^{\mathbf{T}}$ followed by controlled rotations of 20, 30, 40, 50, 60 and 70 degrees around a fixed rotation axis $\mathbf{h} = \frac{1}{14}(1, 2, 3)^{\mathbf{T}}$. We thus, have precise knowledge of the selected points and their correspondences after the motion for error estimation. In order to simulate real world noisy data, each 3D correspondence was corrupted by Gaussian noise with mean 0 and standard deviation 0.125 in one series of experiments and 0.25 in another series of experiments. The estimated parameters of interest for the two algorithms were the rotation axis $\hat{\mathbf{h}}$, the rotation angle $\hat{\theta}$, rotation matrix $\hat{\mathbf{R}}$ and the translation vector $\hat{\mathbf{t}}$. Figures 3 and 4 show comparative results where the solid lines correspond to the data corrupted by noise with deviation 0.125 and the dashed lines correspond to noise with deviation 0.25.

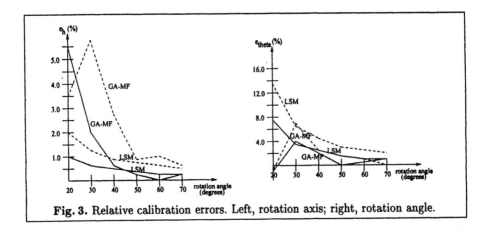

Fig. 3. Relative calibration errors. Left, rotation axis; right, rotation angle.

Figure 3 (left) shows that, although within a level of accuracy that is considered satisfactory, the GA-MF algorithm displays considerable fluctuation in performance for the calibration of the rotation axis. This indicates that the cross product of vectors is sensitive to noise. The LSM algorithm can generally more accurately calibrate the rotation axis especially when data are corrupted by higher levels of noise. In general, the LSM algorithm performs a robust and smooth calibration of the rotation axis as it makes full use of all points information. Figure 3 (right) shows that the GA-MF algorithm can generally more accurately calibrate the rotation angle. Again, the performance of the GA-MF algorithm fluctuates a bit due to the performance variation in calibrating the rotation axis. The overall performance of the GA-MF algorithm is robust and accurate. For both cases if the rotation angle is small, the accuracy of the LSM algorithm seriously decreases.

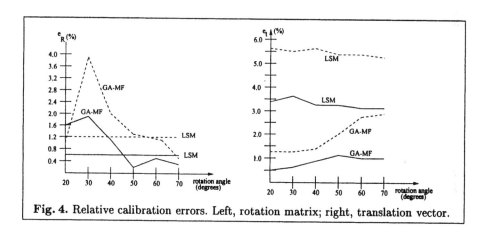

Fig. 4. Relative calibration errors. Left, rotation matrix; right, translation vector.

For large rotation angles, Figure 4 (left) shows that the GA-MF algorithm can more accurately calibrate the rotation matrix. Again, the GA-MF algorithm shows a fluctuating performance due to fluctuations in calibration of the rotation axis and rotation angle. Despite fluctuations, the calibration error is less than 4%. This is so because the GA-MF algorithm makes full use of the property that if the rotation angle is less than 90°, the length of the **RCV** is greater than that of the normal correspondence vector and this renders the rotation parameters more easily extracted. Generally, the performance of the LSM algorithm is smoother than that of the GA-MF algorithm for the calibration of the rotation matrix. Figure 4 (right) shows that the GA-MF algorithm can more accurately calibrate the translation vector than the LSM algorithm. This is so because calibration depends on the determination of the pole and the results show that the pole can effectively extract useful information from noise corrupted data. In general, calibration of the translation vector is more sensitive to noise than calibration of the rotation parameters [5] and the GA-MF algorithm has proved to be particularly adequate to such calibration.

5   Conclusion

In this paper, we investigated the problem of calibrating 3D rigid body motion parameters. Assuming that correspondences between two different coordinate frames have been established, we have proposed a novel method to analyse geometrical properties of reflected correspondence vectors which are synthesised into a single coordinate system. We then proposed a novel algorithm to calibrate motion parameters. For comparison purposes, we also implemented a well known calibration procedure based on the least squares method. An overall analysis of experimental results reveals that the proposed geometrical algorithm is in general more accurate than the least squares based algorithm. While the GA-MF algorithm can more accurately calibrate the rotation angle and the LSM algorithm can more accurately calibrate the rotation axis, we can synthetically use these two methods to counteract noise in real applications. Moreover, we have demonstrated the significance of our geometrical analysis method to the representation and to the development of calibration algorithms for rigid body motion parameters.

References

1. Arun K. S., Huang T. S., and Blostein S. D.: Least-Squares fitting of two 3D point sets. IEEE Transactions on Pattern Analysis and Machine Intelligence, **9** (1987) 698–700
2. Chaudhuri S. and Chatterjee S.: Performance analysis of total least squares methods in three-dimensional motion estimation. IEEE Transactions on Robotics and Automation, **7** (1991) 707–714.
3. Faugeras O. D.: A few steps toward artificial 3-D vision. In: Michael Brady, Robotics Science, The MIT Press, (1989) 39–137.
4. Faugeras O.D. and Hebert M.: The Representation, Recognition, and Locating of 3-D objects. International Journal of Robotics Research, **5** (1986) 27–52.
5. Huang T. S. and Netravali A. M.: Motion and structure from feature correspondence: a review. Proceedings of the IEEE, **82** (1994) 252–268.
6. Huang T. S., Blostein S. D., and Margerum E. A.: Least-squares estimation of motion parameters from 3D point correspondences. in Proceedings of IEEE Conf. Computer Vision and Pattern Recognition, Miami Beach, FL, June 24-26, 1986.
7. Kanatani K.: Analysis of 3D rotation fitting. IEEE Transactions on Pattern Analysis and Machine Intelligence, **16** (1994) 543–549.
8. HC Loguet-Higgins. A computer algorithm for reconstructing a scene from two projections. Nature, **293**(1981) 133–135.
9. Press W. H., Teukolsky S. A., Vetterling W. T., and Flannery B. P.: Numerical Recipes in C: The Art of Scientific Computing. Second edition, Cambridge University Press, 1992.
10. Umeyama S.: Least-squares estimation of motion parameters between two point pattern. IEEE Transactions on Pattern Analysis and Machine Intelligence, **13** (1991) 377–380.

An Automatic Registration Algorithm for Two Overlapping Range Images

Gerhard Roth

Visual Information Technology Group, Building M 50, Montreal Road
National Research Council of Canada, Ottawa, Canada K1A 0R6
E-mail: Gerhard.Roth@iit.nrc.ca; Phone: 613-993-1219; Fax: 613-952-0215

Abstract. This paper describes a method of automatically performing the registration of two range images that have significant overlap. We first find points of interest in the intensity data that comes with each range image. Then we perform a tetrahedrization of the 3D range points associated with these 2D interest points. The triangle pairs of these tetrahedrizations are then matched in order to compute the registration. The fact that we have 3D data available makes it possible to efficiently prune potential matches. The best match is the one that aligns the largest number of interest points between the two images. The algorithms are demonstrated experimentally on a number of different range image pairs.

1 Introduction

Registration is the process of aligning images so that they are in a common co-ordinate frame. When building geometric models from dense 3D range images [11] it is necessary to acquire images from different viewpoints. The first step in the model building process is the registration of these range images. This registration step currently relies on accurate mechanical positioning devices, or on manual processing. The idea of using the 3D data itself to perform the registration is attractive. The assumption is that individual range images will have a significant overlap. When the registration between two images is correct these overlapping regions should blend together with little error, since they represent the same surface patch.

In practice, this type of data driven registration can be further divided into two subclasses: constrained and unconstrained. In the constrained case the assumption is that the transformation between the two range images is already approximately known. This initial estimate is then refined with an iterative closest point (ICP) algorithm [3]. When there are no prior constraints on the transformation, the standard ICP approach will not always work, since it requires that the two images be relatively close to their final position for convergence to be certain.

The problem of automating range image registration when there are no prior constraints is far from being solved. Some methods first triangulate the entire range image, and then attempt to match the triangulations. The difficulty is that the resolution of the triangulations is selected heuristically [1, 2]. In [13] a method is described for registering 3d points sets, but in the context of medical imagery. The concept of spin images, which measures the compatibility of points in different range images, can be used to perform unconstrained registration [9]. However, spin images are only a point to point compatibility measure. How to effectively use them in a search process is still an open question. In [4], the rigidity constraint is used in conjunction with a random sampling algorithm to find the registration between two range images. This has a running time proportional to the number of 3D points in one of the images. The examples do not use the full 3D range images. Instead the range images are subsampled to be around 5K points. The rigidity constraint is also used in [5]. Here, 3 point tuples are matched using principal curvatures and the Darboux frame. This method is applied to a number of images, but as with other methods in the literature, they are subsampled to be around 4K data points. The computation of curvatures is also known to be very noise sensitive. These two methods [4, 5] are similar to ours in that they use the rigidity constraint. While they do not require texture on the objects being scanned, which we do, they are not able to handle geometrically symmetric objects, which our method can deal with. Our method has been tested on full 3D range images, which contain from 60K to 100K data points. We therefore obtain more accurate results, and still do so in reasonable time. To our knowledge, only our approach has been tested on range images which have not been significantly subsampled.

We propose a new way to automatically register two range images, without any prior knowledge of the transformation between them. It is based on the assumption that for every range image, there is also a corresponding registered intensity image. This intensity images represents the amount of light returned by the reflected laser beam. Since this intensity image is only obtained at the frequency of the laser, it is not the same as a standard broad spectrum intensity image. However, the texture of the objects scanned by the rangefinder is still clearly visible in this intensity image. This texture is not viewpoint invariant, since the actual intensity values change with the local surface normal. We deal with this by finding interest points, and using them in the matching process. Since these interest points are differential operators they are not particularly sensitive to the local intensity values, only to their variations [12]. This registered intensity image has previously been used to perform range image registration. In these cases the feature matching was either done manually [7], or the problem of constrained registration was being solved [14], which is much easier than the problem of unconstrained registration. In this paper we describe a method of using the registered range and intensity images to solve the problem of unconstrained registration. Note that for this to be successful there must be at least 30% to 40% overlap between the two range images in order to guarantee that there are enough common feature points.

Corner-like features in 2D intensity images has been successfully used to compute the fundamental and essential matrix [16]. However, in these cases the matching process requires that the two images be close together in viewpoint. This limits the range of possible correspondences of a feature point in one image to those feature points in the other image that are within a certain number of pixels, usually no more than one quarter of the image size [16]. This means that in practice such methods are not able to handle significant image rotations. Such rotations do not usually occur in most 2D video sequences since the intensity images are acquired at a rate of 30hz. By contrast, range images are normally acquired from substantially different viewpoints. Therefore, an algorithm for registering range images must be able to deal with a significant change in the viewpoint between the images, which is the problem of unconstrained registration,

2 Algorithm Description

The registration algorithm has as input the two range images along with their associated intensity images. The output is the relative 3D transformation between these two images. The algorithm proceeds as follows:

1. Compute 2D interest points in the intensity data of each of the two range images.
2. Compute a 3D Delaunay tetrahedrization of these interest points using their associated 3D value.
3. Find all the compatible triangle pairs between the two range images. The compatibility criterion will be defined later in the paper.
4. For each compatible 3D triangle pair, compute a transformation between the range images.
5. Return the transformation which aligns the largest number of interest points.

We will now describe each step in the registration algorithm in more detail.

2.1 Compute Interest Points

Informally, an interest point is an area of an image where there is a large change in the local intensity. The procedure to compute such points is to find the local maxima of the intensity gradient [12]. An example of a good interest point is a corner, since it is clearly a maximum of the gradient in both directions. By contrast, an edge has a significant intensity gradient only in one direction, and is therefore not a good interest point.

The number of interest points found by this process depends on the threshold value of the interest operator, and also on the texture of the object. We will show in our experiments that successful registration results can be achieved using a wide variety of interest operator thresholds. Of course, we do require that the objects being registered have significant texture.

2.2 Compute Delaunay Tetrahedrization

Each interest point in the intensity data has an associated 3D data point in the range image. Three such interest points define a 3D triangle, and it is clear that matching a single pair of triangles between two range images is sufficient to compute a rigid transformation between them. An obvious, but computationally expensive way to match triangle pairs between the images is to use exhaustive or random search. Our idea is to first compute a Delaunay tetrahedrization of the 3D co-ordinates of the interest points.

The Delaunay tetrahedrization in 3D is both unique and invariant to viewpoint. We compute it using an efficient algorithm [6], which runs in $O(n \log n)$ expected time. We consider only the faces of the Delaunay tetrahedrization, that is the triangles of the tetrahedrization as possible corresponding triangles. This greatly limits the search space, since the triangles in the Delaunay tetrahedrization are a small subset of all the possible triangles that can be created from a set of 3D interest points.

Consider the Delaunay tetrahedrization of the interest points in an overlapping region of the two range images. If these computed interest points are the same in each range image, then the Delaunay tetrahedrization of these 3D points will also be the same. Because of intensity variations between images the interest points will differ somewhat, so the Delaunay tetrahedrizations will also differ. However, it is usually the case that a majority of the interest points are the same in the overlapping regions of each range image. There is therefore very likely to be at least one matching pair of triangles in the Delaunay tetrahedrization of the overlapping regions.

2.3   Finding Corresponding Triangles

Typically, there are in the order of 200 feature points, and around 5,000 triangle faces in the 3D Delaunay tetrahedrization of the interest points. A brute force algorithm to match these triangles is not practical, since there are in the order of 10^7 potential matches. To make this correspondence search practical, we use a number of constraints. We first sort the edges of each triangle in length from largest to smallest. Then, we quantize each of these lengths into k bits so that a triangle can be represented by a $3k$ bit string. If there are a total of t triangles, and the edge lengths are distributed evenly, then there are on the average $t/(2^{3k})$ triangles that have the same bit string. Of course, this is the ideal case. In practice, for the value of k we use in our experiments (usually 5 or 6), the number of triangles that have a given bit string is approximately two orders of magnitude less than the total number of triangles. We only match triangles between the two range images that have approximately the same edge lengths; ie. the same bit string representation of these lengths. This matching process is similar to that used in the process of geometric hashing [15]. However, we represent a triangle as a single bit string which makes the lookup process to find similar triangles simple and efficient.

There are other constraints that can be used to further prune the triangle pairs that have the same edge lengths. Given any range image it is easy to compute a local surface normal at each point in the image. Every interest point and therefore every vertex of the Delaunay tetrahedrization has an associated surface normal. If a pair of triangles has the same edge lengths, we check if they also have approximately the same angle differences between the normals at each triangle vertex. If this is not the case, the triangles are not compatible.

Once a triangle pair passes these compatibility tests the transformation that aligns the two triangles is computed. If this transformation is correct it should also align all the matching interest points in the overlapping regions of the two range images. The degree of overlap can be found by counting the number of interest points in the first range image that are within a given tolerance of an interest point in the second range image. To do this efficiently we use a voxel grid, which has an occupied voxel for every interest point in the second image. The grid resolution is simply set to our tolerance for the matching process. The total compatibility score for a transformation is the number of interest points that are aligned in this fashion.

2.4 Algorithm Summary

The pseudo-code for the entire algorithm is as follows:

```
for each of the two range images
  -find the interest points in the intensity image;
  -obtain the 3D range value for each such interest point;
  -compute the 3D Delaunay tetrahedrization for all these 3D points;
  -compute the bit length key for each triangle in the tetrahedrization;
  -store each triangle in a list associated with that length key;
endfor

for each triangle in one image with a given length key
  -find all the triangles in the second image with the same key;
  -if the surface normal differences of the triangle vertices match
  -then
    -compute the transformation that aligns the two triangles;
    -apply this transformation to all the interest points
        in the first image;
    -count the number of interest points in the second image
        that are close to an interest point the first image;
    -save the transformation that aligns the largest
        number of interest points;
  endif
endfor
```

3 Complexity Analysis

The computation of the interest points is proportional to the number of 3D points in the range image. Assume that l interest points are found in the range image. The time required to compute the Delaunay triangles is $O(l \log l)$. The computational time for these two steps is much less than the time of the matching step. To analyze the complexity of the matching step, assume that we have

m 3D Delaunay triangles in one images, and n in the other. Also, assume that the key length is k bits for each edge. Then the average number of triangles in the first image with a given key is $m/(2^{3k})$, and in the second is $n/(2^{3k})$. All the triangles with matching keys in both images are potential matches, which means that each key can have $mn/(2^{6k})$ potential matching triangle pairs. Each potential matching triangle pair requires $O(m)$ operations to evaluate the number of aligned interest points, that is to evaluate the quality measure. Since there are (2^{3k}) possible bit strings this implies that the total time required on the average to perform the matching is $m^2n/(2^{3k})$. Our analysis assumes that the Delaunay triangles are equally distributed across the entire key set. The actual running time depends on the distribution of the lengths of the sides of the Delaunay triangles. If there are many triangles of almost equal size, then the algorithm will be slower. However, both theoretically and practically the average running time is a low order polynomial function of the number of interest points in both images. The reason for this is that having the 3D data available greatly constrains the matching process.

4   Experimental Results

In our experiments, we attempt to register objects which have significant texture, and that have been scanned from different viewpoints by a laser rangefinder. The rangefinder returns both the range data and three channels of registered color intensity data [11]. We convert the R,G,B color channels to hue, intensity and saturation, and find the interest points only in the intensity component.

The objects in our experiments are a duck and vase. We show in the Figures two range views of each object that have been successfully registered. In part (a) and (b) of all the Figures are the interest points. Parts (c) and (d) of Figure 1 shows a rendering of the edges of the 3D Delaunay triangulation. Parts (e) and (f) of Figure 1, and parts (c) and (d) of Figure 2 show two renderings of the correctly registered 3D range images. In these renderings, each range image is distinctively shaded, which demonstrates that the images have been aligned correctly and accurately. It should also be noted that the initial registration results obtained by this method have been further refined by an ICP algorithm which operates only on the interest points, not the entire image [3].

Note that the two images of the vase could not be registered using just the 3D data, since they are rotationally symmetric. However, they were successfully registered using the intensity features and the 3D co-ordinates of these features. In all the Figures the two range images have significant rotational differences. Since the algorithm makes no restrictions on the transformation between the two images it can still find the correct transformation under these conditions. However, success does depend on their being sufficient texture and overlap in the range images. More experiments are necessary to accurately quantify the performance of our method relative to those in the literature.

In terms of execution time each image pair registration took on the order of thirty seconds to three minutes on a standard SGI Indigo workstation. The required time grows quickly with the number of interest points. Up to 300 interest points still produces a reasonable execution time. Of course, if we add constraints on the transformation, such as a limited degree of rotation, then the execution time decreases significantly. In all these experiments there was very little tuning of the thresholds of the interest operator. Examination of the Figures shows that this operator does not produce identical points of interest in each image. Since we only require a single Delaunay triangle pair to be in correct correspondences, such differences in the interest points between the images can easily be handled by our approach.

5   Conclusions

This paper has presented a method of registering range images by finding interest points in the associated intensity data of each range image. The basic idea is to triangulate these interest points in 3D, and then to match the triangles between the range views. The possible matches are pruned using geometric constraints. The best match is the one that aligns the largest number of interest points between the two views. It is found using an exhaustive search of the matching Delaunay triangles. The exhaustive search is practical when 3D data is available, since the geometric constraints are strong enough to prune many false matches.

Instead of using the intensity data of the laser rangefinder to find features it is possible to use the intensity data of a separate colour digital camera which is co-mounted with the laser rangefinder [8]. In this case a calibration process is used to find the associated range point for each pixel of the color camera. This type of color image differs from the intensity image that comes from the reflected laser light of the rangefinder since it is a broad spectrum image. It is therefore easier to find features in such a separate color image. These features can also be used in our registration algorithm, assuming the 2D and 3D sensors are properly calibrated. We plan to perform experiments using a separate color digital camera in the future. However, in this paper we showed that even when using only the intensity data associated with the 3D laser rangefinder image, that it is still often possible to successfully perform reliable unconstrained registration.

Using only the interest points, and furthermore using only the triangles in Delaunay tetrahedrization computed from these interest points to do the matching greatly reduces the computational requirements of the algorithm. We have demonstrated this method on a number of experimental examples. We are able to register range images without assuming any prior knowledge of the transformation between these images. To use the automatic registration algorithm there should be sufficient texture in both images, along with at least 30% to 40% overlap between them. This degree of overlap is not difficult to ensure in practice, especially given the advantages inherent in automating the registration process. The next step will be to make the method practical for objects that have no natural texture by using projected texture.

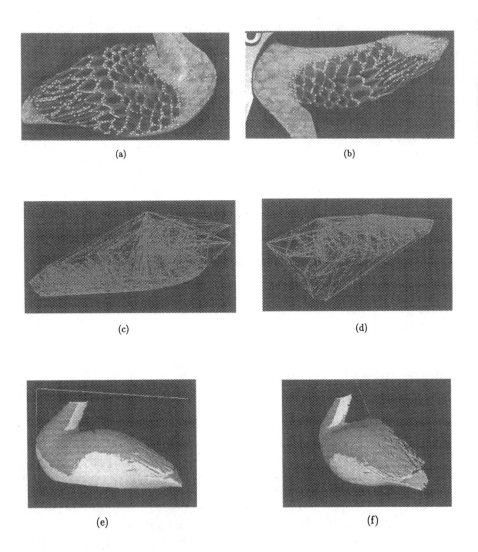

(a)   (b)

(c)   (d)

(e)   (f)

Fig. 1. (a) and (b) Feature points in the duck intensity. (c) and (d) Delaunay tetra-hedrization of the 3D feature points. (e) and (f) The two duck range images registered together.

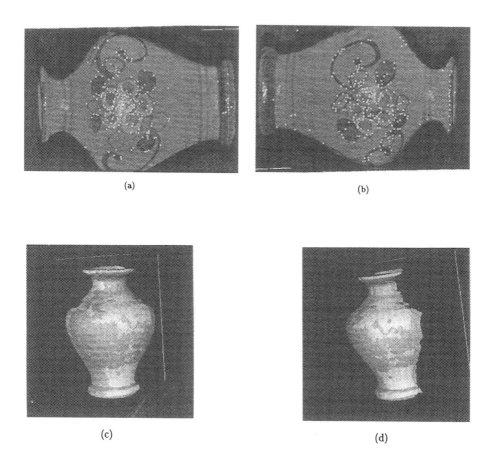

(a)

(b)

(c)

(d)

Fig. 2. (a) and (b) Feature points in the vase intensity. (c) and (d) The two vase range images registered together.

References

1. Ashbrook, A.P., R.B. Fisher, and C. Roberstson and N. Werghi (1998). Finding surface correspondences for object recognition using geometric histograms. In Computer Vision-ECCV98, Freiburg, Germany, pp. 674-686.
2. Bergevin, R., D. Laurendeau, and D. Poussart (1995, January). Registering range views of multipart objects. Computer Vision and Image Understanding **61**(1), 1–16.
3. Besl, P. and N. McKay (1992, Feb.). A method for registration of 3-d shapes. IEEE Trans. on Pattern Analysis and Machine Intelligence **14**(2), 239–256.
4. Cheng, C.-S., Y.-P. Hung, and J.-B. Chung (1998). A fast automatic method for registration of partially overlapping range images. In International Conference on Computer Vision, Bombay, India, pp. 242–248.

5. Chua, C.S. and R. Jarvis (1996). 3D free-form surface registration and object recognition. International Journal of Computer Vision, **17**(1), 77-99.

6. Clarkson, K., K. Mehlhorn, and R. Seidel (1993). Four results on randomized incremental constructions. In Computational geometry, theory and applications, pp. 85–121.

7. Gagnon, E., J.-F. Rivest, M. Greenspan, and N. Burtnyk (1996, June). A computer assisted range image registration system for nuclear waste cleanup. In IEEE Instrumentation and Measurement Conference, Brussels, Belgium, pp. 224–229.

8. Jasiobedzki97, P. (1997) Fusing and guding range measurements with colour video images. In International Conference on Recent Advances in 3D Digital-Imaging and Modelling, Ottawa, Canada, pp. 339–347.

9. Johnson, A. (1997) Surface matching by oriented points. In International Conference on Recent Advances in 3D Digital-Imaging and Modelling, Ottawa, Canada, pp. 121-129.

10. Masuda, T. and N. Yokoya (1995, May). A robust method for registration and segmentation of multiple range image. Computer Vision and Image Understanding **61**(3), 295–307.

11. Rioux, M. (1994). Digital 3-d imaging: theory and applications. In Videometrics III, International Symposium on Photonic and Sensors and Controls for Commercial Applications, Volume 2350, pp. 2–15. SPIE.

12. Smith, S. and J. Brady (1997, May). Susan - a new approach to low level image processing. International Journal of Computer Vision, pp. 45–78.

13. Thirion, J. (1996). New feature points based on geometric invrariants for 3d registration. International Journal of Computer Vision **18**(2), 121–137.

14. Weik, S. (1997) Registration of 3d partial surfaces using luminance and depth information. In International Conference on Recent Advances in 3D Digital-Imaging and Modelling, Ottawa, Canada, pp. 93–100.

15. Wolfson, H. (1990). Model based object recognition by geometric hashing. In Computer Vision-ECCV90, pp. 526-536.

16. Zhang, Z. (1998, March). Determining the epipolar geometry and its uncertainty: a review. International Journal of Computer Vision **27**(2).

On Registering Front- and Backviews of Rotationally Symmetric Objects*

Robert Sablatnig and Martin Kampel

Pattern Recognition and Image Processing Group,
Institute for Computer Aided Automation,
Vienna University of Technology,
Treitlstraße 3/183/2, A-1040 Vienna, Austria
{sab,kampel}@prip.tuwien.ac.at, http://www.prip.tuwien.ac.at

Abstract. Every archaeological excavation must deal with a vast number of ceramic fragments. The documentation, administration and scientific processing of these fragments represent a temporal, personnel, and financial problem. We are developing a documentation system for archaeological fragments based on their profile, which is the cross-section of the fragment in the direction of the rotational axis of symmetry. Hence the position of a fragment (orientation) on a vessel is important. To achieve the profile, a 3d-representation of the object is necessary. This paper shows an algorithm for registration of the front and the back views of rotationally symmetric objects without using corresponding points. The method proposed uses the axis of rotation of fragments to bring two range images into alignment.

1 Introduction

Ceramics are one of the most widespread archaeological finds and are a short-lived material. This property helps researchers to document changes of style and ornaments. Therefore, ceramics are used to distinguish between chronological and ethnic groups. Furthermore, ceramics are used in the economic history to show trading routes and cultural relationships. At excavations a large number of ceramic fragments, called sherds are found. These fragments are photographed, measured, drawn (called *documentation*) and classified. The purpose of *classification* is to get a systematic view of the material found.

Up to now documentation and classification have been done manually which means a lot of routine work for archaeologists and a very inconsistent representation of the real object. First, there may be errors in the measuring process. (Diameter or height may be inaccurate), second, the drawing of the fragment should be in a consistent style, which is not possible since a drawing of an object without interpreting it is very hard to do.

---

* This work was supported in part by the Austrian Science Foundation (FWF) under grant P13385-INF.

Because the conventional method for documentation is unsatisfactory [14], the search for automatic solutions began early (see [7],[21],[16] for some examples). None of the prototype systems developed could satisfy the requirements of the archaeologists since the amount of work for the acquisition was not reduced. Therefore, we developed an automated 3d-object acquisition system with respect to archaeological requirements [15]. With the help of the 2.5d-range images [11] achieved from the acquisition system, a 3d-object model has to be constructed in order to determine the profile.

Archaeological pottery is assumed to be rotationally symmetric since it was made on a rotation plate. With respect to this property the axis of rotation is calculated using a Hough inspired method [22]. To perform the registration of the two surfaces of one fragment, we use a-priori information about fragments belonging to a complete vessel: both surfaces have the same axis of rotation since they belong to the same object. In this first step we concentrate on the registration of the front- and back-view of one fragment which is significantly different from registering the surfaces of different fragments of one object in order to reconstruct the object out of its pieces.

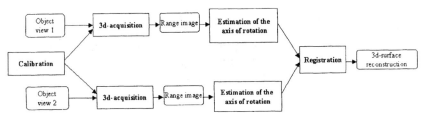

Fig. 1. Overview on 3d-reconstruction from two object views.

Figure 1 gives an overview of a 3d-surface reconstruction from two object views and also shows the structure of this paper. The first step consists of sensing the front- and backside of the object (in our case a rotationally symmetric fragment) using the calibrated 3d-acquisition system (Section 2). The resulting range images are used to estimate the axes of rotation, shown in Section 3. Section 4 presents the proposed registration method for the surface reconstruction and results are presented in Section 5. We conclude the paper with a discussion of the results and give an outlook on future work.

2 Acquisition System

The acquisition system consists of a LCD640 projector and a CCD camera. In order to analyze and process digital images it is important to know the position of the sensor in a reference co-ordinate system. In the process of calibration the parameters to describe the geometrical model are estimated.

The acquisition method for estimating the 3d-shape of a fragment is shape from structured light, which is based on active triangulation. A very simple technique to achieve depth information with the help of structured light is to scan a scene with a laser plane and to detect the location of the reflected stripe. The depth information can be computed out of the distortion along the detected profile.

In our acquisition system the stripe patterns are generated by a computer controlled transparent Liquid Crystal Display (LCD 640) projector. The projector projects stripe patterns onto the surface of the objects. In order to distinguish between stripes they are binary coded. The camera grabs gray level images of the distorted light pattern at different times. With the help of the code and the known orientation parameters of the acquisition system, the 3d-information of the observed scene point can be computed. This is done by using the triangulation principle. The image obtained is a 2D array of depth values and is called a range image.

3 Estimation of Axes of Rotation

The approach exploits the fact that surface normals of rotationally symmetric objects intersect their axis of rotation. The basis for this axis estimation is a dense range image provided by the range sensor. If we have an object of revolution, like an archaeological vessel made on a rotation plate, we can suppose that all intersections n_i of the surface normals are positioned along the axis of symmetry a.

This assumption holds [8] for a complete object or even for its fragment. For each point on the object the surface normal has to be computed. A planar patch of size $s \times s$ can be fitted to the original data using the Minor Component Analysis [13], which minimizes the distance between the points of the surface and the planar patch in an iterative manner in order to compute the optimal value of the normal and discard outliers.

The axis of rotation a is determined using a Hough inspired method [22]. For each point on the object, the surface normals n_i are computed using Minor Component Analysis. In order to determine the axis of rotation a all surface normals n_i are clustered in a 3d Hough-space: All the points belonging to a line n_i are incremented in the accumulator. Hence the points belonging to a large number of lines (like the points along the axis) will have high counter values. All the points in the accumulator with a high counter value are defined as maxima.

In the next step the line formed by the maxima has to be estimated. There are different techniques to solve this problem, in our case the PCA or Principal Component Analysis [12] is used. We have an a-priori knowledge: the maxima points are distributed according to a line (the axis of rotation). The PCA will determine the axis of maximal variance, which is in fact the axis of rotation. The accumulator maxima are taken as candidate points for the estimation of the axis of rotation. Using this technique outliers introduced by noisy range data or discretization errors, can be avoided, since in Hough-space wrong data points are in the minority and do not build a maximum.

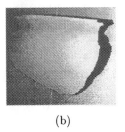

(a)   (b)

Fig. 2. Intensity image (left) and range image (right) of a fragment with rotational axis of the front-view.

Figure 2a shows the intensity image for a front-view of a fragment, Figure 2b shows the result on the range image with the estimated rotational axis. (black regions in Figure 2b indicate points where no range information is available due to occlusion of light stripes).

4 Range Image Registration

The task of building full 3d-models of general objects is difficult, since there is no a-priori knowledge about the shape of the object. One simple method is to use a calibrated turntable upon which the camera is fixed, as described in [17]. Even though the turntable method described above is good at creating 3d-models, there is still the question of getting the bottom of the object sitting on the turntable. So the bottom and the top of the object needs to be scanned in and then registered.

One of the most commonly used algorithms for registering is the *Iterative Closest Point (ICP)* algorithm, based on Besl & Mckay [2]. The basic algorithm consists of calculating the closest points on the surface (which generally is a triangulated surface). Next, the transformation is calculated and applied. The process of calculating the closest points is repeated until the termination criteria are met. Random sampling, as well as least median of squares estimator can be used as termination criteria. Prediction can also be used to speed up the process of registration (refer to [2] for more details). Further information on registration and integration of multiple range image views can be found in [5], [6], [4], and [18].

Fragments of vessels are thin objects, therefore 3d-data of the edges of fragments are not accurate and this data can not be acquired without placing and fixing the fragment manually. Ideally, the fragment is placed in the measurement area, a range image is computed, the fragment is turned and again a range image is computed. To perform the registration of the two surfaces, we use a-priori information about fragments belonging to a complete vessel: both surfaces have the same axis of rotation since they belong to the same object. Furthermore, the distance of the inner surface to the axis of rotation is smaller than the distance of the outer surface. Finally, both surfaces should have approximately the same profile; i.e. the thickness of the fragment should be constant in the average.

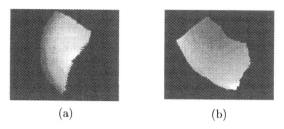

(a)   (b)

Fig. 3. The two views of a fragment.

Figure 3 shows the front- and back- view of a fragment. The goal of the registration is to find the transformation that relates these two views to one another, thus bringing them into alignment so that the two surfaces represent the object in 3d [3]. Since there are no corresponding points, we use a model-based approach. No point to point correspondences are required to determine the interframe transformation needed to express the points from each view in a common reference co-ordinate system [19].

We register the range images by calculating the axis of rotation of each view (Figure 4a) and bringing the resulting axes into alignment (Figure 4b).

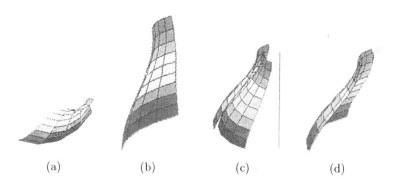

(a)   (b)   (c)   (d)

Fig. 4. Registration steps using synthetic data.

Knowing the surface normals of all surface patches we transform them into a common reference co-ordinate system. In the next step we have to align the surfaces of the objects to avoid intersecting surfaces (Figure 4c). Next the correct match is calculated by iteratively minimizing the error δ which expresses the mean deviation to a standard distance of the two surfaces in the direction of the rotational axis. Finally, the intensity images of both surfaces can be mapped onto the registered 3d-object in order to display the fragment with its original properties. Figure 4d shows the result for synthetic range data with 50 surface points for each view. The computed distance between the inner and the outer surface is 0.42mm with variance of 0.059mm. The registration error is small

(δ=0.13mm, the mean square errors between the original and the computed axes are 0.26mm and 0.31mm respectively).

5   Results

We tested our method on synthetic range images of a synthetic fragment (thickness 1.5mm) with approx. 7000 surface points each where we had a registration error δ=0.016mm. In comparison to the previous results the registration error is smaller since there are more surface points and therefore the computation of the rotation axis is much better.

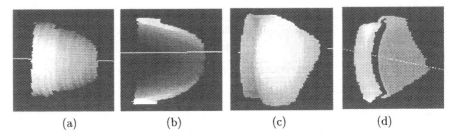

(a)   (b)   (c)   (d)

Fig. 5. Front- and back-view and their axis of rotation of a flowerpot (a,b) and an archaeological fragment (c,d).

To find out if the method is working on real data we used a totally symmetric small flowerpot with known dimensions and took a fragment which covered approximately 25% of the original surface. The range images of the front- and back-view consisted of approximately 10.000 surface points each (Figure 5a,b). The mean distance d between the surfaces is 5.64mm and the registration error δ=1.42mm. The distribution of the registration error δ for the flowerpot is shown in Figure 6a. The registration error increases towards the top of the pot, because of the irregularity of the distance between the surfaces at that region since the flowerpot has an edge (upper border) where inner and outer surface are not parallel.

Figure 5c and d show the front-view, back-view and the axis of rotation of a real archaeological fragment. Registration tests with this fragment resulted in registration errors of approximately δ=1.7mm and a mean distance of d=5.8mm. Figure 6b shows the distribution of δ of a registered archaeological fragment. Marginal peaks are caused by shadow regions of the back-view (see (Figure 5d) at the border of the fragment, where either no range data is processed or the range information is unreliable. The increase of the registration error δ reflects the uneven surface of the fragment.

Table 1 gives an overview of the presented results. It shows the number of points of the back- and front-view and the estimated registration error δ. The increase of δ between the synthetic and real data tests is caused by the error in the determination of the rotational axis.

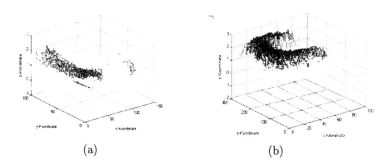

Fig. 6. Distribution of δ for registered flowerpot (a) and archaeological fragment (b).

Data type	# of points(back view)	# of points (front view)	error δ [mm]
synth.	50	50	0.13
synth.	6674	6674	0.016
real	10191	9619	1.42
real	31298	37176	1.72

Table 1. Results of the registration process.

Further problems that arise with real data are symmetry constraints, i.e. if the surface of the fragment is too flat or too small, the computation of the rotational axis is ambiguous (worst case: sphere) which results in sparse clusters in the Hough-space which indicate that the rotational axis is not determinable.

6 Conclusion and Outlook

We have proposed a prototype system for registering the front- and back-view of rotationally symmetric objects from range data. The work was performed in the framework of the documentation of ceramic fragments. For this kind of objects pair-wise registration techniques fail, since there are no corresponding points in the range images. We demonstrated a technique that computes and uses the axis of rotation of fragments belonging to the same vessel to bring two views of a scene into alignment.

The method has been tested on synthetic and real data with reasonably good results. It is part of continuing research efforts to improve the results from various range images since the technique depends on the rotational symmetry of the objects.

References

1. M. Altschuler and B. Altschuler. Measuring Surfaces Space-coded by a Laser-projected Dot Matrix. In *Imaging Applications for Automated Industrial Inspection, SPIE*, volume 182, 1979.

2. P.J. Besl and N.D. McKay. A Method for Registration of 3-D Shapes. *IEEE Trans. on Pattern Analysis and Machine Intelligence*, 14(2), 1992.

3. B. Bhanu. Reprsentation and Shape Matching of 3d Objects. *IEEE Trans. on Pattern Analysis and Machine Intelligence*, 6(3), 1984.

4. Y. Chen and G. Medioni. Object Modelling by Registration of Multiple Range Images. *Image and Vision Computing*, 10(3):145–155, April 1992.

5. C. Dorai, G. Wang, A.K. Jain, and C. Mercer. From Images to Models: Automatic 3D Object Model Construction from Multiple Views. In *13th. International Conference on Pattern Recognition, Vienna*, volume I, Track A, pages 770–774, 1996.

6. C. Dorai, G. Wang, A.K. Jain, and C. Mercer. Registration and Integration of Multiple Object Views for 3D Model Construction. *IEEE Trans. on Pattern Analysis and Machine Intelligence*, 20(1):83–89, January 1998.

7. I. Gathmann. ARCOS, ein Gerät zur automatischen bildhaften Erfassung der Form von Keramik. *FhG - Berichte*, (2):30–33, 1984.

8. Radim Halíř. Estimation of Rotation of Fragments of ArchaeologicalPotter. In W. Burger and M. Burge, editors, *21 th Workshop of the Oeagm*, pages 175–184, Hallstatt, Austria, May 1997.

9. S. Inokuchi, K. Sato, and F. Matsuda. Range-imaging System for 3-d Object Recognition. In *7th. International Conference on Pattern Recognition, Montreal*, pages 806–808, 1984.

10. A. Kak and K. Boyer. Color Encoded Structured Light for Rapid Active Ranging. *IEEE Trans. on Pattern Analysis and Machine Intelligence*, 9:14–28, 1987.

11. D. Marr. *Vision*. Freeman, 1982.

12. E. Oja. *Subspace Methods of Pattern Recognition*. John Wiley, 1983.

13. E. Oja, L. Xu, and C.Y. Suen. Modified Hebian Learning for Curve and Surface Fitting. *Neural Networks*, 5(3):441–457, 1992.

14. C. Orton, P. Tyers, and A. Vince. *Pottery in Archaeology*, 1993.

15. R. Sablatnig and C. Menard. Computer based Acquisition of Archaeological Finds: The First Step towards Automatic Classification. In P. Moscati/S. Mariotti, editor, *Proceedings of the 3rd International Symposium on Computing and Archaeology, Rome*, volume 1, pages 429–446, 1996.

16. C. Steckner. Das SAMOS Projekt. *Archäologie in Deutschland*, (Heft 1):16–21, 1989.

17. R. Szeliski. Image Mosaicing for Tele-Reality Applications. In *2nd. Workshop on Applications of Computer Vision, Sarasota*, pages 44–53, 1994.

18. K. Sakaue T. Masuda and N. Yokoya. Registration and Integration of Multiple Range Images for 3D Model Construction. In *13th. International Conference on Pattern Recognition, Vienna*, volume I, Track A, pages 879–883, 1996.

19. B.C. Vemuri and J.K. Aggarwal. 3D Model Construction from Multiple Views using Range and Intensity Data. In *Proceedings of IEEE Conference on Computer Vision and Pattern Recognition*, pages 435–437, 1986.

20. F. Wahl. A Coded Light Approach for 3dimensional (3D) Vision. Technical Report 1452, IBM, 1984. Research Report RZ.

21. P. Waldhaeusl and K. Kraus. Photogrammetrie für die Archäologie. In Kandler M., editor, *Lebendige Altertumswissenschaft: Festgabe zur Vollendung des 70. Lebensjahres von Hermann Vetters dargebracht von Freunden, Schülern und Kollegen*, pages 423–427, Wien, 1985.

22. S. Ben Yacoub and C. Menard. Robust Axis Determination for Rotational Symmetric Objects out of Range data. In W. Burger and M. Burge, editors, *21 th Workshop of the Oeagm*, pages 197–202, Hallstatt, Austria, May 1997.

Registration of Range Images Based on Segmented Data

Bojan Kverh and Aleš Leonardis

Faculty of Computer and Information Science
University of Ljubljana
Tržaška 25
Ljubljana, Slovenia
tel: (386)-61-1768-381
e-mail: {bojank,alesl}@razor.fer.uni-lj.si

Abstract. We present a new method for registration of range images, which is based on the results we obtain from the segmentation process. To obtain the first set of points needed for registration, we use set of points from first range image. The novelty is how we obtain the second set of points. To obtain the second set we project the first set of points onto geometric parametric models obtained in the second range image. Then we compute the transformation between the two sets of points. The results have shown a significant improvement in precision of the registration in comparison with traditional approach.

Keywords: registration, segmentation, range images.

1 Introduction and motivation

Range image registration is considered an important part of any vision system that requires data acquisition of parts that are not visible from any single viewpoint. There is no 3D sensor system available on the market that would obtain 3D data of the complete surface by a single scan. To extract complete information about the surface of an object we need to scan it from several viewpoints and then merge the data together. This requires that we compute the transformation between data sets obtained from different viewpoints. A rough approximation of the transformation can be obtained by involvement of a human operator, from positioning devices, e.g., a turntable, or automatically by detecting corresponding features in both sets. For some tasks, like reverse engineering, the accuracy of this approximate transformation has to be further improved by one of the refinement methods. In this paper we present a new refinement method which is based on segmented data, i.e., descriptions obtained by Recover-and-Select paradigm. Refinement methods usually compute the transformation either between two sets of points [4, 10] or between a set of points and a set of some surface

* This work was supported by a grant from the Ministry of Science and Technology of Republic of Slovenia (Project J2-0414-1539-98).

elements, e.g., triangles [9]. For each point from the first set, these methods find a correspondent point or surface element in the second set. Transformation between two sets of corresponding points is then computed by methods usually based on least squares. However, due to scanner sampling it is impossible to find the exact corresponding point or surface element for each point in the first set. This situation is illustrated in Fig. 1, which clearly shows that if the refinement is performed in this way, small errors will affect the computed transformation. Therefore we propose a novel approach which alleviates this problem.

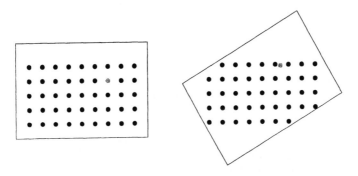

Fig. 1. An example of two corresponding points. Gray point in the right image is not captured by the scanner in the second view.

Registration of all range images is achieved by repeatedly registering pairs of range images. In each step, 3D points from both range images are first segmented into descriptions by the Recover-and-Select paradigm [5, 1, 2]. This gives us geometric parametric models which approximate each segmented region. Let us assume that we have an initial approximation of the transformation. Then, instead of computing the transformation between the two sets of 3D points, we can use one set of points and their correspondent *projections* to geometric parametric models that were obtained for the other set of points. This process can be iterated until the transformation error reaches some predefined value. In this way we cancel the effect of scanner sampling (since the projection of a point to a geometric model really corresponds with the point). The point that corresponds to a particular point from the first set, doesn't need to be captured by the scanner since it is computed from the segmented data.

The rest of the paper is organized as follows: in section 2 we introduce descriptions obtained by the Recover-and-Select segmentation process, in section 3 we present the Projection Iterative Closest Point (PICP) algorithm, geometric parametric models and projections of points onto the models are presented in section 4, while sections 5 and 6 contain experimental results and discussion, respectively.

2 Descriptions

Our method to refine the transformation between two partially overlapping range images assumes that both of them are segmented with Recover-and-Select paradigm [5, 1, 2]. The set of 3D points is segmented into regions. Each region though has its own geometric parametric model that fits best data in the region. The region and its model form a description. To express model M quality of fit to the region R, we use Eq. 1, known as description error,

$$Err = \frac{1}{|R|} \sum_{\mathbf{x} \in R} d(\mathbf{x}, M). \tag{1}$$

Function $d(\mathbf{x}, M)$ calculates distance of $\mathbf{x}$ to the model M.

The Recover-and-Select paradigm can use the following geometric models: planes, superquadrics, spheres, cylinders, cones, and tori. An example of segmented data is shown in Fig. 2. Different gray levels denote different regions.

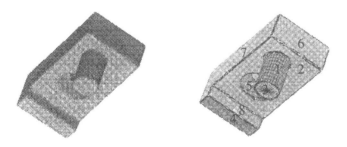

Fig. 2. Generated intensity image of an object and its range image segmentation. An index of description is displayed on each region.

Geometric parametric models in descriptions shown in Fig. 2 are planes (descriptions 2, 3, 4, 6, 7, 8, and 9), cylinder (description 1) and cone (description 5). In our transformation refinement method we use these models to compute projections of points to their surfaces.

3 PICP algorithm

We call our new transformation refinement method Projected Iterative Closest Point (PICP) algorithm, because it is similar to ICP algorithm [4]. ICP algorithm in its simplest implementation computes the transformation between two sets of points X and Y from different coordinate systems. The correspondent point to a point $\mathbf{x} \in X$ is $\mathbf{y}$ which satisfies the condition 2

$$\mathbf{y} = \arg\min_{t \in Y} d(\mathbf{x}, \mathbf{y}). \tag{2}$$

Function d represents a measure of distance between two points. ICP algorithm then computes the transformation that minimizes the sum of distances between transformed points from set X and their corresponding points from the set Y. Several iterations of ICP algorithm can be performed to obtain more accurate transformation.

PICP algorithm was developed to avoid the sampling effect. Using information obtained by segmentation stage and the initial approximation of the transformation between viewpoints from which the point sets have been scanned, we are able to determine pairs of corresponding descriptions. Description D_1 from the first set of points, consisting of region R_1 and model M_1, corresponds to description D_2 from the second set of points, consisting of region R_2 and model M_2 if conditions 3, 4 and 5 are satisfied.

$$\frac{1}{|R_2|} \sum_{\mathbf{x} \in T(R_2)} d(\mathbf{x}, M_1) < d_{tr}, \tag{3}$$

$$\frac{1}{|R_1|} \sum_{\mathbf{y} \in R_1} d(\mathbf{y}, M_2) < d_{tr}, \tag{4}$$

$$\exists \mathbf{x} \in R_1 \wedge \exists \mathbf{y} \in T(R_2) : d(\mathbf{x}, \mathbf{y}) < d_{tr}. \tag{5}$$

T denotes initial approximation of the transformation, function d denotes a measure of distance between two points, or between a point and geometrical parametrical model, while d_{tr} represents a threshold value. For each pair of corresponding descriptions conditions 3 and 4 assure that points from the region of one description are close enough to the geometric model of the other description after the initial approximation of transformation has been applied to one of the sets of points. Condition 5 requires that points from regions of both descriptions are close enough, otherwise we have no information if descriptions really correspond to each other. In Fig. 3 two different segmented range images are shown.

When the pairs of corresponding descriptions are identified, PICP algorithm starts to construct sets X and Y. It starts with $X = Y = \emptyset$. For each pair of corresponding descriptions it adds points from the region of description from the first range image to set X and projections of the same points to geometric model of description from the second range image to set Y. After all pairs of corresponding descriptions have been taken into consideration, sets X and Y are completed. PICP algorithm then uses a method based on quaternions [8] to compute the parameters of the transformation.

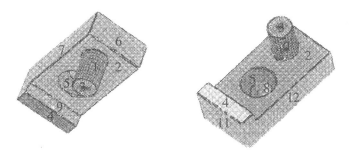

Fig. 3. Two segmented range images. Corresponding descriptions have same indices in both images.

4 Geometric parametric models and projections of points to them

In this section we present geometric parametric models used in the Recover-and-Select based segmentation stage. Also, we show how to compute coordinates of a point that has been projected onto a particular model.

4.1 Projection of a point to a plane

Planes are represented by parameters a, b, c, d and consist of points $\mathbf{t} = (x, y, z)$ that satisfy equation $ax + by + cz + d = 0$. The projection ($\mathbf{t_{proj}}$) of a point to a plane can be computed as shown in Eq. 6.

$$\mathbf{t_{proj}} = \mathbf{t} - d_1\mathbf{n}, \tag{6}$$
$$\text{where} \quad d_1 = \mathbf{nt} + d,$$
$$\mathbf{n} = (a, b, c) \quad \text{and}$$
$$a^2 + b^2 + c^2 = 1.$$

4.2 Projection of a point to a sphere

Spheres are represented by parameters a, b, c, r and consist of points $\mathbf{t} = (x, y, z)$ that satisfy equation $(x-a)^2 + (y-b)^2 + (z-c)^2 - r^2 = 0$. The projection ($\mathbf{t_{proj}}$) of a point to a sphere can be computed as shown in Eq. 7.

$$\mathbf{t_{proj}} = \mathbf{t} - d_2(\mathbf{t} - \mathbf{c}), \tag{7}$$
$$\text{where} \quad d_2 = \frac{d_1 - r}{d_1},$$
$$\mathbf{c} = (a, b, c) \quad \text{and}$$
$$d_2 = \|\mathbf{t} - \mathbf{c}\|.$$

4.3 Projection of a point to a cylinder

Cylinders are represented by parameters $\rho, \phi, \theta, \alpha, k$ [6,7]. Parametrisation of a cylinder is shown in Fig. 4. Values ϕ and θ are used to compute vector $\mathbf{n}$ ($\mathbf{n} = (\cos\phi\sin\theta, \sin\phi\sin\theta, \cos\theta)$), which is parallel to a normal vector of a tangent plane to a cylinder surface and orthogonal to a vector on the cylinder axis ($\mathbf{a}$). The projection ($\mathbf{t_{proj}}$) of a point to a cylinder can be computed as shown in Eq. 8.

$$\mathbf{t_{proj}} = \mathbf{t} - (\|\mathbf{t} - \mathbf{v}\| - 1/k)\mathbf{r}, \tag{8}$$

where $\quad \mathbf{r} = \dfrac{\mathbf{t} - \mathbf{v}}{\|\mathbf{t} - \mathbf{v}\|} \quad$ and

$$\mathbf{v} = (\rho + 1/k)\mathbf{n} + ((\mathbf{t} - (\rho + 1/k)\mathbf{n}) \cdot \mathbf{a})\mathbf{a}.$$

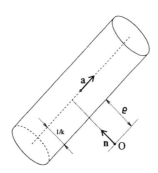

Fig. 4. Parametrisation of a cylinder

4.4 Projection of a point to a cone

Parametrisation of a cone is similar to the parametrisation of a cylinder. A cone has parameters $\rho, \phi, \theta, \sigma, \tau, k$[6,7]. The parametrisation of a cone is shown on Fig. 5. The difference is that here vectors $\mathbf{n}$ and $\mathbf{a}$ are not necessarily orthogonal. Vector $\mathbf{n}$ is computed in the same way as for cylinder, while $\mathbf{a}$ has components $(\cos\sigma\sin\tau, \sin\sigma\sin\tau, \cos\tau)$[6,7]. The projection ($\mathbf{t_{proj}}$) of a point to a cone can be computed as shown in Eq. 9.

$$\mathbf{t_{proj}} = \mathbf{t} - (\|\mathbf{t} - (\rho + 1/k)\mathbf{n} - z\mathbf{a}\| - d)\mathbf{r} \tag{9}$$

where $\quad z = \mathbf{a} \cdot (\mathbf{t} - (\rho + 1/k)\mathbf{n})$,

$$\alpha = \arccos(\mathbf{a} \cdot \mathbf{n}),$$

$$d = \frac{1}{k\sin\alpha} - z\tan\alpha \quad \text{and}$$

$$\mathbf{r} = \frac{\mathbf{t} - (\rho + 1/k)\mathbf{n} - z\mathbf{a}}{\|\mathbf{t} - (\rho + 1/k)\mathbf{n} - z\mathbf{a}\|}.$$

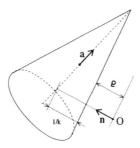

Fig. 5. Parametrisation of a cone

4.5 Projection of a point to a torus

A torus has parameters $\rho, \phi, \theta, \sigma, \tau, k, s$[6, 7]. Parameters $\phi, \theta, \sigma, \tau$ denote vectors $\mathbf{n}$ and $\mathbf{a}$, while both radii of a torus can be computed with $\frac{1}{k}$ and $\frac{1}{s}$. Parametrisation of a torus is shown on a torus cross-section in Fig. 6. The projection $(\mathbf{t_{proj}})$ of a point to a torus can be computed as shown in Eq. 10.

$$\mathbf{t_{proj}} = \mathbf{t} - (\|\mathbf{t} - \mathbf{c} - \mathbf{v}\| - 1/k)\mathbf{r}, \tag{10}$$

where
$$\mathbf{c} = (\rho + 1/s)\mathbf{n} - |1/s - 1/k| \cos(\mathbf{a} \cdot \mathbf{n})\mathbf{a},$$

$$\mathbf{v} = |1/s - 1/k| \sin(\mathbf{a} \cdot \mathbf{n}) \frac{(\mathbf{a} \times (\mathbf{t} - \mathbf{c})) \times \mathbf{a}}{\|(\mathbf{a} \times (\mathbf{t} - \mathbf{c})) \times \mathbf{a}\|} \quad \text{and}$$

$$\mathbf{r} = \frac{\mathbf{t} - \mathbf{c} - \mathbf{v}}{\|\mathbf{t} - \mathbf{c} - \mathbf{v}\|}.$$

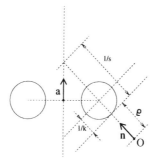

Fig. 6. Parametrisation of a torus

5 Experimental results

We used PICP algorithm in a reverse engineering application [3]. We scanned an object from four viewpoints and obtained four range images (RI_1, RI_2, RI_3, RI_4 shown in Fig. 7). These images were then segmented. After that, we registered images RI_1 and RI_2 and merged them together. At first step we registered the first and the second range image and merged them. In this way we obtained range image RI_{12}. Then we registered images RI_{12} and RI_3 and merged them to obtain the range image RI_{123}. Finally we registered images RI_{123} and RI_4 and merged them to obtain the final image RI_{1234}. When merging two segmented range images to obtain a new range image, we do the following:

1. if description D_1 from the first segmented range image, does not have a corresponding description in the second segmented range image, we insert D_1 into the new range image,
2. if description D_2 from second segmented range image, does not have a corresponding description in the first segmented range image, we insert D_2 into the new range image,
3. if descriptions D_1 and D_2 from different range images correspond and if R_1 is region of D_1 and R_2 is region of D_2, we form a new region $R = R_1 \bigcup R_2$ and fit new model M to region R. We insert description containing R and M into the new range image.

Let us think for the moment of the segmented range image which was generated by merging together two other range images. If some description D was formed by merging together two corresponding descriptions D_1 and D_2, it is obvoius that if the registration stage did not find a good transformation between the two range images, description error (Eq. 1) of D is significantly greater than description error of D_1 and D_2. Similarly, if the transformation found by the registration stage is accurate, description error of D should be similar to description errors of D_1 and D_2. Table 5 shows description errors for three descriptions shown in Fig. 2. Descriptions that were merged from two corresponding descriptions are not much worse in terms of description errors, so it can be concluded that PICP gives satisfactory results.

Description	RI_1	RI_2	RI_{12}	RI_3	RI_{123}	RI_4	RI_{1234}
cylinder (index 1 in Fig. 2)	0.091	0.063	0.1263	0.1112	0.1754	0.1366	0.1986
cone (index 5 in Fig. 2)	0.1284	/	0.1284	0.1215	0.1285	0.1423	0.1746
planar (index 2 in Fig. 2)	0.0950	/	0.0950	0.0980	0.1038	0.1037	0.1056

Table 1. Comparison of description errors between scanned and merged range images

We compared PICP algorithm with ICP in terms of description errors after merging and it turned out that after the same number of steps, PICP always achieves better precision. Descriptions obtained when ICP was used, had errors

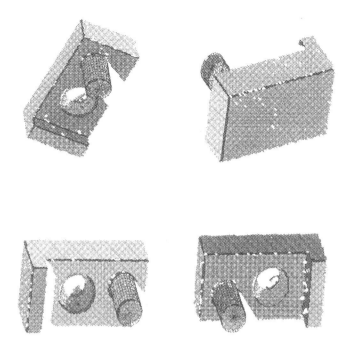

Fig. 7. Four segmented range images

that were 40% to 300% greater than descriptions obtained when we used PICP. PICP has also very good time performance since its complexity is only $O(n)$ [3].

6 Discussion

We presented a new method, PICP algorithm, for registration of range images. It is used to refine the initial approximation of the transformation between range images that were previously segmented. PICP's performance is good in terms of accuracy and required computational time. PICP has some drawbacks as well. The most significant is, that it cannot be used for objects with arbitrary free-form surfaces since our approach requires a segmentation into appropriate models which is currently limited to planes, spheres, cylinders, cones, and tori. Our future work will be directed towards including other types of models. We also intend to compare PICP algorithm with other known refinement algorithms ([9], [10]).

Acknowledgements

We thank Tomaš Pajdla for discussions about our method and Aleš Jaklič for providing a critical overview of our paper.

References

1. P. J. Besl and N. D. McKay. A method for registration of 3-D shapes. *IEEE Transactions on PAMI*, 14(2):239–256, February 1992.
2. B. Kverh. Segmentation and registration of range images as a tool for modelling objects. Msc. thesis, Faculty for Computer and Information Science, University of Ljubljana, 1998.
3. B. Kverh, A. Jaklič, A. Leonardis, and F. Solina. Recover-and-select paradigm on triangulated data. In *Proceedings of the Sixth Electrotechnical and Computer Science Conference ERK'97*, volume B, pages 241–244, September 1997.
4. B. Kverh and A. Leonardis. Using recover-and-select paradigm for simultaneous recovery of planes, second order surfaces and spheres from triangulated data. In *Proceedings of the Computer Vision Winter Workshop CVWW'98*, pages 26–37, February 1998.
5. A. Leonardis, A. Gupta, and R. Bajcsy. Segmentation of range images as the search for geometric parametric models. *International Journal of Computer Vision*, 14:253–277, 1995.
6. G. Lukács, A. D. Marshall, and R. R. Martin. Geometric least-squares fitting of spheres, cylinders, cones and tori. In R. R. Martin and T. V. ady, editors, *RECCAD Deliverable Documents 2 and 3 Copernicus Project No. 1068 Part I, Report on Basic Geometry and Geometry Model Creation with further contributions on Data Acquisition and advanced material on Merging and Applications*, GML 1997/5, Computer and Automation Institute, Hungarian Academy of Sciences, Budapest, 1997.
7. G. Lukács, A. D. Marshall, and R. R. Martin. Faithful least-squares fitting of spheres, cylinders, cones and tori for reliable segmentation. In *ECCV98 - Proceedings of the 5th European Conference on Computer Vision*, volume I, pages 671–686, June 1998.
8. A. D. Marshall and R. R. Martin. *Computer Vision, Models and Inspection*, volume 4 of *World Scientific Series in Robotics and Automated Systems*. World Scientific, September 1991.
9. T. Masuda, K. Sakaue, and N. Yokoya. Registration and integration of multiple range images for 3-D model construction. In *Proceedings of the 13th International Conference on Pattern Recognition*, pages 879–883, August 1996.
10. T. Pajdla, V. Hlaváč, and D. Večerka. Semi-automatic building of complete 3D models from range images. In R. R. Martin and T. Varady, editors, *RECCAD Deliverable Documents 2 and 3 Copernicus Project No. 1068 Part II, Report on Basic Geometry and Geometry Model Creation with further contributions on Data Acquisition and advanced material on Merging and Applications*, GML 1997/5, Computer and Automation Institute, Hungarian Academy of Sciences, Budapest, 1997.

Automatic Grid Fitting for Genetic Spot Array Images Containing Guide Spots

Norbert Brändle[1], Hilmar Lapp[2], and Horst Bischof[1]

[1] Vienna University of Technology
Pattern Recognition and Image Processing Group
Treitlstr. 3/1832, A-1040 Vienna
{nob,bis}@prip.tuwien.ac.at
[2] Novartis Forschungsinstitut
Genetics
Brunner Str. 59, A-1230 Vienna
hilmar.lapp@pharma.novartis.com

Abstract. In the domain of biotechnology array-based methods are used to gain rapid access to genetic information based on the signals of the individual array elements (spots). For an automated analysis of the spots it is necessary to fit a grid to the spots in the digital image in order to represent the array distortions that may occur in the course of the experiment. In order to make the grid fitting problem tractable in a certain class of experiments spot arrays contain a sub-grid of guide spots with a known signal characteristic. We present an automatic grid fitting method for spot array images containing guide spots. Our approach uses simple image processing methods and takes into account prior knowledge inherent in the imaging process.

Keywords: Genomics, Spot Array Images, Grid Fitting

1 Introduction

Rapid access to genetic information is central to the revolution taking place in molecular genetics [2, 7]. Biological systems read, store and modify genetic information by molecular recognition. Because each DNA strand carries with it the capacity to recognize a uniquely complementary sequence through base pairing (A↔T, C↔G) [8], the process of recognition, or *hybridization*, is highly parallel, as every nucleotide in a large sequence can in principle be queried at the same time. Thus, hybridization can be used to efficiently analyze large amounts of nucleotide sequence. The primary approaches include array-based technologies that can identify specific expressed gene products on high density formats, including filters, microscope slides and microchips [2].

Common to all array-based approaches is the necessity to analyze digital images of the array. The ultimate image analysis goal is to automatically assign a quantity to every array element giving information about the hybridization signal (*quantification*). Figure 1 shows a typical image generated in the course

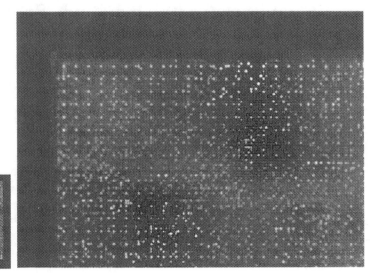

Fig. 1. Oligonucleotide Fingerprint Spot Image. The left image shows the entire spot image. The white rectangle indicates the region belonging to the zoomed right image

of a oligonucleotide fingerprint (ONF) experiment: The high density medium is a filter (nylon membrane) comprising a total of 57600 cDNA [8] spots which were spotted in different spotting cycles by a robot arm carrying a matrix of needles. Detailed information about the spotting procedure can be found in [4, 1]. The intensity of every spot corresponds to the amount of label remaining after hybridizing a liquid containing the labelled probes and subsequently washing off probe not bound to the genetic material. For details about the physical imaging process refer to [3].

Let the spot array in the digital image be represented as a grid $\mathcal{G}$ which is represented as a set of *nodes* in $\{1 \ldots I_G\} \times \{1 \ldots J_G\}$, with I_G as the grid height and J_G as the grid width. Let the spot image be represented as a $M \times N$ matrix $\mathbf{S}$ with pixel coordinates (m, n) or Cartesian spatial coordinates (x, y) with the origin at the upper left corner with x increasing to the right and y increasing downward. For a successful quantification of the hybridization signals it is necessary to assign to every node $g \in \mathcal{G}$ an element of the set of image locations $\mathcal{L} = \{(x, y) \mid x, y \in \mathbb{R}, x, y \geq 0\}$ with sub-pixel accuracy. We call this location assignment function $\mathrm{L} : \mathcal{G} \times \mathbf{S} \to \mathcal{L}$ *grid fitting* [1]. The grid fitting must cope with the following problems:

- **Distortions**: The spot grid in the image is only approximatively regular and rectangular due to anisotropic shrinking and expansion of the membrane,

---

[1] In order to simplify the notation we write $\mathrm{L}((i, j))$ instead of $\mathrm{L}((i, j), \mathbf{S})$, since we are dealing with a single image $\mathbf{S}$.

inaccuracies of the spotting robot, optical distortions, bent needles and other factors.

- **Rotations**: The grid need not be aligned with the image coordinates because the membrane is put manually into the imaging device.
- **Outliers**: Not every node of the grid necessarily contains luminescent spots because the intensity of the hybridization signal to be measured can be low or even zero.

In order to cope with inaccuracies of the spotting robot the grid is usually divided into subunits to represent the nature of the spotting cycles: A grid $\mathcal{G}$ consists of $F_1 \times F_J$ equally sized *fields* $\mathcal{F} \subset \mathcal{G}$ and one field $\mathcal{F}$ consists of $I_F \times J_F$ equally sized blocks $\mathcal{B} \subset \mathcal{F}$. When a large number of outliers is expected from the hybridization experiment (in ONF images for example only 5–20% of all spots can be expected to hybridize) so-called *guide spots* are used: A spot at a certain position within a block contains DNA material which has always a strong hybridization signal irrespective of the hybridization probe. Formally, the guide spot grid $\mathcal{G}^\star$ is a set of nodes in $\{1 \ldots I_{GS}\} \times \{1 \ldots J_{GS}\}$ belonging to guide spots with I_{GS} as the vertical number and J_{GS} as the horizontal number of guide spots. The guide spot grid consists of $F_1 \times F_J$ equally sized *guide spot fields* $\mathcal{F}^\star \subset \mathcal{G}^\star$. The grid in the image of Fig. 1 consists of 3×2 fields, one field consists of 16×24 blocks, each block having a dimension of 5×5 spots with one guide spot at the block center.

One solution of the grid fitting problem was already presented by Hartelius [5] in his Ph.D. thesis. It involves an image rotation in order to align the grid with the image coordinates, block finding and spot finding via Markov Random Fields (MRF) and Simulated Annealing [12]. His algorithm is computationally demanding and is semi-automatic since the user must provide the locations of the corner nodes of the grid. In this paper we present a fully automatic grid fitting approach for spot images containing guide spots and a comparison with the MRF approach of Hartelius. Our approach takes into account different constraints inherent in the imaging process and involves simple image processing operators. Quantification is a problem in its own right and is not presented here (see e.g. [10]). The paper is structured as follows: Section 2 describes the extraction of potential guide spot locations. Section 3 describes the fitting of the locations to the grid nodes and in Section 4 experimental results are presented.

2 Guide Spot Detection

The most reliable grid information lies in the guide spots. Potential guide spot locations in the spot image are detected with the help of two digital filter operators and a local maximum search. The detected guide spot locations are used in a subsequent step to determine the translation of the guide spot grid locations with respect to the origin of the spatial coordinate system. The following prior knowledge is assumed to be available (Distance [pixels] = Distance [mm] / Resolution [mm]): A spot has a theoretical dimension of $M_s \times N_s$ pixels and a block $\mathcal{B}$ has a theoretical dimension of $M_B \times N_B$ pixels.

2.1 Linear Matched Filter

The $M_s \times N_s$ matched filter (MF) $\mathbf{m}$ [9] is constructed by forming an average template from a number of sample guide spots [1, 9]. The spot image $\mathbf{S}$ is filtered with the MF $\mathbf{m}$ in the following way: If $\mathbf{s}_{[m,n]}$ denotes an image patch of the theoretical spot size around any pixel (m, n) of $\mathbf{S}$ rearranged as a $M_s \cdot N_s$-dimensional vector, the image patch $\mathbf{s}_{[m,n]}$ is first normalized to the local intensity mean value $\mu_{\mathbf{s}_{[m,n]}}$ as

$$\tilde{\mathbf{s}}_{[m,n]} = \mathbf{s}_{[m,n]} - \mu_{\mathbf{s}_{[m,n]}} \cdot \mathbf{1}. \tag{1}$$

The response value at (m, n) is then the dot product

$$\mathbf{R}^{M}[m,n] = \tilde{\mathbf{s}}_{[m,n]} \cdot \mathbf{m} \tag{2}$$

corresponding to the statistical covariance between the image patch and the MF. High response values indicate potential guide spot locations.

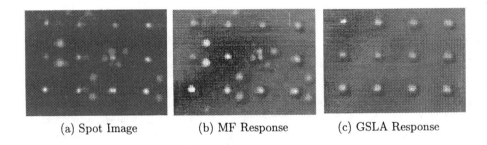

(a) Spot Image   (b) MF Response   (c) GSLA Response

Fig. 2. MF Response and GSLA Response for a Part of a Spot Image

Figure 2b shows the MF response image $\mathbf{R}^{M}$ of the spot image in Fig. 2a. The guide spots and non-guide spots are very similar in shape and intensity. Therefore they cannot be distinguished solely on the basis of the matched filter response.

2.2 Guide Spot Location Amplification (GSLA)

The main idea to distinguish guide spots from other spots is to amplify the locations of potential guide spots by considering the MF response values at the theoretical guide spot 8-neighborhood locations. We formally define the set $\mathcal{T}_{[m,n]}$ which includes the response value $\mathbf{R}^{M}[m, n]$ and the response values of the theoretical guide spot neighborhood locations of (m, n) in $\mathbf{R}^{M}$ as

$$\mathcal{T}_{[m,n]} = \{\mathbf{R}^{M}[m + k, n + l] \mid k \in \{0, M_B, -M_B\} \wedge l \in \{0, N_B, -N_B\}\}. \tag{3}$$

The GSLA response value $\mathbf{R}^{A}[m, n]$ is computed as

$$\mathbf{R}^{A}[m,n] = \text{median}(\mathcal{T}_{[m,n]}). \tag{4}$$

The effect of this nonlinear filter is to amplify those spots which have a high response at the theoretical guide spot neighborhood. Since this is a very unlikely event for non-guide spots, this filter enhances the guide spot locations. The only drawback is that the locations of the grid corner will be suppressed due to the median operation.

2.3 Maximum Search

After having applied the GSLA filter to the MF response image it is very likely that maximum values in $\mathbf{R}^A$ indicate locations of guide spot centers. The strategy of the maximum search is to extract a set of guide spot locations $\mathcal{L}^\star \subset \mathcal{L}$ from a region of interest (ROI) of $\mathbf{R}^A$. The ROI is defined by the *prior guide spot locations* $L_P(\mathcal{G}^\star)$ which are an initial estimate of the guide spot grid locations: Since the spot image border (the connected region not belonging to the physical spot filter) always has significantly lower intensity values than the filter region, it is possible to estimate the locations of the corner nodes of $\mathcal{G}^\star$ with the help of projections of the intensity values onto the axis of the image coordinate system [1]. The inter-guide spot distances of $L_P(\mathcal{G}^\star)$ are defined by M_B and N_B. The potential guide spot locations are searched in the ROI with a local maximum search algorithm [1].

3 Grid Fitting

Due to the possible rotations of the grid the J_{GS} locations of the first row of the guide spot grid need not necessarily correspond to the locations in $\mathcal{L}^\star$ with the J_{GS} lowest y-coordinates. The main principle of the grid alignment $A : \mathcal{L}^\star \times S \to \mathcal{G}^\star$ of the potential guide spot locations is as follows: Transform the prior guide spot locations (see Sect. 2.3) with initial estimates of the grid rotation angle and the grid translation vector such that the transformed prior guide spot locations are near to the detected guide spot locations. The complete grid fitting function is divided into the following steps: 1. Estimate a *global rotation angle* of the grid [2] from the GSLA image $\mathbf{R}^A$. 2. Estimate a *global translation vector* of the grid locations with respect to the prior guide spot locations. 3. Apply the estimated global rotation and translation to the prior guide spot locations resulting in *transformed prior guide spot locations* and assign the detected guide spot locations to the grid nodes. 4. Determine locations for the nodes in $\mathcal{G}^\star$ lacking a valid location assignment. 6. Determine locations for all the nodes in the global grid $\mathcal{G}$.

3.1 Global Rotation Estimation

A global rotation estimation of the grid locations can be achieved by the Radon transform [6] which is based on the projections of the image along specified directions. We use a Cartesian coordinate system with its origin at the image center.

---

[2] The term "global" indicates that the rotation refers to the complete spot grid rather than to a field or a block.

The discrete Radon transform $\mathbf{R}^\mathrm{T}$ of a GSLA response image $\mathbf{R}^\mathrm{A}$ is a $R \times C$ matrix with the following dimensions: The number of rows R corresponds to the number of angles for which the Radon Transform is computed. It depends on the quantization $\Delta\theta$ of the projection angles and the interesting maximum rotation θ_M which is determined by the maximum possible rotation of the physical filter in the imaging device. The grid rotation angle $\theta(r)$ belonging to a row index $r \in \{1 \ldots R\}$ is

$$\theta(r) = -\theta_\mathrm{M} + r\Delta\theta - 1. \tag{5}$$

The number of columns C of $\mathbf{R}^\mathrm{T}$ is the size of the spot image diagonal $(C = \sqrt{M^2 + N^2})$ in order to be able to store the projection at $\theta = 45°$.

Fig. 3. Part of the Radon Transform including the Rotation Estimation

The global rotation estimate $\theta(\mathrm{r})$ is in the row r of the Radon transform $\mathbf{R}^\mathrm{T}$ having the highest projection values. If $\mathcal{M}_r$ denotes the set of the J_GS highest projection values in the row r of $\mathbf{R}^\mathrm{T}$ the estimated rotation angle is found in the row with the highest median of $\mathcal{M}_i$:

$$r = \mathrm{argmax}\{\mathrm{median}(\mathcal{M}_i)\} \tag{6}$$

Figure 3 shows a part of a discrete Radon transform for the GSLA response image of the spot image in Fig. 1 with $\Delta\theta = 0.2°$ and $\theta_\mathrm{M} = 10°$. In the leftmost part of the image the values $\mathrm{median}(\mathcal{M}_i)$ for every row are visualized as horizontal bars. Note that the row with the true rotation angle has the highest pixel values but large intervals of low pixel values. The other rows have lower pixel values, but they are distributed over the columns. A simple horizontal addition of the projection values would therefore not lead to reliable rotation estimations.

3.2 Global Translation Estimation

The global translation $\mathbf{t}'$ of the guide spot grid locations with respect to the origin of the spatial image coordinate system is given by the location $\mathrm{L}((1,1))$ of the upper left node of $\mathcal{G}^\star$. It is not guaranteed by the guide spot detection that $\mathrm{L}((1,1))$ will be in the set of detected guide spot locations $\mathcal{L}^\star$ (See Sect. 2.2). We compute $\mathrm{L}((1,1))$ as the intersection point of two straight lines fitted via least squares error minimization [11]: The first straight line model is fitted to the locations $\mathcal{R}^\star$ belonging to the first row of $\mathcal{G}^\star$ and the second straight line model is fitted to the locations $\mathcal{C}^\star$ belonging to the first column of $\mathcal{G}^\star$. The location sets $\mathcal{R}^\star$ and $\mathcal{C}^\star$ can be determined from $\mathcal{L}^\star$ with the knowledge of the estimated grid rotation angle θ [1]. The translation vector $\mathbf{t}$ with respect to the prior guide spot locations is obtained by subtracting from $\mathbf{t}'$ the prior guide spot location $\mathrm{L}_\mathrm{P}((1,1))$ rotated by $-\theta$.

3.3 Alignment of Detected Locations

Applying a rotation by the angle θ and a subsequent translation $\mathbf{t}$ to the prior guide spot locations $L_P(\mathcal{G}^\star)$ yields a set of *transformed prior guide spot locations* $L_T(\mathcal{G}^\star)$ which are expected to be near the true guide spot locations of the spot image. If a detected guide spot location (x, y) is in a $M_B \times N_B$ window W (see Sect. 2) around a transformed prior guide spot location $L_T((i, j))$, the location $L_T((i,j))$ is then assigned to the node (i, j) of $\mathcal{G}^\star$. If no location is found, then $L((i, j)) = $ nil.

3.4 Guide Spot Grid Parameters

The guide spot grid $\mathcal{G}^\star$ with its aligned locations is modeled in a Cartesian coordinate system (x, y) with its origin in the image center: For every row i of an individual guide spot field $\mathcal{F}^\star$ the parameters a_{r_i} and b_{r_i} of a straight line model $y = a_{r_i} + b_{r_i}x$ are fitted (in the least squares sense) to the locations of the row nodes having a valid location (i.e. nodes with $L((i, j)) \neq$ nil). Similarly, for every column j of $\mathcal{F}^\star$ the straight line parameters a_{c_j} and b_{c_j} are fitted to the valid column node locations (in the coordinates system with swapped x and y). If $L((i, j)) = $ nil), the location $L((i, j))$ is determined by the intersection point of the two fitted straight lines belonging to row i and column j respectively. The guide spot grid $\mathcal{G}^\star$ has a total of $2 * I_F * J_F * F_I * F_J$ parameters.

3.5 Location Assignment to Non-guide Spots

Locations belonging to non-guide spots in a block $\mathcal{B}$ are assigned with the help of the prior knowledge of the theoretical spot distances M_s and N_s and the block rotations. The rotation θ_B of a block $\mathcal{B}$ in row i of a field $\mathcal{F}$ is given by the slope b_{r_i} of the fitted straight line of the row as $\theta_B = \arctan(b_{r_i})$. For a guide spot (i_g, j_g) with a location $L((i_g, j_g)) = (x_g, y_g)$ the locations (x_n, y_n) of the non-guide spots (i_n, j_n) in $\mathcal{B}$ are assigned as follows:

$$\begin{bmatrix} x_n \\ y_n \end{bmatrix} = \begin{bmatrix} x_g \\ y_g \end{bmatrix} + \begin{bmatrix} \cos\theta_B & \sin\theta_B \\ -\sin\theta_B & \cos\theta_B \end{bmatrix} \begin{bmatrix} (i_n - i_g)N_s \\ (j_n - j_g)M_s \end{bmatrix} \tag{7}$$

Equation (7) is valid for guide spots residing in a block center.

4 Experimental Results

We have tested our approach on a set of 350 oligonucleotide fingerprint images (see Fig. 1) with a geometric resolution of 50 pixels/cm and a radiometric resolution of 16 bit. The prior theoretical spot dimensions are 4.5×4.5 pixels. The output of the grid fitting has been compared to the results of the semi-automatic MRF approach of Hartelius [5]. Ground truth data for the 57600 spot locations are not available for high density images of this type. The performance was compared primarily by visually assessing the grid fitting output: If every node of the guide spot grid $\mathcal{G}^*$ has assigned an image location that lies within the image coordinates belonging to the correct guide spot the algorithm is said to have succeeded. The MRF approach succeeded for 78 % of the images, whereas our algorithm succeeded for 100 % of the images in the test set.

We investigated the accuracy of the global rotation estimation as follows: Under the assumption that the estimated global grid rotation angle θ (Sect. 3.1) was correct, the spot image $\mathbf{S}$ was rotated by different angle offsets θ_o to yield a spot image $\mathbf{S'}$. The angle error was computed as $\theta_o - \theta_e$, with θ_e as the estimated grid rotation angle for $\mathbf{S'}$. Figure 4 shows the mean and standard deviation of the error (of 20 images) for a range of θ_o between $-10°$ and $+10°$ with a step size of $0.4°$. The resolution $\Delta\theta$ of the Radon transform was $0.05°$. The plot shows that an angle accuracy of $0.1°$ can be guaranteed irrespective of the rotation θ_o. This accuracy is sufficient, because the whole test set of spot images was analyzed with an angle resolution of $\Delta\theta = 0.2$.

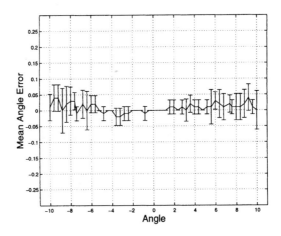

Fig. 4. Mean error and standard deviation of the rotation estimation

Figure 5a shows the upper left part of the fitted grid for the spot image in Fig. 1. The guide spot locations are marked as white crossbars and the locations of the other spots are marked as white dots. The spot locations are initial estimates for the centers of analytical spot models used for the quantification. Note that the locations of the non-guide spots and the guide spot locations that were determined by the intersection of straight lines have sub-pixel accuracy. In the images of Fig. 5 the locations are rounded to the nearest pixel position. Figure 5b shows some spot locations of two adjacent fields $\mathcal{F}_1$ and $\mathcal{F}_2$ of another spot image: The locations of $\mathcal{F}_2$ (the two rightmost guide spot columns) are shifted 3-4 pixels in the vertical direction. Figure 5c shows an image for which the MRF approach fails. The guide spots have a very low hybridization signal and some non-guide spots have a very high hybridization signal. The distortion in the lower left part of the image is a hair. The fitted grid for this image can be

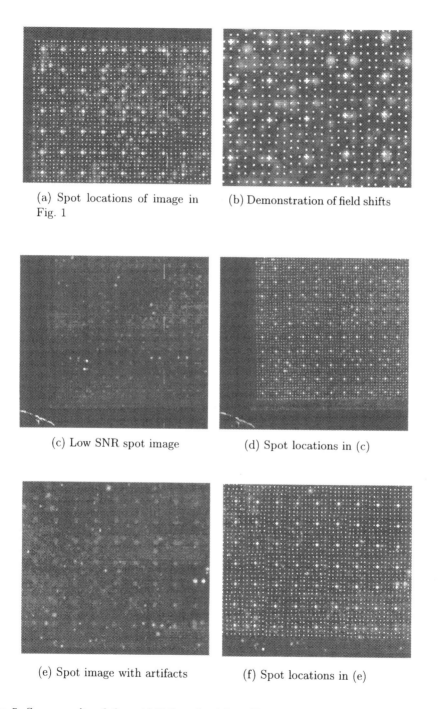

(a) Spot locations of image in Fig. 1

(b) Demonstration of field shifts

(c) Low SNR spot image

(d) Spot locations in (c)

(e) Spot image with artifacts

(f) Spot locations in (e)

Fig. 5. Some results of the grid fitting algorithm. The white crossbars indicate guide spot locations, the white dots indicate locations of the other spots.

seen in 5d. Figure 5e shows another image for which the MRF approach fails. In addition to the low hybridization signal of the guide spots the image contains artifacts: Spots of different sizes may occur in the image which do not belong to hybridization signals. The fitted grid for this image can be seen in Fig. 5f.

5 Conclusion

We presented an algorithm that performs automatic grid fitting for spot array images containing guide spots in the block centers as a safety grid . Guide spots are detected with the help of a matched filter, a nonlinear filter that amplifies the locations of the guide spots and a local maximum search. A grid rotation estimate with a Radon transform helps to assign the detected guide spot locations to the grid nodes. Straight lines are fitted in the least squares sense to the locations of the field rows and columns to fill in the complete grid and to assign block rotations. The algorithm succeeded for the whole test set comprising 350 ONF images of different quality. The approach we have presented is easily extendable to other images (e.g. without guide spots) coming from similar experiments.

References

1. N. Brändle, H. Lapp, and H. Bischof. Fully Automatic Grid Fitting for Genetic Spot Array Images Containing Guide Spots. Technical Report PRIP-TR-058, PRIP, TU Wien, 1999.
2. M. Chee et al. Accessing Genetic Information with High-Density DNA Arrays. Science, 274:610–614, 1996.
3. R. J. Johnston et al. Autoradiography using storage phosphor technology. Electrophoresis, 11:355–360, 1990.
4. S. Meier-Ewert et al. An automated approach to generating expressed sequence catalogues. Nature, 361:375–376, 1993.
5. K. Hartelius. Analysis of Irregularly Distributed Points. PhD thesis, Institute of Mathematical Modelling, Technical University of Denmark, 1996.
6. Anil K. Jain. Fundamentals of Digital Image Processing. Prentice-Hall, 1986.
7. E. Lander. The new genomics: Global views of biology. Science, 274:536–539, 1996.
8. Benjamin Lewin. Genes VI. Oxford University Press, 1997.
9. Shree K. Nayar and Tomaso Poggio, editors. Early Visual Learning. Oxford University Press, 1996.
10. H. J. Noordmans and A. W. M. Smeulders. Detection and Characterization of Isolated and Overlapping Spots. Computer Vision and Image Understanding, 70(1):23–35, 1998.
11. William H. Press et al. Numerical Recipes in C. Cambridge University Press, 1992.
12. Gerhard Winkler. Image Analysis, Random Fields and Dynamic Monte Carlo Methods. Springer Verlag, 1995.

Matching for Shape Defect Detection

Dmitry Chetverikov and Yuri Khenokh

Computer and Automation Research Institute
1111 Budapest, Kende u.13-17, Hungary
Phone: (36-1) 209-6510 Fax: (36-1) 466-7503
E-mail: `csetverikov@sztaki.hu`

Abstract. The problem of defect detection in 2D and 3D shapes is analyzed. A shape is represented by a set of its contour, or surface, points. Mathematically, the problem is formulated as a specific matching of two sets of points, a reference one and a measured one. Modified Hausdorff distance between these two point sets is used to induce the matching. Based on a distance transform, a 2D algorithm is proposed that implements the matching in a computationally efficient way. The method is applied to visual inspection and dimensional measurement of ferrite cores. Alternative approaches to the problem are also discussed.[1]

1 Shape Defect Detection Problem

In industrial shape defect detection, two types of measurements can be distinguished: (a) *Direct, or absolute, measurements*, when a dimension, or another quantity to be measured, is specified with respect to particular shape features (corners, etc.) that are easy to identify and locate. An example of such measurement is obtaining the distance between centroids of two holes. (b) *Relative measurements*, which are specified with respect to a reference shape.

Most of direct measurements are straightforward, as they usually have precise mathematical definitions in terms of images. Such computational definitions are relatively easy to translate into a computer algorithm. Relative measurements are much more challenging, as the definitions are, in fact, implicit. Here, one compares a measured shape to a reference (ideal) shape. For shift- and rotation-invariant comparison, optimal reference position and orientation (pose) of the measured shape is to be found, which requires invariant matching of the two shapes.

The pitfall of matching for defect detection is that no reference pose (e.g., baseline) can be specified *a priori* because defects may deteriorate any part of the shape. It is a typical 'chicken-and-egg' problem. What parts of a measured shape should be considered defective? Usually, it is assumed that those parts of the shape that coincide with the reference, or lie within the tolerance limit, are acceptable. Those parts that lie beyond the tolerance are defective. This means, however, that the defects should be recognized *prior to* the shape comparison whose goal is, in turn, defect recognition itself.

---

[1] This work was supported by the INCO-COPERNICUS grant IC15 CT 96-0742.

Figure 1 illustrates the difficulty being discussed. A measured shape M and a reference shape R are shown. M has defects in the upper parts of the legs. The locations of the defects are unknown for the matching algorithm. Assume one compares R and M by minimizing, under shift and rotation, the maximum local distance between the points of the two shapes. This measure of dissimilarity, which is closely related to the classical Hausdorff distance discussed in section 2.1, is not robust. In the resulting matching, deviations will be found in most parts of the contour: the defects are 'smeared'.

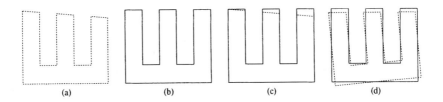

(a)   (b)   (c)   (d)

Fig. 1. (a) Defective measured shape M. (b) Reference shape R. (c) Desired superposition. (d) Superposition obtained using the local distance criterion.

The task is therefore to find the superposition of the measured and the reference shapes that best corresponds to the expectations of a human observer: a 'correct' part should be close to the reference; a 'defective' part should stand out clearly.

A related problem is addressed in robust regression and outlier detection [3], when the parameters of a linear model are estimated based on a set of measured points. However, in this case it is assumed that a majority of these points can be brought into strict correspondence with the reference (model) points, while the outliers are random. In manufactured shapes, both the acceptable and the defective parts of the surface usually have *systematic* deviations from the reference shape. In the acceptable parts, the deviations are within the tolerance limits, while in the defective parts the deviations exceed these limits. In addition, no simple analytical expression for the reference shape can usually be given.

Formally, the general problem of 3D shape matching for defect detection can be stated as follows. Two point sets are given: (a) A *reference point set R*, which represents the 'ideal' shape. R may result either from precise measurement of a physical reference shape, or from computation based on a mathematical description (specification) of the reference shape. R is complete (all parts of surface are represented), dense (every patch is densely covered by the points) and precise (the measurement/computation is precise). (b) A *measured point set M*, which may be incomplete, less dense, noisy, and distorted by defects whose deviation from the reference shape exceeds both noise level and tolerance.

Under these condition, one should find the superposition (matching) of the two point sets, in which the non-defective parts coincide, while the defective parts stand out.

2   Matching for Defect Detection in 2D Shapes

Our approach is based on the notion of the modified *Hausdorff distance* (HD) between two point sets, which is defined in section 2.1. This distance proved to be a general, efficient and robust measure of similarity between two shapes. Its robustness to noisy and incomplete data stems from the global distance between two point sets it operates with: no correspondences between particular points are searched for. A survey of tasks and methods related to the shape matching using the Hausdorff distance can be found in [4].

We apply the HD to a new problem, that of shape defect detection, and present an algorithm for fast computation of the reference pose based on the distance transform [1]. The distance transform (DT) is briefly discussed in section 2.2.

The proposed method first gives an initial guess that approximately normalizes the pose of the measured shape, then uses the modified (median) Hausdorff distance as the precise measure of shape correspondence. The main steps of the method are described in section 2.3.

Several options had been considered to obtain the initial matching. Any local feature based criterion, e.g., the coincidence of the baselines of the shapes, would be neither universal nor stable, because defects may distort that particular local feature. As the centroid is less sensitive to local defects, we finally decided to initially normalize the position by superimposing the centroids of the two shapes. This solution is often used in the industrial machine vision practice.

2.1   The Hausdorff Distance

The basis of our approach is the median Hausdorff distance which is a similarity measure between two arbitrary point sets. The classical Hausdorff distance between two (finite) sets of points, M and R, is defined as

$$h(M, R) = \max_{a \in M} \min_{b \in R} \|a - b\| \tag{1}$$

Here, $h(M, R)$, the directed distance from M to R, will be small when every point of M is near some point of R. This distance is too fragile for practical tasks: for example, a single point in M that is far from anything in R will cause $h(M, B)$ to be large. A natural way to take care of this problem is to replace equation (1) with

$$h^f(M, R) = f^{th}_{a \in M} \min_{b \in R} \|a - b\|, \tag{2}$$

where $f^{th}_{x \in X} g(x)$ denotes the f-th quantile value of $g(x)$ over the set X, for some value of f between zero and one. When $f = 0.5$ one gets the modified *median Hausdorff distance* which we use in our method.

Finally, the modified *mean Hausdorff distance* is defined as

$$h_m(M, R) = \frac{1}{N_M} \sum_{a \in M} \min_{b \in R} \|a - b\|, \tag{3}$$

where N_M is the number of points in M.

Let R be a point set representing the reference shape, M the measured point set. As already mentioned, we assume that R is complete, dense and precise, while M may be incomplete, sparse and noisy. R is obtained either analytically, or as the result of a careful, high-resolution, off-line measurement.

2.2 Using Distance Transform to Minimize HD

Direct computation and minimization of the Hausdorff distance using its definition (2) is very time-consuming. For each relative position of M and R, the number of operations is $O(N^2)$, where N is the number of points. We use the distance transform (DT [1]) of the reference shape to significantly speed up the procedure. A DT converts a binary image to another image in which a pixel value is the distance from this pixel to the nearest nonzero pixel of the binary image.

The distance transform is computed off-line, and only once for each particular reference shape. We apply the 5×5 neighborhood fast chamfer distance transform [1] that gives a good, integer-valued approximation to the exact Euclidean distance. An example of a distance transform is shown in figure 2. Here, the distances are displayed intensity-coded, with large distances shown as light pixels.

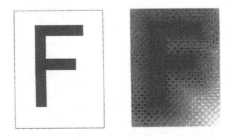

Fig. 2. A shape and its distance transform.

2.3 Main Steps of the Method

The proposed method assumes that the measured shape image has been normalized and binarized. Normalization, in this context, is a geometric correction of an image. Its purpose is twofold: (a) Correct image distortions resulting from the possible deficiencies of the optical setup (camera, lens, viewing geometry). (b) Normalize the size of the image for system calibration (pixel/mm ratio).

The normalization is done by mapping onto the image corners the centroids of four circular markers located close to the corners. This procedure and the subsequent binarization (thresholding) are presented in [6]. The contour is then refined to subpixel accuracy.

The main steps of the proposed matching method are as follows.

Step 1 Find contour points of the reference shape and obtain their DT.

Step 2 Obtain contour points of the measured shape.

Step 3 Compute and superimpose the centroids of the two point sets.

Step 4 Rotate and translate the measured point set with respect to the initial pose. For each relative pose, compute the median HD (2) using the corresponding distance values. To speed up, use a distance limit and discard a pose if a measured point is beyond the limit.

Step 5 Select those relative positions that yield the minimum HD value. (Multiple minima are possible.)

Step 6 Of these, select the one that has the least mean HD (3).

The computational load of this algorithm is $O(N)$ in the worst case, that is, near a matching, either complete or partial. Due to the individual distance check in step 4, most of the non-matching poses are skipped after a few operations.

Step 6 resolves ambiguities arising from the integer approximation of distance. Also, it discards possible false minima when symmetric contours are matched. For such contours, many points may coincide in a wrong orientation because of symmetry, while the rest of them may not coincide at all.

Alternatively to the criterion of step 6, the pose with the largest number of measured points lying within the minimum median HD can be used.

Figure 3 illustrates the matching of a measured contour against a reference contour. In this figure, the measured contour is shown in one of the relative positions, overlaid on the DT of the reference contour. (Step 4.)

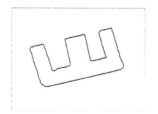 

Fig. 3. Left: The measured contour. Right: The DT of the reference contour with the measured contour overlaid.

3 Tests

3.1 Robustness of Method

Various tests were carried out to study the reliability of the proposed approach. A large number of test contours with different deviations from ideal shape and different missing parts were prepared in order to see how well the algorithm can perform in varying conditions.

The robustness of the algorithm is illustrated in figures 4 and 5. Figure 4 shows that the proposed approach is more robust than both the classical HD and the minimization of the mean Hausdorff distance (3) between the two contours. The classical HD minimizes the maximum distance and results in a wrong matching. Minimizing the mean HD is closely related to the conventional least squares regression. This method also 'smears' the defect.

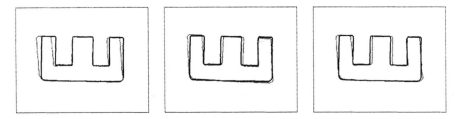

Fig. 4. Left: Proposed method. Center: The classical HD. Right: The mean HD.

Figure 5 demonstrates that our method can cope with incomplete and noisy measurements. Despite these factors, a good result is obtained.

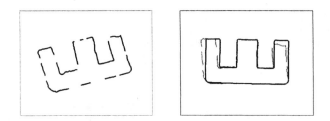

Fig. 5. An incomplete measured contour (left) and its matching.

3.2 Comparison with Reference Measurements

The shape measurement system was applied to a large set (100 pieces) the so-called of E-cores whose dimensions had been manually measured at a ferrite core producing factory using the conventional, manual gauging technique. An example of an E-core is shown in figure 6.

Fig. 6. A defective E-core (bent legs).

For automatic measurement of core dimensions, images of the cores were taken in arbitrary orientations. The measurement requires precise normalization of orientation, which was achieved using the proposed matching algorithm. The algorithm matches an arbitrarily oriented, measured core shape against a model reference shape created in the standard horizontal orientation.

Good correlation between the manual and the automatic measurements was observed. Here, we present the results for one of the dimensions, $A1$. (See [5] for more experimental data.) $A1$ is defined as the average horizontal extension in the normalized orientation. The results are shown in figure 7 where two distributions are plotted. One of them was obtained by the proposed procedure. The other one represents the reference manual measurements done by the core producer. The plots testify that the two measurements are consistent. The mean values of the two distributions are very close, while the spread of the computer-measured values is somewhat larger. The latter effect is partially caused by the discrete approximation of distances in the distance transform.

4   Discussion and Outlook

We have analyzed and formulated the matching problem for shape defect detection in an object represented by contour (2D) or surface (3D) points. A computationally efficient algorithmic solution to this problem for planar shapes has been proposed, implemented and tested.

From the practical point of view, the proposed solution has two drawbacks. One of them is the already mentioned discrete nature of the DT, which limits the accuracy of the approach. The other one is the simplistic, non-compact data structure of the DT. The distance transform can be extended to 3D. However, one would need $O(P^3)$ byte storage space, where P is the size of the DT matrix in each dimension. For 3D laser scanners, a typical resolution required in many tasks is $P \sim 1000$, which would make the method infeasible, at least, for the time being.

Currently, we are working on two alternative solutions that are more easily extendible to the 3D case. One of them is a modification of the robust regression algorithm [3] discussed in section 1. The modification is not straightforward.

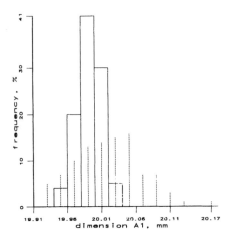

Fig. 7. Results of manual (solid line) and automatic (dashed line) measurements.

In the linear models considered in [3], 3 randomly selected points are sufficient to estimate the parameters of the model, i.e., to determine the 3D surface. In our case, two triples of points, one in R, the other in M, should be repeatedly selected and tested until correspondence is found. In addition, computation of the distance between R and M is more complicated and slower than in an analytical model.

Another alternative solution being considered is based on a flexible, multiresolution representation of a 3D shape. This representation uses the boxing data structure introduced in [2]. It allows to efficiently reduce, at a low resolution, the portion of the 6D parameter space (3 shifts and 3 angles) that should be scanned to refine or discard a candidate matching at a higher resolution.

References

1. G. Borgefors. Distance transforms in arbitrary dimensions. *Computer Vision, Graphics and Image Processing*, 27:321–345, 1984.
2. D. Chetverikov. Fast neighborhood search in planar point set. *Pattern Recognition Letters*, 12:409–412, 1991.
3. P.J. Rousseeuw and A.M. Leroy. *Robust Regression and Outlier Detection.* Wiley Series in Probability and Mathematical Statistics, 1987.
4. W.J. Rucklidge. Efficiently Locating Objects Using the Hausdorff Distance. *International Journal of Computer Vision*, 24(3):251–270, 1997.
5. The SQUASH Consortium. Standard-compliant Quality Control System for High-level Ceramic Material manufacturing. Technical report, EU INCO-COPERNICUS Programme, Project IC15 CT 96-0742, 1998.
6. J. Verestoy and D. Chetverikov. Shape defect detection in ferrite cores. *Machine Graphics and Vision*, 6(2):25–236, 1997.

A Vision Driven Automatic Assembly Unit*

Gernot Bachler, Martin Berger, Reinhard Röhrer, Stefan Scherer, Axel Pinz

Department for Computer Graphics and Vision,
Graz University of Technology,
A-8010 Graz, Austria
{bachler, berger, roehrer, scherer, pinz}@icg.tu-graz.ac.at
WWW home page: http://www.icg.tu-graz.ac.at/

Abstract. The development of a flexible assembly unit is one of the demanding tasks in industrial manufacturing. A higher degree of flexibility is mostly payed by an increasing complexity of the involved hardware. In this paper we present a three-step concept for a vision driven automatic assembly unit. These three steps are robust bin-picking to isolate objects from a pile of unorganized parts, exact pose determination to enable industrial mounting and visual inspection of the final assembling. For robust bin-picking we present a new structured light approach. Experiments show the robust and accurate behavior of the proposed algorithm and motivate the implementation in an industrial system. For exact pose determination, the second step, a pose estimation based on a modified view based approach, followed by a model based refinement is proposed. Initial experiments promise a fast and exact pose determination.

1 Introduction

In industrial production the use of automated assembly does not pay off if there are only a few parts to be produced because mechanical object isolation is expensive and closely tied to particular objects. Optical sensors such as digital cameras allow a more flexible application. Therefore, machine vision plays a more and more important role in increasing the level of automation [1], [8], [16].

We present a three-step concept of an integrated vision and manipulation system for automated industrial assembling (Fig. 1(a)). These steps can be summarized as:

1. Robust bin-picking to isolate a single object from a pile of unorganized parts
2. Exact pose determination of one isolated part to enable gripping for an industrial assembling manipulator
3. Visual inspection of the assembly step and quality control

In this paper we focus on the first two tasks.

---

* This work was supported by M&R Forschungs- und Entwicklungs Ges.m.b.H. and the Austrian 'Forschungs-Förderungs-Fond' under grant 3/13010. We gratefully acknowledge the useful discussions with Dr. Vassili Kravtchenko-Berejnoi during the implementation phase.

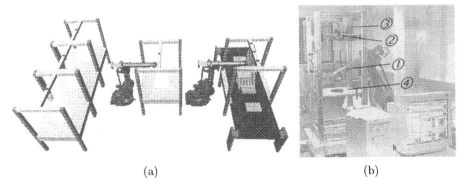

Fig. 1. (a) The proposed three-step concept of the assembly cell; rightmost the robust bin-picking part for different deposit boxes, in the middle the pose estimation task and leftmost the inspection for assembling and quality control. **(b)** The industrial setup for bin-picking consisting of a six degrees-of-freedom robot arm (1), a stereo rig (2), a grid projector (3) and a workplace (4).

To ensure a high degree of flexibility, a vacuum sucker is used for picking various objects. A grip point on a planar surface patch is the only limitation to pick the object. Therefore the bin-picking task is reduced to the identification of planes and the calculation of their 3D coordinates. In Sect. 2 we present a new structured light approach for this task. In most applications structured light is used to solve the stereo correspondence problem (matching) [10], [13], for range data acquisition [9], [12], or to extract shape information along deformed projected lines [2], [5], [11], [14].

Our structured light algorithm combines shape analysis *and* stereo matching. The novel contribution is to analyze the grid pattern of a stereo image pair in 2D and to use the extracted high-level scene description to enable a fast 3D reconstruction of single surface patches. The industrial setup for the bin-picking task is shown in Fig. 1(b)

The diverse algorithms to determine the pose of an object, the second task, can be classified into appearance based approaches [7] and model based approaches [6], [4], [15], [3]. To ensure both, a fast and robust computation, we propose the combination of an adapted eigenspace method and a model fitting technique. The promising initial experiments with the combined pose estimation approach are shown in Sect. 3. Once the exact position is determined, the part can be taken by a special manipulation tool for assembling.

The third step is the inspection of the assembling. This can be a simple completeness detection or a complex scene analysis and is not subject of this paper.

2   Bin-Picking by Robust Plane Detection

As mentioned above, from the visual point of view, the bin-picking task is reduced to 3D plane detection. We propose a fast but coarse *bottom up* computation of plane candidates in a stereoscopic image pair of the scene. The 3D coordinates are obtained by a high level description matching. The obtained 3D estimation of the best plane candidates undergoes a *top down* verification and a refinement step in the original grey image. Fig. 2 outlines the detailed process chain for robust plane detection.

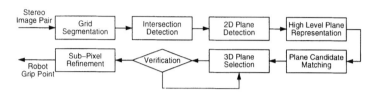

Fig. 2. *Process chain for robust bin-picking*

2.1 Bottom-Up Plane Detection

In order to get a high level description of the planes, the projected grid (Fig. 3(a)) is processed in several steps. We start with a segmentation of the projected grid, proceed with a skeletonization, an intersection detection (Fig. 3(b)), a connecting step and a plane detection by enforcement of perspective projection geometry constraints. The detected planes are stored in a local perspective projection distortion invariant representation. The detected planes are called 'plane candidates' in the scope of high-level plane matching.

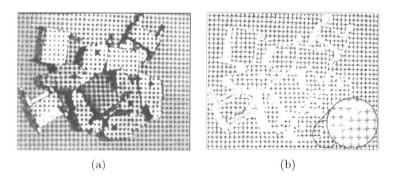

(a)   (b)

Fig. 3. Left image of a pile of parts with the projected grid: **(a)** original grey scale image, **(b)** segmented grid (grey) superimposed with the detected grid intersection points (black).

2.2 Local Perspective-Invariant High-Level Matching

After all plane candidates are identified, corresponding planes in the stereo pair have to be found. Therefore a rectified 'plane image' of each plane is constructed.

Each intersection leads to a point in the plane representation (Fig. 4(a+b)). Correspondence of two planes is found by comparing the plane representation of possible plane pairs. The advantage of using rectified planes is the elimination of distortions in the grid. Fig. 4(c) shows the matching results for the plane pair in Fig. 4(a,b).

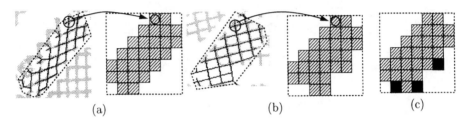

(a) (b) (c)

Fig. 4. High level plane representation: each intersection of a plane is represented as a pixel in the perspective invariant plane representation. Segmentded plane from **(a)** left view, **(b)** right view. **(c)** Matching results for **(a)**-**(b)**, intersection points with no correspondence are marked black.

It has to be mentioned, that a plane representation can flip about 90 degrees if the perspective geometry distortion results in a rotation of the plane in the image which is larger than 45 degrees. This case occurs only, if the tilt angle of this plane is larger than the maximum tilt angle of a plane that can be handled by the vacuum sucker. We therefore call the representation in the sense of its scope locally projective distortion invariant.

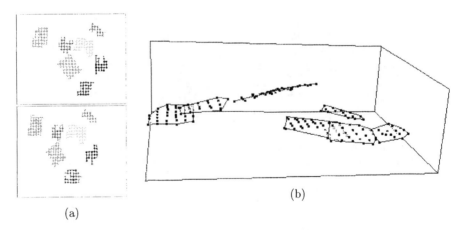

(a)

(b)

Fig. 5. **(a)** Segmented plane candidates from left view (above) and right view (below). **(b)** Computed 3D coordinates of all detected planes of the observed scene

2.3 Top-Down Verification and Refinement

The bottom up strategy is now followed by a top down verification and refinement. False matches can be identified applying a consistency check in 3D. If the reconstructed coordinates exceed given limitations the 2D plane match is classified as false. Fig. 5(b) shows the reconstructed world coordinates of the intersection points of all segmented planes (Fig. 5(a)). The topmost plane should be selected for picking. Our experiments show, that a removal of the topmost object makes it possible to pick more than one object after one reconstruction step of the whole scene. To improve the determined world coordinates all intersection pairs are computed with subpixel accuracy. This is performed by grey level matching in a restricted search area within intersection points.

Once a plane is determined in 3D, a grip point is selected. If there are no special restrictions about the shape of planes (e.g. holes in the distinct surface patch), its center of gravity can be selected as a grip point.

2.4 Experimental Results

The experiments were performed with different kinds of industrial parts[1]. 127 visible planes were manually counted in eight stereoscopic images. 137 planes were totally segmented. 127 (92.7%) of the segmented planes were detected correctly. Some of the correctly segmented planes were split into two ore more representations (28, 20.4%). This is no problem as long as the split up representations are large enough for a grip point selection.

For the computation of 3D planes, 43 plane candidate pairs (86 planes) of the stereoscopic images were selected. Criteria for the selection are the size of a plane candidate and the disparity limitations. 35 planepairs were matched correctly (81.4%). It has to be mentioned, that the top plane candidates (i.e. best candidates for gripping) were always detected correctly.

3 Exact Pose Determination

In this section the methods for determining the exact pose of an isolated object picked out from a bin as described in Sect. 2 are presented. We propose a pose determination method combining an adapted eigenspace method for an initial estimation and a model-based technique for the refinement. The geometric relations between the camera and the workplace (Fig. 7(a)) remain unchanged and are calibrated. Therefore the number of eigenspace parameters can be reduced to one: the rotation angle ϕ around the z-axis of the workplace.

Each object is learned separately, obtaining a number of trajectories in the eigenspace equal to the number of stable poses suggested by the ground plane constraint (see Fig. 6). Each trajectory is parameterized by the rotation angle ϕ. Missing views are interpolated using cubic splines.

---

[1] an online video is available at `http://www.icg.tu-graz.ac.at/~Research/M+R/`

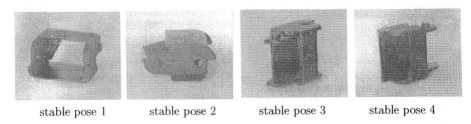

stable pose 1   stable pose 2   stable pose 3   stable pose 4

Fig. 6. Four stable poses of an industrial part

A set of trajectories for four stable poses is shown in Fig. 7(b). We observed a maximum error for the angle ϕ of ± 3 degrees. This relatively poor result is due to our sample width of ten degrees and needs no improvement since we are interested just in an estimation. Additionally to the stable pose

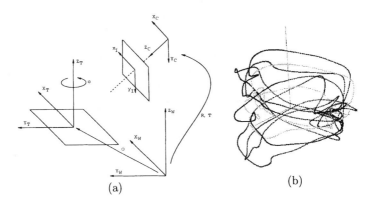

(a)   (b)

Fig. 7. (a) Geometric relations between world coordinate system (X_W, Y_W, Z_W), workplace coordinate system (X_T, Y_T, Z_T), camera coordinate system (X_C, Y_C, Z_C), and image coordinate system (x_I, y_I). **(b)** The first three coordinates of a set of trajectories in eigenspace with interpolated samples

and the rotation angle ϕ we propose a method to obtain a translation vector $\mathbf{V} = (V_x, V_y, V_z)^T$ (Fig. 8) for the object in the workspace coordinate system. The eigenspace parameter ϕ is object centered. There is no a priori information about the translation of an object on the workplace. A CAD-model in the stable pose (detected with the eigenspace method) is virtually put at the center of the workplace, rotated by the angle ϕ and transformed into the camera coordinate system (Fig. 8(a)). From those points which define the bounding rectangle of the projected CAD-model we obtain a 2D - 3D correspondence for four points. The bounding rectangle of the projected CAD-model is mapped onto the bounding rectangle of the captured object. The coordinates of the image points defining its bounding rectangle are now used to estimate the translation vector. The ground plane constraint suggests the z-coordinate of the vector $\mathbf{V}$ to be zero, since the

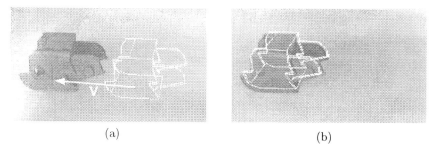

(a)   (b)

Fig. 8. The CAD-model projected onto the image of an object in an unknown pose using: **(a)** only the stable pose and the rotation parameters; **(b)** the stable pose, rotation *and translation* parameters found by the proposed method

object lies on the xy-plane of the workplace coordinate system. This allows to solve the overdetermined system (8 equations, 2 unknowns $\mathbf{V}_x$ and $\mathbf{V}_y$) by some standard method taking into account possible outliers. Hence a translation vector is computed, the CAD-model can be virtually moved on the workplace by $\mathbf{V} = (V_x, V_y, 0)$ and projected again using the set of estimated pose parameters ϕ and $\mathbf{V}$. Fig. 8(b) shows the result for an industrial part. The parameters estimated by the proposed method are input to a model-fitting technique. For fast and robust convergence a good initial pose estimation is needed. The results of our approach are promising and the adaptation of a model-fitting technique for the special requirements in the described assembly cell is subject to future work.

4 Conclusion and Future Work

We presented a three-step concept for an automated vision driven assembly cell. Bin-picking is already implemented in an industrial environment. Our experiments show a robust and accurate behaviour of the proposed method. This system is currently under heavy industrial testing. The initial experiments of the pose determination task promise to be a useful extension to the bin-picking module. The advantages of the proposed concept are:

1. The subdivision into three independent modules enables a flexible, fast, robust and accurate handling of various industrial objects.
2. No information about a single object is needed for the bin-picking task.
3. One universal manipulator, a vacuum sucker, can be used for picking objects of any shape.
4. The pose determination task can be easily adapted to new parts by simply substituting the CAD model and the eigenspace representation.

Future work will focus on the optimization of the bin-picking task, the industrial implementation of the pose determination module and the development of a concept for the inspection module.

References

[1] Ezzet Al-Hujazi and Arun Sood. Range Image Segmentation with Applications to Robot Bin-Picking Using a Vacuum Gripper. *IEEE Transactions on Systems Man and Cybernetics*, 20(6):1313–1324, November/December 1990.

[2] Z. Chen, S.Y. Ho, and D.C. Tseng. Polyhedral Face Reconstruction and Modeling from a Single Image with Structured Light. *IEEE Transactions on Systems Man and Cybernetics*, 23(3):864–872, May/June 1993.

[3] Steven Gold, Anand Rangarajan, Chien-Ping Lu, and Suguna Pappu. New Algorithms For 2D and 3D Point Matching: Pose Estimation and Correspondence. *Pattern Recognition*, 31(8):1019–1031, August 1998.

[4] R. M. Haralick, H. Joo, C. Lee, X. Zhuang, V. G. Vaidya, and M. B. Kim. Pose Estimation from Corresponding Point Data. *IEEE Transactions on Systems Man and Cybernetics*, 19(6):1426–1446, November-December 1989.

[5] Gongzhu Hu and George Stockman. 3-D Surface Solution Using Structured Light and Constraint Propagation. *IEEE Transactions on Pattern Analysis and Machine Intelligence*, 11(4):390–402, April 1989.

[6] David G. Lowe. Three-Dimensional Object Recognition from Single Two-Dimensional Images. *Artificial Intelligence*, 31(3):355–395, March 1987.

[7] Hiroshi Murase and Shree K. Nayar. Visual Learning and Recognition of 3-D Objects from Appearance. *International Journal of Computer Vision*, 14:5–24, 1995.

[8] Krisnawan Rahardja and Akio Kosaka. Vision-Based Bin-Picking: Recognition and Localization of Multiple Comlex Objects Using Simple Visual Cues. In *Proceedings of the International Conference on Intelligent Robotics and Systems*, Osaka, Japan, November 1996. IEEE / RSJ.

[9] Martin Rutishauser and Frank Ade. From Vision to Action: Grasping Unmodeled Objects from a Heap. In *Intelligent Robots and Computer Vision XIV, SPIE's Photonic East Symposium*, volume 2588, pages 375–386. The international Society for Optical Engineering, October 1995.

[10] J. Salvi, J. Batlle, and E. Mouaddib. A robust-coded pattern projection for dynamic 3D scene measurement. *Pattern Recognition Letters*, 19(11):1055–1065, September 1998.

[11] N. Shrikhande and G. Stockman. Surface Orientation from a Projected Grid. *IEEE Transactions on Pattern Analysis and Machine Intelligence*, 11(6):650–655, June 1989.

[12] Marjan Trobina and Aleš Leonardis. Grasping Arbitrarily Shaped 3-D Objects from a Pile. In *International Conference on Robotics and Automation*, volume 1, pages 241–246. IEEE, May 1995.

[13] R.J. Valkenburg and A.M. McIvor. Accurate 3D measurement using a structured light system. *Image and Vision Computing*, 16:99–110, 1998.

[14] Y.F. Wang, A. Mitiche, and J.K. Aggarwal. Computation of Surface Orientation and Structure of Objects Using Grid Coding. *IEEE Transactions on Pattern Analysis and Machine Intelligence*, 9(1):129–137, January 1987.

[15] P. Wunsch and G. Hirzinger. Registration of CAD-Models to Images by Iterative Inverse Perspective Matching. In *Proc. International Conference on Pattern Recognition*, pages 78–83, 1996.

[16] Billibon H. Yoshimi and Peter Allen. Closed-Loop Visual Grasping and Manipulation. In *Proceedings of Image Understanding Workshop*, pages 1353–1360, Palm Springs, California, February 1996.

Scene Reconstruction from Images *

Václav Hlaváč
in cooperation with
Tomáš Pajdla, Radim Šára, Tomáš Svoboda, Martin Urban, and Tomáš Werner

Center for Machine Perception, Faculty of Electrical Engineering
Czech Technical University
Karlovo náměstí 13, 121 35 Prague 2, Czech Republic
hlavac@cmp.felk.cvut.cz, http://cmp.felk.cvut.cz

The CAIP'99 invited lecture overviews the recent development in the scene reconstruction in the context of the Prague research contribution to it by the Center for Machine Perception.

3D Scene Reconstruction from (uncalibrated) 2D images
The second half of nineties witnessed a qualitative move from stereovision that remained for a long time in a one hundred old photogrammetric framework providing relation between two views only. The new impulse was the discovered trilinear and quadrilinear relation among views in projective geometry. Another impulse was the transition from Euclidean using calibrated cameras to projective reconstruction where uncalibrated cameras are sufficient. The observations of the scene provides extensive number (tenth of) images that yield qualitatively better results than before.

The topic will be overviewed from the perspective of own achievements: (a) interpolation from two images [10], (b) relation among interpolation, extrapolation and reconstruction of a full 3D model [13], (c) projective reconstruction from three uncalibrated images [11], (d) introduction of the oriented projective reconstruction [14, 12], (e) selection of an optimal set of reference images [1], (f) search for correspondences if observer just translates or rotates [12], (g) the attempt to generalise a dense sequence correspondence to more general cases [9], (h) autocalibration from uncalibrated views [8, 4].

Omni-directional vision
Epipolar geometry and egomotion estimation algorithm for *central panoramic cameras* was developed and presented [6, 3, 7]. Design and image formation for newly defined central panoramic cameras have been studied [5]. New approach to mobile robot localization has been proposed [2]. The approach relies on a visual map comprising panoramic images which are represented in a rotationally invariant manner.

References

1. Václav Hlaváč, Aleš Leonardis, and Tomáš Werner. Automatic selection of reference views for image-based scene representations. In *Proceedings of the European*

---

* This work was supported by the MSMT VS96049, GACR 102/97/0480, GACR 102/97/0855, and CVUT IG 309809603 grant.

Conference in Computer Vision, volume 1 of *Lecture Notes in Computer Science, No. 1064*, pages 526–535, Heidelberg, Germany, April 1996. Springer.

2. Tomáš Pajdla. Robot localization using shift invariant representation of panoramic images. Technical Report K335-CMP-1998-170, FEE CTU, FEL ČVUT, Karlovo náměstí 13, Praha, Czech Republic, November 1998.

3. Tomáš Pajdla, Tomáš Svoboda, and Václav Hlaváč. Robot motion estimation from panoramic images. In *First International Workshop on Robot Perception for Autonomous Aerial Vehicles*, June 1998. to appear.

4. Tomáš Pajdla and Martin Urban. Camera calibration from bundles of parallel lines. Research Report K335-CMP/99/180, Czech Technical University, Prague, Karlovo nam. 13, 121 35 Prague, Czech Republic, January 1999.

5. Tomáš Svoboda, Tomáš Pajdla, and Václav Hlaváč. Central panoramic cameras: Design and geometry. In Aleš Leonardis and Franc Solina, editors, *Proceedings of Computer Vision Winter Workshop in Gozd Martuljek, Slovenia*, pages 120–133, Ljubljana, Slovenia, February 1998. IEEE Slovenia Section, IEEE Slovenia Section.

6. Tomáš Svoboda, Tomáš Pajdla, and Václav Hlaváč. Epipolar geometry for panoramic cameras. In Hans Burkhardt and Neumann Bernd, editors, *the fifth European Conference on Computer Vision, Freiburg, Germany*, number 1406 in Lecture Notes in Computer Science, pages 218–232, Berlin, Germany, June 1998. Springer.

7. Tomáš Svoboda, Tomáš Pajdla, and Václav Hlaváč. Motion estimation using central panoramic cameras. In Stefan Hahn, editor, *IEEE International Conference on Intelligent Vehicles*, pages 335–340, Stuttgart, Germany, October 1998. Causal Productions.

8. Martin Urban, Tomá Pajdla, and Václav Hlaváč. Camera self-calibration from multiple views. Research Report CTU-CMP-1998-169, CMP FEE CTU, Karlovo nám. 13, 121 35 Prague, Czech Republic, September 1998.

9. Jan Vydržel and Václav Hlaváč. Tracking correspondences in dense motion sequences. Research Report K335-1998-154, CMP FEE CTU, FEL ČVUT, Karlovo náměstí 13, Praha, Czech Republic, January 1998.

10. T. Werner, R.D. Hersch, and V. Hlaváč. Rendering real-world objects using view interpolation. In *Proc. 5th International Conf. on Computer Vision*, pages 957–962, Boston, USA, June 1995. IEEE Computer Society Press.

11. T. Werner, T. Pajdla, and V. Hlaváč. Efficient rendering of projective model for image-based visualization. In *Proceedings of the 14th International Conference on Pattern Recognition, Brisbane, Australia*, pages 1705–1707, Los Alamitos, California, August 1998. International Association for Pattern Recognition, IEEE Computer Society.

12. Tomáš Werner. *Image-Based Visualization of Real 3D Scenes*. PhD thesis, Faculty of Eletrical Engineering, Czech Technical University, FEL ČVUT, Karlovo náměstí 13, Praha, Czech Republic, 1998.

13. Tomáš Werner, Tomáš Pajdla, and Václav Hlaváč. Efficient 3-D scene visualization by image extrapolation. In Hans Burkhardt and Bernd Neumann, editors, *Proc. 5th European Conf. Computer Vision*, volume 2, pages 382–395, Berlin, Germany, June 1998. Springer Verlag.

14. Tomáš Werner, Tomáš Pajdla, and Václav Hlaváč. Oriented projective reconstruction. In M. Gengler, M. Prinz, and E. Schuster, editors, *Pattern Recognition and Medical Computer Vision: 22-nd Workshop of the Austrian Association for Pattern Recognition (ÖAGM/IAPR)*, pages 245–254, Wien, Austria, May 14–15 1998. Österreichische Computer Gesselschaft.

NIIRS and Objective Image Quality Measures

K. J. Hermiston, D. M. Booth

Defence Evaluation and Research Agency
St Andrews Road
Malvern, WR14 3PS, England
kjhermiston@dera.gov.uk dmbooth@dera.gov.uk

Abstract. The introduction of image distortion during compression is of widespread concern, to the extent that the nature and size of the distortion may influence the choice of codec. The ability to quantify the distortion for particular applications is therefore highly desirable, particularly when options for new compression standards (such as JPEG 2000) are being considered. We report on the performance of several degradation measures that have been evaluated on optical imagery having previously undergone compression by the JPEG, wavelet and VQ codecs. The results show the relationship between the subjective National Imagery Interpretability Rating Scale (NIIRS) and several numerical image quality measures. An insight is provided into factors influencing NIIRS evaluation.

Introduction

In recent years we have witnessed a continued growth in digital image processing applications. As data volumes have increased, bandwidth and storage capacity have, in many cases, been placed at a premium, and attention has shifted towards image compression. The suitability of a particular type of compression algorithm is frequently application dependent. For example, when consistency between the reconstructed and original image is paramount then a lossless approach is usual. However, the reduction in data volume achieved by lossless techniques is comparatively low (usually in the order of 2 to 1). Lossy techniques can achieve greater data reduction by weakening constraints on reconstructed image fidelity. The manner in which the information can be sacrificed has given rise to the multitude of algorithms occupying this category.

Lossless compression is assessed on the amount of data reduction achieved and on the complexity and execution time of the algorithm. The assessment of lossy compression algorithms is more difficult because information loss must be quantified in some way. Usually numerical measures, such as Mean Squared Error (MSE) and Peak Signal to Noise Ratio (PSNR), are used in conjunction with the resultant number of bits per pixel to produce the familiar bit-rate distortion graphs. However, it has been reported that MSE and PSNR do not correlate well with a subjective assessment of image quality. Recent studies of the human visual system (HVS) have spawned various image quality measures[1] and compression schemes[2][3] that model aspects of human perception. Even though a complete model of the HVS is beyond current capability, simple HVS weightings have been included in scalar image quality measures, and have reportedly improved correlation under subjective assessment.

In the military domain, the adoption by NATO of the National Imagery Interpretability Rating Scale (NIIRS) for the assessment of reconnaissance imagery has provoked interest. NIIRS has developed under the auspices of the Imagery Resolution Assessments and Reporting Standards (IRARS) committee. It consists of integral levels ranging from 0-9. Higher values indicate a capability to support higher and more detailed analysis. The assignment of NIIRS level is a task-driven activity to detect, distinguish between, and identify specific imaged objects. If the particular object does not appear in the image, the IA must imagine it to do so and assess the image accordingly. The overhead associated with training Imagery Analysts (IAs) to be proficient at NIIRS rating can be considerable. In response to this, and requests for less military orientated criteria, IRARS issued the Civil NIIRS[4], together with agricultural, cultural and natural definitions which were quantised to tenths of a rating based on statistical studies amongst NIIRS rated IAs.

In parallel with our research into image quality, the emerging JPEG 2000 compression standard has provided a stimulus to assess the effects of current JPEG and future wavelet based codecs on NIIRS. The current JPEG standard algorithm has been incorporated into many existing military image file formats such as NITFS[5], STANAG 7023, STANAG 4545/NSIF[6], with JPEG 2000 already earmarked for inclusion in NITFS[7] on its issue. It is within this context that our study is scoped.

This paper presents the results collected from experiments into the subjective and numerical evaluation of digital imagery which has undergone lossy compression and decompression, thereby incurring distortion. The intent is to identify numerical measures that might be used to automatically and intelligently apply compression within imagery intelligence (IMINT) systems to satisfy the NIIRS requirements of end-users on distributed intelligence networks.

Assessment Study

The study used digital versions of the calibration images associated with the Civil NIIRS Scale (Figures 1-16). Each of the sixteen images was sampled onto 512*512, 8 bit pixels, and was compressed separately using the JPEG[8], wavelet[9] and VQ[10] codecs. The compression ratios were 2, 5, 10, 15, 20, 30, 40 and 50:1 for the JPEG and wavelet algorithms. Algorithmic constraints led to VQ compression rates of 2, 4, 9, 16, 20, 30, 42 and 49:1. These provided a sufficiently close correspondence with the JPEG and wavelet compression rates to facilitate direct comparison.

The scalar quality measures[11] assessed in this study are listed in Table 1. All of the measures are bivariate, that is, they measure the differences between corresponding samples in the original image, f , and the (reconstructed) compressed image, f'. The study also evaluated graphical image quality measures, such as histograms and Hosaka plots[12], against the results. However, the dimensional inconsistency between the graphical measures and the NIIRS rating did not facilitate correlation. Of particular note in Table 1 are the measures that incorporate a degree of local correlation with adjacent samples and those which apply a recent and simplistic HVS model into their calculations. They include the HVS versions of Normalised Absolute Error (NAE,13), Normalised Mean Squared Error (NMSE,16), L2 (18,19,20).

Following a numerical analysis of compressed imagery, a subjective assessment of the same imagery was organised with the co-operation of active and NIIRS qualified Image Analysts from military intelligence centers across England. Seven Image Analysts were asked to NIIRS rate the suite of compressed imagery using the March 1996 release of the IRARS Civil NIIRS Reference Guide. The images were systematically presented to the IAs in sequential order of increasing original NIIRS level and of increasing compression. The assigned absolute NIIRS ratings from each IA were recorded and used to form an average rating for each image at each level of compression using each of the three compression algorithms.

During the NIIRS rating exercise, IAs consistently indicated that NIIRS calibration images 5a, 6a, 7a, 8b and 9a were unsuitable for NIIRS assessment in view of the spare features or unusual subject matter which made correspondence with NIIRS definitions difficult. These images were pruned from the study to avoid unnecessary bias caused by inaccurate NIIRS scoring.

With both objective and subjective assessment of imagery complete, the average NIIRS ratings were evaluated for each image at each level of compression and correlated, using the Pearson correlation coefficient, with the corresponding objective measures.

Results and Discussion

Graphical results obtained from the NIIRS assessment of imagery are shown in Figures 17 to 19 illustrating the degradation under JPEG, wavelet and VQ compression respectively. Figure 17 details the average NIIRS rating of JPEG compressed imagery over the considered interval of compression. It is interesting to note the slight deviations from the ideal of the NIIRS ratings at the compression ratio of 2:1 (which is nearly lossless). This is indicative of the inherent inaccuracies of subjective analysis amongst a diverse analyst population. Most images show a similar decline in NIIRS rating as the compression ratio increases, with the rate of degradation increasing slightly at low bit rates. Image 9b shows an unusually steep decline at high compression. During assessment, IAs remarked that this image was sparsely populated with features and the field of view was narrow compared with aerial reconnaissance imagery which they were more used to analysing. Both these properties made image 9b particularly difficult to assess under high compression against a list of criteria in the NIIRS definitions which were not present in the scene and presented difficulty in imagining them in the scene as NIIRS recommends. Image 4b also presented some uncharacteristic behaviour by maintaining an almost uniform level of average NIIRS rating across the spectrum of compression. Again, the exemplar image depicts a scene consisting mainly of large regions of wooded areas with a small farm building as the centre of focus for the IA. The large areas of uniform texture retained their characteristics sufficiently under compression to subjectively reduce the NIIRS degradation.

Figure 18 represents the average NIIRS rating of imagery under wavelet compression. It is interesting to observe the very close similarities between subjective assessment of JPEG and wavelet compressed images. The patterns of NIIRS degradations show a steady trend downwards and again images 4b and 9b show the effects of spare features in subjective image assessment. It is of interest to note the slightly improved NIIRS assessment of images 8a, 7b and 6b at low bit rates using wavelet compression than that obtained by JPEG.

Figure 19 illustrates the degradation of NIIRS under vector quantisation compression. Again, at first sight, the general pattern of decay in NIIRS reflects the overall pattern shown by JPEG and wavelet compression. However, closer examination reveals that there is a faster decay in images 8a, 7b and 6b between compression ratios of 4:1 to 50:1. It is thought that this decay is brought about by the transition to smaller more detailed features in the high NIIRS example imagery. The inherent blocking characteristics which appear at low compression in vector quantised images quickly affect the fine structure of such features influencing the subjective assessment. Unlike JPEG blocking artefacts, which remain at constant 8*8 pixel size, regardless of compression ratio, VQ compression was realised by increasing the block size as the compression increases and maintaining the same size of codebook. The boundaries between blocks become discontinuous quickly as the compression ratio increases. The VQ NIIRS levels also display significantly lower values at the most compressed level of 49:1 than those obtained using JPEG and wavelets.

Figure 20 depicts the average correlation of NIIRS ratings and numerical measures across all images. It is apparent that the entire library of numerical measures used in this study were better able to reflect the NIIRS degradation in wavelet and VQ compressed imagery than they were with the JPEG compressed imagery. This is in part due to the ability of wavelet and VQ compression to encode the imagery to produce an optimally minimised MSE image on reconstruction. The wavelet encoder produces embedded codefiles which ensure that recovered bits minimise optimally the MSE of the reconstructed image when decoded. Since MSE has been shown to provide a good correlation with NIIRS evaluation in this study, the use as a controlling parameter in the wavelet codec has been shown to provide a higher degree of correlation with the NIIRS assessment.

In common with wavelet compression, VQ compression uses MSE to optimise the codebook generation during compression. At each level of compression, the algorithm optimises a codebook collection of model blocks using the Generalised Lloyd Algorithm (GLA) to minimise the MSE of the reconstructed image.

In considering the results presented in Figure 20, it would appear that the following numerical image quality measures provide very good correlation with NIIRS ratings when applied to JPEG and particularly to both VQ and wavelet compressed imagery.

Mean Square Error (1)
Image Fidelity (8)
PMSE (10)
Normalised MSE (14)

The unexpectedly good performance of MSE with NIIRS degradation appears to negate much of the criticism of its low correlation with subjective assessment. These results support recent studies from academia[11] which found that the latter three of the above measures were amongst the best metrics in their subjective/numerical image quality correlation project.

During the subjective assessment of imagery by IAs, a number of points arose which are deemed worthy or relating. When corresponding image features to NIIRS definitions, opinion was expressed that UK analysts are more familiar with European terrain. The NIIRS level 2 criteria to identify road patterns, like clover leafs, on major highway systems, does not take account of the typically larger scale of clover leafs in the US than those found in the UK and Europe. Indeed many frequent image features such as house spacing (NIIRS level 3 criteria) and car sizes (NIIRS level 6 criteria) vary considerably between the UK and the US. It was suggested that such criteria might contain regional variations in future issues of the NIIRS definitions.

During the post assessment discussion, the IAs were asked for their preference of compression. Their view was based on the blocking artefacts present from the JPEG and VQ compression codecs at high compression and the blurring inherent in the wavelet compression at high ratios. Five IAs indicated their first preference was for the JPEG algorithm with their second preference for wavelets. Two IAs indicated they favoured the wavelet compression with JPEG as their second choice. All IAs indicated VQ compression as their last preference. This is supported by the average NIIRS ratings as depicted in Figures 17, 18 and 19 which indicate less visual disortion using JPEG than wavelets except at low bit rates. The IAs expressed concern about the appearance of localised artefacts when viewing highly compressed wavelet compressed imagery and indicated that the artefacts could easily be misidentified as genuine image features in reconnaissance imagery. The IAs also indicated that they could, to a limited extent, look through the blocking, apparent in highly compressed JPEG imagery depending on where the block boundary was placed relative to the image feature of interest. It was generally recognised that interpreting compressed imagery would become easier with experience of the artefact characteristics that a particular compression algorithm can produce.

Conclusions

This paper has characterised the degradation of imagery under compression across the spectrum of interpretable NIIRS levels. The influence of image-feature content within the imaged scene and field of view at higher NIIRS levels on the accuracy of NIIRS rating has been demonstrated. A number of scalar, image quality measures have been shown to provide accurate correlation with the perceived subjective degradation of imagery. These measures may offer the potential of automatic application of compression to satisfy end-user NIIRS requirements. However, the value and robustness of HVS weighted scalar measures has not been proven in this study.

References

1 W. Osberger, N. Bergmann, *An Automatic Image Quality Assessment Technique Incorporating Higher Level Perceptual Factors*, submitted to Signal Processing, available at internet URL http:// www.scsn.bee.qut.edu.au /~wosberg/pub.htm.

2 M. G. Ramos, S. S. Hemami, *Activity selective SPIHT coding* to be published in the Proceedings of SPIE 11th International Symposium of Electronic Imaging January 1999.

3 I. Rabinovitch, A. N. Venetsanopoulos, *High Quality Image Compression Using the Wavelet Transform* Proc ICIP 1997 p283-286.

4 IRARS, NIIRS: Civil NIIRS Reference Guide, March 1996, http://www.fas.org/irp/imint/niirs_c/index.html.

5 MIL-STD-188-198A *Joint Photographic Experts Group (JPEG) Image Compression for the National Imagery Transmission Format Standard*, December 1993. http://164.214.2.51/ntb/ntb_docs/status.html#baseline.

6 STANAG 4545, Edition 1, Ratification Draft 2, NATO Secondary Imagery Format(NSIF) http://www.nato.int/docu/standard.html, June 1998.

7 National Imagery and Mapping Agency (NIMA) National Imagery Transmission Format Standard-Image Compression Users Handbook Final Draft 4th June 1998. http://164.214.2.51/ntb/ntb_docs/status.html#baseline.

8 IJPEG internet site: ftp://ftp.net.com.com/pub/tg/tgl/jpg version 6 of the IJPEG codec.

9 http://www.cs.dartmouth.edu/~gdavis/wavelet/wavelet.html version 0.3 of the Baseline Wavelet Transform Coder Construction Kit. 01/29/97.

10 ftp://isdl.ee.washington.edu/pub/VQ/code/stdvq Vector Quantisation Source code.

11 A. M. Eskicioglu, P. S. Fisher. *Image Quality Measures and their Performance*, IEEE Trans. on Communications, Vol 43, No. 12, December 1995.

12 P. M. Farrelle. Recursive Block Coding for Image Data Compression. Springer-Verlag Books. pp104-108.

Figure 1 NIIRS image 1

Figure 2 NIIRS image 2

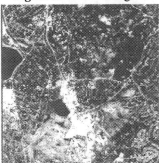

Figure 3 NIIRS image 3a

Figure 4 NIIRS image 3b

Figure 5 NIIRS image 4a

Figure 6 NIIRS image 4b

Figure 7 NIIRS image 5a

Figure 8 NIIRS image 5b

Figure 9 NIIRS image 6a

Figure 10 NIIRS image 6b

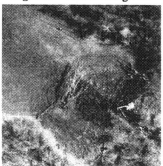

Figure 11 NIIRS image 7a

Figure 12 NIIRS image 7b

Figure 13 NIIRS image 8a

Figure 14 NIIRS image 8b

Figure 15 NIIRS image 9a

Figure 16 NIIRS image 9b

Table 1. Scalar Image Quality Measures

Image Quality Measure	Formula
1 MSE	$\sum\limits_{i=1}^{M}\sum\limits_{j=1}^{N}[f(i,j)-f'(i,j)]^2 / MN$
2 PSNR	$10.0 * \log_{10}\dfrac{255^2}{MSE}$
3 Average Distance	$\sum\limits_{i=1}^{M}\sum\limits_{j=1}^{N}[f(i,j)-f'(i,j)] / MN$
4 Structural Content	$\sum\limits_{i=1}^{M}\sum\limits_{j=1}^{N}f(i,j)^2 / \sum\limits_{i=1}^{M}\sum\limits_{j=1}^{N}f'(i,j)^2$
5 Normalised Cross Correlation	$\sum\limits_{i=1}^{M}\sum\limits_{j=1}^{N}f(i,j)f'(i,j) / \sum\limits_{i=1}^{M}\sum\limits_{j=1}^{N}f(i,j)^2$
6 Correlation Quality	$\sum\limits_{i=1}^{M}\sum\limits_{j=1}^{N}f(i,j)f'(i,j) / \sum\limits_{i=1}^{M}\sum\limits_{j=1}^{N}f(i,j)$
7 Maximum Difference	$\max\{|f(i,j)-f'(i,j)|\}$
8 Image Fidelity	$1-\left[\sum\limits_{i=1}^{M}\sum\limits_{j=1}^{N}[f(i,j)-f'(i,j)]^2 / \sum\limits_{i=1}^{M}\sum\limits_{j=1}^{N}f(i,j)^2\right]$
9 Laplacian MSE*	$\sum\limits_{i=2}^{M-1}\sum\limits_{j=2}^{N-1}\left[O(f(i,j))-O(f'(i,j))\right]^2 / \sum\limits_{i=2}^{M-1}\sum\limits_{j=2}^{N-1}\left[O(f(i,j))\right]^2$
10 Peak MSE	$\dfrac{1}{MN}\sum\limits_{i=1}^{M}\sum\limits_{j=1}^{N}\left[f(i,j)-f'(i,j)\right]^2 / \left[\max(f(i,j))\right]^2$
11/12/13 Normalised Absolute Error*	$\sum\limits_{i=1}^{M}\sum\limits_{j=1}^{N}|O(f(i,j))-O(f'(i,j))| / \sum\limits_{i=1}^{M}\sum\limits_{j=1}^{N}|O(f(i,j))|$
14/15/16 Normalised MSE*	$\sum\limits_{i=1}^{M}\sum\limits_{j=1}^{N}\left[O(f(i,j))-O(f'(i,j))\right]^2 / \sum\limits_{i=1}^{M}\sum\limits_{j=1}^{N}\left[O(f(i,j))\right]^2$
17/18/19/20/21 Lp Norm*	$\left[\dfrac{1}{MN}\sum\limits_{i=1}^{M}\sum\limits_{j=1}^{N}|f(i,j)-f'(i,j)|^p\right]^{\frac{1}{p}}$ $\qquad$ p = 1,2,3

* Note that for LMSE $O(f(i,j)) = f(i+1,j)+ f(i-1,j)+ f(i,j+1)+ f(i,j-1)-4\ f(i,j)$. For NAE, NMSE and L_2-norm $O(f(i,j))$ is defined in 3 ways: 1) $O(f(i,j)) = f(i,j)$ 2) $O(f(i,j)) = f(i,j)^{1/3}$ 3) $O(f(i,j)) = H((u^2+v^2)^{1/2})f(i,j)$ where u and v are co-ordinates in the DCT transform domain,

$r = (u^2+v^2)^{1/2}$ and $H(r) = \begin{cases} 0.05 e^{r^{0.554}} & \text{for } r < 7 \\ e^{-9[|\log_{10} r - \log_{10} 9|]^{2.3}} & \text{for } r \geq 7 \end{cases}$

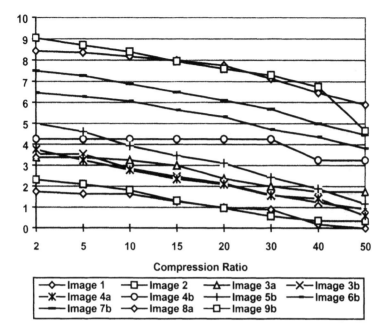

Figure 17 NIIRS degradation under JPEG compression

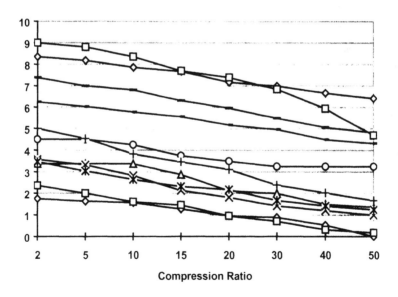

Figure 18 NIIRS degradation under Wavelet compression

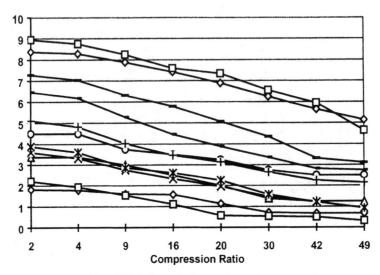

Figure 19 NIIRS degradation under VQ compression

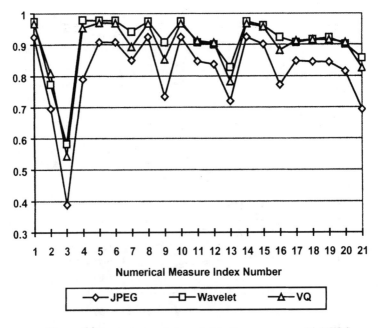

Figure 20 Average correlation of objective measures with NIIRS

Pyramidal Seeded Region Growing Algorithm and Its Use in Image Segmentation

Zoltan Tomori, Jozef Marcin, Peter Vilim

Institute of Experimental Physics, Slovak Academy of Sciences,
Watsonova 47, 04353 Kosice, Slovak Republic
tomori@saske.sk

Abstract. Improvement of "Seeded Region Growing" (SRG) segmentation algorithm based on the pyramidal representation of image is described. Segmentation starts from the proper coarse level of pyramid using seed points chosen by the operator. Segmented contours are projected to the level below. On each subsequent level, SRG algorithm is applied only to pixels inside the window of variable size near the projected contour which leads to the linear dependence of execution time on the image size. Implementation exploiting the graphic user interface allows various forms of the interactive control of image segmentation.

1 Introduction

Segmentation is one of the most challenging problems of digital image processing. Methods based on the principles of watershed algorithm [1] where gradient image can be viewed as a topography achieved the popularity during this decade. Boundaries of segmented regions are determined by ridges which guaranties that the segmented region contour will be always closed.

The similar principle was exploited in "Seeded Region Growing" algorithm (SRG) [2]. The seed points value represents the initial value of each growing region. Regions grow "in parallel" and points where they touch each-other are labeled as boundary points. The parallel grow of regions is simulated by the sequential algorithm with some drawbacks like that the order of scanned pixels has influence on the final shape of segmented boundary. These drawbacks were analyzed in [3] and "Improved Seeded Region Growing" (ISRG) algorithm was proposed in which the results of segmentation are independent on the order of pixel evaluation. However, ISRG doesn't improve SRG from the point of view of execution time. Therefore, our goal was mainly to improve the speed of segmentation. This attempt resulted in an algorithm called "Pyramidal Seeded Region Growing" (PSRG) which is the subject of this contribution. It is based on the hierarchical (pyramidal) representation of image.

Pyramid is such a data structure where the original image represents the base of pyramid (level 0) and each subsequent level lying above the base is the coarser version of the previous one. The simplest form is pyramid 2x2, where each point with coordinates (i,j) of level k is the average of corresponding 4 points from level $k-1$.

$$P_{k,i,j} = \frac{1}{4} \sum_{r=0}^{1} \sum_{s=0}^{1} P_{k-1,2i+r,2j+s} \quad . \tag{1}$$

Top level is represented by single pixel (root). Fig. 1 shows selected levels of pyramid of thermoelectrophoresis image magnified to their original size. Among many algorithms exploiting pyramidal representation, the most valuable ones (from the point of view of speed) are these allowing top-down flow, where the preliminary computation is performed on upper (coarse) levels and results are progressively improved "downwards" the pyramid.

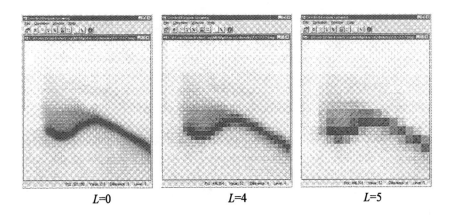

| $L{=}0$ | $L{=}4$ | $L{=}5$ |

Fig. 1. Selected pyramid levels ($L{=}0$, 4, 5) of the thermoelectrophoresis image. Image on each level L is proportionally enlarged to keep the size of displayed level constant.

2 Pyramidal Seeded Region Growing Algorithm (PSRG)

Let DoSRG is original SRG algorithm applied to the image on level k of the pyramid. Let SSL is sequentially sorted list of pixels ordered according to the similarity value

$$\delta(x) = |g(x) - \text{mean}(A_i)| \quad . \tag{2}$$

where $g(x)$ is the gray level of the pixel x and A_i is i-th growing region connected with pixel x. (See [2] for details).

```
Algorithm DoSRG(k)
    Label seed points
    Put neighbors of seed points into the SSL
    while the SSL is not empty do
        Remove first point Y from SSL
        if all neighbors of Y which are already
        labeled (with other than the boundary
        label) have the same label then
            Set Y to this label
            Update mean of corresponding region
            Add neighbors of Y which are neither
            already set nor already in the SSL to
            the SSL according to their similarity
        otherwise
            Flag Y with the boundary label
    endwhile
end.
```

Our pyramidal version of algorithm (PSRG) can be described as follows:

```
Algorithm PSRG()
    Build the pyramid of intensity images P
    Select the proper starting level SL
    DoSRG(SL)
    for k=SL downto 1 do
        Project labels from level k to k-1
        Update all regions
        for each pixel Y on level k do
            if Y is labeled as border then
                for each pixel X from window Y ± w
                    Unlabel X
                    Update running mean of region
                endfor
            endif
        endfor
        for each pixel Y on level k do
            if Y is unlabeled then
                Test its neighbors
                if neighbor X of Y has a region label
                    Unlabel X
                    Put X into SSL according to δ(X)
                    Update running mean of region
                endif
            endif
        endfor
        DoSRG(k-1)
    endfor
end.
```

Pyramidal approach to SRG algorithm is based on the successive refinement of the objects contours segmented on the coarser representation of the image. Only pixels in the vicinity of contours segmented on the previous (coarser) level are evaluated. Procedure DoSRG is applied just to this reduced set of pixels. This process is repeated on each pyramid level until the full-resolution image is reached. Subroutine DoSRG is either original SRG described above or the improved one (ISRG) published in [3].

The general problem of any pyramidal edge detection algorithm is the deformation of contour when performed on coarse level. Coarse-to-fine tracking is therefore a kind of operation "reverse" to pyramid building when algorithm tries to find the exact edge points positions near the contour predicted by the coarse level. The question is how large this fine-search window should be. In some algorithms is this size given by the type of pyramid (e.g. overlapped pyramid 4x4 in [4] ,[5]). Calculations in [6] showed that the contour deformation between two successive levels is within ± 2 pixels range assumed that each coarse level is blurred version of the previous one using Gaussian filter with $\Delta\sigma = 1$.

Non-overlapped pyramid 2x2 has been exploited in PSRG and the search area in the vicinity of predicted contour is determined by the area consisting of pixels unlabeled by an "unlabeling window". Each pixel outside this area takes label from its parent. There exists the built-in conflict in PSRG between the degree of image smoothing (start level) and the precision of final detection. More coarse levels of pyramid concentrate the global information. It depends on our requirements how much we will force algorithm to respect their estimation.

It is desired to suppress unnecessary details of contour shape (along with the noise) and then the size of unlabeling window $w=4$ pixels is sufficient (2 sons + 2 pixels from both sides). On the other side, such limit leads to the poor detection of objects with elongated parts (Fig. 2a) and therefore the variable unlabeling window size is necessary (Fig. 2b). The criterion for the determination of unlabeling window size should reflect the homogeneity of corresponding region. If this region is inhomogeneous enough, border pixels can be projected correctly and unlabeling window can remain small, otherwise its size is increased until the corresponding area becomes inhomogeneous.

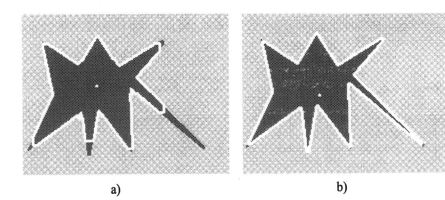

a)   b)

Fig. 2. Effect of unlabeling window size on objects with elongated parts. a) constant size 4x4 b) algorithm controlled variable size. White spot inside the dark region is seed point

3 Interactive Approach to Segmentation Using PSRG

A number of practical problems of image segmentation can be efficiently solved only with the assistance of human operator. An interactive principle is included on all versions of SRG algorithm, because the initial step is the manual enter of seeds. Our goal was to allow some other forms of interactive pre-requisite, intermediate and post-processing manipulation.

The most important step in the pre-processing phase is the selection of proper initial level which can be supported as follows:

- operator can control the displayed level of pyramid using keyboard arrows,
- the cursor size is changed proportionally to the pixel size on current level,
- when operator moves cursor on some level, the information about the underlying region homogeneity is shown continuously. This helps the operator to set the seed position which represents homogeneous region with the prospect of growth.

Post-processing interactive correction can perform additional segmentation of area which was detected as one object but evidently consists of more objects. Additional seed points can refine the original segmentation. If the gradient between two objects is too low then algorithm is very often unable to locate border according to operator's expectation. However, operator can permit post-segmentation growth of preferred object. Ambiguous area can be interactively erased and SRG is applied to the erased area again. Seed points are kept unchanged, but similarity criterion is changed as follows:

$$\delta_i(x) = c_i \left| g(x) - \text{mean}(A_i) \right| \ . \tag{3}$$

where c_i is user supplied coefficient. Smaller weights should be used for objects with preferred growth. The result of post-processing correction of segmented area is shown in Fig. 3.

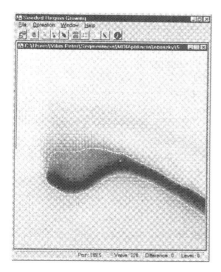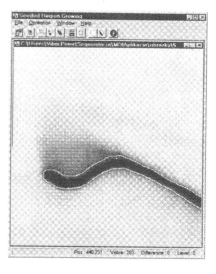

Fig. 3. Result of the region weight change in the post-processing phase

4 Implementation

Programs were written using Borland C++ 5.02 exploiting Object Windows Library for GUI control and Borland container classes library which allows simple programming of data structures like SSL (Sequentially Sorted List).

Algorithms has not been optimized from the point of view of speed yet. Optimization of operations with SSL (memory management) which consumes most of operation time should substantially improve execution time in Table 1.

PSRG has been implemented in two versions as a stand-alone Windows program and as a plug-in module of free image processing program ImageTool (developed at the University of Texas, San Antonio, available from [7].

5 Discussion and Conclusions

This contribution deals with improvement of "Seeded region growing" algorithm accomplished by its pyramidal version with interactive control options.

Pyramidal approach leads to O(log n) execution time as can be expected in algorithm based on hierarchical top-down principle. Table 1 compares execution time starting from various levels of hierarchical image representation. It implies that the execution time in PSRG depends linearly on image size and in the case of traditional SRG (equal to PSRG starting from level 0) is this dependency quadratic.

However, this rule is true only for the relatively small number of objects. In fact, execution time is proportional to the sum of perimeters of all contours detected on the pyramid. For many small regions this time can be higher then in original SRG. This limit is less substantial if we take into account that SRG itself usually requires manual enter of seed points which automatically limits its exploitation for images containing many objects.

There are also another factors influencing PSRG speed. Noise-free image allows immediate processing of pixels near the growing region without the need to keep them in the SSL. Short SSL list means efficient sorting, which is the most time consuming part of algorithm. Reduced number of gray levels has the similar positive effect to the SSL length. Elongated objects (or its parts) requires larger unlabeling window which has the negative influence to algorithm speed, too.

Fig. 4 shows the results of segmentation for 256 x 256 image from Table 1. The same seed positions was used for all cases. Only the minor differences between results achieved by PSRG and SRG algorithms are visible.

Proper attention has been paid to the selection of starting level. Very coarse starting level leads to the destructive loss of details and finally to wrong segmentation, unnecessary fine starting level decreases the efficiency of PSRG. The general rule for the optimal selection of start level cannot be found because problem is task-dependent and image-dependent. Some auxiliary information (e.g. Interest measures in [4]) can help us to make a decision and we implemented them into the interactive part of algorithm. Therefore, interactive supervision is not only a question of program "look", but represents the way to accomplish successful segmentation.

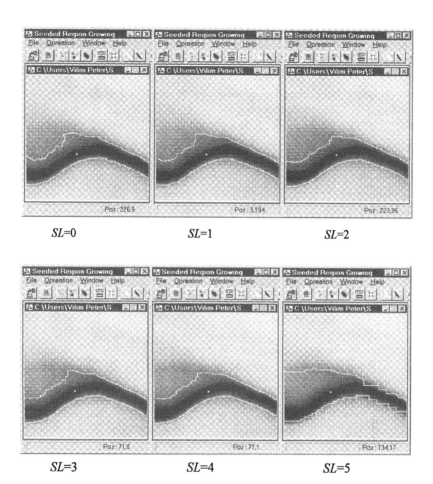

Fig. 4. Results of segmentation for PSRG started on different *SL (start level)*. If *SL* = 0, then PSRG is reduced to original SRG algorithm without exploitation of pyramid.

	SL = 0	SL = 1	SL = 2	SL = 3	SL = 4	SL = 5
256 x 256	12.5	1.5	1.0	0.52	0.45	0.32
512 x 512	184	16.9	5.6	4.6	2.6	0.86

Table 1. Dependence of the execution time on *SL (start level)*. Time is given in [s], *SL* value is equal to the number of hierarchy levels used in PSRG. *SL*=0 corresponds to traditional SRG without exploitation of pyramidal image representation.

If SRG represented the novel approach to segmentation, ISRG improved the precision keeping the speed almost constant, PSRG presented here augments both above mentioned methods and increases its speed. In addition, PSRG forces contours to keep the approximate shape given by coarse level which can partly remove unnecessary details (see low-gradient part of image on Fig. 4). If details are desired, PSRG should start from less-smoothed level.

This work was supported by the grant # 2/5142/98 of Slovak Grant Agency VEGA.

References

1. Meyer, F – Beucher, S.: Morphological segmentation. J. Visual Communication and Image Representation 1 (1990) 21-46.
2. Adams, R – Bischoff, L.: Seeded region growing. IEEE Trans on Pattern Analysis and Machine Intelligence PAMI-16 (1994) 641-647.
3. Mehnert, A. – Jackway, P.: An improved seeded region growing algorithm. Pattern Recognition Letters 18 (1997)1065-1071.
4. Gross, A. D. – Rosenfeld, A.: Multiresolution object detection and delineation. Computer Vision, Graphics and Image Processing 39, (1987) 102-115
5. Tomori, Z.: Border detection of the object segmented by the "pyramid linking" method. IEEE Transactions on Systems, Man and Cybernetics SMC-25 (1995) 176-181.
6. Bergholm, F.: Edge focussing. IEEE Trans on Pattern Analysis and Machine Intelligence PAMI-9 (1987) 726-741.
7. ftp://maxrad6.uthscsa.edu.

Multi-image Region Growing
for Integrating Disparity Maps

Uğur M. Leloğlu[1,2] and Uğur Halıcı[2]

[1] TUBİTAK-BİLTEN,
ODTÜ, 06531, Ankara, Turkey
tlf: +90 312 210 13 10, fax: +90 312 210 13 15
`leloglu@image.bilten.metu.edu.tr`
[2] Computer Vision and Neural Networks Research Group
Middle East Technical University, 06531, Ankara, Turkey
`halici@rorqual.cc.metu.edu.tr`

Abstract. In this paper, a multi-image region growing algorithm to obtain planar 3-D surfaces in the object space from multiple dense disparity maps, is presented. A surface patch is represented by a plane equation and a set of pixels in multiple images. The union of back projections of all pixels in the set onto the infinite plane, forms the surface patch. Thanks to that hybrid representation of planar surfaces, region growing (both region aggregation and region merging) is performed on all images simultaneously. Planar approximation is done in object space by linear least square estimation using all data points of the region under question in all images. Linear edge segments detected on colour images are used for constraining the region growing during the region aggregation phase as well as for detection of borders of surface patches. Experimental results on disparity maps obtained from high-resolution aerial images of urban areas demonstrate the performance of the algorithm.

1 Introduction

Reaching at meaningful 3-D surface descriptions from a depth map (either from a laser scan camera or from a stereo pair) is one of the challenging problems of computer vision. An important phase of this 3-D reconstruction process is the segmentation of the image into regions which correspond to continuous surfaces. Once the image is segmented, each surface can be modelled by parametric surface models or by implicit functions.

When multiple depth images are available, we have more information about the underlying structure: First of all, some surfaces that cannot be seen in an image may be seen in another one. Using sufficient number of views with various viewing angles, we may even have a complete description of all visible surfaces. Secondly, more measurements are available, so parametric surfaces are statistically more accurate. Besides, a point to which a disparity value cannot be assigned in a stereo pair due to weak texture or matching ambiguities can have a disparity value in another pair, so we have less "holes" to guess for.

However, the introduction of multiple views makes the problem more complicated due to several reasons. If the images are segmented, the fusion of surface patches becomes difficult, because the segmentations, in general, are not stable across views. An alternative is to try to fit surfaces to a cloud of 3-D data points formed by measurements from all views. But in that case, useful information about the connectivity of the data points is discarded.

In this paper, we present a method to obtain 3-D surfaces from multiple registered dense disparity maps using region growing (both region aggregation and region merging) in multiple images simultaneously. As a result, segmented regions are consistent across all views. By means of a hybrid representation, we can benefit from intensity edges while we perform 3-D reconstruction in the object space.

This paper is organised as follows: In the next section, some related work is summarised. In section 3, the multi-image region growing algorithm is described. Section 4 shows some of the experimental results on disparity maps obtained from real images and Section 5 concludes the paper.

2   Related Work

Region growing is used frequently for range image segmentation [11], sometimes with the contribution of edge detection[13]. Algorithms for range image segmentation are not always directly applicable to disparity maps. Even in area-based stereo, there are large regions where a disparity value cannot be assigned due to weak texture, occlusion or matching ambiguities. Some other common problems are false matches, disparity quantisation and artefacts caused by large correlation windows.

[1] uses region growing and edge detection for segmentation from stereo. Using the segmentation results, some of the intensity edges are classified as depth discontinuities. Then all boundary line segments belonging to the same region are linked so as to form a closed contour.

Another approach to disparity map segmentation is to accept colour image segmentation as initial disparity map segmentation. After planar approximation in each segment, region merging is applied considering the angle between normals of two neighbouring regions, the height difference along the border separating them and their colorimetric similarity[4]. A similar region merging method is presented in [8]. Disparity map is segmented into planar regions using histogram-based clustering, then some of the small regions are merged to neighbouring large regions depending on the orientation angles and average surface distance.

For integrating range data from multiple sources, there are several approaches. One idea is first to detect depth discontinuities in each range image and then to fuse the resulting 3-D models (generally triangulated meshes) in the object space [6], [12]. This method is used when dense and accurate depth data is available.

In integration of disparity maps, another possibility is to track matches in disparity maps obtained from successive images in a video sequence [9]. As a result, outliers are detected and, by the help of Kalman filtering, the noise is reduced. Small surface gaps are interpolated using a parametric surface model. This method yields very successful results, but it requires large number of sequential images where the distances between successive viewpoints are much smaller than the average scene depth.

Another approach is to project all data points into the object space and then to fit surfaces to resulting 3-D point clouds [2], [5], [7]. In that case, we lose the connectivity of 3-D points in the images and we cannot benefit from edges.

3 Multi-image Region Growing

3.1 Surface Representation

Man made environments contain mostly planar surfaces. For that reason, we approximate the surfaces with planar surface patches in 3-D. This polyhedral world assumption may be too strong, but, the region growing algorithm presented can be extended to higher-order surfaces as done in [11]. A planar patch is defined by the equation

$$a\,x + b\,y + c\,z + d = 0 \tag{1}$$

describing an infinite plane in the object space and a set of pixels on views. The union of intersections of the infinite plane with the volume defined by each pixel.

3.2 The Algorithm

We assume that we have multiple disparity maps of the same area as well as the intensity images. As long as the calibration parameters are known, the camera locations and orientations may be arbitrary.

A block diagram of the algorithm is depicted in Figure 1. Each step of the algorithm is explained in detail in the following subsections.

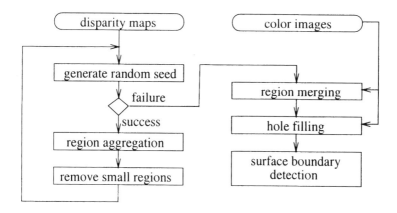

Fig. 1. Block diagram of the region growing and 3-D reconstruction algorithm

3.3 Choice of Seed Points

Each iteration begins with the choice of a seed point at a random coordinate of a random image. If the point does not satisfy the following conditions, another seed point is chosen:

- The seed point should not be on or adjacent to an edge point.
- The seed point should not be on or adjacent to a pixel that is not assigned a disparity value.

If a valid or successful (i.e. leading to a large region) seed point cannot be found after a certain number of trials, the region aggregation phase is terminated.

3.4 Region Aggregation

First, an initial plane is calculated using the seed point and its 8-neighbours. At each image, we consider the neighbouring pixels of the region that has a disparity value and that are not yet assigned to any region. If the distance of the corresponding 3-D point to the plane is smaller than a threshold, then that pixel is added to the current region. In that case, the projections of the point where the line and the point intersect on the other images are also calculated. If a close depth value is found, that pixel is also included in the region.

During region aggregation, the region is not allowed to cross edges. Edges may correspond to surface-reflectance discontinuities or illumination discontinuities as well as depth or orientation discontinuities. In the first case, the use of edges may divide a surface into several partitions, which is undesired, but such regions are merged later. On the other hand, this constraint prevents generation of wrong planar approximations when the seed point is near a roof edge (orientation discontinuity) or a jump edge (depth discontinuity). Besides, it prevents the region to invade into neighbours several pixels, depending on the angle between the surfaces, if the region cannot find a gap to leak into the other side.

After each pass, the planar approximation is updated. The approximation is realised by least squares estimation that minimise the smallest distance from the plane to points belonging to the region in all views. Since outliers are not added to the set of points, we do not need to use complicated methods for planar approximation like least-median-of-squares[4].

When the region cannot grow any more, the expansion of that region is terminated. If the region is too small (less than 50 pixels in our implementation) it is removed, because such regions generally correspond to very rough surfaces that should be modelled using higher order surfaces.

Note that the segments generated may have holes, they may not even be connected in some views. This is not a weakness but a virtue of the algorithm. Because, we know that unconnected parts of a region actually belong to the same surface since they are connected in some other view.

3.5 Region Merging

Since depth maps may be oversegmented because of edges that constrain region aggregation or because of noise, two regions that satisfy the following requirements are merged:

- The two regions must be neighbours in at least one of the images.
- If both regions are large (i. e. larger than 1500 pixels),
  - The angle between the normals should be small[4].
  - The distance of the centre of gravity of each region to the plane of the other should be small.
- If only one of the regions is large, the distance of the centre of gravity of the smaller region to the plane of the larger one should be small.
- Each surface patch defines a mapping among the pixels in each view. Sum-of-squared-differences (SSD) is calculated on the intensity images for both regions. Then we assume that the regions are merged and we do a planar approximation for the merged region. The SSD calculated for the merged region should not be greater than the sum of SSDs of the two regions.

3.6 Hole Filling

Because of missing data, there may be holes in the regions. There are two possibilities for such a hole. Either it is actually part of the region but could not be included due to missing depth values or noise, or there is another surface whose projection is inside the hole. To decide which is the case, each hole is tested to see if it satisfies the following conditions:

- The average of squared differences between the plane and available data points inside the hole should be small.
- The SSD of intensity images is evaluated over the hole assuming that the hole is part of the region. The SSD should be small.

If the hole corresponds to another surface (like a chimney on a roof) then it is very likely that at least one of the above conditions is violated.

3.7 Surface Boundary Detection

Two different mechanisms are employed to detect surface boundaries precisely. Firstly, in each image, the intersections of the planes corresponding to neighbouring regions are calculated and these 3D lines are projected to the image plane[8]. Any matching 2D line segment is assumed to be the line separating the regions. This way, roof edges are detected.

Secondly, each 2D line segment in each image is tested to see if it corresponds to a jump edge in that image. For that purpose, we use a method similar to that of [1]. At each side of the edge, along a strip, several pixels away from the edge, the histogram of regions is calculated (see Figure 2). The peak of the histogram is assumed to correspond to the main region at that side of the edge. If the height difference of the planes along the line is large [4], then the line is assumed to be a jump edge in that image and to be part of the boundary of the upper surface.

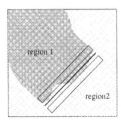

Fig. 2. Strips at two sides of an edge segment.

4 Results

For the experiments, we have used disparity maps obtained from high-resolution aerial images of urban areas generated by a multi-view stereo algorithm [10]. Figure 3a and Figure 3b show one of the images and the corresponding disparity map, respectively. Two such disparity maps of the same area are used to generate the 3-D model. For detecting linear edge segments in the colour images FEX program is used [3].

Figure 3c depicts the results of pixel aggregation. The segmentation map after region merging and hole filling is shown in Figure 3d. Figure 4 shows detected discontinuities projected on the original colour image.

5 Conclusions

In this paper, a new method for segmenting multiple disparity maps simultaneously and for obtaining 3-D surface models, is presented. Planes are fitted to each segment and, using the linear edge segments detected on colour images, most of the surface boundaries are also detected. The performance of the algorithm is demonstrated on real images.

Further research is going on for detecting closed boundaries of all surface patches and for generating visually pleasant VRML models.

References

1. L.-H. Chen and W.-C. Lin. Visual surface segmentation from stereo. *Image and Vision Computing*, 15:95–106, 1997.
2. P. Fua. From multiple stereo views to multiple 3-d surfaces. *International Journal of Computer Vision*, 24(1):19–35, 1997.
3. C. Fuchs and Stephan Heuel. Feature extraction. In W. Förstner, editor, *Proc. of Third Course in Digital Photogrammetry*, Bonn, Germany, 1998.
4. S. Girard, P. Guérin, M. Roux, and H. Maître. Building detection from high resolution colour images. In *International Symposium on Remote Sensing, Europhoto98*, Barselona, Spain, 1998.
5. A. A. Goshtasby. Three-dimensional model construction from multi-view range images: Survey with new results. *Pattern Recognition*, 31(11):1705–1714, 1998.

Fig. 3. a) *up left:* One of the three views of the same area (1024×1024). Ground resolution is approximately 8cm×8cm. Courtesy of Eurosense. b) *up right:* Corresponding disparity map. Darker points are closer to the camera. Unassigned pixels are white. c) *down left:* Segmentation results after region aggregation. Each segment has a different random color. d) *down right:* Segmentation results after region merging and hole filling.

Fig. 4. Discontinuities projected on the original color image.

6. A. Hilton, A. J. Stoddart, J. Illingworth, and T. Windeatt. Implicit surface-based geometric fusion. *Computer Vision and Image Understanding*, 69(3):273–291, 1998.
7. H. Hoppe, T. DeRose, T. Duchamp, J. McDonald, and W. Stuetzle. Surface reconstruction from unorganised points. In *Computer Graphics 26: SIGGRAPH'92 Conference Proceedings*, volume 26, pages 71–78, 1992.
8. R. Koch. Surface segmentation and modeling of 3-d polygonal objects from stereoscopic image pairs. In *International Conference on Pattern Recognition 96*, Vienna, Austria, 1996.
9. R. Koch, M. Pollefeys, and L. Van Gool. Multi viewpoint stereo from uncalibrated video sequences. In *Proc. ECCV'98, LNCS*, Freiburg, Germany, 1998. Springer-Verlag.
10. U. M. Leloğlu, M. Roux, and H. Maître. Dense urban dem with three or more high-resolution aerial images. In D. Frisch, M. Englich, and M. Sester, editors, *GIS - Between Visions and Applications*, pages 347–352, Stuttgart, Germany, 1998. ISPRS.
11. G. Maître, H. Hügli, and J. P. Amann. Range image segmentation based on function approximation. In A. Gruen and E. Baltsavias, editors, *Close Range Phogrammetry Meets Machine Vision, SPIE vol. 1395*, pages 275–282, 1990.
12. C. Schütz, T. Jost, and H. Hügli. Free-form 3d object reconstruction from range images. In *The Proceeding of Virtual Systems and Multimedia, VSMM97*, Geneva, Switzerland, 1997.
13. D. Zhao and X. Zhang. Range-data-based object surface segmentation via edges and critical points. *IEEE-PAMI*, 6(6):826–830, 1997.

Contrast Enhancement of Gray Scale Images Based on the Random Walk Model

Bogdan Smolka   and   Konrad W. Wojciechowski

Dept. of Automatics Electronics and Computer Science
Silesian University of Technology
Akademicka 16 Str, 44-101 Gliwice, Poland
bsmolka@peach.ia.polsl.gliwice.pl

Abstract In this paper a new approach to the problem of contrast enhancement of gray scale images is presented. The algorithms introduced here are based on a model of a virtual particle, which performs a random walk on the image lattice. It is assumed, that the probability of a transition of the walking particle from a lattice point to a point belonging to its neighbourhood is determined by the Gibbs distribution, defined on a specified neighbourhood system.

In this work four algorithms of contrast enhancement are presented. The first algorithm traces the visits of the walking particle and determines their relative frequencies. The second operator assigns to each lattice point the probability of a stationary Markov chain, generated by the trajectory of the randomly walking particle. The third algorithm is based on a concept of a jumping particle and the last one uses the information contained in the statistical sum of the Gibbs distribution of the transition probabilities.

1 Introduction

Contrast enhancement is one of the main fields of the low-level image processing. The aim of the methods of image enhancement is the improvement of the visual appearance of an image and its conversion to a form better suited for analysis by a computer vision system. The commonly used techniques can be divided into [1–6] :

• *spatial domain operations* (convolutions with high-pass filter masks, unsharp masking, inverse contrast ratio mapping, local adaptive contrast enhancement, histogram transformations) and

• *frequency domain operations* (high-pass filtering through the manipulation of the Fourier transforms, homomorphic filters, Wiener filters).

The techniques of image enhancement perform quite well on images with uniform spatial distribution of gray values. Difficulties arise when the object and background assume broad range of gray tones or when the background brightness is strongly nonhomogeneous.

The choice of a particular enhancement method is a difficult task, since there exists a whole variety of different algorithms. Additionally, it is difficult to compare the different methods, as there is no generally accepted criterion of image quality. Moreover, the choice of a particular method must be adequate to the character of the image deterioration.

This situation promotes the development of new universal methods, which could be applied independently of the way in which the image quality was degraded. For this reason, new algorithms of contrast enhancement, based on a concept of a randomly walking particle have been developed.

2   Random Walk Model

Let the image be represented by a matrix B of size $N_r \times N_c$, $B = \{B(i,j), i = 1, 2, \ldots, N_r, \ j = 1, 2, \ldots, N_c\}$ and let the distance between the points (i,j) and (k,l) be determined by the metric $\rho_{(i,j),(k,l)} = \max\{|i-k|, |j-l|\}$. Assuming that the neighbourhood of a given point (i,j) consists of points (k,l), which satisfy the condition $\rho_{(i,j),(k,l)} \leq \lambda$, let us assume that the the virtual particle can visit its neighbours or stay in its temporary position. Figure 1 shows the particle's trajectory for the neighbourhood system defined by $\lambda = 1$ (eight-neighbourhood) and $\lambda = 2$. In this work it was assumed, that the particle moves on the image

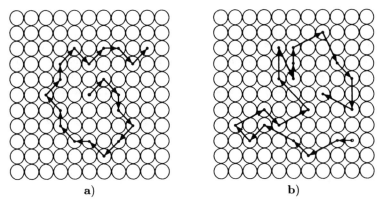

a)   b)

Fig. 1. Random walk in case of two neighbourhood systems: $\lambda = 1$ a) and $\lambda = 2$ b).

lattice with probabilities derived from the Gibbs distribution formula

$$
p_{ij,kl} = \begin{cases} \dfrac{\exp\left\{\frac{V(i,j)-V(k,l)}{T}\right\}}{\sum\limits_{m=-\lambda}^{m=\lambda} \sum\limits_{n=-\lambda}^{n=\lambda} \exp\left\{\frac{V(i,j)-V(i+m,j+n)}{T}\right\}} & : \ \rho_{(i,j),(k,l)} \leq \lambda \\[4mm] 0 & : \ \rho_{(i,j),(k,l)} > \lambda \end{cases} \quad , \quad (1)
$$

where $V(i,j)$ is the potential energy of (i,j) and T plays the role of the temperature of the statistical system. In this work $V(i,j)$ is represented by the gray scale value of the image point (i,j).

Let $p_{ij,kl}^{(n)}$ denote the transition probability, that the virtual particle starting from the point (i,j) visits in the n-th step the point (k,l). Then the transition matrix Π is regular, since every point of the image lattice is accessible from any other point in fewer than $N_r + N_c$ steps. For the eight-connectivity system the matrix Π is is also aperiodic and there exists a limiting matrix Π^* which satisfies $\lim_{n \to \infty} \Pi^n = \Pi^*$.

If the positive elements of the Π matrix are denoted by • and the zero-elements by ○, then the transition matrix for the eight-connectivity system on a toroidal lattice takes the form

$$
\Pi = \begin{bmatrix}
XX\,0\,0\,0\,0\cdots0\,0\,0\,0\,0\,X \\
XXX\,0\,0\,0\cdots0\,0\,0\,0\,0\,0 \\
0\,XXX\,0\,0\cdots0\,0\,0\,0\,0\,0 \\
\cdots\cdots\cdots\cdots\cdots\cdots\cdots\cdots \\
\cdots\cdots\cdots\cdots\cdots\cdots\cdots\cdots \\
0\,0\,0\,0\,0\,0\cdots0\,0\,XXX\,0 \\
0\,0\,0\,0\,0\,0\cdots0\,0\,0\,XXX \\
X\,0\,0\,0\,0\,0\cdots0\,0\,0\,0\,XX
\end{bmatrix}, \quad
X = \begin{bmatrix}
\bullet\,\bullet\,\circ\,\circ\,\circ\,\circ\cdots\circ\,\circ\,\circ\,\circ\,\circ\,\bullet \\
\bullet\,\bullet\,\bullet\,\circ\,\circ\,\circ\cdots\circ\,\circ\,\circ\,\circ\,\circ\,\circ \\
\circ\,\bullet\,\bullet\,\bullet\,\circ\,\circ\cdots\circ\,\circ\,\circ\,\circ\,\circ\,\circ \\
\cdots\cdots\cdots\cdots\cdots\cdots\cdots\cdots \\
\cdots\cdots\cdots\cdots\cdots\cdots\cdots\cdots \\
\circ\,\circ\,\circ\,\circ\,\circ\,\circ\cdots\circ\,\circ\,\bullet\,\bullet\,\bullet\,\circ \\
\circ\,\circ\,\circ\,\circ\,\circ\,\circ\cdots\circ\,\circ\,\circ\,\bullet\,\bullet\,\bullet \\
\bullet\,\circ\,\circ\,\circ\,\circ\,\circ\cdots\circ\,\circ\,\circ\,\circ\,\bullet\,\bullet
\end{bmatrix}. \quad (2)
$$

The shape of the transition matrix also guarantees the existence of the limiting distribution in this case, since the main diagonal and the two adjacent ones consist of positive elements only and by the theorem of Frechet this matrix is regular [7, 8].

3 New Algorithms of Image Enhancement

3.1 Method of Evaluating of the Frequency of Visits

Using the 8-neighbourhood system, the transition probability (1) becomes [9, 10]

$$
P_{ij,kl} = \begin{cases}
\dfrac{\exp\{\beta[V(i,j)-V(k,l)]\}}{\displaystyle\sum_{m=-1}^{1}\sum_{n=-1}^{1}\exp\{\beta[V(i,j)-V(i+m,j+n)]\}} & : \ \rho_{(i,j),(k,l)} \leq 1 \\[4mm]
0 & : \ \rho_{(i,j),(k,l)} > 1
\end{cases}
, \quad \beta = \frac{1}{T}. \quad (3)
$$

Let the particle starts its movement from each lattice point (i,j) and let the area $D_{i,j}^{\mu}$ on which the particle can perform the walk be reduced to a square of $n = (2\mu + 1)^2$ points (k,l) for which $\rho_{(i,j),(k,l)} \leq \mu$. Additionally let the area D^{μ} has the topological structure of a torus.

Then the virtual particle moving on the D^{μ} visits its points and after a large enough number of steps, a limiting frequency of visits will be established.

Since it is assumed that the movement of the particle represents an evolution of a Markov chain, the particle after some steps "forgets" from which point it started its walk and its movement can not be distinguished from the trajectory of a particle, which started from a point (k,l), which is in the neighbourhood relation with (i,j).

In order to enhance the image contrast using the concept of a random walk, to each image point the relative frequency of visits $\Gamma(i,j)$ was assigned

$$
\Gamma(i,j) = \frac{\displaystyle\sum_{\substack{(k,l),(i,j) \\ \rho_{(k,l),(i,j)} \leq \mu}} \eta_{(k,l),(i,j)}}{\displaystyle\sum_{\substack{(m,n),(k,l) \\ \rho_{(m,n),(k,l)} \leq \mu,\ \rho_{(k,l),(i,j)} \leq \mu}} \eta_{(k,l),(m,n)}}, \quad (4)
$$

where $\eta_{(k,l),(i,j)}$ is the frequency of visits of (i,j) by a particle which started from the point (k,l).

In such a way, from the original image B a map of the frequencies of visits of each point can be obtained. The resulting image depends strongly on the temperature T and the parameter μ in (3). The decreasing of the temperature leads to a stronger enhancement of the image details and even small image features can be made visible. The parameter μ is also very important, as it should be large enough to enable the visualization of the local image structures.

The results of image enhancement using this algorithm are shown in Fig. 3 b). Starting from each point of the image lattice, the particle performed 1000 steps at $\beta = 10$ and $\mu = 15$.

3.2 Method of the Limiting Vector

Let again the particle starts it walk from point (i,j) and let its movement be restricted to the area $D_{i,j}^{\mu}$ with the centre at (i,j). The transition probabilities for the eight-neighbourhood system are given by the Gibbs distribution (3) and the random walk can be viewed as an evolution of an aperiodic, regular and ergodic Markov chain.

The second algorithm consists of assigning to each lattice point (i,j) the value of the limiting probability $p_{i,j}^{*} = \lim_{n\to\infty} p_{kl,ij}^{(n)}$, $\rho_{(k,l),(i,j)} \leq \mu$, where $p_{kl,ij}^{(n)}$ is the probability of reaching in the n-th step the state (k,l) when starting from (i,j). The limiting probabilities were obtained using the numerical algorithm of von Mises [11] performing seven iteration. The results of this algorithm are shown in Fig. 3 c) using the same parameters β and μ as in the previous algorithm.

3.3 Model of a Jumping Particle

Because there exist an infinitely many trajectories joining two points of an image lattice and the calculations of the limiting probabilities of a Markov chain are time consuming, a drastic simplification of the rules of the particle movement were introduced.

Let the particle can move from a given point (i,j) directly to a point (k,l) for which $\rho_{(i,j),(k,l)} \leq \lambda$. In this way the points (i,j) and (k,l) can be declared as neighbours, with respect to the metric ρ with parameter λ. The probability, that the point (i,j) is visited in one step by particles jumping from the neighbouring points is

$$P_{i,j}^{\lambda} = \frac{1}{(2\lambda+1)^2} \sum_{\substack{(k,l) \\ \rho_{(i,j),(k,l)} \leq \lambda}} \frac{\xi_{kl,ij} \exp\{\beta[V(k,l) - V(i,j)]\}}{Z(k,l)}, \quad \xi_{kl,ij} = \frac{1}{1 + \rho_{(i,j),(k,l)}}, \quad (5)$$

where $\xi_{kl,ij}$ is a variable decreasing with the growing distance between image points. Assigning to each lattice point (i,j) the probability $P_{i,j}$ from (5) leads to an image with improved contrast. The effect of the algorithm is shown in Fig. 3 c). The choice of parameter λ is once again very important. Taking small values of λ results in enhancing very subtle details, whereas for large λ values the algorithm generates an image in which only relatively large features are visible. The influence of the λ value on the result of image enhancement depicts Fig. 2.

415

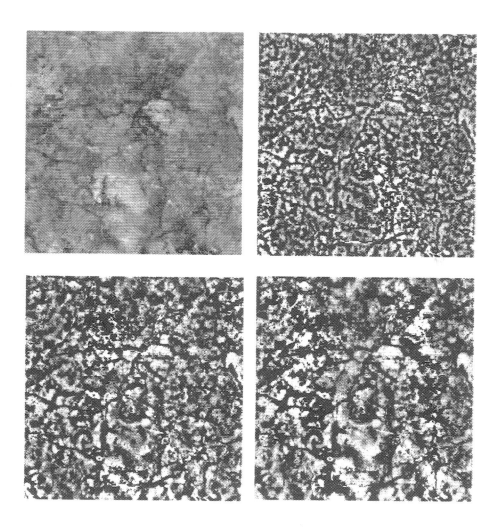

Fig. 2. Influence of the parameter λ on the of contrast enhancement: **upper left**) marble stone surface, **upper right**) contrast enhancement at $\lambda = 5$, **bottom left**) $\lambda = 10$, **bottom right**) $\lambda = 15$, ($\beta = 10$ in all three cases).

3.4 Iterative Method of Contrast Enhancement

Let once more the particle at the point (i,j) can visit its neighbours, that is points (k,l) for which $\rho_{(i,j),(k,l)} \leq \lambda$. Then the probability, that the particle will **not** jump to its neighbours, but stay in the temporary position is equal to

$$
P_{ij,ij} = \begin{cases} \dfrac{\xi_{ij,ij} \exp\{\beta[V(i,j)-V(i,j)]\}}{\displaystyle\sum_{m=-\lambda}^{\lambda} \sum_{n=-\lambda}^{\lambda} \xi_{(i,j),(i+m,j+n)} \exp\{\beta[V(i,j)-V(i+m,j+n)]\}} & : \rho_{(i,j),(k,l)} \leq \mu \\[6pt] 0 & : \rho_{(i,j),(k,l)} > \mu \end{cases} . \quad (6)
$$

In this way the non-zero probabilities are equal to $Z(i,j)^{-1}$, where Z is the statistical sum - the denominator of (6).

The assignment of the probability $P_{ij,ij} = Z(i,j)^{-1}$ to each lattice point leads to an enhanced image, which can be processed in an iterative way. This procedure is especially useful when applied to images with nonhomogeneous illumination. Figure 4 shows the successive iterations of the algorithm performed on a badly illuminated test image. As can be seen, the iterations lead to a binary image with maximal contrast.

4   Conclusions

In this work, four methods of contrast enhancement of gray scale images are presented. The main contribution of this work is the development of a quite new approach to the problem of contrast enhancement. Its essence is the combination of the random walk model and the Gibbs distribution property of the virtual particle walking on the image lattice. Although a very simple model of the potential energy was assumed, the obtained results are encouraging. The authors plan to compare the results obtained with the known methods of image enhancement using different measures of contrast and image quality and to verify the algorithms using a more sophisticated model of the image potential.

References

1. **Gonzalez R.C.** : An overview of image processing and pattern recognition techniques, in Handbook of Geophysical Exploration, Pattern Recognition and Image Processing, edited by F. Aminzadeh, Geophysical Press, **20**, 1987
2. **Pratt W.K.** : Digital Image Processing, New York, John Willey & Sons, 1991
3. **Gonzalez R.C.   Woods R.E.** : Digital Image Processing, Reading MA, Addison- Wesley, 1992
4. **Yaroslavsky L.   Murray E.** : Fundamentals of Digital Optics, Birkhäuser, Boston, 1996
5. **Klette R.   Zamperoni P.** : Handbuch der Operatoren für die Bildverarbeitung, Vieweg Verlag, Braunschweig, Wiesbaden, 1992
6. **Zamperoni P.** : Methoden der digitalen Bildsignalverarbeitung, Vieweg, Braunschweig, 1991
7. **Romanowskij W.I.** : Discrete Markov Chain, (in russian), Moscow, 1949
8. **Kai Lai Chung** : Markov Chains with Stationary Transition Probabilities, Springer Verlag, Berlin, Heidelberg 1967
9. **Smolka B.   Wojciechowski K.** : A new method of texture binarization, Lecture Notes on Computer Science, **1296**, 629-636, 1997
10. **Smolka B.   Wojciechowski K.** : Contrast enhancement of badly illuminated images, Lecture Notes on Computer Science, **1296**, 271-278, 1997
11. **Böhm W.   Gose G.   Kahmann J.** : Methoden der Numerischen Mathematik, Vieweg Verlag, Braunschweig, Wiesbaden, 1985

417

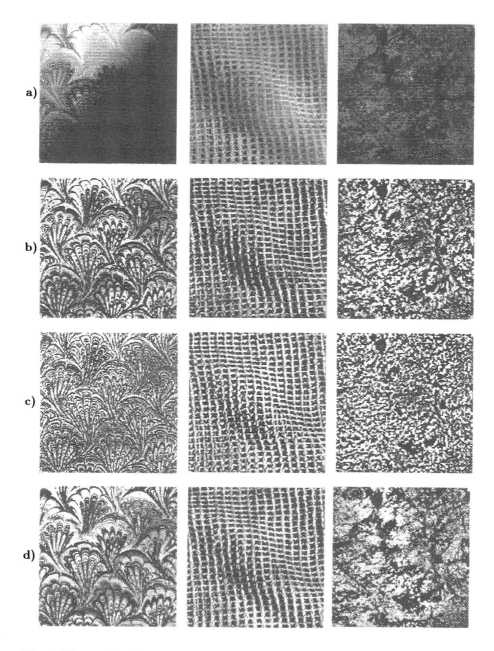

Fig. 3. The results of the presented image enhancement algorithms : **a)** test images (marble paper, fabric and satellite image), **b)** images obtained using the method of evaluating of the relative frequencies of visits, **c)** results achieved with the algorithm of finding the limiting probability of a Markov chain, **d)** images obtained using the model of a jumping particle ($\mu, \lambda = 15$, $\beta = 10$).

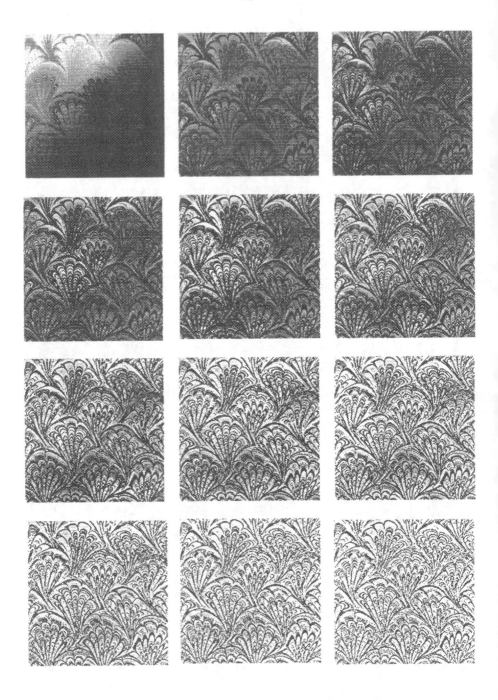

Fig. 4. The effect of the iterative procedure of image enhancement (successive iterations from upper left to bottom right, $\lambda = 15$, $\beta = 10$).

Skeletonization on Projection Pictures

Attila Fazekas

Kossuth Lajos University, Debrecen P.O.Box 12, H-4010, Hungary

Abstract. The template method is one of the most often used group of image transformations in digital image processing. In this paper we try to reveal the possibility of template methods on the perpendicular projections of 2D binary pictures. We examined the theoretical background of the template methods on projection pictures. As a result of this we created an algorithm, which accomplishes this template method. To illustrate the theoretical results, a special template method, the thinning method is implemented.

Keywords. Skeletonization, template method, projection picture

1   Introduction

One of the most often used group of image transformations in digital image processing is the so called template method. This method is commonly used both in preprocessing and for obtaining specific features.

This is especially true in the case of 2D binary pictures. To prove this statement let us refer to the thinning templates, which are used for skeletonization. The literature on this topic is quite rich [6,10]. Besides the thinning on the classical pixel-pixel type picture representations, investigations also include the picture representation based on the quadtree [2], run-length [8], and contour-coding [1,5].

It often happens in medical digital image processing that instead of the digital pictures only their projections are known. These projections are considered as the representations of the original pictures if the unique reconstructibility is assured. The following question arises: if the projections are considered as representations, how can the template methods, especially the thinning method, be applied on these picture representations.

This paper deals with the possibility of template methods on the perpendicular projections of 2D binary pictures. We present the theoretical background of the template method on projection pictures, and as a result of this an algorithm has been created, which provides template methods on projection pictures. In the last chapter we implement a special template method, the thinning method, to illustrate the theoretical results.

2 Basic Definitions, Notations

This chapter contains some well-known concepts of digital topology wich will be used later on.

Definition 1. A set $X \subseteq \mathbb{Z}^2$ is called a digital set and its elements are called points. A function f defined on a digital set X with values $\{0, 1, \ldots, n - 1\}$ ($n \in \mathbb{N}$) is called an n-valued digital picture. If $n = 2$ then f is a binary picture.

Definition 2. Let f be a binary picture defined on the digital set X. The set $F = \{p \mid f(p) = 1, p \in X\}$ is the foreground or object of the binary picture f and the set $B = \{p \mid f(p) = 0, p \in X\}$ is the background of f.

Definition 3. Let X be a digital set and $p_{(x,y)} \in X$ its arbitrary element. The $\mathcal{N}_4(p)$ 4-neighbourhood of the point p are those points of X whose (x', y') coordinates satisfy the equality $|x - x'| + |y - y'| = 1$. The $\mathcal{N}_8(p)$ 8-neighbourhood of the point p are those points of X which differ from p and whose coordinates satisfy the following two inequalities at the same time: $|x - x'| \leq 1$, $|y - y'| \leq 1$.

Definition 4. Let p and q be two points of the digital set X. The n-path from p to q is defined as the sequence of points $p = p_0, p_1, \ldots, p_k = q$ ($k \in \mathbb{N}$), where the elements of the sequence belong to X and p_i is the n-neighbour of p_{i-1} ($1 \leq i \leq k$).

Definition 5. A digital set X is n-connected if there is an n-path between any two points p and q in X.

Definition 6. The "n-path from p to q" relation is an equivalence relation on the digital set X and the classes induced by this are called the n-components of X.

Definition 7. Let p' and p'' be any two points of the $\mathbb{Z}^2$, so that $\mathrm{Pr}_i(p') \leq \mathrm{Pr}_i(p'')$ ($i = 1, 2$). The ith coordinate of the point p is indicated by $\mathrm{Pr}_i(p)$. The digital set W is called a window with size $(\mathrm{Pr}_1(p'') - \mathrm{Pr}_1(p') + 1) \times (\mathrm{Pr}_2(p'') - \mathrm{Pr}_2(p') + 1)$ if
$$W = \{p \mid \mathrm{Pr}_i(p') \leq \mathrm{Pr}_i(p) \leq \mathrm{Pr}_i(p'')\}.$$

3 Reconstructibility of Binary Matrices

Let f be a binary picture defined on the window X with size $m \times n$. In this case f can be represented as a binary matrix, element (i, j) of which is equal to 1 if $f((k + i - 1, l + j - 1)) = 1$ holds and 0 otherwise, where k is the horizontal and l is the vertical coordinate of the upper left corner.

    This pixel-pixel type of binary picture representation allows us to characterize those cases when the perpendicular projections of the binary pictures can be considered the representation of the original pictures. In this chapter we are going to investigate it with the help of the results on reconstructibility of binary matrices.

Definition 8. Let $X = (x_{ij})$ be a binary matrix with size $m \times n$. On the row projection of the matrix X we mean the vector $R(X) = R = (r_1, r_2, \ldots, r_m)$ and on the column projection we mean the vector $C(X) = C = (c_1, c_2, \ldots, c_n)$, where

$$r_i = \sum_{j=1}^{n} x_{ij}, \qquad c_j = \sum_{i=1}^{m} x_{ij}.$$

The class of binary matrices with row and column projections R and C is denoted by $\mathcal{X}(R, C)$.

Definition 9. We say that a matrix X can be uniquely reconstructed from its projections if $X' \in \mathcal{X}(R, C)$ implies $X = X'$.

Theorem 10. (see in [11]). *A binary matrix $X \in \mathcal{X}(R, C)$ can be uniquely reconstructed if and only if it does not contain switching components*
$$X_1 = \begin{pmatrix} 1 & 0 \\ 0 & 1 \end{pmatrix} \text{ and } X_2 = \begin{pmatrix} 0 & 1 \\ 1 & 0 \end{pmatrix} \text{ as a minor.}$$

Using this theorem we can see that the perpendicular projections of binary pictures defined on a window X can be considered as the reconstruction of the picture f if and only if the binary matrix can be uniquely reconstructed from its projections with the pixel-pixel picture representation defined by f.

It is possible to generalize the unique reconstruction problem. In this case we can prescribe the elements of the uniquely reconstructible binary matrix.

Definition 11. Let $P = (p_{ij})$ and $Q = (q_{ij})$ be two binary matrices with size $m \times n$. We say $Q \geq P$ if $q_{ij} \geq p_{ij}$ for all positions $(i, j) \in \{1, 2, \ldots, m\} \times \{1, 2, \ldots, n\}$. The class of $\mathcal{X}_P^Q(R, C)$ then can be defined as:

$$\mathcal{X}_P^Q(R, C) = \{X \mid X \in \mathcal{X}(R, C), \ P \leq X \leq Q\}.$$

It is clear that if $P = \{0_{ij}\}$ (zero matrix) and $Q = \{1_{ij}\}$ then we have the class $\mathcal{X}(R, C)$ again. It is easy to prove [4] that instead of investigating the class $\mathcal{X}_P^Q(R, C)$, as a consequence of the transformation $\mathcal{X}_0^{Q-P}(R - R(P), C - C(P))$, it is sufficient to investigate $\mathcal{X}_0^{Q-P}(R - R(P), C - C(P))$, which is then shortly called class $\mathcal{X}^{Q-P}(R - R(P), C - C(P))$.

In this class the switching chain plays the role of the switching component [4].

Definition 12. We say that the binary matrix $X \in \mathcal{X}^Q$ has a switching chain if there is a sequence of different free positions of X,

$$\{(i_1, j_1), (i_1, j_2), (i_2, j_2), (i_2, j_3), \ldots, (i_p, j_p), (i_p, j_1)\},$$

such that

$$x_{i_1 j_1} = x_{i_2 j_2} = \ldots = x_{i_p j_p} = 1 - x_{i_1 j_2} = 1 - x_{i_2 j_3} = \ldots = 1 - x_{i_p j_1}$$

$(p \geq 2)$. The position (i, j) is said to be free if the corresponding matrix element is not prescribed by Q, i.e. $q_{ij} = 1$.

According to this, the unique reconstructibility is characterized as it follows.

Theorem 13. (see in [4]) *A binary matrix $X \in \mathcal{X}^Q(R, C)$ can be uniquely reconstructed if and only if X has no switching chain.*

4 The Theoretical Background of the Template Methods on Projection Pictures

For template processes on projections of uniquely reconstructable binary matrices, first we have to provide, that the fitting of a template should be determined only by analysing the projections of the template and of the original picture. The following theorem provides the theoretical background.

Theorem 14. *Let $X = \begin{pmatrix} X_1 & X_2 & X_3 \\ X_4 & X_5 & X_6 \\ X_7 & X_8 & X_9 \end{pmatrix}$, $P = \begin{pmatrix} P_1 & P_2 & P_3 \\ P_4 & P_5 & P_6 \\ P_7 & P_8 & P_9 \end{pmatrix}$, and $Q = \begin{pmatrix} Q_1 & Q_2 & Q_3 \\ Q_4 & Q_5 & Q_6 \\ Q_7 & Q_8 & Q_9 \end{pmatrix}$ be binary matrices with size $m \times n$. Let X_i, P_i, and Q_i $(i = 1, 2, \ldots, 9)$ be submatrices with the same size in X, P, and Q, respectively. Furthermore, let $\overline{X} = \begin{pmatrix} X_1 & X_2 & X_3 \\ X_4 & 0 & X_6 \\ X_7 & X_8 & X_9 \end{pmatrix}$, $X^0 = \begin{pmatrix} 0 & 0 & 0 \\ 0 & X_5 & 0 \\ 0 & 0 & 0 \end{pmatrix}$, and $X^{0\prime} = \begin{pmatrix} 0 & 0 & 0 \\ 0 & X_5' & 0 \\ 0 & 0 & 0 \end{pmatrix}$ be binary matrices with size $m \times n$. Let X_5 and X_5' be binary matrices with the same size. The projections of binary matrices $X, \overline{X}, X^0$, and $X^{0\prime}$ are denoted by $R(X), R(\overline{X}), R(X^0)$, and $R(X^{0\prime})$ and $C(X), C(\overline{X}), C(X^0)$, and $C(X^{0\prime})$. We assume that X is uniquely reconstructible in the class $\mathcal{X}_P^Q$, and X_5' is a non-zero uniquely reconstructible matrix. A matrix X with size $m \times n$ is uniquely reconstructible from projections $R(X) - R(X^{0\prime})$ and $C(X) - C(X^{0\prime})$ in the class $\mathcal{X}_{\overline{P}}^{\overline{Q}}$ if and only if $X_5' \leq X_5'$, where $\overline{Q} = \begin{pmatrix} Q_1 & Q_2 & Q_3 \\ Q_4 & 0 & Q_6 \\ Q_7 & Q_8 & Q_9 \end{pmatrix}$ and*

$$\overline{P} = \begin{pmatrix} P_1 & P_2 & P_3 \\ P_4 & 0 & P_6 \\ P_7 & P_8 & P_9 \end{pmatrix}.$$

Proof of theorem. Suppose that the matrix X' is uniquely reconstructible from projections $R(X') = R(X) - R(X^{0\prime})$ and $C(X') = C(X) - C(X^{0\prime})$ in the class $\mathcal{X}^{\overline{Q}}_{\overline{P}}$. Since the matrix $X, \overline{X}, X^0$ and $X^{0\prime}$ uniquely reconstructible from their projections, the following equations hold:

$$R(X') = R(X) - R(X^{0\prime})$$
$$C(X') = C(X) - C(X^{0\prime})$$
$$R(X) = R(\overline{X}) + R(X^0)$$
$$C(X) = C(\overline{X}) + C(X^0)$$

From these equations we have

$$X = X' + X^{0\prime}$$
$$X = \overline{X} + X^0.$$

According to the construction of the matrix X^0 and $X^{0\prime}$ if some element of the matrix $\overline{X}$ is equal to 1, then the corresponding element of the matrix X' is also equal to 1. This feature of the matrix $\overline{X}$ and X' implies that the matrix $X' - \overline{X}$ is a binary matrix. This result and the above equations imply that the matrix $X^0 - X^{0\prime}$ is also a binary matrix. As a consequence, if one of the elements of the matrix $X^{0\prime}$ is 1, than the corresponding element of the matrix X^0 is also 1. Thus $X^{0\prime} \leq X^0$, so $X'_5 \leq X_5$.

Conversely. We assume that $X'_5 \leq X_5$. It is easy to see, that the matrix $X' = X - X^{0\prime}$ is a binary matrix. We prove the fact that the reconstruction of the matrix X' from projections $R(X) - R(X^{0\prime})$ and $C(X) - C(X^{0\prime})$ is unique.

Suppose that the matrix $X'' \neq X'$ can be reconstructed from the projections $R(X) - R(X^{0\prime})$ and $C(X) - C(X^{0\prime})$. Then the equations $X' = X - X^{0\prime}$ and $X'' = X - X^{0\prime}$ imply the equation $X' + X^{0\prime} = X'' + X^{0\prime}$. Since the matrix $X^{0\prime}$ is uniquely reconstructible from its projections, the equation $X' = X''$ also holds. This is a contradiction. The uniquely reconstruction of the matrix X' in the class $\mathcal{X}^{\overline{Q}}_{\overline{P}}$ also holds. In order to prove this, it is enough to realize that the matrices $X - X^{0\prime}$ and $X - X^0$ can only differ in those elements which have not been prescreibded in matrices $\overline{P}$ and $\overline{Q}$. □

In the following step we are going to determine how the unique reconstructibility can be held during template processing. This theorem has been published in paper [3] as a joint work with G.T. Herman and S. Matej.

Theorem 15. *Let* $X = \begin{pmatrix} X_1 & X_2 & X_3 \\ X_4 & x & X_6 \\ X_7 & X_8 & X_9 \end{pmatrix}$ *and* $X' = \begin{pmatrix} X_1 & X_2 & X_3 \\ X_4 & x' & X_6 \\ X_7 & X_8 & X_9 \end{pmatrix}$ *are binary matrices, furthermore let* x *be the element of matrix* X *with coordinates* (i,j). *Let the equation* $x' = 1 - x$ *be true. If the binary matrix* X *is uniquely reconstructible in class* $\mathcal{X}_P^Q$ *then* X' *can be uniquely reconstructed in the class* $\mathcal{X}_{P'}^{Q'}$, *where* P' *and* Q' *differ from binary matrices* P *and* Q *only with the element with coordinates* (i,j), *which can be defined as* $p_{ij} = q_{ij} = x'$.

Proof of theorem. It is easy to verify that in a uniquely reconstructible matrix X, in class $\mathcal{X}_P^Q$, it can happen that by changing one element a switching chain can be constructed, which can damage the uniqueness of the reconstructibility. In order to avoid this the new value of the element changed should be fixed in the matrices P and Q. This means that the matrix X' can be uniquely reconstructed in class $\mathcal{X}_{P'}^{Q'}$. □

5 Projection Template Processing

On the basis of our result we can construct a template method on projection pictures. The idea is the following: First we investigate the fitting of the template in a point of the picture using Theorem 14. Using the notations of this theorem, the fact that there is a matrix, which is uniquely reconstructible from the projections $R(X) - R(X^{0\prime})$ and $C(X) - C(X^{0\prime})$ in the class $\mathcal{X}_{\overline{P}}^{\overline{Q}}$, means that all the 1's found in the template X_5' fit in the 1's in X_5. Then we check, whether there is such a matrix which can be uniquely reconstructed from the projections $R(X^c) - R(X^{0\prime c})$ and $C(X^c) - C(X^{0\prime c})$ in the class $\mathcal{X}_{\overline{P^c}}^{\overline{Q^c}}$. If there is, then it means that the 0's found in the template X_5' will fit in the 0's in X_5.

These two investigations together mean the exact fitting of template X_5'. If the template fits in the point ij of the binary matrix, then we can modify it as well as the jth element of R, and the ith element of C on the projection pictures. Then we modify the required elements of sets P and Q based on the Theorem 15. This will guarantee the unique reconstructibility of matrix.

6 Thinning Method on Projection Pictures

Now as an example let us investigate an algorithm which executes the template method in the case of 3×3 templates. We are going to use this algorithm for our thinning algorithm.

```
Program Thinning;
Var P,Q:Array[1..1,0..m] Of Byte;
  PC,QC:Array[1..1,0..m] Of Byte; /* The complement of P and Q */
  TP,TQ:Array[1..1,0..m] Of Byte; /* Temporary P and Q */
TPC,TQC:Array[1..1,0..m] Of Byte /* Temporary PC and QC */
  R,RC:Array[0..l+1] Of Integer; /* The row projections */
  C,CC:Array[0..m+1] Of Integer; /* The column projections */
TR,TRC:Array[0..l+1] Of Integer; /* Temporary R and RC */
TC,TCC:Array[0..m+1] Of Integer; /* Temporary C and CC */
RM,RMC:Array[0..l+1] Of Integer;
CM,CMC:Array[0..m+1] Of Integer;
      SR:Array[1..k,1..3] Of Integer; /* Row projections of the templates */
      SC:Array[1..k,1..3] Of Integer; /* The same for columns */
     SRC:Array[1..k,1..3] Of Integer;
     SCC:Array[1..k,1..3] Of Integer;
     Fit:Boolean;
Begin
R[0]:=0;R[m+1]:=0;C[0]:=0;C[n+1]:=0; /* Set the border */
RC[0]:=1;RC[m+1]:=1;CC[0]:=1;CC[n+1]:=1; /* Set the border's complement */
Fit:=True;
While Fit Do
  Begin
  Fit:=False;
  For h:=1 To k Do
    Begin
    TP:=P;TQ:=Q;TPC:=PC;TQC:=QC;
    RM:=R;CM:=C;RMC:=RC;CMC:=CC;
    For i:=1 To m Do
     For j:=1 To l Do
       Begin
       TR:=R;TC:=C;TRC:=TR;TCC:=TC;
       R[j-1]:=TR[j-1]-SR[h,1];RC[j-1]:=TRC[j-1]-SRC[h,1];
       R[j]:=TR[j]-SR[h,2];RC[j]:=TRC[j]-SRC[h,2];
       R[j+1]:=TR[j+1]-SR[h,3];RC[j+1]:=TRC[j+1]-SRC[h,3];
       C[i-1]:=TC[i-1]-SC[h,1];CC[i-1]:=TCC[i-1]-SCC[h,1];
       C[i]:=TC[i]-SC[h,2];CC[i]:=TCC[i]-SCC[h,2];
       C[i+1]:=TC[i+1]-SC[h,3];CC[i+1]:=TCC[i+1]-SCC[h,3];
       If (R,C) uniquely reconstructible in $\mathcal{X}_{TQ}^{TP}$ Then

          If (RC,CC) uniquely reconstructible in $\mathcal{X}_{TQC}^{TPC}$ Then
            Begin
            Fit:=True;
            P[i,j]:=0;Q[i,j]:=0;PC[i,j]:=1;QC[i,j]:=1;
            RM[j]:=RM[j]-1;CM[i]:=CM[i]-1;
            RMC[j]:=RMC[j]+1;CMC[i]:=CMC[i]+1;
            End;
         R:=TR;C:=TC;RC:=TRC;CC:=TCC;
         End;
       R:=RM;C:=CM;RC:=CMC;CC:=CMC;
       End;
     End;
  End.
```

This algorithm deletes a point of a binary picture if it fits on at least one template. This subcycle should be applied as long as there are any changes on the picture. If we use templates during the thinning procedure, which belong to two or more subcycles then the algorithm should be used with the templates belonging to the adequate subcycle. If we would like to delete every point in a subcycle at the same time on which at least one template fits then we have to use two picture planes avoiding the case when a previously deleted point would change the neighbours of the examined point.

For thinning we can use the templates in [9]. It is easy to see that these templates fulfill the requirements of being uniquely reconstructible. Fig. 1a shows the original object on which the positions "•" represent the "1" elements, the "o" represent the "0" elements, while the "*" represent those "0" elements which are prescribed to get unique reconstructibility. On Fig. 1b the skeleton is marked with "•", and the positions marked by number i are deleted in the ith subcycle.

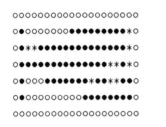

Fig. 1a. The original image.

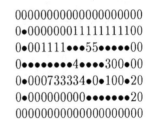

Fig. 1b. The skeleton of the original image.

References

1. C. Arcelli, Pattern thinning by contour tracing, *Computer Graphics and Image Processing* **17**, 130–144 (1981).
2. L. A. Breene, Quadtrees and hypercubes: Grid embedding strategies based on spatial data structure addressing, *The Computer Journal* **36**, 562–569 (1993).
3. A. Fazekas, G. T. Herman and S. Matej, On processing binary pictures via their projections, *International Journal of Imaging Systems and Technology* **9**, 99–100 (1998).
4. A. Kuba, Reconstruction of unique binary matrices with prescribed elements, *Acta Cybernetica* **12**, 57–70 (1995).
5. P. C. K. Kwok, A thinning algorithm by contour generation, *Communications of the ACM* **31**, 1313–1324 (1988).
6. L. Lam and S.-W. Lee and C. Y. Suen, Thinning methodologies – A comprehensive survey, *IEEE Transactions on Pattern Analysis and Machine Intelligence* **14**, 869–885 (1992).
7. L. Latecki, U. Eckhardt and A. Rosenfeld, Well-composed sets, *Computer Vision and Image Understanding* **61**, 70–83 (1995).
8. L. Ji, J. Piper and J.-Y. Tang, Erosion and dilation of binary images by arbitrary structuring elements using elements using interval coding, *Pattern Recognition Letters* **9**, 201–209 (1989).
9. W. K. Pratt and I. Kabir, Morphological binary image processing with a local neighborhood pipeline processor, *Proceeding of Computer Graphics Tokyo '84*, 321–343 (1985).
10. C. Ronse, A topological characterization of thinning, *Theoretical Computer Science* **43**, 31–41 (1986).
11. H. J. Ryser, Combinatorial properties of matrices of zeros and ones, *Canadian Journal of Mathematics* **9**, 371–377 (1957).

Computing the Intrinsic Camera Parameters Using Pascal's Theorem

Bodo Rosenhahn, Eduardo Bayro-Corrochano* **
Institut für Informatik und Praktischer Mathematik
Christian-Albrechts-Universität zu Kiel
Preußerstrasse 1-9, 24105 Kiel, Germany
bro@ks.informatik.uni-kiel.de, edb@ks.informatik.uni-kiel.de

Abstract. The authors of this paper adopted the characteristics of the image of the absolute conic in terms of Pascal's theorem to propose a new camera calibration method. Employing this theorem in the geometric algebra framework enables the authors to compute a projective invariant using the conics of only two images which expressed using brackets helps us to set enough equations to solve the calibration problem.

1 Introduction

The computation of the intrinsic camera parameters is one of the most important issues in visual guided robotics. One important method of selfcalibration is based on the image of the absolute conic and it requires as input only information about the point correspondences [5, 4, 1].

In this paper we re establish the idea of the absolute conic in the context of Pascal's theorem and we get different equations than the Kruppa equations [4, 1]. Although the equations are different they rely on the same principle of the invariance of the mapped absolute conic. The consequence is that we can generate equations so that we require only a couple of images whereas the Kruppa equation method requires at least three images [4]. As a prior knowledge the method requires the translational motion direction of the camera and the rotation about at least one fixed axis through a known angle in addition to the point correspondences. The paper will show that although the algorithm requires the extrinsic camera parameters in advance it has the following clear advantages: It is derived from geometric observations, it does not stick in local minima in the computation of the intrinsic parameters and it does not require any initialization at all. We hope that this proposed method derived from geometric thoughts gives a new point of view to the problem of camera calibration.

In this paper we are modelling the properties of the projective space $\mathcal{P}^3$ using the geometric algebra $\mathcal{G}_{1,3,0}$ and that of the projective plane $\mathcal{P}^2$ using $\mathcal{G}_{3,0,0}$. Next, we will briefly outline the basic operations of projective geometry within the geometric algebra which are used in the following sections. For a complete introduction of geometric algebra and algebra of incidence in computer vision the reader should consult [3].

---

* This work was granted by DFG(So-320-2-1).
** New Adress: Eduardo Bayro-Corrochano, Centro de Investigacion en Matematicas, A.C., Apartado Postal 402, 360000-Guanajuato, Gto. Mexico

In an n-dimensional vector space with an orthonormal basis $\{e_1, \ldots, e_n\}$ we define an 2^n dimensional space consisting of $e_1^{m_1} \ldots e_n^{m_n} (m_j = 0, 1)$, where e_i^0 is interpreted as the identity, and define a product on these basis vectors which fulfils $e_1^{m_1} \ldots e_r^{m_r} e_{r+1}^{m_{r+1}} \ldots e_n^{m_n} = -e_1^{m_1} \ldots e_{r+1}^{m_{r+1}} e_r^{m_r} \ldots e_n^{m_n} (m_r = m_{r+1} = 1)$. The element $I = e_1 \ldots e_n$ is called unit pseudoscalar and any pseudoscalar can be written as $P = \alpha I$ where α is a scalar. If I^{-1} denotes the inverse of I, so that $II^{-1} = 1$, then the magnitude of P relative to I will be called **bracket** and is defined by $PI^{-1} = \alpha \equiv [P]$. The bracket of the maximum grade multivector $a_1 \ldots a_n$ is determined by $[a_1 \ldots a_n] \equiv [a_1 \wedge a_2 \wedge \ldots \wedge a_n] = (a_1 \wedge a_2 \wedge \ldots \wedge a_n) I^{-1}$ and can be taken as a definition of the determinant. The dual A^* of an r-vector A is defined by $A^* = AI^{-1}$. In a geometric algebra of an n-dimensional vector space one can define the **join** for the linearly independent r-vector A and s-vector B by $J = A \wedge B$. If A and B are not linearly independent the join is not given simply by the wedge but by the commonly spanned subspace. If A and B have a common factor (i.e. there exists a k-vector C such that $A = A'C$ and $B = B'C$ for some A', B') then we can define the 'intersection' or **meet** of A and B as $A \vee B$ where $(A \vee B)^* = A^* \wedge B^*$. Thus the dual of the meet is given by the join of the duals.

The paper is organized as follows: Section two presents a new method for computing the intrinsic camera parameters based on Pascal's theorem. Section three is devoted to the experimental analysis and section four to the conclusion part.

2   Camera Calibration Using Pascal's Theorem

This section presents a new technique in the geometric algebra framework for computing the intrinsic camera parameters. In this context we use the ideas of Maybank, Faugeras and Luong [5, 4] to enforce an epipolar line defined by the epipole or fundamental matrix and a point $(1, \tau, 0)^T$ to be tangential to the image of the absolute conic and analyse Pascal's theorem in this context. Consider the points a, a', b, b' and c' lying on a conic, we can compute the following bracket expression

$$[c'ab][c'a'b'] - \rho[c'ab'][c'a'b] = 0 \quad \Rightarrow \rho = \frac{[c'ab][c'a'b']}{[c'ab'][c'a'b]} \tag{1}$$

for some $\rho \neq 0$. This equation is well known and represents a projective invariant which has been used often in real applications of computer vision. Now consider c as some other point placed on the conic, we can get a conic equation fully represented in terms of brackets

$$[cab][ca'b'][ab'c'][a'bc'] - [cab'][ca'b][abc'][a'b'c'] = 0. \tag{2}$$

For further details about describing a conic by brackets and points placed on the conic the reader should consult [2]. Using the collinearity constraint, we can write the geometric formulation of Pascal's theorem

$$\underbrace{((a' \wedge b) \vee (c' \wedge c))}_{\alpha_1} \wedge \underbrace{((a' \wedge a) \vee (b' \wedge c))}_{\alpha_2} \wedge \underbrace{((c' \wedge a) \vee (b' \wedge b))}_{\alpha_3} = 0. \tag{3}$$

This theorem proves that the three intersecting points of the lines which connect opposite vertices of a hexagon circumscribed by a conic are collinear ones. Since equation (3) fulfills a property of any conic, it should also be possible to compute the intrinsic camera parameters using this equation. In geometric algebra a conic can be described by the points lying on the conic. Furthermore, the image of the absolute conic can be described by the image of the points lying on the absolute conic. Let us choose for all computations the following imaginary points lying on the absolute conic

$$\boldsymbol{A}_0 = \begin{pmatrix} 1 \\ i \\ 0 \\ 0 \end{pmatrix}, \boldsymbol{B}_0 = \begin{pmatrix} i \\ 1 \\ 0 \\ 0 \end{pmatrix}, \boldsymbol{A}'_0 = \begin{pmatrix} i \\ 0 \\ 1 \\ 0 \end{pmatrix}, \boldsymbol{B}'_0 = \begin{pmatrix} 1 \\ 0 \\ i \\ 0 \end{pmatrix}, \boldsymbol{C}'_0 = \begin{pmatrix} 0 \\ i \\ 1 \\ 0 \end{pmatrix}, \quad (4)$$

where $i^2 = -1$. The projected points of $\boldsymbol{A}_0, \boldsymbol{B}_0, \boldsymbol{A}'_0, \boldsymbol{B}'_0, \boldsymbol{C}'_0$ can be described by $\boldsymbol{a} = K[R|t]\boldsymbol{A}_0 = KR\boldsymbol{A}, \boldsymbol{b} = K[R|t]\boldsymbol{B}_0 = KR\boldsymbol{B}, \boldsymbol{a}' = K[R|t]\boldsymbol{A}'_0 = KR\boldsymbol{A}', \boldsymbol{b}' = K[R|t]\boldsymbol{B}'_0 = KR\boldsymbol{B}', \boldsymbol{c}' = K[R|t]\boldsymbol{C}'_0 = KR\boldsymbol{C}'$, where

$$\boldsymbol{A} = \begin{pmatrix} 1 \\ i \\ 0 \end{pmatrix}, \boldsymbol{B} = \begin{pmatrix} i \\ 1 \\ 0 \end{pmatrix}, \boldsymbol{A}' = \begin{pmatrix} i \\ 0 \\ 1 \end{pmatrix}, \boldsymbol{B}' = \begin{pmatrix} 1 \\ 0 \\ i \end{pmatrix}, \boldsymbol{C}' = \begin{pmatrix} 0 \\ i \\ 1 \end{pmatrix}. \quad (5)$$

In addition the rotated points $R^T \boldsymbol{A}, \ldots, R^T \boldsymbol{C}'$ also lie at the conic. Using the rotated points, the image of the absolute conic can be described by $\boldsymbol{a} = K\boldsymbol{A}, \boldsymbol{b} = K\boldsymbol{b}, \boldsymbol{a}' = K\boldsymbol{A}', \boldsymbol{b}' = K\boldsymbol{b}', \boldsymbol{c}' = K\boldsymbol{C}'$. To use $\boldsymbol{a}, \ldots, \boldsymbol{c}'$ in the bracket notation of conics we suppose an orthonormal basis of the image plane, $B_1 = \{e_1, e_2, e_3\}$ and formulate $\boldsymbol{X}$ as a linear combination of B_1, i.e. $\boldsymbol{X} = \sum_{i=1}^{3} x_i e_i$ and introduce an operator $\underline{K}e_i = \underline{K}(e_i) = \sum_{j=1}^{3} e_j k_{ji}$, with k_{ji} the elements of the matrix of intrinsic camera parameters, K.

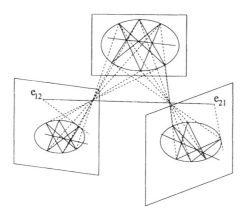

Fig. 1. Pascal's theorem in the conic images

The application of the principle of duality allows us to set $\boldsymbol{c} = KK^T \boldsymbol{l}_c$ for a point $\boldsymbol{c}$ on the image of the absolute conic. This point depends on the intrinsic camera parameters and a line $\boldsymbol{l}_c$ tangent to the image of the absolute conic computed in terms of the epipole and a point $\boldsymbol{q}$ lying at infinity, $\boldsymbol{l}_c =$

$(e_{12} \wedge q)I^{-1} = (p_1 e_1 + p_2 e_2 + p_3 e_3) \wedge (e_1 + \tau e_2)I^{-1}$. The point c can be described by using the adjoint operator $\overline{K}$ of $\underline{K}$. The above expression for the point c can thus be formulated as $c = \underline{K}\overline{K}l_c$.

Now we simplify equation (3) and get the bracket equation of the α's

$$([a'bc']c - [a'bc]c') \wedge ([a'ab']c - [a'ac]b') \wedge ([c'ab']b - [c'ab]b') = 0$$

$$\Leftrightarrow \underbrace{([A'BC'](\overline{K}l_c) - [A'B(\overline{K}l_c)]C')}_{\alpha_1} \wedge \underbrace{([A'AB'](\overline{K}l_c) - [A'A(\overline{K}l_c)]B')}_{\alpha_2} \wedge$$

$$\underbrace{([C'AB']B - [C'AB]B')}_{\alpha_3} = 0. \tag{6}$$

The computation of the intrinsic parameters is done first if the intrinsic parameters remain stationary under camera motions and second if these parameters change.

2.1 Computing stationary intrinsic parameters

We consider one camera motion. The involved projective transformations are $P_1 = K[I|0]$ and $P_2 = P_1 \begin{pmatrix} R_1 & t_1 \\ 0_3^T & 1 \end{pmatrix}^{-1} = P_1 D^{-1}$. The 4×4-matrix D describes the extrinsic camera parameters. The optical centres of the cameras are $C_1 = (0,0,0,1)^T$ and $C_2 = DC_1$. In geometric algebra we use instead the notations $\underline{P_1}$, $\underline{P_2}$, $C_1 = e_4$ and $C_2 = \underline{D}C_1$. Thus we can compute their epipoles as $e_{21} = \underline{P_2}C_1$, $e_{12} = \underline{P_1}C_2$.

Next, we show by means of an example that the homogeneous coordinates of the points $\alpha_1, \alpha_2, \alpha_3$ are entirely independent of the intrinsic parameters. This condition is necessary for solving the problem. We choose a camera motion given by

$$[R_1|t_1] = \begin{pmatrix} 0 & -1 & 0 & | & 2 \\ 1 & 0 & 0 & | & -1 \\ 0 & 0 & 1 & | & 3 \end{pmatrix}. \tag{7}$$

For this motion the epipoles are $e_{12} = (2k_{11} - k_{12} + 3k_{13})e_1 + (-k_{22} + 3k_{23})e_2 + 3e_3$ and $e_{21} = (k_{11} + 2k_{12} - 3k_{13})e_1 + (2k_{22} - 3k_{23})e_2 - 3e_3$. By replacing e_{12} in the equation (6), we can make explicit the $\alpha's$

$$\alpha_1 = ((-3 + 3i)k_{11}\tau)e_1 + (3k_{11}\tau - ik_{12}\tau + ik_{22} + 2ik_{11}\tau - 3k_{12}\tau + 3k_{22})e_2 +$$
$$(ik_{11}\tau + 3k_{12}\tau - 3k_{22} + ik_{12}\tau - ik_{22})e_3$$

$$\alpha_2 = (-3ik_{11}\tau - 2k_{12}\tau + 2k_{22} - 2k_{11}\tau)e_1 + (-6i(k_{12}\tau - k_{22}))e_2 +$$
$$(-3k_{11}\tau - 4ik_{12}\tau + 4k_{22} + 2ik_{11}\tau)e_3$$

$$\alpha_3 = (1 - i)e_1 + (1 - i)e_2 + 2e_3. \tag{8}$$

Note that α_3 is fully independent of $\underline{K}$.

According to Pascal's theorem these three points are lying on the same line. Thereto, by replacing these points in Pascal's equation (3), we get the following second order polynomial in τ

$$(-40ik_{12}^2 - 52ik_{11}^2 + 16ik_{11}k_{12})\tau^2 + (-16ik_{11}k_{22} + 80ik_{12}k_{22})\tau - 40ik_{22}^2 = 0. \tag{9}$$

Solving this polynomial and choosing one of the solutions which is nothing else than the solution for one of the two lines tangent to the conic, we get

$$\tau := \frac{16ik_{11}k_{22} - 80ik_{12}k_{22} + 24\sqrt{14}k_{11}k_{22}}{2(-40ik_{12}^2 - 52ik_{11}^2 + 16ik_{11}k_{12})}. \tag{10}$$

Now considering the homogeneous representation of the intersecting points

$$\alpha_i = \alpha_{i1}\mathbf{e}_1 + \alpha_{i2}\mathbf{e}_2 + \alpha_{i3}\mathbf{e}_3 \sim \frac{\alpha_{i1}}{\alpha_{i3}}\mathbf{e}_1 + \frac{\alpha_{i2}}{\alpha_{i3}}\mathbf{e}_2 + \mathbf{e}_3, \tag{11}$$

and the case of exactly orthogonal image axis, $k_{12} = 0$, we get

$$\alpha_{11} = \frac{2i - 3\sqrt{14} + 10 + 2i\sqrt{14}}{2 + 3i\sqrt{14} + 16i + 2\sqrt{14}} \quad \alpha_{12} = 26\frac{i}{2 + 3i\sqrt{14} + 16i + 2\sqrt{14}}$$

$$\alpha_{21} = \frac{(1 + i)(-2i + 3\sqrt{14})}{-5i + \sqrt{14} - 13} \quad \alpha_{22} = -\frac{-11 + 3i\sqrt{14} - 3i - 2\sqrt{14}}{-5i + \sqrt{14} - 13}. \tag{12}$$

The coordinates are indeed independent of the intrinsic parameters. After this illustration which helps for the understanding we will get now the coordinates using any camera motion. For that let us define $s = s_1\mathbf{e}_1 + s_2\mathbf{e}_2 + s_3\mathbf{e}_3 = [I|0]\underline{DC}_1$. Using this value, the epipole is $e_{12} = \underline{K}[I|0]\underline{DC}_1 = \underline{K}s$. Note that in this expression the intrinsic camera parameters are separate from the extrinsic ones. Similar as above using the general camera motion and the epipole value, the coordinates for the intersecting points read

$$\alpha_{11} = -\frac{(-s_3s_1s_2 + is_3\sqrt{s_3^2(s_1^2 + s_2^2 + s_3^2)} - is_3^3 - is_3s_1^2 + s_1\sqrt{s_3^2(s_1^2 + s_2^2 + s_3^2)} - is_2s_3^2)}{(-is_3s_1s_2 - s_3\sqrt{s_3^2(s_1^2 + s_2^2 + s_3^2)} - s_3^3 - s_3s_1^2 + is_1\sqrt{s_3^2(s_1^2 + s_2^2 + s_3^2)} + s_2s_3^2)}$$

$$\alpha_{21} = \frac{-2s_3(s_3^2 + s_1^2)}{-is_3s_2s_1 - s_3\sqrt{s_3^2(s_1^2 + s_2^2 + s_3^2)} - s_3^3 - s_3s_1^2 + is_1\sqrt{s_3^2(s_1^2 + s_2^2 + s_3^2)} + s_2s_3^2}$$

$$\alpha_{12} = \frac{(-1 - i)(is_1s_2 + \sqrt{s_3^2(s_1^2 + s_2^2 + s_3^2)})s_3}{-is_3s_1s_2 - s_3\sqrt{s_3^2(s_1^2 + s_2^2 + s_3^2)} + s_3s_1^2 + s_3^3 + s_1\sqrt{s_3^2(s_1^2 + s_2^2 + s_3^2)} - is_2s_3^2}$$

$$\alpha_{22} = \frac{i(is_3s_1s_2 + s_3\sqrt{s_3^2(s_1^2 + s_2^2 + s_3^2)} + is_1\sqrt{s_3^2(s_1^2 + s_2^2 + s_3^2)} + s_2s_3^2 + is_3s_1^2 + is_3^3)}{-is_3s_1s_2 - s_3\sqrt{s_3^2(s_1^2 + s_2^2 + s_3^2)} + s_3s_1^2 + s_3^3 + s_1\sqrt{s_3^2(s_1^2 + s_2^2 + s_3^2)} - is_2s_3^2}. \tag{13}$$

Note that also in the general case the intrinsic parameters are totally cancelled out. These invariance properties can be used to obtain equations which depend on the four unknown intrinsic camera parameters. For this we suppose point correspondences between two cameras and the motion parameters. First, the values of the invariant homogeneous α_i can be calculated by the known motion and formulas (13). Second, we calculate the epipole from the point correspondences, calculate $\overline{K}l_c$ and solve an equation system for τ similar to (10) to achieve a polynomial depending on the intrinsic camera parameters which must be equal to the calculated values of the α_i's by (13). Thus, we should find another set of equations to solve the problem. The way to do that is simply to consider the second camera with its epipole e_{21}. Since we are assuming that the intrinsic parameters remain constant, we can consequently gain a second set of four equations depending again of the four intrinsic parameters from the second epipole.

The interesting aspect here is that we require only one camera motion to find a solvable equation system. Other methods gain for each camera motion only a couple of equations, thus they require at least three camera motions to solve the problem [4].

2.2 Computing non–stationary intrinsic parameters

As a difference we compute now the line l_c using the fundamental matrix and a point lying at infinity of the second camera which is equal to the cross product of the second epipole with the point at infinity, namely $l_c = \overline{F}(e_1 + \tau' e_2)$.

Now similar as in previous case we will use an example for facilitating the understanding. We will use the same camera motion given in equation (7) and suppose orthogonal image axis.

Similar as above we compute the α's and according Pascal's theorem we gain a polynomial similar as equation (9). We select one of both solutions of the gained second order polynomial and substitute it in the homogeneous coordinates of the α's given by

$$\alpha_{11} = -\frac{i(-5i - 4 + i\sqrt{14})}{5i + 2 + 2i\sqrt{14}} \qquad \alpha_{21} = \frac{-2 + 3i\sqrt{14}}{5i + 2 + 2i\sqrt{14}} \tag{14}$$

$$\alpha_{12} = \frac{10 - 10i}{-4i - 2 + 3i\sqrt{14} - \sqrt{14}} \qquad \alpha_{22} = -\frac{8 + 6i - \sqrt{14} + 3i\sqrt{14}}{-4i - 2 + 3i\sqrt{14} - \sqrt{14}}. \tag{15}$$

Now we will consider the general motion

$$[R|t] = \begin{pmatrix} r_{11} & r_{12} & r_{13} & t_1 \\ r_{21} & r_{22} & r_{23} & t_2 \\ r_{31} & r_{32} & r_{33} & t_3 \end{pmatrix}. \tag{16}$$

In matrix algebra the fundamental matrix reads

$$F = K^{-T} E K'^{-1} = \begin{pmatrix} k_{11} & 0 & k_{13} \\ 0 & k_{22} & k_{23} \\ 0 & 0 & 1 \end{pmatrix}^{-T} \begin{pmatrix} E_{11} & E_{12} & E_{13} \\ E_{21} & E_{22} & E_{23} \\ E_{31} & E_{32} & E_{33} \end{pmatrix} \begin{pmatrix} k'_{11} & 0 & k'_{13} \\ 0 & k'_{22} & k'_{23} \\ 0 & 0 & 1 \end{pmatrix}^{-1} \tag{17}$$

and in geometric algebra the operator of the fundamental matrix reads $\underline{F} = \underline{K}^{-1} \underline{E} \underline{K}'^{-1}$. The matrix $E = [t]_\times R_1$ is the so called essential matrix. We can compute the α's using this general expression for $\underline{F}$ and get again equations fully independent of the intrinsic parameters. Together with the equations of the α's obtained using the first epipole, the intrinsic parameters can be found by solving a quadratic equation system.

3 Experimental Analysis

In this section we present the test of the method based on Pascal's theorem using firstly simulated images. We will explore the effect of increasing noise in the computation of the intrinsic camera parameters. The experiments with real images show that the performance of the method is reliable.

3.1 Experiments with simulated images

In order to test the performance of our approach we carried out a motion of the camera about the y–axis and a small translation along the three camera axes by increasing noise. For the tests we used exact arithmetic instead of floating point arithmetic. The Table 1 shows the computed intrinsic parameters. The most right column of the table shows the value obtained by substituting these parameters in the polynomial (9) which gives zero for the case of zero noise. The values in this column show that by increasing noise the computed intrinsic parameters cause a tiny deviation of the ideal value of zero. This indicates that the procedure is relatively stable against noise. Note that there are remarkable deviations shown by noise 1.25.

Noise(pixels)	k_{11}	k_{13}	k_{22}	k_{23}	Error
0	500	256	500	256	10^{-8}
0.5	504	259.5	503.5	258	0.004897
1	482	242	485	254	0.011517
1.25	473	220	440	238	0.031206
1.5	517	272	518	266	0.015
2	508	262.5	504	258.5	0.006114

Table 1. Intrinsic parameters by rotation about the y–axis and translation along the three axes by increasing noise.

3.2 Experiments with real images

In this section we present experiments using real images with one general camera motion as shown in Figure 2. The motion was done about the three coordinate axes. We used a calibration dice and for comparison purposes we computed the intrinsic parameters from the involved projective matrices by splitting the intrinsic parameters from the extrinsic ones. The reference values were: First camera $k_{11} = 1200.66$, $k_{22} = 1154.77$, $k_{13} = 424.49$, $k_{23} = 264.389$ and second camera $k_{11} = 1187.82$, $k_{22} = 1141.58$, $k_{13} = 386.797$, $k_{23} = 288.492$ with mean errors of 0.688 and 0.494, respectively. Thereafter using the gained parameters $[R_1|t_1]$ and $[R_2|t_2]$ we computed the $[R|t]$ between cameras which is required for the Pascal's theorem based method. The fundamental matrices were computed using a non-linear method. Using the method of Pascal's theorem with 12 point correspondences unlike 160 point correspondences used by the algorithm with the calibration dice we computed the following intrinsic parameters $k_{11} = 1244$, $k_{22} = 1167$, $k_{13} = 462$ and $k_{23} = 217$. These values resemble quite well to the reference ones and cause an error of $\sqrt{|eqn_1|^2 + ... + |eqn_8|^2}$: 0.00496045 in the error function, where eqn_i are the constraint equations depending on the intrinsic parameters. The difference to the reference values is attributable to inherent noise in the computation and to the fact that the reference values are not exact, too.

Since a system of quadratic equations is to be solved we resort to an iterative procedure for finding the solution. First, we tried the *Newton–Raphson method* and the *Continuation method* [4]. These methods were not practicable enough due to their complexity. We used instead a variable in size window minima search which through the computation ensures the reduction of the quadratic error. This simply approach worked faster and reliable.

In order to visualize how good we gain the epipolar geometry we superimposed the epipolar lines for some points using the reference method and method of Pascal's theorem. In both cases we computed the fundamental matrix in terms of their intrinsic parameters, i.e. $F = K^{-T}([t]_{\times}R)K^{-1}$. Figure 2.r shows this comparison. It is clear that both methods give quite similar epipolar lines and interesting enough it is shown that the intersecting point or epipole coincides almost exactly.

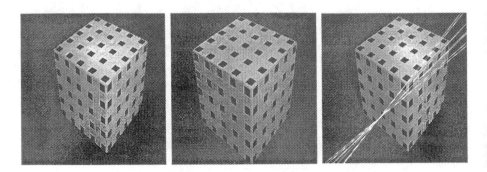

Fig. 2. Scenario and (r.) superimposed epipolar lines using the reference and Pascal's theorem based method.

4 Conclusions

This paper presents a geometric approach to compute the intrinsic camera parameters in the geometric algebra framework using Pascal's theorem. We adopted the projected characteristics of the absolute conic in terms of Pascal's theorem to propose a new camera calibration method based on geometric thoughts. The use of this theorem in the geometric algebra framework allows us the computation of a projective invariant using the conics of only two images. Then, this projective invariant expressed in terms of brackets helps us to set enough equations to solve the calibration problem. Our method requires to know the point correspondences and the values of the camera motion. The method gives a new point of view for the understanding of the problem thanks to the application of Pascal's theorem and it also explains the overseen role of the projective invariant in terms of the brackets. Using synthetic and real images we show that the method performs efficiently without any initialization or getting trapped in local minima.

References

1. Hartley, R. I. 1994. An algorithm for self–calibration from several views. In *Proc. Conference on Computer Vision and Pattern Recognition*, Seatle, WA, pp. 908–912.
2. Hestenes, D. and Ziegler, R. 1991. Projective geometry with Clifford algebra. *Acta Applicandae Mathematicae*, 23: 25–63.
3. Lasenby J. and Bayro-Corrochano E. 1997. Computing 3–D projective invariants from points and lines. In *G. Sommer, K. Daniilidis, and J. Pauli, editors, Computer Analysis of Images and Patterns, 7th Int. Conf., CAIP'97, Kiel, Springer-Verlag*, Sept., pp. 82-89.
4. Luong Q. T. and Faugeras O. D. 1997. Self–calibration of a moving camera from point correspondences and fundamental matrices. *International Journal of Computer Vision*, 22(3), pp. 261–289.
5. Maybank S. J. and Faugeras O. D. 1992. A theory of self–calibration of a moving camera. *International Journal of Computer Vision*, 8(2), pp. 123–151.

Calibration Update Technique for a Zoom Lens

Marina Kolesnik

GMD - German National Research Center for Information Technology
Schloss Birlinghoven, Sankt-Augustin, D-53754 Germany
E-mail: kolesnik@gmd.de

Abstract. A technique for a zoom lens camera is presented that enable to update continuously camera intrinsic parameters while zooming. This is based on the fact that the spatial angle subtended by the optical center and the same two 3-D points in the standard camera coordinate system remains invariant for subsequent image frames on which the two 3-D points are located. The new focal length value is obtained as the solution of the quadratic equation we derive from the angular invariance. We give an error estimate to show how the uncertainty in the localization of corresponding points in the image frames may affect the accuracy of the updated value of the focal length. We report the results of two experiments when sequences of magnified images were taken by fixed and by rotating camera.

1. Introduction

Zoom lens cameras are widespread in surveillance, monitoring, tracking, robotic navigation or video conferencing. A typical scenario for an active zooming camera is the following: First, an area of special interest is located in the image plane. Second, the active camera changes its focal length to increase image resolution for the area of particular interest. Third, camera calibration update is done to carry out further investigations based on the image. The active camera can be tilting (for monitoring, video conferencing or tracking purposes) or mobile, mounted on a robot. In both cases the camera motions can be restricted while changing the focal length without any loss of the general functionality.

The calibration of zoom lens cameras is important to obtain metric information from images. Previous studies followed three major directions: 1) precise calibration for each possible position of the focus and zoom range [1]; 2) calibration update of the camera intrinsic parameters based on pre-calibration and tracking image points [2], [3]; 3) self-calibration of a zoom lens camera [4], [5], [6], [7] based on constraints due to the rigidity of the camera motion. The idea of self-calibration based on the invariance of the absolute conic under rigid transformations was introduced in the pioneering work [8]. A detailed guidance how to determine the principle point by zooming, calibration of lens distortion and focal length as well as some practical aspects are given by Li et al. [1]. Pre-calibrated zoom lens cameras are considered in [2] and [3] where it is noted that the change in focus induces 3x3 affine transformation between the image planes. Consequently, camera calibration is corrected by tracking 3 points while zooming. Several self-calibration approaches are based on applying the rigidity constraint (Kruppa's equations) to moving zoom lens cameras. Enciso et al. [6] perform the self-calibration of a stereo rig with varying intrinsic parameters. It is based on establishing the epipolar geometry and solving the Kruppa's equations. Heyden et al. [7] prove that the self-calibration for a single moving camera is possible when the aspect ratio is known and no skew is present. Pollefeys et al. [5] use linear and non-linear approaches to obtain self-calibration based on absolute quadric. All self-calibration methods rely on computationally extensive linear/non-linear minimization procedures, which make camera parameters consistent with the rigidity of the camera motion. Therefore they are of limited practical use for mobile robots that do not have adequate resources on-board. In addition the self-calibration methods are highly sensitive to noise, that leads to imprecision and inaccuracy.

An interesting attempt to cope with these difficulties is made by Sturm [4]. Interdependence of the internal parameters is introduced to reduce the number of unknowns. The coordinates of the principle point are modeled as polynomial functions of the scale factor. The camera is pre-calibrated by classical method [9] for several zoom positions to obtain the order and the coefficients of the polynomial model. The number of unknowns in the Kruppa's equations is thus reduced to one.

In this paper we do not attempt to present theoretical considerations or new algorithm of camera calibration, rather a practical technique to quickly update camera calibration while zooming. We suppose that the camera is classically calibrated for the original (reference) image frame with the minimum zoom value. Two experiments were conducted with the camera fixed to a tripod while zooming: 1) the camera did not move; 2) the camera performed mostly rotation.

We assume that the optical center of the camera stays at the same position between the two image frames while zooming. Strictly speaking, the camera center moves for a real camera. However, the motion of the center is negligible compared to the distance to object points. Therefore the spatial angle subtended by the optical center and the same two 3-D points in the camera coordinate system remains invariant for both image frames. This angular invariance is used to update the new focal length by tracking a pair of points in the image frames. The camera is modeled as a pinhole with four unknown internal parameters. Lens distortions must be removed before the focal length update is performed. We suppose that zooming changes only camera focal length while leaving the principle point fixed, i.e. the camera center moves along the optical axis.

The paper is organized as follows. The problem is posed in section 2. Its mathematical description is given in section 2.1. The effect of inaccurate localization of the corresponding points is investigated in section 2.2. The analysis of the experiments is presented in section 3. Concluding remarks are made in Section 4.

2. The focal length update

We consider a firmly mounted pinhole camera equipped with a zoom lens. The camera performs sequential image acquisition. Let us consider two subsequent image frames N and N+1. We make the following assumptions concerning the camera and the images:

1   The optical center of the camera does not move between the frames N and N+1.
2   The camera may perform rotation between the frames N and N+1. There should be enough overlap between the frames to establish reliable point correspondences.

3 Optical distortions are removed from both images. The camera internal parameters are known for the image frame N.

These conditions allow us to derive an equation that describes the variation of the focal length between the image frames N and N+1.

2.1. Writing equations

We work in the coordinate system attached to the camera (standard coordinate system). The perspective projection matrix P is modeled by four internal parameters [3]:

$$P = \begin{vmatrix} -\alpha_u & 0 & u_0 \\ 0 & -\alpha_v & v_0 \\ 0 & 0 & 1 \end{vmatrix} \tag{1}$$

where u_0, v_0 are the coordinates of the principle point and α_u, α_v are the image scaling factors for rows and columns, respectively. Let $\alpha_u, \alpha_v, u_0, v_0$ be the known internal parameters for the image frame N. Noting that α_u and α_v are linear functions of the focal length, the parameters $\alpha_u', \alpha_v', u_0', v_0'$ of the frame N+1 are defined as:

$$\{f' = \mu f\} \Rightarrow \{\alpha_u' = \mu\alpha_u, \alpha_v' = \mu\alpha_v, u_0' = u_0, v_0' = v_0\} \tag{2}$$

The focal lengths f, f' correspond to the frames N, N+1, respectively, and μ is the unknown variable. The equation of the optical ray $<C,a>$ defined by the pixel a and the camera optical center C is expressed as:

$$A = \lambda P^{-1}\tilde{a} \tag{3}$$

where $\tilde{a} = [a_1, a_2, 1]$ is the homogeneous coordinate vector of the pixel a, 3-D vector A defines a point on the optical ray $<C,a>$ and λ - varies between $-\infty$ and $+\infty$.

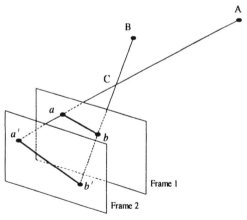

Figure 1. The geometry of the two consecutive images for the camera with zoom lens.

Let us consider the geometry of the two consecutive image frames (Fig.1). We select two 3-D points A and B in the scene and their images a, a' and b, b' on the image frames N and N+1, respectively. The angle defined by the optical rays $<a,C>$ and $<b,C>$ and the angle defined by the corresponding pixels $<a',C>$ and $<b',C>$ are physically the same angle expressed as:

$$\cos\varphi = \frac{AB}{|A||B|} \tag{4}$$

where AB is the scalar product of the 3-D vectors A and B. Combining (3) and (4) for the frames N and N+1 we can write:

$$\frac{(P_1^{-1}\tilde{a})^T (P_1^{-1}\tilde{b})}{\left|P_1^{-1}\tilde{a}\right|\left|P_1^{-1}\tilde{b}\right|} = \frac{(P_2^{-1}\tilde{a}^{\,\prime})^T (P_2^{-1}\tilde{b}^{\,\prime})}{\left|P_2^{-1}\tilde{a}^{\,\prime}\right|\left|P_2^{-1}\tilde{b}^{\,\prime}\right|} \tag{5}$$

where: $\tilde{a}, \tilde{a}^{\,\prime}, \tilde{b}, \tilde{b}^{\,\prime}$ are the homogeneous coordinates of the image pixels a, a', b and b', and P_1 and P_2 are the perspective projection matrices for the frames N and N+1 expressed in (1). Let M be the value of the left-hand side of (5) that does not contain the unknown variable. The expansion of (5) using expression (1) and (2) for matrices P_1 and P_2 yields:

$$\frac{\left(u_0 - a_1^{\,\prime}\right)\left(u_0 - b_1^{\,\prime}\right)}{\mu^2\alpha_u^2} + \frac{\left(v_0 - a_2^{\,\prime}\right)\left(v_0 - b_2^{\,\prime}\right)}{\mu^2\alpha_v^2} + 1 =$$

$$= M \sqrt{\frac{\left(u_0 - a_1^{\,\prime}\right)^2}{\mu^2\alpha_u^2} + \frac{\left(v_0 - a_2^{\,\prime}\right)^2}{\mu^2\alpha_v^2} + 1} \times \sqrt{\frac{\left(u_0 - b_1^{\,\prime}\right)^2}{\mu^2\alpha_u^2} + \frac{\left(v_0 - b_2^{\,\prime}\right)^2}{\mu^2\alpha_v^2} + 1}$$

From this we obtain a quadratic equation for the unknown $\mu_1 = \mu^2$:

$$\mu_1^2 (1 - M^2) + \mu_1 (2\Delta - M^2(\Delta_1 + \Delta_2)) + \Delta^2 - M^2\Delta_1\Delta_2 = 0$$

where:

$$\Delta = \frac{\left(u_0 - a_1^{\,\prime}\right)\left(u_0 - b_1^{\,\prime}\right)}{\alpha_u^2} + \frac{\left(v_0 - a_2^{\,\prime}\right)\left(v_0 - b_2^{\,\prime}\right)}{\alpha_v^2} \tag{6}$$

$$\Delta_1 = \frac{\left(u_0 - a_1^{\,\prime}\right)^2}{\alpha_u^2} + \frac{\left(v_0 - a_2^{\,\prime}\right)^2}{\alpha_v^2}$$

$$\Delta_2 = \frac{\left(u_0 - b_1^{\,\prime}\right)^2}{\alpha_u^2} + \frac{\left(v_0 - b_2^{\,\prime}\right)^2}{\alpha_v^2}$$

By construction the above equation must have a real positive root. Summarizing, we have obtained the following procedure to update calibration for the camera while zooming:

1. Compute correspondences for a pair of selected points in the two image frames N and N+1 taken by the camera with modified focal length.
2. Write down equation (6) for the pair of corresponding points.
3. Solve (6) for the real positive root and update the scaling factors α_u, α_v.

2.2. Error propagation

In this section we investigate how sensitive the solution of equation (6) is to possible errors in the localization of corresponding points. In other words the question we ask is this: assuming an error is made when establishing a correspondence in the frame N+1 for the pair of points in the frame N, how does it affect the value of μ, which is the solution of the equation (6)?

We assume that both corresponding points on the frame N+1 are shifted from their true position due to poor localization. Let d be the maximum shift error for the coordinates of the points a',b'. Rewriting the coefficients in (6) as the functions of d we obtain:

$$\mu_1^2(1-M^2)+\mu_1\,f_1(d)+f_2(d)=0$$

$$f_1(d)=2\Delta-M^2(\Delta_1+\Delta_2) \tag{7}$$

$$f_2(d)=\Delta^2-M^2\Delta_1\Delta_2$$

The Δ,Δ_1,Δ_2 are the following functions of d:

$$\Delta=\frac{\left(u_0-a_1^1-d\right)\left(u_0-b_1^2-d\right)}{\alpha_u^2}+\frac{\left(v_0-a_2^1-d\right)\left(v_0-b_2^1-d\right)}{\alpha_v^2}$$

$$\Delta_1=\frac{\left(u_0-a_1^1-d\right)^2}{\alpha_u^2}+\frac{\left(v_0-a_2^1-d\right)^2}{\alpha_v^2}$$

$$\Delta_2=\frac{\left(u_0-b_1^1-d\right)^2}{\alpha_u^2}+\frac{\left(v_0-b_2^1-d\right)^2}{\alpha_v^2}$$

We introduce a value S to denote the generalized size of the CCD matrix in pixels. Without loss of generality we can select a pair of 3-D points in such a way that the distance between the image points a and b in the frame N $(a'$ and b' in the frame N+1) is of the order of 1/2S. Hence we assume:

$$\alpha_u\sim\alpha_v\sim kS$$

$$(u_0-a_1^1)\sim(v_0-a_2^1)\sim(u_0-b_1^1)\sim(v_0-b_2^1)\sim 1/2\,S$$

We obtain the following assessments for the derivatives of the Δ,Δ_1 and Δ_2 at $d=0$:

$$\left.\frac{\partial\Delta}{\partial d}\right|_{d=0}=\frac{-2u_0+a_1^1+b_1^1}{\alpha_u^2}+\frac{-2v_0+a_2^1+b_2^1}{\alpha_v^2}\sim\frac{2}{k^2S}$$

$$\left.\frac{\partial\Delta_1}{\partial d}\right|_{d=0}\approx\left.\frac{\partial\Delta_2}{\partial d}\right|_{d=0}\sim\frac{2}{k^2S} \tag{8}$$

Using first order Taylor expansion in the small variable d, the above approximations (8) and considering $M\sim 1/2$, we obtain the following assessments for the functions f_1 and f_2:

$$f_1\approx f_1|_{d=0}+d\,\frac{5}{k^2S}\,;\qquad f_2\approx f_2|_{d=0}+d\,\frac{2}{k^4S}\,;$$

Similarly, for the discriminant D of the equation (7) we obtain: $D\approx D|_{d=0}+d\,\dfrac{4}{k^4S}$

The first order Taylor expansion in the small d for μ, which is the solution of (7), gives:

$$\mu\approx\mu_0+d\,\frac{4}{kS}$$

It follows that the error $d=n$ (pixels) in localizing the corresponding point induces the distortion for μ: $\delta\mu\sim4n/kS$. Assuming for instance that $d \sim S/400$ (1-2 pixels), $k\sim1/2$ and the $\mu_o\sim1$ the expected variation for μ is equal to: $\delta\mu\sim2/100\sim 2\%$.

The assessment above is the possible error of one step. If the calibration update is performed N times in the sequence, the expected cumulative error is $Sqrt(N)$ times larger. It is valid under assumption that errors made in each step are independent and identical, i.e. they can be treated as independent probability variables. Summarizing we conclude that the coefficient μ that updates the value of the focal length is reasonably stable with respect to poor localization of the corresponding points within pixels.

3. Experimental results

In our experiment we used a firmly fixed CCD camera that has 680x480 pixels resolution. The camera was equipped with the manually adjustable zoom lens. The focal length of the lens could be optionally changed within the range 4.5~10mm, which corresponds to the ~25°-45° variation in the camera viewing angle. We used a correction matrix to remove optical distortions from the images before the calibration update. We computed the correction matrix with the help of commercially available software package [11] that adjusts distortion parameters to be consistent with arbitrary u_o, v_o, α_u and α_v. Two sequences of 19 images were similarly acquired in the two experiments. The camera did not move during the first sequence and rotated (the optical center was not on the axis of rotation) during the second sequence. The first image frames in both sequences (acquired with the minimal focal length) were used as the reference frames. We applied angular calibration approach [12] to obtain the camera internal parameters for the reference frame. Images with the numbers 2-11 were acquired with gradually increasing focal length. Images with the numbers 12-19 were acquired with the gradually decreasing focal length back to its minimum. Accordingly, the focal length of the last 19th frame eventually coincides with the focal length of the first reference frame. Three distinctive pairs of manually selected points with the pixel accuracy were tracked to update the camera internal calibration. The results of the final focal length update presented in the tables are the average of the three independent calculations applied for the three point pairs. Two different calibration updates were carried out on the basis of each sequence:

1. The focal length for each image frame was updated on the basis of the reference frame 1 by tracking the same pair of points. The results are presented in the columns "independent" of tables 1,2. The last 18th updating coefficients are eventually 1, which is in a good agreement with the data: the first and the last image frames were acquired with practically identical camera calibration. The coefficient 1.042488205 (table 2) varies more from 1. This is due to the fact that the actual values of the spatial angles were a bit distorted because of the motion of the optical center during rotation.

2. The focal length for each image frame was updated sequentially, i.e. on the basis of the previous image frame. The results based on tracking the same pairs of points are shown in the column "Sequential:1" of table 1. Different point pairs were used for the similar update shown in the column "Sequential:2" of table 2. Obviously, by construction of the

experiment, the following equivalence between "independent" update coefficient μ_{ind}^{i} for the frame i, and the respective "sequential" update coefficient μ_{seq}^{i} must hold:

$$\mu_{ind}^{i} = \mu_{seq}^{1} \times \mu_{seq}^{2} \times ... \times \mu_{seq}^{i} \tag{9}$$

The results of the recalculations based on (9) are shown in the columns "Accumulated:1" (for the same point pairs) and "Accumulated:2" (for the different point pairs) of tables 1,2. One can see almost perfect correspondence comparing respective frame numbers for the columns "Independent" and "Accumulated:1" in both tables. The difference between the values in the column "Accumulated:2 " and the correspondent values in the column "Independent" is most significant for the frames with bigger magnification. This may be explained by the fact that bigger spatial angles were used while tracking different point pairs in the sequence. These angles were distorted much more due to the camera rotation when compared to the smaller spatial angles used while tracking the same point pairs in the image sequence. The small overlap (about 25%) between the first and the last image frames in the sequence did not allow us to choose another point pairs. It must be noted that there is no accumulated error for the 17th and 18th coefficients in the column "Accumulated:2", whose values virtually coincide with the respective coefficients in the column "Independent" (table 2).

Frames N->N+1	μ: Independent	μ: Accumulated:1		Frames N->N+1	μ: Sequential:1
1->2	1.049890601	1.049890601		1->2	1.049890601
1->3	1.097537652	1.097557901		2->3	1.045402159
1->4	1.135482723	1.135504318		3->4	1.034573499
1->5	1.207894076	1.207922163		4->5	1.063775931
1->6	1.278697066	1.278721380		5->6	1.058612400
1->7	1.403922381	1.403937068		6->7	1.097922573
1->8	1.538290268	1.538299311		7->8	1.095703893
1->9	1.682791704	1.682793591		8->9	1.093931187
1->10	1.824675178	1.824632956		9->10	1.084288035
1->11	1.880829435	1.880754496		10->11	1.030757715
1->12	1.659055838	1.659060637		11->12	.8821250410
1->13	1.490284159	1.490352203		12->13	.8983108694
1->14	1.380356705	1.380407216		13->14	.9262288558
1->15	1.317993310	1.318039562		14->15	.9548193801
1->16	1.239971917	1.240075530		15->16	.9408484888
1->17	1.158752376	1.158856006		16->17	.9345043729
1->18	1.079144661	1.079240295		17->18	.9312980125
1->19	1.000661276	1.000755912		18->19	.9272781204

Table 1. Results of the focal length update acquired by the still camera while zooming. The same point pairs were tracked in the image sequence to compute the new calibration.

Frames N->N+1	μ: Independent	μ: Accumulated:1	μ: Accumulated:2	Frames N->N+1	μ: Sequential:2
1->2	1.070402008	1.070402008	1.075321241	1->2	1.075321241
1->3	1.116654620	1.116655490	1.115047313	2->3	1.036943446
1->4	1.164474737	1.164473855	1.170720176	3->4	1.049928700
1->5	1.236684749	1.236689917	1.250940626	4->5	1.068522309
1->6	1.310590102	1.310579969	1.321264655	5->6	1.056216920
1->7	1.402384615	1.402360640	1.424322492	6->7	1.077999390
1->8	1.448220039	1.448112147	1.482660128	7->8	1.040958165
1->9	1.539923256	1.539776686	1.584254909	8->9	1.068521962
1->10	1.658449511	1.656176613	1.732064637	9->10	1.093299208
1->11	2.046914720	2.046781367	2.184242151	10->11	1.261062725
1->12	1.691119884	1.671162877	1.771565001	11->12	.8110662092
1->13	1.504400948	1.504561014	1.587320114	12->13	.8959987997
1->14	1.397044734	1.397247668	1.468390930	13->14	.9250754885
1->15	1.285414584	1.285670698	1.358834963	14->15	.9253904630
1->16	1.217636632	1.217937349	1.279259685	15->16	.9414385997
1->17	1.180938083	1.181194907	1.188534749	16->17	.9290801255
1->18	1.074897834	1.075137297	1.079197686	17->18	.9080068440
1->19	1.042488205	1.042934536	1.044412084	18->19	.9677671638

Table 2. Results of the focal length update acquired by the rotating camera while zooming. The same point pairs with relatively small spatial angles (overlap problem) were used in the columns "Independent" and "Accumulated:1". Different point pairs with the spatial angle about half viewing angle of the camera were used in the columns "Sequential:2" and "Accumulated:2".

4. Conclusion

We have presented an efficient approach how to update the value of the focal length for a zoom lens. The method is based on the spatial angular invariance for a particular pair of the image points traced in the subsequent image frames. The positive root of the quadratic equation gives the updating coefficient for the new focal length value. The method substantially relies on tracking the corresponding points in the image frames. It is shown that the method is stable with respect to a possible poor localization of the corresponding points.

There are several advantages of the angle-based focal length update approach. Firstly, it gives simple quadratic equation that is easy to solve compared to the computationally extensive minimization procedures used by the self-calibration methods. Second, limited number of point correspondences that is necessary for calibration update can be reliably established for subsequent image frames. The condition that restricts camera translation between two subsequent image frames while zooming is provided in many applications.

The drawback of the method is the assumption that the principle point of the camera remains fixed while zooming. In fact, zooming can shift the principle point by tens of pixels depending on the lens technology used in the camera [13]. Evidently, by keeping the principle point fixed we might produce a bias in the camera calibration. It is the subject for future research to include in our calibration update the effect of varying principle point.

Our experiments have shown highly stable performance for the sequential recalculations of the focal length. No accumulated error is detected over the sequence of 19 image frames for the rotating camera when different point pairs were used for calibration update. Despite we used manually selected pairs of correspondent points, we are confident that automatic correspondence searching procedure will provide reliable update of the focal length.

Acknowledgements

The author is indebted to Mr. G. Paar (JOANNEUM Research, Austria) for his professional support in removing optical distortions from the original images.

References

1. M.Li and J.M. Lavest: Some Aspects of Zoom Lens Camera Calibration. *IEEE Transactions on Patt. Anal. And Mach. Intell.* Vol.18, No.11, November (1996) 1105-1110.
2. J.L.Crowley, P. Bobet and C. Schmid: Maintaining Stereo Calibration by Tracking Image Points. *CVPR*, (1993) 483-488.
3. R. Enciso, T, Vieville.and A. Zisserman: An affine solution to Euclidean calibration for a zoom lens. *RobotVis Project INRIA BP 93, FR-06902,* Sophia-Antipolis, France.
4. P. Sturm: Self-calibration of a moving zoom-lens camera by precalibration. *Image and Vision Computing.* Vol.15, (1997) 583-589.
5. M. Pollefeys, R. Koch and L. Van Gool: Self-Calibration and Metric Reconstruction in spite of Varying and Unknown Internal Camera Parameters. *Proc. ICCV,* (1998)
6. R. Enciso and T. Vieville: Self-calibration from four views with possibly varying intrinsic parameters. *Image and Vision Computing.* Vol.15, (1997) 293-305.
7. A. Heyden and K.Aström: Euclidian Reconstruction from Image Sequences with Varying and Unknown Focal Length and Principle Point. *Proc. CVPR'97* (1997).
8. S. J. Maybank and O. D. Faugeras: A Theory of Self-Calibration of a Moving Camera. *International Journal of Computer Vision.* 8:2, (1992) 123-151.
9. O. D. Faugeras and G. Toscani: Camera calibration for 3D computer vision. *Proc. 2nd ECCV. Santa Margherita,* May (1992) 563-578.
10. O. Faugeras: Three-Dimensional Computer Vision. A Geometric Viewpoint. MIT Press, Cambridge, MA, (1993) 55-65.
11. H. Kager (ed.): ORIENT, a universal photogrammetric adjustment system. *Reference Manual.* Inst. for Photogrammetry and Remote Sensing, Technical University Vienna, (1995)
12. M. Kolesnik: Using Angles for Internal Camera Calibration. *GMD Report #55,* Sankt Augustin, Germany (1999).
13. R.G.Willson and S.A. Shafer. What is the Center of the Image? *Proc. CVPR'93.* June 15-17, New York (1993)

Which Slightly Different View is the Right One?*

Jasna Maver[1,2]

University of Ljubljana
[1]Faculty of Computer and Information Science,
[2]Faculty of Arts
Aškerčeva 2, 1000 Ljubljana, Slovenia
Jasnam@fri.uni-lj.si

Abstract. In the paper we consider a problem of the next view planning using stereo vision. Our objective is to resolve the ambiguities of occluded regions — parts of the scene which are acquired only in a one of the images of the two stereo views. We assume a general scene. The next view problem is first analyzed under the assumption that the next camera location can lie only in the particular epipolar plane, then we extend the analysis to a general set of possible camera locations. An important feature of the proposed approach is that it does not require 3-D reconstruction of the scene. Experimental results demonstrate the proposed solution.

keywords: the next view planning, view synthesis methods, stereo vision

1 Introduction

Recently there has been a great interest in view synthesis methods, which generate intermediate images of the scene from images acquired at nearby viewpoints. Generation of an intermediate image may not be possible due to occlusions, i.e., parts of the scene which are acquired only in one of the images of nearby viewpoints. To resolve the occlusions the authors in the literature suggest the acquisition of additional images from slightly different viewpoints [1, 7].

Selection of nearby viewpoints can be planned and represented as an exploration task. In the literature, the space which has to be explored can be determined in different ways:

1. By *the bounds of the scene*. The bounds can be defined explicit, by the X, Y, Z coordinates of the bounding surface of the scene, perhaps by a sphere. Another possibility is to define the domain of points which have to be explored in an abstract way, like it is in the work of Kutulakos et al. [4]: The scene to be explored is a complete surface of an object.

* This work was supported by a grant from the Ministry of Science and Technology of Republic of Slovenia (Projects S2-7978-0781-98, J2-0414-1539-98).

2. By *the set of possible viewpoints from which the images of the scene can be acquired.* Usually the camera can not acquire the images of all the surface points of a particular scene, due to limitations of camera locations, e.g., due to obstacles. The set of possible camera locations can also be chosen to serve the proposed method. An example is a method for voxel coloring [5]: The set of possible camera locations are all the points which satisfy the ordinal visibility constraint.

3. By *the set of viewpoints from which the scene is going to be reconstructed or visualized.* For example: The idea of the view interpolation methods is to synthesize only those views lying on the line connecting two reference views.

The scene of our interest are parts of the scene which are acquired only in a one of the images of the two stereo views. We assume a general scene. The problem of the next view selection is first analyzed under the assumption that the next camera location can lie only in the particular epipolar plane, then we extend the analysis to a general set of possible camera locations. An important feature of the proposed approach is that it does not require 3-D reconstruction of the scene.

In the paper we are not solving the correspondence problem, i.e., finding for each point in one image the matching point in the other. We assume that an algorithm for finding corresponding pairs of pixels in two stereo views is available and the algorithm is able to determine occluding points. In [2,3] the authors showed that occlusions can be detected by searching for disparity discontinuities.

In Section 2 we describe the geometry and introduce the assumptions. In Section 3 we formulate the problem and in Section 4 we propose the solution. Experimental results are demonstrated in Section 5.

2 Assumptions and geometry of the camera

Let $\mathcal{R}$ be the retinal plane of the camera with the center C. The camera coordinate system is located at C. Z axis is aligned with camera's optical axis, is orthogonal to $\mathcal{R}$, and points toward $\mathcal{R}$. $\mathcal{R}$ lies at $Z = f$, where f denotes a camera focal length. X and Y coordinates of the camera coordinate system are parallel to x and y axes of $\mathcal{R}$, respectively.

A point M occludes point N in the retinal plane $\mathcal{R}$ if M lies on the line $\langle C, M \rangle \equiv \langle C, N \rangle$ in front of N. By rotating the camera about its origin, i.e., by changing the orientation of $\mathcal{R}$ the point N remains occluded since the ordering of the points M and N on the line $\langle C, M \rangle$ with respect to C stay the same. It follows: A change in camera orientation does not resolve occlusions. Occlusions can only be resolved by changing the camera position.

We assume two stereo views. Retinal planes $\mathcal{R}_1$, $\mathcal{R}_2$ belong to the first and the second camera, respectively. Since the change in camera orientation does not resolve occlusions and since there exists a projective transformation of one image to another which allows the gaze direction to be modified after the photograph is taken [6], we assume that planes $\mathcal{R}_1$ and $\mathcal{R}_2$ are parallel. The center of the first camera is assumed to lie at the origin of the world coordinate system:

$C_1 = (0,0,0)$. The center of the second camera C_2 is located at (T_x, T_y, T_z) (Fig. 1 left). The epipolar plane $\mathcal{E}_{C_1,C_2,P}$, going through the scene point P and the two camera centers C_1 and C_2, intersects $\mathcal{R}_1$ and $\mathcal{R}_2$ along epipolar lines e_1 and e_2, respectively.

Let $p_1 = (x_{p,1}, y_{p,1})$ and $p_2 = (x_{p,2}, y_{p,2})$ denote the projection of a 3D point P in $\mathcal{R}_1$ and $\mathcal{R}_2$, respectively. $P = (X_{P,1}, Y_{P,1}, Z_{P,1}) = (\frac{x_{p,1} Z_{P,1}}{f}, \frac{y_{p,1} Z_{P,1}}{f}, Z_{P,1})$. The coordinates of P in the coordinate frame of the second camera are:

$$(X_{P,2}, Y_{P,2}, Z_{P,2}) = (X_{P,1} - T_x, Y_{P,1} - T_y, Z_{P,1} - T_z).$$

Hence, the projection of P in $\mathcal{R}_2$ is

$$\begin{pmatrix} x_{p,2} \\ y_{p,2} \end{pmatrix} = \begin{pmatrix} f\frac{X_{P,1}-T_x}{Z_{P,1}-T_z} \\ f\frac{Y_{P,1}-T_y}{Z_{P,1}-T_z} \end{pmatrix} = \begin{pmatrix} \frac{x_{p,1} Z_{P,1}-T_x f}{Z_{P,1}-T_z} \\ \frac{y_{p,1} Z_{P,1}-T_y f}{Z_{P,1}-T_z} \end{pmatrix}. \tag{1}$$

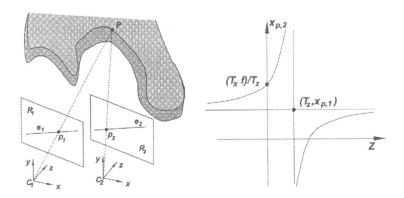

Fig. 1. left) Stereo geometry of two identical parallel cameras. right)An image of $x_{p,2}$. A similar image can be drown for $y_{p,2}$.

By moving point P on the line $\langle C_1, p_1 \rangle$, its projection in $\mathcal{R}_2$ changes by the homography (1). The homography (1) is monotone with regard to Z coordinate (Fig. 1 right). An interesting fact is that the point at infinity[1] P_∞ always projects to the point $(x_{p,1}, y_{p,1})$ no matter where is the new center of the camera located.

3 Points that are visible only in one of the stereo views

We begin by considering situations in which one part of the scene is visible only in a one of the two stereo views. Let points of the interval (p_2, q_2) on e_2 in $\mathcal{R}_2$ have no corresponding matches on e_1 in $\mathcal{R}_1$ (see all three cases of fig. 2). The points p_2 and q_2 correspond to p_1 and q_1 on e_1, respectively.

---

[1] The term "at infinity" is often used instead of "ideal". *Ideal point of a line*—A fictitious point representing the direction of a straight line; it is common to the line and to all parallel lines.

447

We require the existence of an empty triangle $\triangle C_1 C_2, S$ (Fig. 2). This practically means that between the two camera centers C_1, C_2, in the scene, there does not lie any object. The point S is the intersection of the line $\langle C_2, p_2 \rangle$ with the line $\langle C_1, q_r \rangle$. q_r is the right most point on e_1 (looking at e_1 from C_1) which stereo pair is either p_2 or one of the point on the left side of p_2 on e_2 (looking at e_2 from C_2). Fig. 2 depicts three different situations. The point Q is the intersection of the line $\langle C_1, q_r \rangle$ with the line $\langle C_2, q_2 \rangle$.

The role of the first and the second retinal plane are interchangeable.

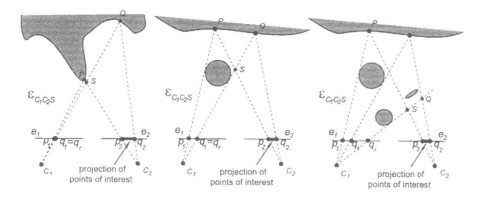

Fig. 2. Different stereo situations.

3.1 Domain of occluded points

To the interval (p_2, q_2) on e_2 in $\mathcal{R}_2$ projects the area of the epipolar plane $\mathcal{E}_{C_1, C_2, S}$ inside the sector determined by the three points: S_∞, C_2, Q_∞ (Fig. 3).

Observation 1: The triangle $\triangle(S, C_2, Q)$ is empty.

Proof: Let in $\triangle(S, C_2, Q)$ there exist a set of object points and let O include all of them. Some points of O project in $\mathcal{R}_2$ to the interval (p_2, q_2). Let us denote them as $O_{(p_2, q_2)}$. Points of (p_2, q_2) do not have stereo pairs therefore, all the points of $O_{(p_2, q_2)}$ are occluded in $\mathcal{R}_1$. Since we require $\triangle(C_1, C_2, S)$ to be empty, the points of $O_{(p_2, q_2)}$ can only be occluded in $\mathcal{R}_1$ by the points from O. Let o_c be the closest point of O to $\mathcal{R}_2$. o_c is also the closest point of $O_{(p_2, q_2)}$ to $\mathcal{R}_2$. Hence, o_c is either not occluded in $\mathcal{R}_1$ or $\triangle(C_1, C_2, S)$ is not empty.

In accordance with the Observation 1 the domain of occluded points $\mathcal{D}_{S, C_2, Q}$ is the area inside the sector $\angle(S_\infty, C_2, Q_\infty)$ above the line $\langle C_1, q_r \rangle$. It is determined by four points: S_∞, S, Q, Q_∞ (Fig. 3).

4 Possible camera locations

Our task is to compute the next camera location C_N to acquire the second image of object points from domain $\mathcal{D}_{S, C_2, Q}$ whose projection is (p_2, q_2). Since

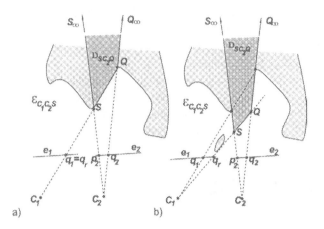

Fig. 3. Domain of occluded points $\mathcal{D}_{S,C_2,Q}$ a) the case where $q_1 = q_r$ b) the case where $q_1 \neq q_r$.

the points of our interest can lie anywhere in $\mathcal{D}_{S,C_2,Q}$, we first have to determine camera locations where none of the object points outside $\mathcal{D}_{S,C_2,Q}$ occludes none of the points of $\mathcal{D}_{S,C_2,Q}$.

4.1 C_N is constrained to lie in epipolar plane $\mathcal{E}_{C_1,C_2,S}$

By selecting the next camera location C_N in $\mathcal{E}_{C_1,C_2,S}$ we preserve the epipolar line belonging to $\mathcal{E}_{C_1,C_2,S}$. The scene points which may potentially occlude $\mathcal{D}_{S,C_2,Q}$ lie on $\mathcal{E}_{C_1,C_2,S}$. We define the obstacle points to be the scene points outside $\mathcal{D}_{S,C_2,Q}$.

None of the obstacle points occludes none of the point of $\mathcal{D}_{S,C_2,Q}$ if the scene is empty in the area from C_N to $\mathcal{D}_{S,C_2,Q}$ for all optical lines of C_N which intersect $\mathcal{D}_{S,C_2,Q}$, hence, the area between $\mathcal{D}_{S,C_2,Q}$ and the two "tangents" t_1, t_2 to $\mathcal{D}_{S,C_2,Q}$ intersecting at C_N has to be empty.

The bounding lines of $\mathcal{D}_{S,C_2,Q}$: $\langle S, Q \rangle$, $\langle C_2, Q \rangle$, and $\langle C_2, S \rangle$ divide the region outside $\mathcal{D}_{S,C_2,Q}$ into six parts (Fig. 4).

- *Area I:* Area I is the intersection of points lying above the line $\langle C_1, Q \rangle$ with the points lying below the line $\langle C_2, S \rangle$. $t_1 = \langle C_N, S_\infty \rangle$, $t_2 = \langle C_N, S \rangle$, hence $t_1 \parallel \langle C_2, S_\infty \rangle$ (Fig. 4 b).
- *Area II:* Area II is the intersection of three half-planes: points lying above the line $\langle C_2, Q \rangle$, points lying below the line $\langle C_2, S \rangle$, and points lying below the line $\langle S, Q \rangle$. $t_1 = \langle C_N, S_\infty \rangle$, $t_2 = \langle C_N, Q \rangle$, hence $t_1 \parallel \langle C_2, S_\infty \rangle$ (Fig. 4 c).
- *Area III:* Area III is the triangle $\triangle(S, C_2, Q)$. This area is of special importance. C_N can be placed anywhere in the triangle $\triangle(S, C_2, Q)$ since $\triangle(S, C_N, Q) \subset \triangle(S, C_2, Q)$ and in accordance with Observation 1, $\triangle(S, C_2, Q)$ is empty. $t_1 = \langle C_N, S \rangle$, $t_2 = \langle C_N, Q \rangle$ (Fig. 4 d).

- *Area IV:* Area IV is the intersection of three half-planes: points lying below the line $\langle C_2, Q \rangle$, points lying above the line $\langle C_2, S \rangle$, and points lying below the line $\langle S, Q \rangle$. $t_1 = \langle C_N, S \rangle$, $t_2 = \langle C_N, Q_\infty \rangle$, hence $t_2 \parallel \langle C_2, Q_\infty \rangle$ (Fig. 4 c)
- *Area V:* Area V is the intersection of points lying below the line $\langle C_2, Q \rangle$ with the points lying above the line $\langle S, Q \rangle$. $t_1 = \langle C_N, Q \rangle$, $t_2 = \langle C_N, Q_\infty \rangle$, hence $t_2 \parallel \langle C_2, Q_\infty \rangle$ (Fig. 4 b)
- *Area VI:* Area VI is the intersection of points lying below the line $\langle C_2, P \rangle$ with the points lying below the line $\langle C_2, Q \rangle$. $t_1 = \langle C_N, S_\infty \rangle$, $t_2 = \langle C_N, Q_\infty \rangle$, hence $t_1 = \parallel \langle C_2, S_\infty \rangle$ and $t_2 \parallel \langle C_2, Q_\infty \rangle$ (Fig. 4 d)

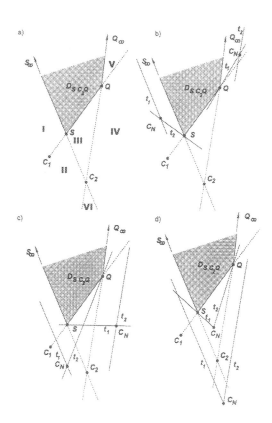

Fig. 4. a) Areas I, II, III, IV, V, and VI. b) Positioning C_N in area I and area V. c) Positioning C_N in area II and IV. d) Positioning C_N in area III and area VI.

The areas I, II, IV, V, and VI all require an empty scene from C_N to infinity along the one (in the case of areas I, II, IV, V) or the two (in the case of area VI) borders of $\mathcal{D}_{S,C_2,Q}$. Only the area III does not introduce any constraint on the scene outside $\mathcal{D}_{S,C_2,Q}$. By selecting the new camera position in $\triangle(S, C_2, Q)$ we move closer to the $\mathcal{D}_{S,C_2,Q}$ therefore, due to expansion, the objects in $\mathcal{D}_{S,C_2,Q}$ which are closer to the camera occlude the objects behind. In general to acquire the second image of points projecting to (p_2, q_2) we need more than one view.

4.2 The new camera location C_N lies outside $\mathcal{E}_{C_1,C_2,S}$

Selecting C_N outside $\mathcal{E}_{C_1,C_2,S}$ seems to be much harder problem to analyze, since points of $\mathcal{D}_{S,C_2,Q}$ after repositioning the camera do not project to the same epipolar line.

Usually we are not interesting in how to move the observer to resolve only the occlusions of one epipolar line, but primarily, how to move the observer to resolve occlusions belonging to at least one occluded region.

Let us now analyze an occluded region $\mathcal{O} = \cup_i(o_{2_i}, q_{2_i})_{e_{2_i}}$ (Fig. 5a). Such occluded region can be detected on one side of the occluding contour o_2; $o_2 = \cup_i o_{2_i}$. A bunch of optical lines going through C_2 and occluded points of $\mathcal{O}$ determine a conic volume $\mathcal{V}_{C_2,\mathcal{O}}$. Points inside $\mathcal{O}$ are connected along the epipolar lines e_{2_i}. To each interval $(o_{2_i}, q_{2_i})_{e_{2_i}}$ belongs the domain of occluded points which can be computed in the same way as it is described in Section 3 for the interval (p_2, q_2). Hence, for each $(o_{2_i}, q_{2_i})_{e_{2_i}}$ there exists the line $\langle S_i, Q_i \rangle$. All the lines $\langle S_i, Q_i \rangle$ determine a bounding surface $\mathcal{B}_{C_2,\mathcal{O}}$ (Fig. 5 b). In accordance with the Observation 1 for each interval $(o_{2_i}, q_{2_i})_{e_{2_i}}$, there exists an empty triangle $\triangle(S_i, C_2, Q_i)$. All the triangles $\triangle(S_i, C_2, Q_i)$ form an empty conic volume $\mathcal{V}_{empty}$ with the vertex at C_2 and bounded by $\mathcal{B}_{C_2,\mathcal{O}}$ and the surface formed by a set of lines passing through the border of $\mathcal{B}_{C_2,\mathcal{O}}$ and intersecting at C_2. The domain of occluded points $\mathcal{D}_{C_2,\mathcal{O}}$ is a part of the volume $\mathcal{V}_{C_2,\mathcal{O}}$ above $\mathcal{B}_{C_2,\mathcal{O}}$.

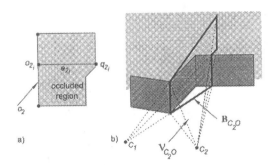

Fig. 5. a) Occluded region $\mathcal{O}$. b) Surface $\mathcal{B}_{C_2,\mathcal{O}}$.

For each point of $\mathcal{O}$ the corresponding point at infinity projects to the same point no matter where the new camera center is located. Hence, the same is true for the complete region $\mathcal{O}$: Points at infinity belonging to $\mathcal{O}$ project to the same region $\mathcal{O}$ in $\mathcal{R}_N$ no matter where C_N is located.

The problem is analogous to the problem analyzed for positioning C_N in the epipolar plane $\mathcal{E}_{C_1,C_2,S}$. C_N is a good camera location if the volume formed by parts of the lines connecting C_N with all the surface points of $\mathcal{D}_{C_2,\mathcal{O}}$ is empty.

Points are divided into two sets: Into the points which lie inside the conic volume $\mathcal{V}_{empty}$ and into the points outside $\mathcal{V}_{C_2,\mathcal{O}}$.

Again, only the camera locations in $\mathcal{V}_{empty}$ do not introduce any constraints on the scene outside $\mathcal{D}_{C_2,\mathcal{O}}$.

4.3 Feature points versus uniform regions

Stereo algorithms can recover the shape of areas where there are sufficient feature points to establish correspondences. In the areas of uniform intensity the depth values are than interpolated. Hence, in practice, to determine the depth of uniform regions between feature points, it is enough to acquire the second image of feature points.

Staying in the epipolar plane $\mathcal{E}_{C_1,C_2,F}$

Let ξ be a feature point of our interest and let at F be a feature event in the scene which causes ξ to be a feature point on (p_2, q_2) (Fig. 6). Let ξ_l and ξ_r be the closest left and the closest right feature on (p_2, q_2) to ξ, respectively and let features at ξ_l and ξ_r correspond to feature events at F_l and F_r, respectively. The boundaries which determine the allowable space for the next camera locations have to be computed assuming the worst case situation: The point projecting to ξ lies at infinity: $F = F_\infty$, the points projecting to the two neighboring features lie on the line $\langle S, Q \rangle$: $F_l = F_{l_{\langle S,Q \rangle}}$, $F_r = F_{r_{\langle S,Q \rangle}}$. Hence, $F_{l_{\langle S,Q \rangle}}$ occludes F_∞ when the optical line $\langle C_N, F_{l_{\langle S,Q \rangle}} \rangle$ becomes parallel to the optical line $\langle C_2, \xi \rangle$. Let us denote this line as $\langle C_{N_{limit}}, F_{l_{\langle S,Q \rangle}} \rangle$. Similarly, $F_{r_{\langle S,Q \rangle}}$ occludes F_∞ when the optical line $\langle C_N, F_{r_{\langle S,Q \rangle}} \rangle$ becomes parallel to the optical line $\langle C_2, \xi \rangle$. Again, we denote this line as $\langle C_{N_{limit}}, F_{r_{\langle S,Q \rangle}} \rangle$. Hence, in $\mathcal{E}_{C_1,C_2,F}$ C_N can be selected anywhere between the two parallel lines $\langle C_{N_{limit}}, F_{l_{\langle S,Q \rangle}} \rangle$ and $\langle C_{N_{limit}}, F_{r_{\langle S,Q \rangle}} \rangle$ where $\langle C_{N_{limit}}, F_{l_{\langle S,Q \rangle}} \rangle \parallel \langle C_2, \xi \rangle \parallel \langle C_{N_{limit}}, F_{r_{\langle S,Q \rangle}} \rangle$. The camera can not be positioned on the line $\langle C_2, \xi \rangle$ since for C_N lying on $\langle C_2, \xi \rangle$ there is no parallax for F.

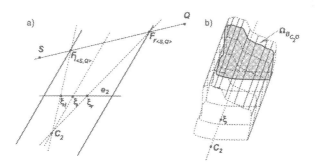

Fig. 6. Acquisition of feature points. a) Allowable locations when staying in $\mathcal{E}_{C_1,C_2,F}$. b) Allowable locations when moving outside $\mathcal{E}_{C_1,C_2,F}$.

Moving outside the epipolar plane $\mathcal{E}_{C_1,C_2,F}$

Let ξ be the feature point of our interest. Let $\Omega = \cup_i \omega_i$ be the set of neighboring features around ξ in the occluded region $\mathcal{O}$. Again, to compute the possible camera locations, we assume the worst case situation: The point projecting to ξ lies at infinity, the points projecting to the neighboring features of Ω lie on $\mathcal{B}_{C_2,\mathcal{O}}$, let denote them as $\Omega_{\mathcal{B}_{C_2,\mathcal{O}}}$. The solution is a cylindrical volume formed by lines parallel to line $\langle C_2, \xi \rangle$ going through the points of $\Omega_{\mathcal{B}_{C_2,\mathcal{O}}}$. Again, the camera can not be positioned on the line $\langle C_2, \xi \rangle$ due to the same reason as above.

5   Experimental results

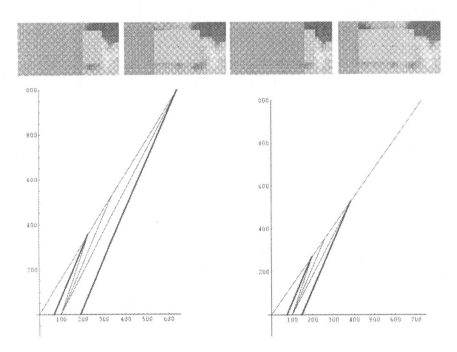

Fig. 7. Top row: Two pairs of stereo images. Bottom row: Images depict possible camera locations (bounded by the bold lines) for epipolar plane $\mathcal{E}_{C_1,C_2,F_2}$. C_1 is located at $(0,0)$ and C_2 at $(100,0)$ (axes are denominated in pixel units).

The experiment demonstrates that the size of the possible displacement of the camera to acquire the slightly different view does not depend only on the distance between the features in the occluded region but also

- on the distance of the occluded feature to the occluding border
- and on the distance of the obstacle to the camera center C_1.

We generate two pairs of stereo images. The scene contains two planes occluding a sky with clouds in the background (Fig. 7). Let us denote by $\mathcal{A}_1$ the darker plane and by $\mathcal{A}_2$ the lighter plane. $\mathcal{A}_1$ and $\mathcal{A}_2$ are parallel to image planes $\mathcal{R}_1$ and $\mathcal{R}_2$. The distance between the left and the right camera center is 100 pixel units. $\mathcal{A}_2$ was in both examples located at $Z = 10f$. $\mathcal{A}_1$ is in the first example located at $Z = 2.857f$ and in the second at $Z = 1.857f$, $f = 100$ pixel units. In both examples the plane $\mathcal{A}_1$ occludes in the left image three black points lying on the lighter plane $\mathcal{A}_2$. Let us denote the black points from left to right as ξ_1, ξ_2, and ξ_3. The distances $d_{\xi_2,\xi_1}, d_{\xi_3,\xi_2}$ between the points ξ_2 and ξ_1 and the points ξ_3 and ξ_2 respectively, are equal. The feature point of our interest is ξ_2. Let us denote by F_2 the point in the scene which projects to ξ_2.

The area of possible camera locations in $\mathcal{E}_{C_1,C_2,F_2}$ is limited by borders—bold lines in the bottom images of fig. 7 which are determined as described in the previous subsection. The border b_{ξ_3} which is due to the feature ξ_3 is further away from C_2 than it is the border b_{ξ_1} which is due to the feature ξ_1, even though the distances $d_{\xi_2,\xi_1}, d_{\xi_3,\xi_2}$ are equal. In the second example (right stereo pair of images in fig. 7) we move the plane $\mathcal{A}_1$ closer to C_1. The movement of $\mathcal{A}_1$ closer to C_1 causes the borders b_{ξ_1}, b_{ξ_3} (bold lines in the right image of the bottom row of fig. 7) of area of possible camera locations to move closer to C_2.

6 Conclusions

View synthesis methods generate intermediate images of the scene from images acquired at nearby viewpoints. The generation of the complete intermediate image may not be possible due to occlusions: It is not possible to generate the intermediate image of the scene which is acquired only in one of the nearby images or is not acquired in any of the images of nearby viewpoints. In the paper we analyze the former case. The results of the analysis can be used in planning the set of nearby viewpoints. In the future we will concentrate on the following problems:

- how to group the feature points in all occluded regions to optimize the set of nearby views,
- we plan to extend the analysis to panoramic images.

References

1. S. E. Chen and L. Williams. View Interpolation for Image Synthesis. Proc. SIG-GRAPH 93. In *Computer Graphics* 1993, pages 279–288.
2. D. Geiger, B. Ladendorf, and A. Yuille. Occlusions and Binocular Stereo. *International Journal of Computer Vision*, 14:211-226,1995
3. J. J. Little and W. E. Gillett. Direct Evidence for Occlusion in Stereo and Motion. In *Proceedings of First European Conference on Computer Vision*, pages 336–340, 1990.
4. K. N. Kutulakos and C. R. Dyer Global Surface reconstruction by Purposive Control of Observer Motion. Tech. Rep. 1141 Computer Sciences Department, University of Wisconsin—Madison, April 1993. Available via ftp from ftp.cs.wisc.edu.
5. S. M. Seitz and C. R. Dyer. Photorealistic Scene Reconstruction by Voxel Coloring. In *Proceedings on Computer Vision and Pattern Recognition*, pages 1067–1073, June 1997, Puerto Rico.
6. S. M. Seitz and C. R. Dyer. View Morphing. In *Proceedings of Computer Graphics*, pages 21–30, Annual Conference series, 1996.
7. T. Werner, R.D. Hersch, and V.Hlaváč. Rendering Real-World Object Without 3-D Model In *Proceedings of 6th International Conference, CAIP'95*, pages 146–153, 1995.

Convex Layers: A New Tool for Recognition of Projectively Deformed Point Sets

Tomáš Suk and Jan Flusser *

Institute of Information Theory and Automation
Academy of Sciences of the Czech Republic
Pod vodárenskou věží 4, 182 08 Praha 8, Czech Republic
suk@utia.cas.cz, flusser@utia.cas.cz

Abstract. A new more efficient algorithm for recognition of projectively deformed point sets is presented. It supposes the type of projective transformation, where the orientation of a triangle is preserved.

First, convex layers of two compared sets are computed and left-tangents are found. If their structures are different, then an advanced algorithm is used. It creates the new structure of one set according to the other set. If the numbers of the points in the convex layers differ from each other, then the suitable points are moved to other layers, and if the left-tangents do not link the corresponding points, then their structure is corrected. The suitable combination of optimal structure correction criteria was investigated by a numerical experiment, the angle criteria were found as the best.

Finally, five-point cross ratios of sequences of the points from both sets are computed and compared.

1   Introduction

One of the important tasks in image processing and computer vision is recognition of objects on images captured under different viewing angles. If the scene is planar, then the distortion between two frames can be described by *projective transform*

$$x' = (a_0 + a_1 x + a_2 y)/(1 + c_1 x + c_2 y)$$
$$y' = (b_0 + b_1 x + b_2 y)/(1 + c_1 x + c_2 y) \,, \tag{1}$$

where x, y are the coordinates in the first frame and x', y' are the coordinates in the second one.

The scope of this contribution is limited to the cases, where objects on images can be described by finite sets of points and its projective transforms have th same number of points, e.g. vertices of polygons. The algorithm does not solve cases, when some points are occluded on some images.

Under (1) the straight line

$$1 + c_1 x + c_2 y = 0 \tag{2}$$

---

* This work has been supported by the grant No. 102/98/P069 of the Grant Agency of the Czech Republic.

is not mapped into the second plane (or it is mapped into the infinity). Each triangle not intersected by the straight line (2) is transformed by (1) with the same orientation of the vertices. It implies some additional theorems dealing with topology of the set, e.g. the convex hull is preserved during the projective transform in that case. This property can be used for construction of very efficient algorithms for recognition of point sets. While the computing complexity of the invariants [2] in the general case of the projective transform when the points can lay in both parts of the plane is $O(n^5)$, there was published an algorithm for recognition of projectively deformed point sets in [1], which utilizes preserving triangle orientation, with computing complexity $O(n \log n)$.

Concisely, it consists of a few steps:

1. Compute convex layers of both sets.
2. Compute left-tangents.
3. Acquire left-list of points.
4. If the structures of the sets are the same, compute five-point cross ratios and declare the sets as equivalent if and only if the cross ratios are equivalent.

The convex layers of a set are obtained by computing convex hull of the set, removing the points of the hull and computing convex hull of the rest of the set. It is repeated while some points remain in the set.

From a point of the outer layer we can draw the two tangents to the adjacent inner convex layer. While looking at the inner hull from the point, we can classify these tangents as left-tangents and right-tangents (Fig. 1 a). We must choose one of them, but the other can be used as well. The left-tangents were chosen here.

The system of the left-tangents with fictitious point in the center of the innermost convex layer forms a tree. The tree can be transcribed into a sequence of the points by depth-first search of the tree. The sequence is called left-list. The transcription is ambiguous, we can choose any point in the innermost hull as the beginning. If the innermost hull contains one point only, the same situation occurs in the next convex layer. We must consider each of these sequences and find the pair of sequences, one from the first set and the other from the other set, with the least deviation of the cross ratios. If the deviation is zero, then the sets are precisely equivalent, else the deviation can be considered as the measure of the degree of difference between the sets.

The main disadvantage of this algorithm is its high sensitivity to noise. If some three points are almost collinear, then small perturbations in their coordinates can affect the structures of the convex hulls and left-tangents and the sets are declared as completely different. This paper describes a new algorithm which is more robust to noise in point coordinates.

The simplest projective invariant is a five-point cross ratio [3, 4, 5 and 6]:

$$\varrho(A, B, C, D, E) = \frac{P(A, B, C)P(A, D, E)}{P(A, D, C)P(A, B, E)} \tag{3}$$

where $P(A, B, C)$ is the area of the triangle with vertices A, B and C. The point A is called the common point of the cross ratio.

Since the set of n points has $2n$ degrees of freedom and the projective transform has 8 parameters, therefore the set can have only

$$m = 2n - 8 \tag{4}$$

independent invariants to projective transform. In [1] there are only n invariants computed for n points, we propose to take m invariants, which describe the set completely. If we have a sequence of n points, we can compute the cross ratio of the first point as the common one and following 4 points, then the second point as the common one and following 4 points and so on until we take the fifth point from the end as the common one and last 4 points. We obtain $m/2$ invariants. Now we take the sequence in the reverse order, it means the last point as the common one and 4 preceding points, then last but one as common and preceding 4 points and so on until the fifth point as common one and first 4 points. So each point is used at least once as the common one in m cross ratios.

Since the cross ratio can be infinity or near infinity and it can lead to instability, we use the arc tangent of the cross ratio instead of the mere cross ratio.

2   Basic Algorithm

Our algorithm determines, if two point sets are equivalent up to projective transform. It can be outlined as follows

1. Compute convex layers of both sets.
2. Compute left-tangents.
3. Obtain left-list of points.
4. Find the degree of symmetry of the left-list.
5. If the structures of the sets differ from each other, use advanced algorithm from the next section.
6. For each degree of symmetry compute invariants of the corresponding sequence.
7. The minimum distance of two sets of the invariants in Euclidean feature space expresses the measure of equivalence of the point sets.

The convex layers can be computed either the Chazelle's fast algorithm [7] with the computing complexity $O(n \log n)$ or much simpler Jarvis' march [8] with running time $O(n^2)$. If we know the convex layers, then the left-tangents can be found with the computing complexity $O(n)$.

The tree is created from the left-tangents and transcribed into the left-list by the depth-first search of the tree (Fig. 1 c). There are the convex layers outlined by the thick line and the left-tangents by the thin line in the Fig. 1 b.

The left-list of this set can be

$$C_1, B_1, A_1, A_2, C_2, B_2, A_3, B_3, A_4, C_3, B_4, A_5.$$

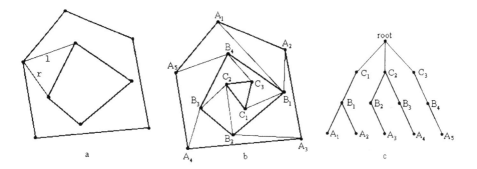

Fig. 1. a: Left-tangent (l) and right-tangent (r). b: The convex layers (thick line) and the left-tangents (thin line). c: The tree from the left-tangents

From all possible sequences is used that one with the greatest lexicographic order. If we want to find which of two sequences has greater lexicographic order, we find the first points from the beginnings, whose convex depths differ from each other, and the sequence with the greater convex depth of the point is lexicographically greater. E.g. the sequence of the convex depths of the left-list from the Figs. 1 b,c is

$$3, 2, 1, 1, 3, 2, 1, 2, 1, 3, 2, 1.$$

We can use all cyclic permutations beginning by 3, but we use the sequence

$$3, 2, 1, 3, 2, 1, 1, 3, 2, 1, 2, 1$$

because it is lexicographically greatest – the fourth member of this permutation is the greatest.

In case of only one point in the innermost convex layer, this point is ever first in the left-list and it is omitted from the cyclic permutations.

The degree of symmetry is the number expressing how many cyclic permutations of a sequence is the same as the original sequence. For instance, the degree of symmetry of the sequence

$$3, 2, 1, 1, 3, 2, 1, 1, 3, 2, 1, 1$$

equals 3. The degree of symmetry of a sequence without any symmetry equals 1. If the degree of symmetry of the left-list depths is greater than 1, then we must consider the corresponding number of the cyclic permutations.

The structures of both sets are identical, if the sequence of the convex depths of the points in the left-lists are the same, otherwise the advanced algorithm must be used. It is described in the next section.

The minimum distance of two sets of the invariants in m-dimensional Euclidean feature space expresses the measure of equivalence of the point sets. The sets are exactly equivalent, it means they can be transformed by projective transform one into another, if and only if their distance equals zero. The number of structural changes can serve as the auxiliary criterion of similarity.

3 Advanced Algorithm

The aim of this algorithm is to arrange the structure of one point set according to the structure of the other set, what is refining with respect to the basic algorithm, which cannot change the structure.

The advanced algorithm consists of two steps, translations of points into different convex layers (1,2) and arrangement of left-tangents (3,4). First, one set is labeled as reference and the other one as input.

Data from the basic algorithm:

n = number of points in one set
$cd_1[i]$ = convex depth of the i-th point of the input set
$cd_2[i]$ = convex depth of the i-th point of the reference set
nd = convex depth of of the reference set
$np[i]$ = number of points with convex depth = i in the reference set
$lt_1[i, k]$ = left-tangent of the k-th point in the i-th layer (input set)
$lt_2[i, k]$ = left-tangent of the k-th point in the i-th layer (reference set)
Left-tangent means its order in its convex layer.

Advanced algorithm:

1. for i:=1 to n
   if($cd_1[i] < cd_2[i]$)
   $point\_down(cd_1[i], cd_2[i])$
   if($cd_1[i] > cd_2[i]$)
   $point\_up(cd_1[i], cd_2[i])$
   endfor(i)

2. for i:=1 to n
   if the i-th point was moved in Step 1
   $inner\_tangent(lt_1, i)$
   endfor(i)

3. for j:=1 to $np[1]$ //loop via points of the first layer
   for i:=1 to $nd - 1$ //loop via other layers
   for k:=1 to $np[i]$ //loop via points of the layer
   $no := (lt_1[i, k] - lt_2[i, k]) \bmod np[i]$
   $sh[no] := sh[no] + 1$ //compute histogram of the layer rotations
   endfor(k)
   Find maximum $maxr$ of the histogram sh.
   If there is more identical maxima, compute sum of left-tangent criteria for each of them.
   $noerl := noerl + np[i] - maxr$ //number of changes of the left-tangents
   endfor(i)
   $minnoerl$:=minimum number of changes $noerl$
   If there is more identical minima, we consider that with less sum of the left-tangent criteria.
   $nchm[i]$:=position of the minimum
   endfor(j)

4. for $i:=1$ to nd
   for $j:=1$ to $np[i]$
   $lt_1[i,j] := lt_2[i,(j+nchm[i]) \bmod np[i]]$
   endfor(j)
   endfor(i)

The procedures $point\_down(i,j)$ and $point\_up(i,j)$ search, what point from the i-th layer is the best candidate for translation to the j-th layer. The procedure $point\_up(i,j)$ examines triangles from three adjacent points in the source layer, the procedure $point\_down(i,j)$ examines triangles from a point in the source layer and two corresponding points in the destination layer [1]. A number of both the layer criteria and the tangent criteria was tested, see the section Numerical Experiments.

The procedure $inner\_tangent(lt,i)$ compute a new left-tangent of the translated i-th point.

The aim of the next part of the algorithm is to find the optimal way, how to copy left-tanget numbers from the reference set to the input one. The criterion is the whole number of changes and if there is more than one identical minima, we use another auxiliary left-tangent criterion.

4 Numerical Experiments

The following two numerical experiments were carried out. The goal of the first experiment was to determine, what layer and left-tangent criteria are the best.

A set of 16 points was created by a pseudo-random generator. They are uniformly distributed from 0 to 511. Independent zero-mean Gaussian noise with gradually increasing standard deviation up to 30 was added to the coordinates of all points and the original set was compared with the noisy set by our algorithm. The experiment was repeated 100 times for each standard deviation and for each of 7 combinations of the criteria. The following combinations of the criteria and the whole number of correct results can be seen in Tab. 1. The criteria are illustrated in Fig. 2.

We can see the success rate of the 7th combination is the best one. The success rate of the 6th combination is only slightly worse.

The goal of the second experiment was the comparison of the advanced algorithm with mere base algorithm.

Two basic sets of 16 uniformly distributed points were created again. One of them was corrupted by noise like in the previous case. The noisy set was compared both with the original set and with the other set by our algorithm.

---

[1] The points are corresponding, when the point in the source layer lies in the triangle from two adjacent points in the destination layer and the centroid of the innermost layer, e.g. the pair of the points A_3 and A_4 is corresponding to the point B_2 in the Fig. 1 b.

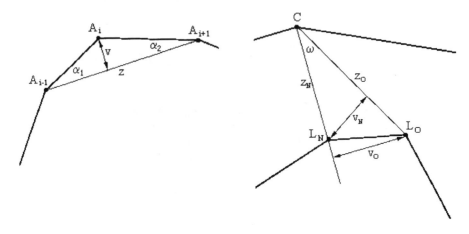

Fig. 2. The criteria of the point translations to another layer (left) and those of the new left-tangent (right)

10 basic sets were created for each standard deviation and the experiment was repeated 10 times for each pair of the basic sets and for each standard deviation. The average deviations in the feature space can be seen in Fig. 3 left. The solid line expresses comparison with the original set and the dashed line with the other set. The number of the correct results can be seen in Fig. 3 right. The solid line expresses the number of correct results of the advanced algorithm and the dashed line that of the basic algorithm.

Table 1: The combinations of the criteria

No.	layer criterion	left-tangent criterion	correct
1	distance v	old distance v_O	739
2	relative distance v/z	relative old distance v_O/z_N	724
3	distance v	new distance v_N	740
4	relative distance v/z	relative new distance v_N/z_O	723
5	triangle area $P = vz/2$	triangle area $P_l = v_O z_N/2 = v_N z_O/2$	746
6	angle maximum of α_1 and α_2	angle ω	854
7	angle sum $\alpha_1 + \alpha_2$	angle ω	860

We can see that the advanced algorithm yields good results in case of significantly greater noise than the mere basic algorithm without possibility of structure corrections.

5 Conclusion

In this contribution, we improved the original method for recognition of projectively deformed point sets [1]. The simpler evaluation of symmetric sets was

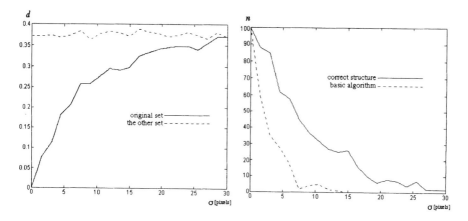

Fig. 3. The average deviations d in the feature space (left) and the number n of the correct results (right) with respect to the noise standard deviation σ

used instead of relatively complicated searching substrings in the string and more suitable number of invariants was computed. The new method of structure correction was established. The correspondence between the points of both sets can be found in case of need. The behavior of the method was demonstrated by numerical experiments, one of them showed the angle criteria of the moving point are the most suitable.

References

1. Rao N. S. V., Wu W., Glover C. W.: Algorithms for Recognizing Planar Polygonal Configurations Using Perspective Images, In IEEE Trans. on Robotics and Automation, Vol. 8 (1992), No. 4, pp. 480-486
2. Suk T., Flusser J.: The Features for Recognition of Projectively Deformed Images, In Proc. ICIP'95, Washington (1995), Vol. III, pp. 348-351
3. Reiss T. H.: Recognition Planar Objects Using Invariant Image Features, Lecture Notes in Computer Science, Vol. 676, Springer (1993)
4. Lenz R., Meer P.: Point configuration invariants under simultaneous projective and permutation transformations, In Pattern Recognition, Vol. 27 (1994), No. 11, pp. 1523-1532
5. Maybank S. J.: Classification Based on the Cross Ratio, In proc. of The Darpa-Esprit Workshop on Applications of Invariants, Azores, Portugal (1993), pp. 113-132
6. Sinclair D., Boufama B., Mohr R.: Independent Motion Segmentation and Collision Prediction for Road Vehicles, In COPR94 (1994), pp. 958-961
7. Chazelle B.: On the Convex Layers of a Planar Set, In IEEE Trans. on Information Theory, Vol. IT-31 (1985), No. 4, pp. 509-517
8. Preparata F. P., Shamos M. I.: Computational Geometry, Springer (1985)

Measurement of Ski-Jump Distances by the Use of Fuzzy Pattern Comparator

Veselko Guštin[1], Aleš Lapajne[1], Rober Kodrič[1], Tomislav Žitko[2]

University of Ljubljana
[1]Faculty of Computer and Information Science,
[2]Faculty of Electrical Engineering,
Tržaška 25, 1000 Ljubljana
E-mail: Veselko.Gustin@fri.uni-lj.si

Abstract

A measurement of the ski-jump distance using the computer programme JumpIt is presented. The technique we use is based mainly on the recording of consecutive images in the computer memory in real-time. We propose an algorithm based on off-line recursive filtering, and fuzzy pattern comparison of the bit mask of two sets of windows: the learning one and the testing one. The result of this comparison should give us the point of landing of the ski jumper. The length is calculated from the calibrated ski jumping hill.

1 Introduction

The ski-jump may be divided into three phases: the inrun and take-off, the free flight and preparation for landing, the landing and outrun. The most interesting phase for us, the spectators, is surely the jump or the flight and landing, or touch down, i.e. what interests us here is the length of the jump. For the trainer, however, other phases may be much more important [3].

Today, this measurement is carried out mainly in the "classical" manner, with subjective visual observation, either at the very landing-point, or by means of video (recorder) replays. The International Ski Federation (FIS) uses expensive computer equipment for the measurement, which is, unfortunately, a very tight-kept secret.

In Chapter 2 of this paper we will first describe the computer hardware system for measuring distance, and in Chapter 3 we will describe by type the programming procedures which are necessary to determine the length of the jump. Then, in Chapter 4, we will present a comparison between the results measured by our system and the "classically" measured results. In the Conclusion, we will mention certain findings, which we obtained from testing our equipment on site.

2 Computer Hardware and Video Equipment

We are using a personal computer and the frame grabber, produced by Data Translation [5,6]. It receives the signal with either a b/w or colour camera;

this enabled us - by using, amongst others, the appropriate *JumpIt* programme written in Visual C++ - to store in real-time the consecutive frames (film) within the available computer memory. The main memory then made it possible to display the recorded images up to 150 frames, i.e. up to 6 seconds of filming. One b/w incoming frame **S** is an array of $I \times J$ picture elements (pixels) or bytes, representing 8-bit grey levels. From our frame grabber we get the array of $I = 768$, and $J = 576$ pixels, so that the entire database for one image requires 442,368 bytes. An ordinary b/w or colour camera is positioned in such a way that, while capturing the information, its position remains unaltered, and that it covers approximately 40-50 metres of the descent from the ski jumping hill. The proposed final hardware solution with 3 cameras is shown in Fig. 1.

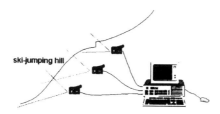

ski-jumping hill

Figure 1: Proposed (final) hardware solution

3 Computer Software

3.1 Manual Measurement at the Landing-Point

We used the principle of linear interpolation in order to calculate the points between two shifts. We drew out separately each of the shifts between two points which we had previously marked on the "live image", either with a small flag or a striped pole; then we had already transferred the image into the computer memory. We marked the "live" distance at every 10 m point. It proved that the introduction of this method was sufficiently accurate and that there were almost no errors, i.e. the margin of error was less than 0.5 m. The calibrated ski jumping hill is shown in Fig. 2.

Once the system has been calibrated it is ready for measuring. Hand directed control is essential; when the ski jumper is observed, the starting of the computer recording of frames begins.

Depending on the size of the ski jumping hill, we limit the number of windows used, particularly if we are taking measurements on smaller ski-jumps. Thus the operator chooses from a range between 3 seconds (75 frames) to 6 seconds (150 frames) depending upon how long he intends to follow the jumper.

As the longest flight on 180-m ski jump hill was approximately 7 seconds, we considered it sufficient to store 3/4 of that "maximal" time. After the recording

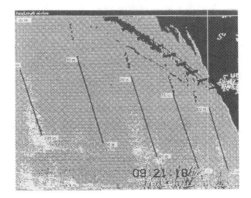

Figure 2: Calibrated ski jumping hill from 60 m to 100 m with a camera positioned at the out-run and with marked observation area S'.

we had to match the frame with the landing point of the ski jumper. This is done mainly in the same manner as by video recorder. When the frame with the just-landed ski-jumper is located, we put the cursor in the right position (between the two feet), and the length is displayed in the computer monitor.

3.2 Proposed Algorithm for System Measurement

Once we have successfully completed the initial phase for the manual measurement of the ski jump distance, we plan to make the measured distance of the ski-jump length fully automated. Therefore we upgrade our *JumpIt* program with:
- fuzzy detection of the ski-jumper in the field of observation,
- recursive filtering,
- fuzzy comparison and analysis of the jump, and
- calculation of the distance.

We decided to choose the principle of the fuzzy pattern comparator (FPC). This treats the patterns as fuzzy sets and calculates a (hamming) distance measure between multiple sets [7]. The threshold value defines the "membership" function which determines how two patterns match.

3.2.1 Fuzzy Detection of the Ski Jumper Within the Area of Observation

However, since the "hand directed" waiting for the ski-jumper - who should first appear at the edge of the area of observation - is a "demanding and enduring" job, we preferred to use the principle of motion detection by fuzzy filtering. The area of observation S' is an array of $(I^* \dots I^{**}) \times (J^* \dots J^{**})$ b/w pixels (or bytes). In Fig. 2. this area is positioned in the right part of the picture, in the last 1/8 vertical strip of the image, and is selected by the operator in the calibration phase.

a) The images come in from the video source and we calculate for the first incoming frame the amount of pixels u_1 inside the observation area $s'_{m',n'} \in$ $\mathbf{S}'$, and $\mathbf{S}' \subset \mathbf{S}$:

$$u_1 = \sum_{m=0}^{\frac{(I^{**}-I^*)}{m}} \sum_{n=0}^{\frac{(J^{**}-J^*)}{n}} \mathbf{S}_{I^*+m\cdot k, J^*+n\cdot l}$$

where m, n are integer numbers, and $I^*+m\cdot k$, $J^*+n\cdot l$ are the co-ordinates within the observation area $\mathbf{S}$, k and l are the step of the grid within the array, usually 2, 4, 8, ... In our case, with the camera positioned in the out run the operator had to take $I^* = 672, I^{**} = I, J^* = 1, J^* = J$. We then repeated the calculation for each next frame and obtained a set of summarised points on the grid within the observation area:

$$u_1, u_2, u_3, \ldots, u_p, u_{p+1}, \ldots, u_P.$$

b) Once it had been shown that the difference between the two consecutive calculations

$$u_p \gg u_{p-1},$$

or

$$|u_{p-1} - u_p| > \delta_u$$

was greater than the previously prescribed threshold δ_u, we then commenced with the procedure of sequential recording of the frames into the memory (film). Also, in this range of vision no third object should appear, e.g. a spectator, measurement adjudicator, or snow-leveller. The aim of the programme is not to separate the one from the other. We preferred to use the third camera only for this purpose. The threshold - once the procedure has been set in motion for receiving data from the camera - is decided in the measuring phase of the pre-initial jumps and is dependent on various factors, such as the camera position, distance from the focal figure, weather conditions, visibility, and illumination on the landing slope.

3.2.2 Recursive Filtering

Since it is necessary to separate the dynamics from the background, we used the same filter which we had previously used [1,2] and which is also to be found in other applications [4].

a) From each incoming frame $\mathbf{S}$, we calculate the filtered image $\mathbf{F}$ by the recursive formula,

$$f_{i,j}(t+1) = \frac{c_1 \cdot f_{i,j}(t) + c_2 \cdot s_{i,j}(t)}{c_1 + c_2},$$

for $i = 1 \ldots I$, and $j = 1 \ldots J$, and $c_1 \gg c_1$. Here $s_{i,j}$ represents the pixels from the live picture $\mathbf{S}$ from the camera, therefore $s_{i,j} \in \mathbf{S}$ and

$f_{i,j}$ are pixels within the filtered image $\mathbf{F}$, therefore $f_{i,j} \in \mathbf{F}$, hence the background. The two matrixes are of the same dimensions, thus

$$|\mathbf{S}| = |\mathbf{F}|.$$

The constants c_1 and c_2 determines the relation between live and filtered picture in the recursive filter.

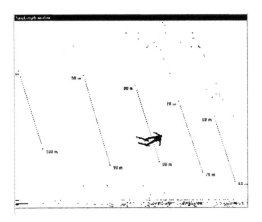

Figure 3: Black-and-white filtered bit mask image $\mathbf{M}$.

b) In order to achieve the fastest and simplified calculation, it is reasonable to select

$$c_1 + c_2 = 4, 8, 16, \ldots .$$

The division, however, is a time-consuming action, which can be simply substituted by shifting to the right by a specified number of places. In our example, we found that $c_1 = 3$, and $c_2 = 1$ gave good results and division by 4 is substituted by a 2-places shift to the left.

c) If we subtract the live image $\mathbf{S}$ from the filtered image $\mathbf{F}$, and we consider that:

$$\texttt{if } (|f_{i,j}(t+1) - s_{i,j}(t+1)| > \delta'_m) \quad m_{i,j} = \text{"1"} ;$$
$$\texttt{else } m_{i,j} = \text{"0"} ;$$

for $i = 1 \ldots I$ I and $j = 1 \ldots J$, then we get $\mathbf{M}$, where $m_{i,j} \in \mathbf{M}$, is the matrix of the binary image (bit mask) with the moving object, which is of the same dimension as the incoming image, thus $|\mathbf{S}| = |\mathbf{M}|$, where δ'_m is the treshold.

Filtered in this way, the image provides us with the basis for a fuzzy comparator, which can be used to determine the course of the flight to the landing and outrun. The binary image thus obtained is presented only in two colours: black ("0" or 00h) and white ("1" or FFh). (See Fig. 3.). The number of levels we take for the bit mask image is related to transformation function c_t. We prefer the same number of levels in the (input) learning matrices and in the testing one, or in our case, two.

3.2.3 Fuzzy Comparator and Locating the Frame of the Landing

From among the many algorithms, which deal with the analysis of both 2D and 3D images, we chose the analysis of binary images $\mathbf{M}$ that we got from the recursive filtering. We had already recognised the method in [1,2], for the identification of signs from hand-written numbers. This method is otherwise regarded as being among the "naive" methods, yet it proved to be successful and fast, for it gave us the results in just a few seconds.

a) If we look at the recorded images $\mathbf{S}$ we will find the sequence of the ski jumper in flight position (F), in landing (E) position, and in landed position (L). If we write this sequence by letters, (see Fig. 4) we get the set: where we take one part for learning. It is our goal to find the frame r,

$$\overbrace{F_1, F_2, F_3, \ \ldots \ , F_{r-3}, E_{r-2}, E_{r-1},}^{input\ testing\ patterns\ from\ the\ pre-initial\ and\ all\ other\ jumps\ \mathbf{S}} \underbrace{L_r, L_{r+1}, L_{r+2}, L_{r+3}, \ \ldots \ L_{R-2}, L_{R-1}, L_R}_{learning\ patterns\ from\ the\ pre-initial\ jumps\ \mathbf{S}_p} ,$$

Figure 4: Set of input patterns from pre-initial jumps.

i.e. the landed position of the jumper. What we are first interested in is the sequence $\rho = r \ldots R$ of the so-called *learning pattern matrixes* of the moving ski jumper

$$x_{i',j'} \in \mathbf{X}(\rho),$$

where $i' = 1 \ldots I'$ and $j' = 1 \ldots J'$, which we are able to compute from the set of filtered bit map images of

$$\mathbf{M}(r), \mathbf{M}(r+1), \ldots \mathbf{M}(R) \text{ and } \mathbf{X}(\rho) \in \mathbf{M}(\rho).$$

How do we get the learning pattern matrixes $\mathbf{X}(\rho)$? We determine in advance the data file for the learning recordings of landings, which we arrange by the pre-initial jumper making a descent of the ski jumping hill. Then, off-line, we calculate the filtered images $\mathbf{F}$. We then scan in binary images $\mathbf{M}(\rho)$ for each group of neighbouring pixels, which we put in a rectangle, i.e. possible moving object matrix. Usually there can be found few such candidate matrices. We can then get the initial co-ordinates of the first moving object matrix inside the area of observation, when two conditions are fulfilled. First, the number of pixels must be greater of some δ_p, and second, once we find the first object (ski jumper). This method is to be taken into account since it is very useful in making the calculation faster, as it helps to check only a smaller part of the array $\mathbf{M}$.

Thus we get the following set of matrixes from the pre-initial jump:

$$x_{i',j'}(\rho) \in \mathbf{X}(\rho),$$

for $\rho = 1 \ldots R$, and $i' = 1 \ldots I'$ or $j' = 1 \ldots J'$ are the co-ordinates within $\mathbf{X}(\rho) \subset \mathbf{M}(\rho)$. The pattern in the matrix is adjusted below, on the right and left edge of the window. We would use the "stretch" function for this purpose. We then say that the figure has been normalised. In our case we did some experiments with $I' = 20$ and $J' = 30$. It is the operator who, from the set of pre-initial jump images $\mathbf{M}(\rho)$, has to select the starting r and final R.

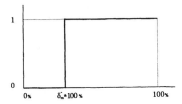

Figure 5: Class 0 transformation function c_t for values $\delta'_m \cdot 100\%$ taken between 0% and 100%, and generalised to 1.

b) From the sequence of learning pattern matrixes, $x_{i',j'}(\rho) \in \mathbf{X}(\rho)$, $\rho = r \ldots R$, we are able to create the *cumulative* learning matrix $\mathbf{W}'$ as:

$$w'_{i',j'} = \left(\frac{1}{R-r} \sum_{\rho=r}^{R} x_{i',j'}(\rho) \right) \cdot 100\%$$

for $i' = 1 \ldots I'$ and $j' = 1 \ldots J'$, where $w'_{i',j'} \in \mathbf{W}'$. The matrixes $\mathbf{X}(\rho)$ and matrix $\mathbf{W}'$ are of the same dimensions, thus $|\mathbf{X}(\rho)| = |\mathbf{W}'|$! The cumulative learning matrix $\mathbf{W}'$ was generalised [2] in class 0 in such a way as defines the transformation function c_t, which we see in Fig. 5.

The generalisation through class 0 function (c_t) gives us the generalised cumulative matrix $\mathbf{W}$:

$$w_{i',j'} = c_t(w'_{i',j'})$$

for $i' = 1 \ldots I'$ and $j' = 1 \ldots J'$, where $w_{i',j'} \in \mathbf{W}$ and $|\mathbf{W}'| = |\mathbf{W}|$, and t=0. We prefer the calculation with integer values because of the faster execution time.

c) Once the system was learned, we began with measuring the distance. We introduce here the algorithm of *best matching* as the "fuzzy spatial cross-correlation" $m_k(\rho)$ between two matrixes, the generalised cumulative learning and testing one.

We programmed the algorithm, so that a pixel $x_{i',j'}(\rho)$ in the testing matrix is compared with neighbouring elements of the cumulative learning matrix W:

$$m_k(\rho) = \sum_{i''=-\rho}^{+\rho} \sum_{j''=-\tau}^{+\tau} (w_{i'+i'',j'+j''} \equiv_f x_{i,j}(\rho))$$

for $i' = 1\ldots I'$ and $j' = 1\ldots J'$, and for all testing patterns, thus $\rho = 1\ldots R$, where $x_{i',j'}(\rho)$, is an element from a set of testing pattern matrixes $X(\rho)$ of the last recorded jump, W is the cumulative learning matrix, and $w_{i'+i'',j'+j''} \in W$, while $\equiv_f$ stands for the fuzzy equivalence function [1], and is defined as

$$W \equiv_f X(\rho) \Rightarrow max\left(min\left(W, X(\rho)\right), min\left(1 - W, 1 - X(\rho)\right)\right)$$

The values ρ, τ are integer numbers, define the width and the height of the neighbouring area, and $k = 1\ldots K$, where

$$K = (|-\sigma| + 1 + \sigma) \cdot (\tau|| + 1 + \tau).$$

d) As a result of the fuzzy comparison, for each compared matrix $m_k(\rho)$ we obtain a set of K coefficients. The best matching $m_k(\rho)$ of each of the compared matrixes is obtained when we extract the maximal one, thus:

$$m(\rho) = MAX(m_k(\rho)),$$

for $k = 1\ldots K$, and $\rho = 1\ldots R$. From the sequence of the best matching values $m(\rho)$, we then calculated the difference between two coefficients in series,

$$|m(\rho) - m(\rho - 1)| = \epsilon,$$

for $\rho = 1\ldots R$. When we find that $\epsilon > \delta_r$, then we get $\rho = r$, i.e. the frame number r of the just-landed ski jumper.

In Fig. 6 we see the cumulative learning bit mask matrix W and a part of the set of input testing matrixes $W(\rho)$, and the result of comparison between both.

3.2.4 Calculation of the Distance

We now take the calibrated frame r from the set of frames S for the calculation of the ski jumper distance. The program, which has to be learned by the operator, could be able to locate the position of skies in the observation frame, and to mark the specific point as when done manually. The position of the cursor gives us the length of the ski jump.

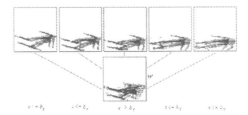

Figure 6: Fuzzy comparison between input testing matrixes $\mathbf{X}(\rho)$ (above) and the learning cumulative matrix $\mathbf{W}$ (below).

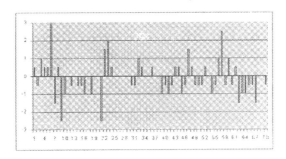

Figure 7: Deviation (in metres) between the official results of the measurers, and our results at the second series of (70) jumps.

4 Results

We performed the last manual test of the system with measurements from ski jumpers for the Slovenian MIP Cup, 17 Jan. 1999 at Planica, on a 90-metre ski jumping hill. We observed a landing area of 60 to 100 m. The camera was placed at approximately 10 m to the side of the outrun. The results can be seen in Fig. 7. We notice immediately a deviation in series of a distance of up to 2 m, to the disadvantage of the official measurers.

5 Conclusion

It is planed that this system will be expanded in order to achieve better and more reliable automatic detection of the ski jumper. We also wish to further improve the method of measurement. For the present, it would seem that the greatest mistake we can make is only in the length between the two recordings (fields). We also wish to complete the system by addition of the starting-list, assessment, and the archive storage of frames (3 or 5 pictures around the just-landed ski-jumper). We would also wish to be granted by FIS a licence for ski-jump mesurements.

6 Acknowledgements

We should like to thank the Slovenian Ski Association who made it possible for us to co-operate in the measurements. Considerable interest was also shown in the *JumpIt* system for the electronic measurement of ski-jumps. Our thanks also go to the firm *AVI inženiring* for setting up the video equipment and for giving the initial stimulus for the realisation of the project.

References

[1] V. Guštin, J. Virant. Pattern Recognition with Fuzzy Neural Network. *The Euromicro Journal, Microprocessing and Microprogramming*, Vol. 40, No.10-12, pp. 935-938, 1994.

[2] V. Guštin. Realization of a Fuzzy Boolean Neural Network Classifier. *Euromicro Journal, Microprocessing and Microprogramming*, Vol. 40, No.1, pp. 23-32, 1994.

[3] Wolfram Mueller, et al. Dynamics of Human Flight on Skis: Improvements in Safety and Fairness in Ski Jumping, *J. Biomechanics*, Vol. 29, No. 8, pp. 1061-1068, 1996.

[4] Phillip M. Ngan. Motion Detection in Temporal Clutter. *http://www.vision.irl.cri.nz/mv_ team/isa/9698/poster.html*, 1997.

[5] -, Data Translation Imaging Product Catalogue, 1997.

[6] -, Frame Grabber SDK, Data Translation, 1997.

[7] -, Fuzzy Pattern Comparator Application System, American NeuraLogix, Inc.

2D Motion Analysis by Fuzzy Pattern Comparator

Mirjana Bonković, Darko Stipaničev, Maja Štula

LaRIS-Laboratory for Robotics and Intelligent Systems
Faculty of Electrical Engineering, Mechanical Engineering and Naval Architecture
UNIVERSITY OF SPLIT
R.Boskovica bb, 21000 SPLIT, C r o a t i a
Web: http://zel.fesb.hr/laris
e-mail: mirjana@fesb.hr, dstip@fesb.hr

Abstract.

The motion analysis plays an important role in the broad range of applications, such as visual feedback control, scene reconstruction and similar. In this paper the technique for 2D-motion analysis based on displacement vector determination is described. In our approach we have used the device called Fuzzy Pattern Comparator (FPC) and develop the fuzzy algorithm based on novel, cylindrical fuzzy sets of directions. Their introduction made the calculation of displacement vector direction and magnitude quite simple. The method was verified experimentally and used in final positioning of the robot arm.

1   Introduction

Displacement vector field [1], [2] which can be defined as a mapping between one frame from an image sequence and the proceeding or following frame is of the great importance in motion analysis, for applications such as visual feedback control [3] or scene reconstruction [4]. Existing techniques for image matching roughly fall into two categories: continuous and discrete.

In the *continuous approaches*, the interframe motion is approximated by motion velocity and the displacement vectors are given as optical flow [2].

The *discrete approach* techniques treat the images as samples of the scene taken at discrete times. Scene features of interest such as high gray value variations, edges, lines or intensity patterns, are extracted from the images and matched appropriately. The procedure is usually based on sophisticated techniques for overlapping recognized shapes in a sequence of analyzed images [5], [6].

The concept of fuzzy sets fits very naturally in the framework of discrete image matching techniques, so we have used for object displacement vector determination the fuzzy correlation method based on Fuzzy Pattern Comparator (FPC) board.

The procedure had two phases:

- the learning phase, and
- the working phase.

In the *learning phase* the robot was positioned up, down, up-left, up-right, down-left and down-right according to its desired, final position (the camera's central point) and appropriate images were captured and saved in external pattern memories.

In the *working phase,* when the robot has entered in the camera's image field, the displacement vectors of its motion were determinate in real time, taking into account the differences between its image and the images of its displacements, saved in the external pattern memories. The displacement vectors directions and magnitudes were calculated by fuzzy algorithm based on cylindrical fuzzy sets of direction. These values were used as input to robot control algorithm which have moved the robot gripper toward its desired, final position (the camera's central point).

2 System Overview

The system consists of the robot arm, video camera, FPC board and a personal computer (486, 66MHz) as Fig.1a). shows. The robot arm (type MICROBOT TeachMover) was equipped with motorized jaw-type gripper, which was in our experiments always in horizontal position. The camera was mounted in front of the robot arm as Fig1b). shows. The main control task of this visual feedback control system was to guide the robot arm until the gripper takes the predefined central position in the camera's image.

The used equipment was more or less standard one, except the Fuzzy Pattern Comparator (FPC).

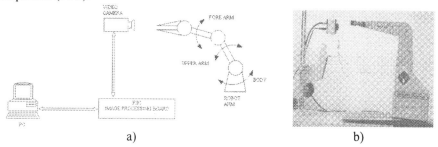

a)   b)

Fig. 1. a) Experimental system setup   b) Photo of the experimental system

FPC is s specialized device particularly designed to provide very fast comparison of frames of data. The FPC utilizes fuzzy logic's method of comparing data in order to determine the similarities or differences in groups of data.

Data is entered serially into the device, where it is formatted into bit fields and compared, one field at a time, with data stored in external pattern memories. In our case the input data where the live image from the video camera. Each input image frame was stored in the first pattern memory and compared with six predefined pattern images stored in external pattern memories during the learning phase. Each pattern memory has a register where results of comparison were stored. The smallest value indicates that the input data due the most clearly matched to its appropriate stored pattern.

3 The Learning Procedure

During the learning phase, robot arm is moved in front of the camera. Its six characteristic position (Fig.2.) viewed by the fixed camera (up, right-up, right-down, down, left-down and left-up) were captured and stored in six FPC board external pattern memories. In working phase a comparison was made between the data from a new image of a scene with data stored in the external pattern memories. Each difference between pixels with the same xy coordinate from stored and captured images has increased the value of the register accompanied to that pattern memory. The final result was an accumulation of errors across the all image elements. Consequently, if two images differ from each other in all the image elements, the value of the register accompanied to the pattern memory have been 192x132=25344 because the input image was digitalised in 192x132 pixels.

Contrary, for two identical images the register will have zero value. These facts were used for displacement vector determination described in the following section.

4 Displacement Vector Determination

The displacement vector is determined by the three-step procedure. The first one is *calculation of the degrees of fulfillment* between the input scene and stored displaced images in memories. Each displace image has its own degree of fulfillment, which is reciprocal to the error sums. If an error sum is equal to zero, then the stored pattern is identically equal to the compared one. Therefore, the degree of fulfillment has maximum value, which is equal to 1.

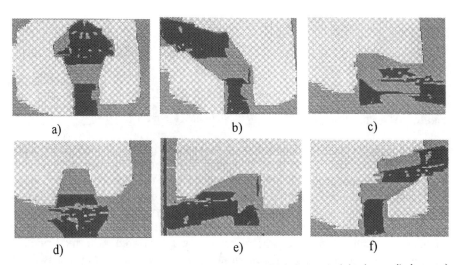

Fig. 2. Images of the predefined displacements: a) up, b) left-up, c) right-down, d) down, e) left-down and e) right-up

Contrary, the highest value of an error sum means the weakest match between captured and stored patterns, resulting with zero degree of fulfillment. Consequently, the overall degree of fulfillment is a 6-element vector $\bar{\lambda} = [\lambda_1 \ \lambda_2 \ \lambda_3 \ \lambda_4 \ \lambda_5 \ \lambda_6]$, whose elements are calculated using the equations

$$if \qquad E_i = max \ Err \qquad \lambda_i = 0 \qquad (1)$$

$$else \ if \qquad E_i = min \ Err \qquad \lambda_i = 1 \qquad (2)$$

$$else \ \lambda_i = \frac{E_i - minErr}{maxErr - minErr} \qquad (3)$$

where maxErr is the maximal error sum and minErr is minimal error sum between input image and displaced images stored in pattern memories. In our case minErr was 0 and maxErr = 25344. The degree of fulfillment λ_1 corresponds to the displace image "up", λ_2 to "right-up", λ_3 to "right-down", λ_4 to "Down", λ_5 to "left-down", λ_6 to "left-up".

The second step is *determination of the displacement vector output fuzzy set*. For this purpose we have introduced cylindrical fuzzy sets of direction depicted on Fig.3.

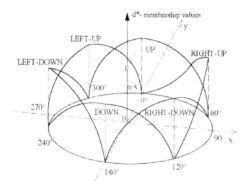

Fig. 3. Cylindrical fuzzy sets of direction.

As Fig.3. shows the definition set of cylindrical fuzzy sets of direction is a circle of directions from 0° to 360°. Membership functions have their maximal value 1 at 0, 60, 120, 180, 240 and 300 degrees for "up", "right-up", "right-down", "down", "left-down" and "left-up" directions.

The output displacement vector fuzzy set is determined using the fuzzy composition

$$d^* = \bigvee_i (\lambda_i \wedge d_i^*) \qquad (4)$$

Where λ_i is the degree of fulfillment for the i-th displacement image (up, right-up, etc...) and d_i^* is its corresponding fuzzy sets of direction. If, for example, the captured robot arm position is identically equal to the displacement image called right-up, its overall degree of fulfillment will be

$$\vec{\lambda} = \begin{bmatrix} 0 & 1 & 0 & 0 & 0 & 0 \end{bmatrix} \tag{5}$$

and its corresponding output fuzzy sets are shown on Fig.4a). Similarly if the robot arm on the input image is positioned somewhere between "left-down" and "left-up" as image 6a) shows, the overall degree of fulfillment will be

$$\vec{\lambda} = \begin{bmatrix} 0 & 0 & 0 & 0 & 0.2 & 0.8 \end{bmatrix} \tag{6}$$

and corresponding output displacement vector fuzzy set is shown on Fig.4b).

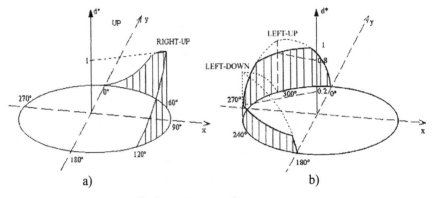

a)   b)

Fig. 4. The shape of the output displacement vector fuzzy sets:

a) for $\vec{\lambda} = \begin{bmatrix} 0 & 1 & 0 & 0 & 0 & 0 \end{bmatrix}$

b) for $\vec{\lambda} = \begin{bmatrix} 0 & 0 & 0 & 0 & 0.2 & 0.8 \end{bmatrix}$

The final, third stage is *the interpretation of the output displacement vector fuzzy set* which results in displacement vector direction and magnitude. The displacement vector is defined inside the circle of directions, which is the definition set of cylindrical fuzzy sets of direction. The displacement vector always starts from the circle center and its end point (x_d, y_d) is defined by the center of gravity of the output displacement vector fuzzy set, calculated by formulas:

$$X_d = \frac{\sum_{d=0}^{360} d * d^*(d)\cos(d)}{\sum_{d=0}^{360} d^*(d)} \tag{7}$$

$$y_d = \frac{\sum\limits_{d=0}^{360} d * d^*(d)\sin(d)}{\sum\limits_{d=0}^{360} d^*(d)} \tag{8}$$

$$|d| = \sqrt{x_d^2 + y_d^2} \tag{9}$$

$$d_\varphi = atan\frac{y_d}{x_d} \tag{10}$$

where d*(d) is the value of the output displacement vector fuzzy set membership function for direction d. |d| is the magnitude of the displacement vector and d_φ is its direction. Fig.5a). shows the example for the output displacement vector fuzzy set from the Fig.4b).

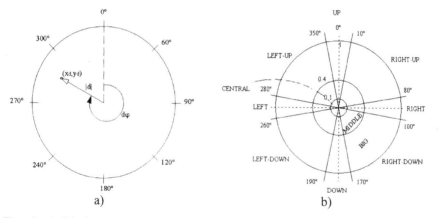

a)   b)

Fig. 5. a) Displacement vector for output displacement vector fuzzy set for Fig.4b). b)Definition of linguistic terms used in linguistic if-then rules (left-up, left-down, down, right-down, right, right-up, up, central, middle and big)

5   Control of Robot Arm

The robot arm used in experiment was of RRR type (educational robot MICROBOT TechMover) schematically shown in Fig.1a). For its 2D movement according to camera's optical center, only two robot segments were moved: robot BODY and ELBOW, because the workspace was limited by the camera's field of view. The space viewed by the camera was divided in sections as Fig.5b). shows. There were eight direction formed by four lines crossing through center "S" and three magnitude sections (central, middle and big) formed by three concentric circles. Based on these lines and circles, two linguistic variables were introduced: distance from the center of the image (MAGNITUDE) and displacement position (DIRECTION). The control algorithm was defines by simple linguistic if-then rules:

If DIRECTION is *left*, then move robot segment BODY to the *right*.

If DIRECTION is *left-up*, then move robot segment BODY to the *right*, and robot segment ELBOW to the *down. etc.*

All together 8 linguistic rules were used. Also, as the robot arm was controlled by the step motors, it was possible to control the number of steps for faster achieving the central position and three additional rules were introduced:

If MAGNITUDE is CENTRAL then STOP.

If MAGNITUDE is MIDDLE then NUMBER_OF_STEPS is 1.

If MAGNITUDE is BIG then NUMBER_OF_STEPS is 4.

In each iteration step, the if part of rules were tested and appropriate control action (incremental movement of robot segments) were applied.

6 Experimental Results and Conclusions

The robot positioning experiment confirmed our theoretical expectations. Using our FPC based image displacement vector determination system the robot was controlled in real time. Fig.6. and 7. shows one of experiments. For the robot starting position shown on Fig.6a) the displacement vector direction and magnitude were 200° and 0.4. Fig.6b) shows the final image. Although quite big oscillations in direction determination were noticed (mostly introduced by the unstable camera capturing characteristic) the arm was very precisely positioned from starting to the final position.

a)   b)

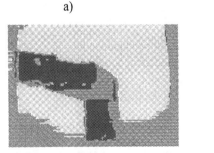 

Fig. 6. a) The starting image b) The image of final positioned robot arm

This system insensibility is mostly due to fuzzy algorithms used in displacement vector detection. Fig.7. shows the measurement results:

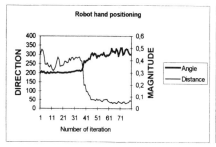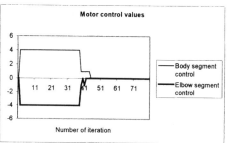

Fig. 7. Experiment results of fuzzy robot arm positioning based control: Displacement vector direction and magnitude and motor control values. (no. of steps in each iteration).

Our experiment shows that quite inexpensive equipment could be used for not so simple control tasks based on image pattern recognition. The limitation of the system was 4-bit image digitalization, as Fig.2. and Fig.6. shows. In the future we intend to use more precise digitalization unit and to improve our system introducing the possibility for robot arm orientation recognition. The second task could be achieved using additional pattern memories. Also, we intend to break the limits in robot workspace limitations, and develop the system for 3D vector displacement determination.

References

[1] Hans-Helmut Nagel, W. Enkelmann: An Investigation of Smoothness Constraints for the Estimation of Displacement Vector Fields from Image Sequences, in book **Computer Vision Principles**, pp. 498-526, Eds. R. Kasturi and R. C. Jain, IEEE Computer Society Press, 1991

[2] B. K. P. Horn: **Robot Vision**, MIT Press, 1981.

[3] Peter I. Corlee, **Visual Control of Robots**, John Wiley&Sons Inc., 1996

[4] J. J. Weng: 3D Motion Analyses from Image Sequences Using Point Correspondences, from **Handbook of Pattern Recognition and Computer Vision**, pp. 395-441, Eds. C. H. Chen, L. F. Pau, P. S. P. Wang, Worlds Scientific, Publishing Company, 1993.

[5] Bernd Jahne: **Digital Image Processing**, Springer-Verlog, 1991.

[6] A. K. Jain: **Fundamentals of Digital Image Processing**, Prentice Hall International Editions, 1989.

[7] **Fuzzy Set Comparator** (MD1210), Micro Devices, USA, 1989.

Acknowledgements

This work was supported by Ministry of Science and Technology of Republic of Croatia through Main Project: *"Complex System Control by Intelligent Methods"* and its Minor Projects.

Image Retrieval System Based on Machine Learning and Using Color Features

Janez Demšar, Dragan Radolović and Franc Solina

janez.demsar@fri.uni-lj.si, franc.solina@fri.uni-lj.si

Faculty of Computer and Information Science, University of Ljubljana
Tržaška cesta 25, SI-1000 Ljubljana, Slovenia

Abstract. We describe an interactive system for content based image retrieval. The system presents the user with 15 randomly selected images from the database. The user grades the images with one of five possible grades (YES, yes, neutral, no, NO) according to what he is looking for. The system returns the first 15 images with the highest probability of YES grade. The attributes used are a combination of color features. Three different machine learning techniques are compared.

Keywords: content based image retrieval, machine learning, color features

1 Introduction

The progress of computer industry has turned a cheap personal computer from an enhanced typewriter and number crunching machine into a powerful multimedia device. With availability of cheap, yet fast mass storage media, this led to foundations of large collections of images, audio and video files. Contrary to the textual databases, multimedia collections are much harder to organize into searchable libraries, which decreases their usefulness [1].

The simplest way to organize a multimedia collection is to manually define a hierarchy of 'themes' and divide the objects onto the defined groups. To find a particular object, the user steps down the hierarchy to the desired subgroup and then browses the acquired objects. Although the idea is straightforward, it has certain drawbacks. It requires a well defined and extendible structure and a great amount of work for manual classification of each object, including the objects that arrive later. In addition, such a structure is normally not disjunctive, i.e. one and the same object (or a subtheme) falls into different groups (or themes), which can be confusing. Our personal experience using such commercial collections is that searching through the structure can be slow and inefficient.

The alternative is to do the things automatically. In systems of this kind, objects are normally not organized in any kind of hierarchy and, in principle, no keywords or similar descriptions are given manually. Instead, certain simple attributes are extracted automatically and the user queries by specifying the values of attributes. Attributes can be given directly, as the user input, or indirectly, by showing the system one or more images with the contents we are looking for.

Searching algorithm which compares the values of attributes of the images in the collection with the required values can range from a simple distance measuring to sophisticated artificial intelligence methods.

The second approach is seldom used. The 'manual' approach is reliable in the sense that we always know what to expect, while the alternative might perform better but it might also not work at all. Our opinion is, however, that the amount of multimedia data is becoming infeasible for the manual handling and that the automatic methods must be explored and improved.

In this work we limited ourselves to collections of images. Also, our intention is not to develop a super-fast and super-accurate image retrieving system but merely to explore the usability of machine learning methods in this area and find the solutions for overcoming the encountered drawbacks. We will first describe the related work of other researchers and briefly introduce some previous method that incorporates a machine learning algorithm. After discussing the major difficulties, we shall present a new method for querying image databases.

2   Related Work

Existing systems for content based image retrieving (CBIR) generally use attributes that are manually or automatically extracted from images and then stored and managed in conventional database systems [2]. Besides that, precalculated attributes are often too domain specific or too general. Chabot System [3] for example, integrates a relational database containing keyword and other conventional data with color analysis technique to allow searching by keywords and dominant colors. It allows queries as "*mostlyOrange* and *someBlue*" which should, presumably, describe images of sunset over seas and lakes. The problem with this approach is in finding the right combination of attributes; query must be often refined. Also, those queries do not seem to describe the content of the image accurately enough.

Searching of an image database requires from the user to select or grade some initial images according to their likeliness to the searched images or to set some boundaries for the values of attributes. Attributes used in this context as well as the distances between the attributes must be fairly simple and fast for computation. Different types of attributes can be used. The most popular are *Color attributes* as they can be computed fast and in a straightforward manner. Several types of color attributes and their combinations can be used (histograms, moments, primary colors, averages etc.) which can also be computed separately in predefined parts of the image. Color attributes are not sensitive to location, rotation, scale and resolution. On the average they give good results but they miss images which are to a human observer very similar but of different colors. *Texture* [4] is somewhat more difficult to define and compute than color and is more sensitive to resolution. *Shape* (composition, structure) is much more difficult to define and compute than color attributes. Even if one can reliably recover shape attributes the definition of similarity or the distance among different shapes poses a very difficult problem. But since the human perception of similarity of images can not be reduced only to color and texture, this area of research is very important. Due to computational complexity the current shape attributes used in the framework of image retrieval are limited mostly only to edges, corners and interest points [5]. For shape attributes it is particularly more difficult to attain location, rotation and scale invariance.

Over the Web several commercial products and research systems for content based image retrieval can be tested. *QBIC (Query by image content)* is an IBM product [6] which is based mostly on color, color layout and texture attributes. *VIRAGE* [7] uses also composition and structure. *MetaSEEK* [8] combines the previous two search systems with the home grown *Vseek* using color and texture.

3 Machine Learning for Image Retrieving

The idea of using machine learning tools for image retrieval is not new. Our work is based on [9], where the ID3 algorithm is run to learn from example and counter example images, and the resulting classifier acts as an image query. We describe the method as an illustration of a straightforward application of machine learning techniques and of the related problems.

ID3 [10] is a simple learning algorithm from the "top-down induction of decision trees" family, which recursively divides the examples into groups and further into ever smaller subgroups using values of features as criteria, until it creates a 'clean' or almost clean subgroups. For CBIR, images are divided into subgroups until (almost) all images in a subgroup correspond to the same class (*wanted* or *not wanted*). The result of such a learning is a tree structure with each internal node containing a criteria for obtaining the branch that a particular image goes into, and the leaves containing groups of images of the same kind.

The attribute set used in [9] was simple, mostly describing proportions of basic colors in the image or in the central area of the image. The adaptation to the basic algorithm was a search for informative colors. The system used a local optimization to find colors which could be used as attributes in a decision tree. In the case of querying for images of faces it usually found a color which could be recognized as the skin color to be the most informative and used it in decisions of type 'images with less than 10% of this color do not represent faces'.

Experiments were run on a small collection of 167 images, of which 67 were images of a human face and the rest had different contents. Images were classified manually. Randomly chosen 70% of images (119 images) were given to the learning algorithm as a learning data with which it has built a decision tree. The learned ability to distinguish between images of faces and other images was then tested on the remaining examples. Although proportions of correctly classified images were encouraging, the method has never been put into practical use.

1. Machine learning algorithms can be seen as advanced statistical algorithms and, as always, a small sample means unreliable results. ID3 is quite sensitive to this. The described system has 119 examples as learning data. Would the user be prepared to find and manually classify such a great number of examples to perform a query?

2. Another problem with the approach was that ID3 classified each image as having the desired contents or not having it. Estimating the probabilities of having the desired contents would be more appropriate approach since it would enable the image browser to sort the images and present them to the user with the 'best' images first, instead of presenting only the images which the classifier guesses to have the desired contents and hiding the others.

3. The same holds for the user part. User should not be forced to classify each image as being "good" or "bad". Instead, he must be given a chance to grade the images according to how close they are to the desired image.

4. The system seems to rely on its ability to find informative colors. The proportion of, say, red or blue color in the image is probably far less informative than the proportion of the skin color. Thus, using an approximation for the skin color was essential and the search for informative colors was absolutely required. On the other hand, this search is quite slow.

4 Improvements

It is obvious that the described method requires major modifications. We have tested two algorithms besides ID3, and introduced example weights and class probabilities to soften the classification. We have also added some attributes.

4.1 Learning Algorithms

To solve the first problem, we reconsidered the chosen learning algorithm. ID3 is a strong learning algorithm that presents the gathered knowledge in a 'brain-compatible' form. In many cases, we are interested in the obtained decision tree and its explanation, and do not use it to classify unseen cases. The image retrieval problem is, however, of a different kind. Although we can admire the interpretability of the trees derived by the described system, the user does not really care about it. A more primitive learning algorithm that is able to learn with less examples and can estimate probabilities of classes should be used instead of ID3. In practice it does not matter whether the knowledge is 'readable' or not, since it is improbable that an occasional user would like to understand or even modify it. We decided to try out three different algorithms, ID3, k-nearest neighbors and naive Bayesian classifier.

Instead of classifying *ID3* estimated probabilities of classes. This was achieved by stopping the tree induction before the subgroups were clean and, when "classifying", returning the relative frequency of the node examples corresponding to certain class as the probability of that class.

K-nearest neighbors uses a distance measure (Euclidean distance, Manhattan distance, or some other) to find the k nearest neighbors of an example which is being classified. The algorithm usually selects the most frequent class among the neighbors as a prediction for the example's class. In our case, we are interested in probabilities of classes so the system returns relative frequency of the class as an estimate for the probability. Examples are also weighted by their distance from the reference example.

Naive Bayes classifier is based on Bayesian probability formula. Supposing the independence of attributes, the probability that example E is in class r_k is $P(r_k|E) = P(r_k) \prod_{i=1}^{n} \frac{P(r_k|v_i)}{P(r_k)}$, where $v_1, v_2, \ldots, v_n$ are values of attributes for E and the probabilities $P(r_k|v_i)$ and $P(r_k)$ are estimated from the learning set.

Estimating the probabilities enables the system to rank the images. The user is given the same opportunity. Instead of deciding for or against an image, he assigns grades. The grades are converted to weights of examples. The image with greater positive or negative grade is given a greater weight.

The last problem from the previous section was partially eliminated by using more robust learning algorithms and by introducing some new color attributes. The attributes were defined and extracted by Dragan Radolović [11].

4.2   Color Attributes for Image Query

Color histograms are of high dimension and therefore it is difficult to compute the distance between them. The first and second moment of color histograms [12] are much more compact and easier to compare. The **first and second moment of color histogram** is the average RGB color and its dispersion. In our system they are computed on the whole image and in the central part (middle three fifths) of the image. **Compactness of colors** measures the proportions of pixels of "mostly red", "mostly green", "mostly blue" and "gray" colors surrounded by pixels of a similar color. The pixel of color (r, g, b) is "mostly red" if $r - \max(g, b) > \delta$, "mostly green" and "mostly blue" are defined similarly, all remaining pixels are "gray". **Proportions of basic colors** are proportions of pixels of "mostly red", "mostly green", "mostly blue" and "other" colors.

ID3 discretizes the attributes by finding such boundary that informativity of the obtained binary attribute is maximal. For Naive Bayesian classifier, attributes are discretized on five intervals with approximately equal number of examples. K-nearest neighbors normalizes the values to interval $[0, 1]$.

5   Implementation

The color attributes are precalculated for all images in the database. The computation of all color attributes for 1000 images takes less than 10 minutes on a PC. This set of attributes serves as input to the learning part of the system. The learning part of the new system was done by our general machine learning system ML*. ML* is a modular system which incorporates all of the listed learning algorithms, all of them also support example weighting.

First, the system presents the user a certain number (15) of randomly selected images and asks him to grade them (Fig. 1). Each image can be given one of five grades, with the lowest meaning that the image is completely different from what he looks for and the highest meaning that the image is of exactly the right type. The user is not required to classify all the images.

Fig. 1. An example of a query for human faces. The user gave the highest grade to the two faces and the second highest to the image of a group of people. All other images have the lowest grade.

The data posted, grades are converted to classes and weights. The images having the middle grade are skipped. The lower two grades are converted to 'NO' class, with the lowest having weight 1 and the other 0.5. The higher two grades correspond to 'YES' class with the highest having weight 1 and the other 0.5. The precalculated attributes together with the just constructed class and weight values are given to the learning algorithm. The obtained classifier is used to estimate the probabilities of 'YES' class for all other images in the database, which have not been presented to the user yet. The fifteen images with the highest probability of 'YES' class are presented to the user as an answer to his query and examples for its refinement.

If the query was unsuccessful the user can grade the presented set of images to point at the good and the bad examples again. The images are added to the previous examples. The learning algorithm re-learns with the new examples and a new selection of fifteen images is presented to the user again. In the case of a satisfactory answer, the user can refine the query by being more selective when

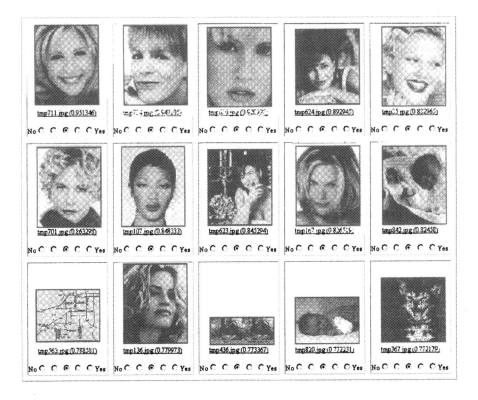

Fig. 2. Answer of the query from Figure 1. Only three of fifteen images are complete misses, all other images represent human faces.

assigning the good grades. The system decreases the weights of images of past queries and thus gives the new examples a bigger importance. The user can count on that and request a larger concept in the beginning and narrow it (become more strict) later, without the good grades from the first rounds of the process interfering in the later rounds.

Finally, if the query was a complete success, the user can request the next fifteen closest images using exactly the same classifier as before.

6   Results

Instead of statistical analyses which can distract the attention from actual usability of the system, we tested the system "visually" by using it to retrieve images from a database containing 1000 images.

Figures 1 and 2 show an example of a query for images of faces and the answer. Note that from the fifteen random images that were initially chosen by the system, only two presented a positive and one half-positive example. The answer is relatively accurate; only three of fifteen images are complete misses.

As expected, the *k-nearest neighbors* method was by far the best of the three learning methods tested. ID3 proved to be unable to handle such a small learning set with a great number of attributes; also, its probability estimation method assigned just a few different probabilities to the images so that many images had the same probability, which is impractical. *Naive Bayes classifier's* poor performance was probably due to the strong correlations between the attributes. K-nearest neighbors method also proved to be efficient in retrieving images of faces, animals, landscapes, cityscapes and similar. It is also fast enough, although the other two methods were a bit faster.

7   Conclusions and Further Work

As mentioned, our goal was to adopt the general machine learning methods for the use in image retrieving systems. Different methods were examined and small adaptations were made to incorporate them in a web based search engine. Experiments show that the most promising method for now is the *k-nearest neighbors*, especially for its ability to work with smaller example sets than the smarter methods. This is not surprising, the fact is that most of working systems for image retrieval already use a simplified version of this method. Its performance could be further improved by refining probability estimation function and how it is influenced by the learning examples of different classes at different distances from the example which is being classified.

Although the system is actually able to retrieve images belonging to simple "concepts", like those mentioned above, the concepts that can be distinguished from each other are much to wide. For example, an image retrieving system is expected to be able to retrieve not just "images of faces" but at least "images of female faces" if not even "images of faces of middle-aged blondes with green eyes, round glasses and not too much make up". It is obvious that the given color attributes do not describe images precisely enough. Therefore, the future work shall focus mostly on searching for new image describing features.

References

1. Ramesh Jain. Visual information management. *Communications of the ACM*, 40(12): 31–32, 1997.
2. V. N. Gudivada, V. V. Raghavan. Content-Based Image Retrieval Systems. *Computer*, 28:18–12, 1995.
3. Ogle V. E. (1995) Chabot: Retrieval from a Relational Database of Images, In *Computer 28*, pp. 40-48.
4. F. Liu, R. Picard. Periodicity, directionality, and randomness: world features for image modeling and retrieval. *IEEE Transactions on Pattern Analysis and Machine Intelligence*, 18(7):722–733, 1996.
5. C. Schmid, R. Mohr. Local grayvalue invariants for image retrieval. *IEEE Transactions on Pattern Analysis and Machine Intelligence,* 19(5):530–535, 1997.
6. http://wwwqbic.almaden.ibm.com/
7. http://www.virage.com/virdemo.html

8. http://ctr.columbia.edu/metaseek/

9. J. Demšar, F. Solina. Using machine learning for content-based image retrieving. In *Proceedings of the 13th International Conference on Pattern Recognition*, volume IV, pages 138–142, Vienna, Austria, August 1996.

10. J. T. Quinlan. Induction of decision trees, In *Machine Learning 1*, pp. 81-106. Kluwer Academic Publishers, 1986.

11. Dragan Radolović. *Image database queries based on color information*, B.Sc. Thesis (in Slovene). Faculty of Computer and Information Science, University of Ljubljana, 1998.

12. M. Stricker, A. Dimai. Spectral covariance and fuzzy regions for image indexing. *Machine vision and applications*, 10(2):66–73, 1997.

Face Parts Extraction Window Based on Bilateral Symmetry of Gradient Direction

Shunji Katahara and Masayoshi Aoki

Faculty of Engineering, Seikei University
3-3-1 Kichijoji-kitamachi, Musashino-shi, Tokyo, 180-0001, Japan
{katahara, masa}@ee.seikei.ac.jp

abstract. We propose a simple algorithm to determine face parts extraction window in face image. We utilize bilateral symmetries between and within face parts. We also use knowledge about size and locationship of face parts. First, we examine bilateral symmetries around vertical orientation edge, then obtain symmetry measures. The symmetry measures are projected onto y-axis to produce histogram of the measures. We estimate height of face parts regions by frequency of the histogram. Face parts region, which contains maximum frequency of the histogram, becomes a candidate of face parts region that includes eyes and eyebrows. Secondly, the measures that exist within the height of the face parts region are projected onto x-axis to estimate width of face parts region. We determine face parts extraction windows by the estimated height and width. Finally, we detect irises in the candidate of face parts region that includes eyes and eyebrows, using circular mask.

1 Introduction

Face image processing is recently used for personal identification, facial expression detection, drowsiness detection and so on[1, 2, 3]. The important face parts such as eyebrows, eyes, nose and mouth are used to express facial feature. Active contour model, deformable template model, local smoothness of image density, color information of image and knowledge about shape and locationship of face parts are used to detect face parts[4, 5, 6, 7]. Reisfeld describes symmetry transform to find facial feature points[8].

When application of face image recognition is limited to the specific purpose, in the case which one face is always clearly obtained, algorithm for extracting face parts becomes extremely simple by using symmetry. We focus bilateral symmetries between and within face parts. It is possible to determine face parts extraction windows by utilizing bilateral symmetry of face parts. It is also possible to estimate horizontal positions of irises by bilateral symmetry between face parts in the face parts extraction windows.

We detect symmetry measure on face parts using bilateral symmetry of gradient directions. We determine width and height of face parts extraction windows from histogram of symmetry measures, and detect irises in the window.

2 Symmetry detection

2.1 Edges for symmetry detection

We define 8-kinds of gradient direction as shown in Fig. 1 to detect bilateral symmetry. In order to examine bilateral symmetry of shape of the object by gradient direction, we use some of a pair of horizontal and vertical orientation edges represented in Fig. 1.

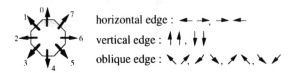

horizontal edge : ◄ ─►, ─► ◄─

vertical edge : ▲▲, ▼▼

oblique edge : ↘ ↗, ↗ ↘, ↗ ↖, ↘ ↙

Fig. 1. Direction and combination of edge

2.2 Symmetry measure

Bilateral symmetry around gradient direction at each point of interest is detected, and symmetry measure is obtained as described below (see Fig. 2). We examine direction of gradient at the same distance from point of interest. If both directions of gradient at the same distance from point of interest have bilateral symmetry each other, symmetry measure at this point is incremented.

Horizontal and vertical directions of gradient are encompassed in g(x,y), and bilateral symmetries around horizontal and vertical directions of gradient are separately evaluated in bm(x,y).

Distance from point of interest is horizontally extended until it approaches to the side of the image.

Point of interest horizontally moves from left side to right side of the image.

This procedure in each line described above is repeated from top to bottom line of the image.

$$bm(x,y) = \begin{cases} +1.0 & g(x-i,y) = g(x+i,y) \\ {\scriptstyle(increment\ 1.0)} & i=1,\ \text{maximum distance from } P(x,y) \\ 0 & otherwise \end{cases}$$

where,

$bm(x,y)$: bilateral symmetry measure at $P(x,y)$

$g(x,y)$: direction of gradient at $P(x,y)$

i : horizontal distance from $P(x,y)$

$x \in [i, m-i]$, m : width of image

$y \in [1, n]$. n : height of image

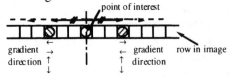

point of interest

gradient direction → ↑ ↓   ← gradient ↑ ↓ direction   row in image

Fig. 2. Symmetry detection

2.2.1 Projection onto x-axis

We obtain histogram of symmetry measure SM(x) by projecting symmetry measure bm(x,y) onto x-axis. We estimate width of face parts region from frequency of histogram SM(x), and we find midline of face from a peak of the frequency of histogram SM(x).

$$SM(x) = \sum_{j=1}^{n} bm(x, j)$$

where,

$SM(x)$: projection of $bm(x, y)$ onto x - axis

$$SM(y) = \sum_{i=1}^{m} bm(i, y)$$

where,

$SM(y)$: projection of $bm(x, y)$ onto y - axis

2.2.2 Projection onto y-axis

We obtain histogram of symmetry measure SM(y) by projecting symmetry measure bm(x,y) onto y-axis. By projecting symmetry measure bm(x,y) around vertical orientation edge onto y-axis, we can estimate height of face parts region from the frequency of histogram SM(y).

Peak of the frequency of the histogram also gives the height of eyes or eyebrows position.

3 Face parts extraction window

3.1 Extraction of effective edge

We obtain edge image by applying edge detection operator to face image(see Fig. 3 and 4). We think that low strength edge does not contain useful information about face parts, therefore we extract effective edge from edge image using discriminant analysis for thresholding(see Fig. 5).

We use template type Robinson operator[9, 10] to detect edge strength and direction.

 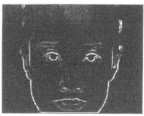 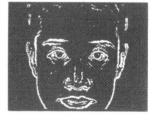

Fig. 3. Original image (320x240 pixel, 8bit gray level)　　**Fig. 4.** Edge image　　**Fig. 5.** Effective edge image

3.2 Midline of face

We examine bilateral symmetry around horizontal orientation edge on effective edge image. Map of symmetry measure bm(x,y) is represented in Fig. 6-a). Projecting symmetry measures onto x-axis produces histogram of symmetry measure SM(x). We find midline of face from a peak of the frequency of the histogram.

If face image is captured in plain background, we can estimate width of face from the frequency of the histogram.

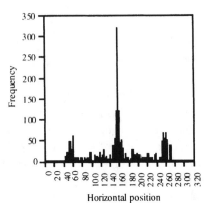

a) Map of bilateral symmetry measure
bm(x,y) around horizontal orientation edge

b) Histogram of symmetry measure SM(x)

Fig. 6. Symmetry measure around horizontal edge

3.3 Height of face parts region

We utilize bilateral symmetry between and within face parts, in order to estimate height of face parts region.

We examine bilateral symmetry around vertical orientation edges on effective edge image. Map of symmetry measure bm(x,y) is represented in Fig. 7-a). Projecting the symmetry measures onto y-axis produces histogram of symmetry measure SM(y).

We estimate height of face parts region using this histogram. We also estimate a face parts region that includes eyes and eyebrows from a peak of the frequency of the histogram. Intersection of midline of face and height of the peak of the histogram around vertical orientation edge nearly represents the center of the face.

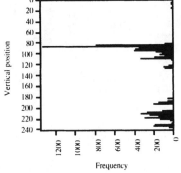

a) Map of bilateral symmetry measure
bm(x,y) around vertical orientation edge

b) Histogram of symmetry measure SM(y)

Fig. 7. Symmetry measure around vertical edge

In order to estimate height of face parts region, we roughly observe the histogram the same as we observe an image by low graduation level and low resolution.

3.3.1 Removing useless symmetry measure

In order to remove useless symmetry measure in the histogram, we requantize the frequency of the histogram at each class in 7-bit depth. This operation is effective to remove the frequency less than 1 % of maximum frequency as useless symmetry measure(see Fig. 8-a)).

3.3.2 Observing image height by low resolution

In order to roughly observe the height of face parts region, we observe height of image by low resolution. We subdivide the image height into every 8 lines for this purpose. If useful symmetry measure is included in the subdivided section, the subdivided section becomes useful class to determine height of face parts region. We finally obtain the number of three candidates face parts region around height(see Fig. 8-b)).

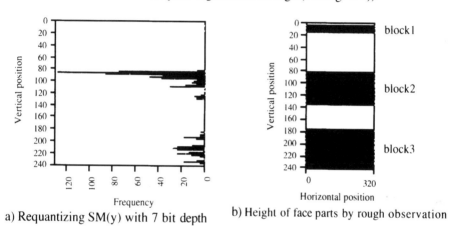

a) Requantizing SM(y) with 7 bit depth   b) Height of face parts by rough observation

Fig. 8. Height of face parts region

3.4 Width of face parts region

We produce another histogram by projecting the symmetry measures that exist within each height of the candidate of face parts region onto x-axis(see Fig. 9-a)~c)).

We estimate width of face parts region by the frequency of the histogram. We examine the frequency of the histogram from left side and from right side to the center to determine width of face parts region. We find a position at which the frequency of the histogram exceeds the value of the useless symmetry decided in 3. 3. 1, respectively. A distance between the positions determines width of the face parts region. Therefore, block-1 does not correspond to face parts region(see Fig. 9-a)).

Remarkable multimodal histogram appears in the face parts region that includes eyes and eyebrows, because these face parts have bilateral symmetry between and within them(see Fig. 9-b)).

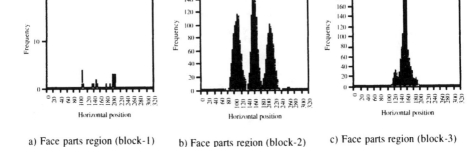

a) Face parts region (block-1)   b) Face parts region (block-2)   c) Face parts region (block-3)

Fig. 9. Histogram of the symmetry measure in each face parts region

3.5 Face parts extraction window

We estimate height of face parts region by projecting symmetry measures along vertical orientation edge onto y-axis.

The symmetry measures along vertical orientation edges that exist inside of the height of face parts region are projected onto x-axis to estimate width of face parts region.

Intersection regions of estimated height and width become candidates of face parts extraction window (see Fig. 10).

Fig. 1 0. Face parts extraction windows

4 Iris detection

4.1 Estimation of iris diameter

By projecting the symmetry measures onto x-axis, we can make sure of remarkable multimodal histogram of face parts region that includes eyes and eyebrows (see Fig. 9-b)). Left and right side peaks arise from bilateral symmetry within face parts, whereas middle of the peaks arises from bilateral symmetry between face parts. Left and right side peaks also give the midline of each face parts individually. We think that the distance between left and right side peaks corresponds to width between eyes.

We show typical size around iris as follows. Iris has a diameter of 12 mm, and width between center of irises is 63 mm ± 7 mm for adult men or 61 mm ± 6 mm for adult women.

We determine diameters for detecting iris as follows.

We prepare some of one-pixel thickness concentric circle masks to detect boundary of iris.

$$\frac{d}{(D_{men})_{max.}} D' \leq d' \leq \frac{d}{(D_{women})_{min.}} D'$$

where,

d' : estimated iris diameter (pixel)

D' : distance between lateral peaks (pixel)

d : human iris diameter (12mm)

D : distance between right and left iris

(men; 63 ± 7mm, women; 61 ± 6mm)

$$sumE(x, y) = \oint_s t \cdot f_e \cdot f_i ds$$

where,

t : concentric circle mask

$$\begin{cases} 1 : on\ circle \\ 0 : otherwise \end{cases}$$

f_e : effective edge

$$\begin{cases} 1 : on\ effective\ edge \\ 0 : otherwise \end{cases}$$

f_i : strength of edge

4.2 Search area for iris detection

Strength of edges that correspond to effective edge on the mask in search area is accumulated to detect boundary of iris. We detect best-matched iris diameter and iris position on the condition that cumulative strength of edges becomes largest.

The mask for detecting right iris moves inside of the candidate of face parts extraction window, which includes eyes and eyebrows. The search area is approximated by the estimated right iris position derived from the symmetry histogram represented in Fig. 9-b), and searched width is determined by estimated iris diameter(see Fig. 11).

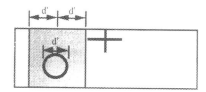

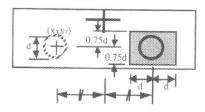

Fig. 11. Search area for right iris detection   **Fig. 12.** Search area for left iris detection

Left iris is detected by the same procedure that right iris has been detected. We use knowledge about symmetry between face parts to detect left iris. By the symmetry between face parts, left iris may be detected same distance from midline. We limit the search area for detecting left iris in upward and downward equivalent to 0.75 of right iris diameter, and in leftward and rightward equivalent to right iris diameter(see Fig. 12).

4.3 Iris detection result

Iris detection results are represented in Fig. 13-a)~d) with face parts extraction window and search area for iris detection.

In Fig. 13-a), irises have been detected as 17 pixel across in diameter, and distance between the center of irises has been detected as 86 pixel length. Detected centers of iris positions are indicated by cross hair. Right iris has been detected at the position (116, 114), and left iris has been detected at (202, 116).

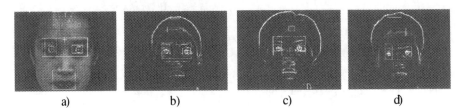

a)   b)   c)   d)

Fig. 1 3. Face parts extraction windows and detected irises

Table 1. Iris detection results

	d(right)	d(left)	D	D/d
(a)	17	17	86	5.1
(b)	13	11	62	5.2
(c)	13	13	65	5.0
(d)	9	11	57	5.7

Table 1 shows detected iris diameters; d(right), d(left), width between irises; D and ratio of D by d; D/d. In Table 1, D/d indicates in the range of 5.0 to 5.7. We think that if D/d is within the range of 4.6 to 6.0, detection has been successfully done.

5 Concluding remarks

1) Midline of face and width of face are detected using bilateral symmetry around horizontal orientation edge.

2) Peak of a frequency of histogram of symmetry measure around vertical orientation edge gives height of eyes or eyebrows position.

3) Intersection of midline of face and height of a peak of histogram of symmetry measures around vertical orientation edges nearly represents center of face.

4) Height of face parts region is estimated from histogram that is produced by projecting symmetry measures along vertical orientation edges onto y-axis. Width of face parts region is estimated from another histogram that is produced by projecting the symmetry measures around vertical orientation edges, which exist within each height, onto x-axis. Face parts extraction windows are determined by the height and the width.

5) Irises are detected from a candidate of face parts extraction window, which includes eyes and eyebrows. In order to detect iris inside of an extraction window, diameter of irises

is estimated from distance between peak positions of multimodal histogram of symmetry measure.

6) Irises as important face parts are detected. Position of irises, diameter of iris and distance between irises give useful information to detect another face parts such as eyebrows, upper eyelids, mouth and so on.

References

1. Huang C. L. and Chen C. W.: Human Facial Feature Extraction for Face Interpretation and Recognition, Pattern Recognition, Vol. 25, No. 12 (1992) 1435-1444
2. Katahara S. and Aoki M.: Eye Closure Detection from Eyelid Motion, Proceedings of the 30th International Symposium on Automotive Technology & Automation, 97ATT031 (1997) 215-222
3. Nakano T. et. al.: Blink Measurement by Image Processing and Application to Warning of Driver's Drowsiness in Automobiles, Proceedings of the 1998 IEEE International Conference on Intelligent Vehicles, Vol. 1 (1998) 285-290
4. Kass M. et. al.: Snake:Active Contour Models, Int. J. Computer Vision (1988) 321-331
5. Xie X. et. al.: On Improving Eye Feature Extraction Using Deformable Templates, Pattern Recognition, Vol. 27, No. 6 (1994) 791-799
6. Chow G. and LI X.: Towards a System for Automatic Facial Feature Detection, Pattern Recognition, Vol. 26, No. 12 (1993) 1739-1755
7. Chen C. and Chiang S. P.: Detection of Human Faces in Colour Images, IEE Proceedings Vision, Image and Signal Processing, Vol. 144, No. 6 (1997) 384-388
8. Reisfeld D. and Yeshurun Y.: Preprocessing of Face Images: Detection of Features and Pose Normalization, Computer Vision and Image Understanding, Vol. 71, No. 3 (1998) 413-430
9. Robinson G.: Edge Detection by Compass Gradient Mask, CGIP 6 (1977) 492-501
10. SPIDER Working Group: SPIDER Users' Manual, Joint System Development Corp. , Tokyo (1983)

Modelling Needle-Map Consistency with Novel Constraints

Philip L. Worthington   Edwin R. Hancock

Department of Computer Science, University of York, UK
{plw,erh}@minster.cs.york.ac.uk

Abstract. This paper addresses two fundamental criticisms of existing global shape-from-shading techniques. Firstly, we introduce a new update framework which ensures data-closeness at all stages whilst iteratively recovering the needle-map by treating the image irradiance equation as a hard constraint. Secondly, we apply the apparatus of robust statistics and novel constraints on needle-map smoothness to accomodate surface orientation discontinuities in a principled fashion.

1   Introduction

The recovery of shape-from-shading (SFS) is a longstanding research goal of the computer vision community. From a computational viewpoint, SFS usually involves solving the image irradiance equation (IIR) to recover a set of surface normals, often referred to as the needle-map. However, since the IIR is under-constrained, additional constraints such as local smoothness of the needle-map must be invoked for reasons of computational tractability.

One of the most popular approaches is to adopt a regularization framework [8, 1, 4] and iteratively recover the needle-map. The regularization functional penalizes departures from the IIR, whilst imposing smoothness constraint on the recovered needle-map. A fundamental obstacle to progress in energy minimization approaches to shape-from-shading is the failure, to-date, to develop a constraint function which is uniquely minimized by a needle-map which closely matches the true needle-map. The search for such a functional is hindered by the regularization framework, which in many cases leads to a stark choice between numerically instability and severe model-dominance. Data-closeness is invariably sacrificed to improve smoothness.

The aim of the work reported in this paper is to provide a novel framework for shape-from-shading in which the dual deficiencies of poor data-closeness and simplistic consistency constraints are redressed. We make two distinct contributions. Firstly, we show how data-closeness can be imposed as a hard constraint. By reformulating the problem in this way, we avoid concerns about numerical stability and parameter selection. This facilitates the second contribution of the paper, which is to experiment with novel needle-map consistency constraints in place of needle-map smoothness. We develop and compare three novel needle-map consistency constraints. The design of these constraints is motivated by the desire to take better account of the *piecewise* smooth nature of most real-world surfaces, using robust adaptive smoothness, gradient consistency and curvature consistency.

2 The Horn and Brooks Algorithm

The local surface normal may be written as $\mathbf{n} = (-p, -q, 1)^T$, where $p = \frac{\partial z}{\partial x}$ and $q = \frac{\partial z}{\partial y}$. For a light source at infinity, we can similarly write the light source direction as $\mathbf{s} = (-p_l, -q_l, 1)^T$. If the surface is Lambertian the reflectance map is given by $R(p, q) = \mathbf{n} \cdot \mathbf{s}$. The normalized image irradiance equation (IIR) [3] states that $E(x, y) = R(p, q)$, where $E(x, y)$ is the image brightness. Alone, the IIR provides insufficient constraint for the unique recovery of the needle-map. An additional constraint, that the needle-map varies smoothly, is usually applied

Smooth surface recovery is posed as a variational problem in which a global error-functional is minimized through the iterative adjustment of the needle map. In the formulation of Brooks and Horn [1], the error functional is defined to be

$$I = \int \int \left(E(x, y) - \mathbf{n} \cdot \mathbf{s} \right)^2 + \lambda \left(\left\| \frac{\partial \mathbf{n}}{\partial x} \right\|^2 + \left\| \frac{\partial \mathbf{n}}{\partial y} \right\|^2 \right) dx dy \qquad (1)$$

where $\frac{\partial \mathbf{n}}{\partial x}$ and $\frac{\partial \mathbf{n}}{\partial y}$ are the directional derivatives of the needle-map in the x and y directions respectively, and λ is a Lagrange multiplier. The magnitudes of these quantities are used to measure the smoothness of the surface, with a large value indicating a highly-curved region. However, it should be noted that a planar surface has $\frac{\partial \mathbf{n}}{\partial x} = \frac{\partial \mathbf{n}}{\partial y} = \mathbf{0}$.

Applying the calculus of variations and discretizing the resulting Euler equation yields the update equation to estimate the needle-map at epoch $k + 1$ using the estimate at epoch k of the form $\mathbf{n}_{i,j}^{(k+1)} = \bar{\mathbf{n}}_{i,j}^{(k)} + \frac{\epsilon^2}{2\lambda} \left(E_{i,j} - \mathbf{n}_{i,j}^{(k)} \cdot \mathbf{s} \right) \mathbf{s}$, where ϵ is the spacing of pixel-sites on the lattice and $\bar{\mathbf{n}}_{i,j}$ is the local 4-neighbourhood mean of the surface normals around pixel position i, j.

3 A Novel Framework for SFS

A common criticism of the Horn and Brooks algorithm, and of similar approaches, is the tendency to oversmooth important surface detail such as surface orientation discontinuities. In this paper, we aim to improve data-closeness, and also improve the modelling of smoothness to accomodate surface orientation discontinuities. We commence by noting that the Horn and Brooks functional (Equation 1) consists of two terms whose aims are different and competing. If the scheme is initialized with the true needle-map, the solution will "walk-away" as brightness error is traded for smoothness. Horn [4] attempts to address this problem by increasing the relative importance of the brightness error over successive iterations. Here, we treat the data-closeness and constraint terms separately, in such a way that data-closeness is always guaranteed; that is, the IIR is treated as a hard constraint. Within this constraint, the task becomes to iterate towards the true needle-map, using minimization of smoothness error or some other constraint functional. The need for λ is removed entirely.

Geometrically, we consider the IIR to define a cone about the light source direction for each surface normal. The normals forming the needle-map can only assume directions defined by this cone, as shown in Figure 1.

3.1 Using the Image Irradiance Equation as a Hard Constraint

The new framework requires us to minimize the constraint functional

$$I = \int \int \psi\left(\mathbf{n}(x,y), \mathbf{N}(x,y)\right) dxdy \tag{2}$$

whilst satisfying the hard constraint imposed by the IIR

$$\int \int (E - \mathbf{n} \cdot \mathbf{s})\, dxdy = 0 \tag{3}$$

where $\mathbf{N}(x,y)$ is the set of local neighbourhood vectors about location (x,y). The function ψ is a localized function of the current surface normal estimates. Clearly, it is possible to incorporate the hard data-closeness constraint directly into ψ, but this needlessly complicates the mathematics. Instead, we choose to impose the constraint after each iteration by mapping the updated normals back to the most similar normal lying on the cone.

If we take the smoothness constraint of Horn and Brooks as an example, then $\psi\left(\mathbf{n}, \mathbf{N}\right) = \left\|\frac{\partial \mathbf{n}}{\partial x}\right\|^2 + \left\|\frac{\partial \mathbf{n}}{\partial y}\right\|^2$. Discretizing ψ directly in terms of forward differences yields $\psi\left(\mathbf{n}_{i,j}, \mathbf{N}_{i,j}\right) = (\mathbf{n}_{i+1,j} - \mathbf{n}_{i,j})^2 + (\mathbf{n}_{i,j+1} - \mathbf{n}_{i,j})^2$. In practice, to maintain symmetry, we take the mean of the forward and backward difference results, since we cannot use the central difference because this does not include the centre normal, $\mathbf{n}_{i,j}$. Doing this and differentiating leads to an update equation of the form $\mathbf{n}_{i,j}^{k+1} = \bar{\mathbf{n}}_{i,j}^{k}$. This is a simple neighbourhood averaging action which if left unchecked will eventually lead to a flat surface. Note that this update equation could be obtained from the Horn and Brooks update equation simply by setting the brightness error to zero, or by using the constraint function $\psi\left(\mathbf{n}, \mathbf{N}\right) = (\mathbf{n} - \bar{\mathbf{n}})^2$.

We now apply the hard constraint provided by the IIR to modify this smoothing action and force it to iteratively search the reduced subspace of solutions fitting the IIR. We opt to do this by rotating the updated normal back to the nearest vector lying on the cone, so the updated process becomes $\mathbf{n}_{i,j}^{k+1} = \Theta \bar{\mathbf{n}}_{i,j}^{k}$, where Θ is a rotation matrix. Figure 1 illustrates the update process, and compares it with the Horn and Brooks update.

3.2 Initialization

The new framework requires an initialization which ensures that the image IIR is satisfied at the outset. This differs from the Horn and Brooks algorithm, which is

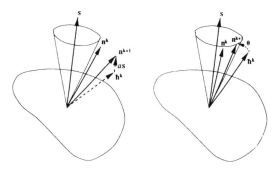

Fig. 1. Comparison of update process for Horn and Brooks (left) and the new framework with the same smoothness constraint (right). Horn and Brooks allows the updated normal to wander away from the cone, sacrificing brightness error for smoothness. The movement in the light source direction is as in the left-hand diagram, where $a = \frac{\epsilon^2}{2\lambda}(E - \mathbf{n} \cdot \mathbf{s})$. In contrast, the new framework forces the brightness error to be satisfied by using the rotation matrix Θ to map the smoothness update to the closest point on the cone.

usually initialized by estimating the occluding boundary normals, with all other normals set to point in the light source direction. We choose to initialize each normal such that its projection onto the image plane lies in the negative image gradient direction. This has intuitive appeal, since the gradient is a first-order estimate of surface tilt in the case of a light source directly overhead [6]. We have also applied this initialization to the Horn and Brooks algorithm, and find that it produces significantly better and faster results than using the occluding boundary.

4   Needle-Map Consistency Constraints

The novel framework developed in the preceding section, incorporating the smoothness constraint of Horn and Brooks, largely frees us from problems of numerical instability and error trade-off. Our second contribution is to develop new methods of enforcing needle-map consistency. This is motivated by our observations that traditional smoothness constraints are ill-suited to the SFS problem. Remarkably little attention has been focused upon alternatives to smoothness constraints [2, 11].

Here, we describe three promising approaches. The first is simply to modify the action of needle-map smoothness to more adequately model the properties of real surfaces. Secondly, we use higher-order consistency between the needle-map and image to ensure satisfaction of the IIR in terms of image and needle-map gradient as well as intensity. Finally, we consider the topographic properties of surfaces to develop consistency constraints based upon surface curvature.

4.1 Robust Adaptive Smoothness Consistency

We consider the robust regularizer developed in [9, 10]. In essence, this assumes that the recovered needle-map should be smooth, except where there is a high probability that a discontinuity is present, in which case the smoothing is reduced. We define the robust adaptive smoothness constraint function as $\psi\left(\mathbf{n}, \mathbf{N}\right) = \rho_\sigma\left(\left\|\frac{\partial \mathbf{n}}{\partial x}\right\|\right) + \rho_\sigma\left(\left\|\frac{\partial \mathbf{n}}{\partial y}\right\|\right)$, where $\rho_\sigma(\eta)$ is a robust kernel defined on the residual η and with width parameter σ. Applying the calculus of variations, discretizing and re-arranging, yields the general update equation

$$\mathbf{n}_{i,j}^{(k+1)} = \Theta\left(\frac{\partial}{\partial x}\rho_\sigma'\left(\left\|\frac{\partial \mathbf{n}}{\partial x}\right\|\right) + \frac{\partial}{\partial y}\rho_\sigma'\left(\left\|\frac{\partial \mathbf{n}}{\partial y}\right\|\right)\right) \tag{4}$$

where $\rho_\sigma'(\|\frac{\partial \mathbf{n}}{\partial x}\|) = \frac{\partial}{\partial \mathbf{n}_x}\left(\rho_\sigma(\|\frac{\partial \mathbf{n}}{\partial x}\|)\right)$ and $\rho_\sigma'(\|\frac{\partial \mathbf{n}}{\partial y}\|) = \frac{\partial}{\partial \mathbf{n}_y}\left(\rho_\sigma(\|\frac{\partial \mathbf{n}}{\partial y}\|)\right)$.

In [10] we showed that the sigmoidal-derivative M-estimator

$$\rho_\sigma(\eta) = \frac{\sigma}{\pi}\log\cosh\left(\frac{\pi\eta}{\sigma}\right) \tag{5}$$

which is a continuous version of Huber's estimator, possesses good properties for handling surface discontinuities.

4.2 Gradient Consistency

Zheng and Chellappa [11] used an intensity gradient constraint in conjunction with integrability. This constraint ensured that the intensity gradient of the image reconstructed from the needle-map matched the gradient of the input image. However, this constraint is not useful in the new framework since the reconstructed image always precisely matches the input due to the application of the IIR as a hard constraint. We therefore attempt to design a constraint which uses the needle-map structure directly.

To formulate a gradient consistency constraint, we wish to ensure that the image gradients in the x and y directions match those of the needle-map. In other words, differentiating the IIR, and assuming the light source direction is constant with respect to x and y, we obtain $\frac{\partial E}{\partial x} = \frac{\partial \mathbf{n}}{\partial x}{\cdot}\mathbf{s}$ and $\frac{\partial E}{\partial y} = \frac{\partial \mathbf{n}}{\partial y}{\cdot}\mathbf{s}$. Incorporating this into the robust regularizer constraint function, we define an adaptive kernel width to be the mean of the exponentials of the local gradient consistency

$$\sigma = \sigma_0\frac{1}{N}\sum_{l\in N}\exp\left(-\left(\left(\frac{\partial E}{\partial x} - \frac{\partial \mathbf{n}_l}{\partial x}{\cdot}\mathbf{s}\right)^2 + \left(\frac{\partial E}{\partial y} - \frac{\partial \mathbf{n}_l}{\partial y}{\cdot}\mathbf{s}\right)^2\right)\right) \tag{6}$$

If the image and needle-map gradients correspond well, σ is large and little smoothing is applied. However, if some of the neighbourhood normals exhibit large deviations from image consistency, this causes the kernel to narrow, reducing the influence of the inconsistent normals. If the mean inconsistency is very large, the kernel narrows to the point where all the neighbourhood normals are in the saturation region of the kernel, causing the algorithm to revert to the Horn and Brooks smoothing process.

4.3 Curvature Consistency

Needle-map smoothness appears to be an over-strong and inappropriate constraint for SFS. This is primarily because real surfaces are more likely to be *piecewise* smooth [7]. If we instead consider surfaces in terms of curvature characteristics, orientation discontinuities usually correspond to sharp surface ruts or ridges. Furthermore, the curvature classes for locations either side of a rut or a ridge should be the most similar classes, either trough or saddle rut for a rut, or dome or saddle ridge for a ridge (Figure 2). This property of smooth variation in class suggests that curvature consistency may be a more appropriate constraint for SFS than smoothness.

The use of a curvature consistency measure was introduced to SFS by Ferrie and Lagarde [2]. They use global consistency of principal curvature directions to refine the surface estimate returned by local shading analysis.

We propose to formulate curvature consistency in terms of the shape index of Koenderink and van Doorn [5]. This is a continuous measure which encodes the same curvature class information as $H - K$ labels, but in an angular representation. It is defined in terms of the principal curvatures κ_1 and κ_2 such that $\phi = \frac{2}{\pi} \arctan \frac{\kappa_2 + \kappa_1}{\kappa_2 - \kappa_1}$. In terms of normals this becomes

$$
\phi = \frac{2}{\pi} \arctan \frac{\left(\frac{\partial \mathbf{n}}{\partial x}\right)_x + \left(\frac{\partial \mathbf{n}}{\partial y}\right)_y}{\sqrt{\left(\left(\frac{\partial \mathbf{n}}{\partial x}\right)_x - \left(\frac{\partial \mathbf{n}}{\partial y}\right)_y\right)^2 + 4 \left(\frac{\partial \mathbf{n}}{\partial x}\right)_y \left(\frac{\partial \mathbf{n}}{\partial y}\right)_x}}
\tag{7}
$$

Figure 2 shows the range of shape index values and the type of curvature which they represent.

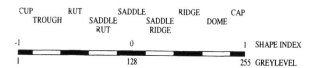

Fig. 2. The shape index scale ranges from -1 to 1 as shown. The shape index values are encoded as a continuous range of grey-level values between 1 and 255, with grey-level 0 being reserved for background and flat regions (for which the shape index is undefined).

Since direct minimization of curvature differences does not yield a unique update equation, we have investigated the use of an adaptive robust regularizer in conjunction with curvature consistency. We use the shape index statistics to adaptively set the width of the robust kernel, σ, to be applied over the neighbourhood. The kernel width determines the level of smoothing applied to the neighbourhood. If the shape index of the neighbourhood varies greatly from the shape index at the current normal, then strong smoothing is applied, whereas an already smoothly-varying shape index pattern receives less attention. Once again, we have $\psi(\mathbf{n}, \mathbf{N}) = \rho_\sigma\left(\left\|\frac{\partial \mathbf{n}}{\partial x}\right\|\right) + \rho_\sigma\left(\left\|\frac{\partial \mathbf{n}}{\partial y}\right\|\right)$ where again $\mathbf{N}$ is the set of local neighbourhood normals used to calculate the finite difference approximations to $\frac{\partial \mathbf{n}}{\partial x}$ and $\frac{\partial \mathbf{n}}{\partial y}$. However, instead of a fixed kernel width, σ, we use the adaptive value

$$\sigma = \sigma_0 \exp\left(-\left(\frac{1}{N}\sum_{l \in \mathbf{N}} \frac{(\phi_l - \phi_c)^2}{\Delta\phi_d^2}\right)^{\frac{1}{2}}\right) \qquad (8)$$

where ϕ_c is the shape index associated with the central normal, $\mathbf{n}_{i,j}$, and $\Delta\phi_d$ is the difference in shape index between the centre values of adjacent curvature classes. The number of neighbourhood normals used in calculating the finite difference approximations to $\frac{\partial \mathbf{n}}{\partial x}$ and $\frac{\partial \mathbf{n}}{\partial y}$ is denoted N, and σ_0 is a reference kernel width which we set to unity. Using the scale of Figure 2, $\Delta\phi_d = \frac{1}{8}$.

5 Experiments

We have performed experiments using synthetic and real-world images, using the SFS schemes developed in the preceding sections. Using synthetic data, we attempt to analyse some of the difficulties inherent in SFS, focusing upon the fundamental point that no existing constraint functional possesses a unique global minimum matching the true surface. We demonstrate by example that none of the constraint functions considered here have this property, but that this does not in itself prevent the new framework generating improved approximations to the true needle-map. This research poses important questions about the design of future constraint functions using the new framework. We also demonstrate the qualitative improvements offered by the new framework in conjunction with the novel constraints when applied to real-world images.

Figure 3 offers an assessment of the quality of the different constraints developed in this paper, using ground-truth data from a synthetically-generated image of two co-joined spheres. The normalized values of several measures are plotted against iteration number for the Horn and Brooks algorithm and each of the new framework schemes. The most important measures are the normal error, as calculated against ground truth, and the value of the constraint function summed over the whole image. We also include the surface reconstruction error, and the average change in normal direction is plotted as a potential termination criterion, but is seen to be ill-suited to the task. All the schemes, including the Horn and Brooks algorithm, are initialized using the novel initialization of Section 3.2.

Since the normal error is an absolute measure unaffected by which constraint function is used, and as all the schemes use the same initialization and so start with identical (and, due to the new initialization, relatively low) values of normal error, we can compare these normalized values directly. Doing so, we see that when the smoothness and gradient consistency constraints are modelled using a robust error kernel, there is a reduction in the normal error to approximately 57% of the initial value, which Horn and Brooks struggles to approach despite the improved initialization and many more iterations.

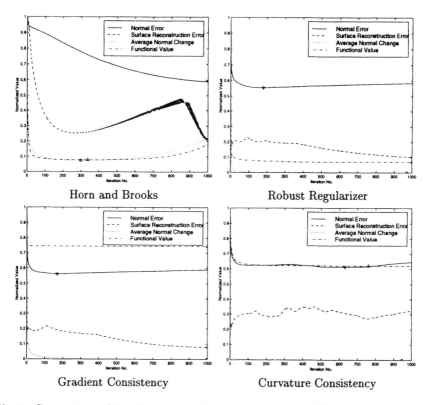

Fig. 3. Comparison of the behaviour of various measures for Horn and Brooks, and the new framework using different constraint functions (see text).

In Figure 4, we illustrate the application of the SFS schemes developed in this paper to real images from the Columbia Object Image Library, which present challenges such as albedo changes, regions of brightness saturation and deviations from the Lambertian assumption. In the first row of Figure 4, the needle-map recovered by the Horn and Brooks algorithm is seen to be extremely smooth, even losing large-scale details such as the wing of the toy duck. The new framework schemes all perform qualitatively well, recovering the wing and neck structure. Similarly, in the second row of Figure 4, the new framework schemes all yield significant qualitative gains over the Horn and Brooks algorithm.

6 Conclusions

We have presented a novel framework for SFS which addresses the problem of model dominance encountered using many traditional iterative algorithms. Specifically, we ensure the fullest use of the data available, namely the input image, by incorporating the image irradiance equation as a hard constraint.

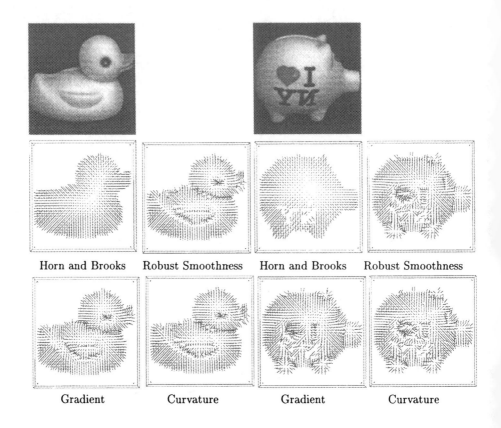

Horn and Brooks	Robust Smoothness	Horn and Brooks	Robust Smoothness
Gradient	Curvature	Gradient	Curvature

Fig. 4. Needle-maps recovered from real images.

Our second contribution is to develop new constraints on needle-map consistency. These range from simple modifications to the traditional smoothness constraint, to topographic curvature and gradient consistency. The new schemes yield significant improvements in needle-map recovery.

Gradient consistency makes only relatively weak assumptions about the surface physics. This is in contrast to the curvature consistency techniques developed here as a result of modelling the topographic properties of surfaces. The slight advantage of the gradient consistency schemes over the more complicated curvature consistency methods suggests that we can obtain good results without recourse to complex arguments about the behaviour of real surfaces.

With the new framework, we have demonstrated the development of principled, novel constraint functions modelling the properties of real surfaces. It is hoped that this research will stimulate the search for still better constraints, possessing global minima which more closely correspond to true surfaces.

References

1. Brooks, M.J. and Horn, B.K.P. (1986) Shape and Source from Shading, *IJCAI,* pp. 932-936, 1985.
2. Ferrie, F.P. and Lagarde, J. (1990)Curvature Consistency Improves Local Shading Analysis, *Proc. IEEE International Conference on Pattern Recognition,* Vol. I, pp. 70-76.
3. Horn, B.K.P. and Brooks, M.J. (eds.), *Shape from Shading,* MIT Press, Cambridge, MA, 1989.
4. Horn, B.K.P. (1990) Height and Gradient from Shading, *IJCV,* Vol. 5, No. 1, pp. 37-75.
5. Koenderink, J.J. and van Doorn, A.J. (1992) Surface Shape and Curvature Scales, *IVC,* Vol. 10, pp. 557-565.
6. Lee, C.H. and Rosenfeld, A. (1985) Improved Methods of Estimating Shape from Shading using the Light Source Coordinate System, *Artificial Intelligence,* Vol. 26, No. 2, pp.125-143.
7. Marr, D.C. (1982), *Vision,* Freeman, San Francisco.
8. Poggio, T., Torre, V. and Koch, C. (1985) Computational Vision and Regularization Theory, *Nature,* Vol. 317, No. 6035, pp. 314-319.
9. Worthington, P.L. and Hancock, E.R. (1997) Needle Map Recovery using Robust Regularizers, *Proc. BMVC,* BMVA Press, Vol. I, pp. 31-40.
10. Worthington, P.L. and Hancock, E.R. (1998) Needle Map Recovery using Robust Regularizers, *IVC,* to appear.
11. Zheng, Q. and Chellappa, R. (1991) Estimation of Illuminant Direction, Albedo, and Shape from Shading, *IEEE PAMI,* Vol. 13, No. 7, pp. 680-702.

Improved Orientation Estimation for Texture Planes Using Multiple Vanishing Points

Eraldo Ribeiro* and Edwin Hancock

Department of Computer Science,
University of York, UK.
{eraldorj, erh}@cs.york.ac.uk

Abstract. Vanishing point locations play a critical role in determining the perspective pose of textured planes. However, if only a single point is available then the problem is undetermined. The tilt direction must be computed using supplementary information such as the texture gradient. In this paper we describe a method for multiple vanishing point detection and, hence complete perspective pose estimation, which obviates the need to compute the texture gradient. The method is based on local spectral analysis. It exploits the fact that spectral orientation is uniform along lines that radiate from the vanishing point on the image plane. We experiment with the new method on both synthetic and real world imagery. This demonstrates that the method provides accurate pose angle estimates, even when the slant angle is large.

1 Introduction

Key to shape-from-texture is the problem of estimating the orientation of planar patches from the perspective foreshortening of regular surface patterns. Conventionally, the problem is solved by estimating the direction and magnitude of the texture gradient [2, 5]. Geometrically, the texture gradient determines the tilt direction of the plane in the line-of-sight of the observer. Broadly speaking, there are two ways in which the texture gradients can be used for shape estimation. The first of these is to perform a structural analysis of pre-segmented texture primitives in terms of the geometry of edges, lines or arcs [7, 3, 8]. The second approach is to cast the problem of shape-from-texture in the frequency domain [4, 1, 9]. The main advantage of the frequency domain approach is that it does not require an image segmentation as a pre-requisite. For this reason it is potentially more robust than its structural counterpart.

Unfortunately, the estimation of the magnitude and direction of the texture gradient is a task of some fragility. An alternative route to the local slant and tilt parameters of the surface is to estimate the whereabouts of vanishing points [6]. When only a single vanishing point is available, texture gradient information is still a prerequisite since the surface orientation parameters can only be determined provided that the tilt direction is known. However, if two or more

---

* Supported by CAPES-BRAZIL, under grant: BEX1549/95-2

vanishing points are available, then not only can the slant and tilt be determined uniquely, they can also be determined more accurately. As a result, searching or back-projection are not required to recover the parameters of planar orientation.

The aims of the current work are twofold. Firstly, we aim to provide a detailed analysis of the properties of the local spectral moments under perspective geometry. We commence by considering the local distortions of the spectral moments. Here our approach is to use the Taylor expansion to make a local linear approximation to the transformation between texture plane and image plane. In other words, we identify the affine transformation that locally approximates the global perspective geometry. With this local approximation to hand, we can apply Bracewell's [11] affine Fourier theorem to compute the local frequency domain distortions of the spectral distribution. Based on this analysis we show that lines of uniform spectral orientation radiate from the vanishing point. Our second contribution is to exploit this property to locate multiple vanishing points. We provide an analysis to show how the slant and tilt parameters can be recovered from a pair of such points.

2   Vanishing Points

We commence by reviewing the projective geometry for the perspective transformation of points on a plane. Specifically, we are interested in the perspective transformation between the object-centred co-ordinates of the points on the texture plane and the viewer-centred co-ordinates of the corresponding points on the image plane as shown in Figure 1. Suppose that the texture plane is a distance h

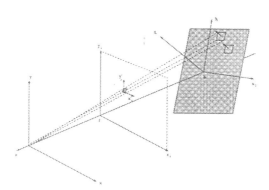

Fig. 1. Perspective projection of a planar surface onto an image plane. The projection of a local patch over the texture plane onto the image plane is also shown.

from the camera which has focal length $f < 0$. Consider two corresponding points that have co-ordinates $\mathbf{X_t} = (x_t, y_t)^T$ on the texture plane and $\mathbf{X_i} = (x_i, y_i)^T$ on the image plane. The perspective transformation between the two co-ordinate systems is $\mathbf{X_i} = T_p\mathbf{X_t}$. We represent the orientation of the viewed surface plane using the slant angle σ and tilt τ angles. This parametrisation is a natural way to model local surface orientation. For a given plane, considering a viewer-centred representation, the slant is the angle between viewer sight line and the normal vector of the plane. The tilt is the angle of rotation of the normal vector around the sight line axis. The elements of the transformation matrix T_p can be computed using the slant angle σ and tilt angle τ in the following manner

$$
T_p = \frac{f\cos\sigma}{h - x_t\sin\sigma}
\begin{bmatrix} \cos\tau & -\sin\tau \\ \sin\tau & \cos\tau \end{bmatrix}
\begin{bmatrix} 1 & 0 \\ 0 & \frac{1}{\cos\sigma} \end{bmatrix}
\tag{1}
$$

The perspective transformation in Equation 1 represents a non-linear geometric distortion of a surface texture pattern onto an image plane pattern. Unfortunately, the non-linear nature of the transformation makes Fourier domain analysis of the texture frequency distribution somewhat intractable. In order to proceed we therefore derive a local linear approximation to the perspective transformation. However, it should be stressed that the global quilting of the local approximations preserves the perspective effects required for recovering of shape-from-texture. With this linear model, the perspective distortion can be represented as T_p^* which is the linear version of the perspective transformation given by 1.

We linearize T_p using a first-order Taylor formula. Let (xo_t, yo_t, h) be the origin or expansion point of the local coordinate system of the resulting affine transformation. This origin projects to the point (xo_i, yo_i, f) on the image plane. We denote the corresponding local coordinate system on the image plane by $\mathbf{X_i'} = (x_i', y_i', f)$ where $x_i = x_i' + xo_i$ and $y_i = y_i' + yo_i$. The linearised version of T_p in equation 1 is obtained through the Jacobian $J(.)$ of $\mathbf{X_i}$ where each partial derivative is calculated at the point $\mathbf{X_i'} = \mathbf{0}$. After the necessary algebra, the resulting linear approximation is $T_p^* = J(\mathbf{X_i})\,|_{(\mathbf{X_i'}=0)} = J(T_p\mathbf{X_t})\,|_{(\mathbf{X_i'}=0)}$. Rewriting T_p^* is terms of the slant and tilt angles we have

$$
T_p^* = \frac{\Omega}{hf\cos\sigma}
\begin{bmatrix} xo_i\sin\sigma + f\cos\tau\cos\sigma & -f\sin\tau \\ yo_i\sin\sigma + f\sin\tau\cos\sigma & f\cos\tau \end{bmatrix}
\tag{2}
$$

where $\Omega = f\cos\sigma + \sin\sigma(xo_i\cos\tau + yo_i\sin\tau)$. Hence, T_p^* depends on the expansion point (xo_i, yo_i) which is a constant.

In order to determine the surface plane normal vector $\mathbf{N} = (p, q, 1)^T$, we need to estimate two different vanishing points in the image plane. Suppose that the two points are $\mathbf{V_1} = (x_{v1}, y_{v1})^T$ and $\mathbf{V_2} = (x_{v2}, y_{v2})^T$. The resulting normal vector components p and q are found by solving the system of simultaneous linear equations

$$
\begin{bmatrix} x_{v1} & y_{v1} \\ x_{v2} & y_{v2} \end{bmatrix}
\begin{bmatrix} p \\ q \end{bmatrix} = -\begin{bmatrix} f \\ f \end{bmatrix}
\tag{3}
$$

The solution parameters, p and q are

$$
p = f\frac{y_{v1} - y_{v2}}{x_{v1}y_{v2} - x_{v2}y_{v1}} \qquad q = f\frac{x_{v2} - x_{v1}}{x_{v1}y_{v2} - x_{v2}y_{v1}}
\tag{4}
$$

Using the two slope parameters, the slant and tilt angles are computed using the formulae $\sigma = \arccos\left(\frac{1}{\sqrt{p^2+q^2+1}}\right)$ and $\tau = \arctan\left(\frac{q}{p}\right)$.

3 Projective Distortion of the Power Spectrum

The Fourier transform provides a representation of the spatial frequency distribution of a signal. In this section we show how local spectral distortion resulting from our linear approximation of the perspective projection of a texture patch can be computed using an affine transformation of the Fourier representation. We will commence by using an affine transform property of the Fourier domain [11]. This property relates the linear effect of an affine transformation A in the spatial domain to the frequency domain distribution. Suppose that $G(.)$ represents the Fourier transform of a signal. Furthermore, let $\mathbf{X}$ be a vector of spatial coordinates and let $\mathbf{U}$ be the corresponding vector of frequencies. According to Bracewell *et al*, the distribution of image-plane frequencies $\mathbf{U_i}$ resulting from the Fourier transform of the affine transformation $\mathbf{X_i} = A\mathbf{X_t} + B$ is given by

$$G(\mathbf{U_i}) = \frac{1}{|det(A)|}e^{2\pi j\mathbf{U_t}^T A^{-1}\mathbf{B}}G[(A^T)^{-1}\mathbf{U_t}] \tag{5}$$

In our case, the affine transformation is T_p^* as given in Equation 2 and there are no translation coefficients, i.e., $\mathbf{B} = \mathbf{0}$. As a result Equation 5 simplifies to:

$$G(\mathbf{U_i}) = \frac{1}{|det(T_p^*)|}G[(T_p^{*T})^{-1}\mathbf{U_t}] \tag{6}$$

In other words, the effect of the affine transformation of co-ordinates T_p^* induces an affine transformation $(T_p^{*T})^{-1}$ on the texture-plane frequency distribution. The spatial domain transformation matrix and the frequency domain transformation matrix are simply the inverse transpose on one-another.

We will consider here only the affine distortion over the frequency peaks, i.e., the energy amplitude will not be considered in the analysis. For practical purposes we will use local power spectrum as the spectral representation of the image. This describes the energy distribution of the signal as a function of its frequency content. In this way we will ignore complications introduced by phase information. Using the power spectrum, small changes in phase due to translation will not affect the spectral information and the Equation 6 will hold. The power spectrum representation of an image may be defined as the square magnitude of its Fourier transform, which is always non-negative by definition. Our overall goal is to consider the effect of perspective transformation on the power-spectrum. In practice, however, we will be concerned with periodic textures in which the power spectrum is strongly peaked. In this case we can confine our attention to the way in which the dominant frequency components transform. According to our affine approximation and Equation 6, the way the Fourier domain transforms locally is governed by

$$\mathbf{U_i} = (T_p^{*T})^{-1}\mathbf{U_t} \tag{7}$$

In the next Section we will use this local spectral distortion model to establish some properties of the projected spectral distribution in the viewer-centred co-ordinate system. In particular, we will show that lines radiating from the vanishing point connect points with identically oriented spectral components. We will exploit this property to develop a geometric algorithm for recovering the image-plane position of the vanishing-point, and hence, for estimating the orientation of the texture-plane.

4 Lines of Constant Spectral Orientation

In this section we focus on the directional properties of the local spectrum distribution. We will show how the uniformity of the spectral angle over the image plane can be used to estimate the vanishing point location and hence compute planar surface orientation.

On the texture-plane the frequency-domain angle of the unprojected spectral component is given by $\beta = \arctan\left(\frac{v_t}{u_t}\right)$. Using the affine transformation of frequencies given in Equation 7, after perspective projection, the corresponding frequency domain angle in the image plane is

$$\alpha = \arctan\left[\frac{v_i}{u_i}\right] = \arctan\left[\frac{u_t f \sin\tau + v_t\,(xo_i \sin\sigma + f\cos\tau\cos\sigma)}{u_t f \cos\tau - v_t\,(yo_i \sin\sigma + f\sin\tau\cos\sigma)}\right] \quad (8)$$

For simplicity, we confine our attention to a rotated system of image-plane coordinates in which the x-axis is aligned in the tilt direction. In this rotated system of coordinates, the above expression for the image-plane spectral angle simplifies to

$$\alpha = -\arctan\left[\frac{(f\cos\sigma + xo_i \sin\sigma)v_t}{yo_i v_t \sin\sigma - f u_t}\right] \quad (9)$$

Let us now consider a line in the image plane radiating from a vanishing point which results from the projection to a family of horizontal parallel lines on the texture plane. This family of parallel lines would be originally described by the spectral component $\mathbf{U_s} = (0, v_t)^T$. After perspective projection this family of parallel lines can be written in the "normal-distance" representation as

$$\mathcal{L} : r = xo_i \cos\theta + yo_i \sin\theta \qquad \forall\,(xo_i, yo_i) \in \mathcal{L} \quad (10)$$

where r is the length of the normal from the line to the origin, and, θ is the angle subtended between the line-normal and the $x-axis$. Since this line passes through the vanishing point $\mathbf{V} = (x_v, y_v) = (-f\cos\sigma/\sin\sigma, 0)^T$ on the image plane, for $f < 0$, r can also be written as

$$r = x_v \cos\theta + y_v \sin\theta = \frac{-f\cos\sigma}{\sin\sigma}\cos\theta \quad (11)$$

Substituting Equation 11 into Equation 10 and solving for yo_i we obtain

$$yo_i = -\left(\frac{f\cos\sigma + xo_i\sin\sigma}{\tan\theta\sin\sigma}\right) \tag{12}$$

By substituting the above expression into Equation 9 and using the fact that $u_s = 0$, after some simplification we find that $\alpha = \theta$, $\forall\,(xo_i, yo_i) \in \mathcal{L}$. As a result, each line belonging to the family $\mathcal{L}$ connects points on the image plane whose the local spectral distributions have a uniform spectral angle α. These lines will intercept at a unique point which is a vanishing point on the image plane.

By using this property we can find the co-ordinates of vanishing points on the image plane by connecting those points which have corresponding spectral components with identical spectral angles. We meet this goal by searching lines for which the angular correlation between the spectral moments is maximum.

5 Experiments

In this section we provide some results which illustrate the accuracy of planar pose estimation achievable with our new shape-from-texture algorithm. This evaluation is divided into two parts. We commence by considering projected Brodatz textures [10] with known ground-truth slant and tilt. The second part of our experimental study focuses on natural texture planes where the ground truth is unknown. In order to give some idea of the accuracy of the slant and tilt estimation process, we back-project the textures onto the fronto-parallel plane. Since the textures are man-made and rectilinear in nature, the inaccuracies in the estimation process manifest themselves as residual skew.

We commence with some examples of projected textures. In this experiment, we have taken three different texture images from the Brodatz album and have projected them onto planes of known slant and tilt. The textures are regular natural textures of almost regular element distribution. Superimposed on the projected textures are the estimated lines of uniform spectral orientation. The values for the estimated orientation angles are listed in the Table 1.

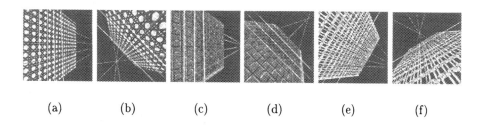

(a)   (b)   (c)   (d)   (e)   (f)

Fig. 2. *Brodatz textures. (a) and (b) D101; (b) and (c) D1; (d) and (e) D20.*

TABLE 1 - Actual × Estimated slant and tilt values (Brodatz Textures)

img	actual (σ)	(τ)	estimated (σ')	t(τ')	abs. error σ'	τ'
(a)	30	0	34.5	0.0	4.5	0.0
(b)	50	225	53.9	223.5	3.9	1.5
(c)	30	0	27.5	0.0	2.5	0.0
(d)	45	45	51.7	44.6	6.7	0.4
(e)	30	-30	30.4	-23.3	0.4	6.7
(f)	60	120	59.6	125.0	0.4	5.0

This part of the experimental work focuses on real world textures with unknown ground-truth. The textures used in this study are two views of a brick-wall, a York pantile roof and the lattice casing enclosing a PC monitor. The images were collected using a Kodak DC210 digital camera and are shown in Figure 3. There is some geometric distortion of the images due to camera optics. This can be seen by placing a ruler or straight-edge on the brick-wall images and observing the deviations along the lines of mortar between the bricks.

Superimposed on the images are the lines of uniform spectral orientation. In the case of the brick-wall images these closely follow the mortar lines. In Figure 4 we show the back-projection of the textures onto the fronto-parallel plane using the estimated orientation angles. In the case of the brick-wall, any residual skew is due to error in the estimation of the slant and tilt parameters. It is clear that the slant and estimates are accurate but that there is some residual skew due to poor tilt estimation.

6 Conclusions

We have described an algorithm for estimating the perspective pose of textured planes from pairs of vanishing points. The method searches for sets of lines that connect points which have identically oriented spectral moments. These lines intercept at the vanishing point. The main advantage of the method is that it does not rely on potentially unreliable estimates texture gradient to constrain the tilt angle. As a result the estimated tilt angles are more accurately determined.

There are a number of ways in which the ideas presented in this paper can be extended. In the first instance, we are considering ways of improving the search for the vanishing points. Specific candidates include Hough-based voting methods. The second line of investigation is to extend our ideas to curved surfaces, using the method to estimate local slant and tilt parameters. Studies aimed at developing these ideas are in hand and will be reported in due course.

References

1. B.J., Super and A.C., Bovik, Planar surface orientation from texture spatial frequencies, *Pattern Recognition*, **28** (1995),729-743,
2. D., Marr, Vision: A computational investigation into the human representation and processing of visual information, *Freeman*, (1982),

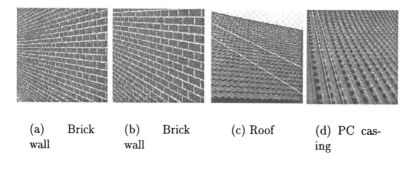

(a)   Brick   (b)   Brick   (c) Roof   (d) PC cas-
wall   wall   ing

Fig. 3. *Real texture images.*

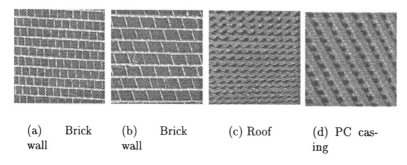

(a)   Brick   (b)   Brick   (c) Roof   (d) PC cas-
wall   wall   ing

Fig. 4. *Back-projected images.*

3. J., Aloimonos and M.J., Swain, Shape from texture, *Biological Cybernetics*, **8**, **3**, (1988), 345-360,

4. J., Krumm and S.A., Shafer, Texture segmentation and shape in the same image, *IEEE International Conference on Computer Vision* (1995), 121-127,

5. J. J., Gibson, The perception of the visual world, *Houghton Miffin* (1950),

6. J.S., Kwon and H.K., Hong and J.S., Choi, Obtaining a 3-D orientation of projective textures using a morphological method, *Pattern Recognition*, **29** (1996), 725-732,

7. K., Ikeuchi, Shape from regular patterns, *Artificial Intelligence*, **22** (1984), 49-75,

8. K., Kanatani and T. Chou, Shape from Texture: General Principle, *Artificial Intelligence*, **38** (1989), 1-48,

9. M.J., Black and R., Rosenholtz, Robust estimation of multiple surfaces shapes from occluded textures, *IEEE International Symposium on Computer Vision* (1995), 485-490,

10. P., Brodatz, Textures: A Photographic Album for Artists and Designers, *Dover* (1966),

11. R.N., Bracewell and K.-Y., Chang and A.K., Jha and Y.-H., Wang, Affine theorem for two-dimensional Fourier transform, *Electronics Letters* , **29**, **3** (1993), 304,

Relating Scene Depth to Image Ratios

Jens Arnspang, Knud Henriksen, and Fredrik Bergholm

Computer Science Department, University of Copenhagen,
Universitetsparken 1, DK-2100 Copenhagen Ø, Denmark
email: arnspang@diku.dk
phone: +45 35 32 14 00   fax: +45 35 32 14 01

Abstract. An alternative to the classic depth from stereo disparities is presented. In this new approach two scene points with different and finite depths are viewed by two identical cameras with parallel optic axes. Both the case of frontal views and of side views, where both cameras are rotated by the same angle is addressed. A simple construction for the vanishing point of the line connecting the two scene points is presented. Both for frontal views and side views it is shown that the relative scene depth of two points equals the reciprocal of the ratio of the image plane distances from the image points to the vanishing point of the line they define. For side views it is furthermore shown how the lens plane separation and ratios of image plane distances to vanishing points directly determine the absolute depths to the scene points. Neither camera focal length, image plane optic center, image coordinate scale nor coordinate disparities are used in the calculations of the absolute scene depths.

1 Introduction

Calculation of scene depth from binocular views, consisting of images from two cameras with a known configuration, is a classic paradigm within computational vision, see Grimson (1981), Ballard and Brown (1982) for an overview. A dominant mainstream of research has concentrated on 4 steps:

1. Initial calibration of the cameras, concerning camera focal length, image plane optic center, base line between camera optic centers, and often also orientation of the cameras.
2. Matching of image points between the two monocular views.
3. Calculation of coordinate disparity for each pair of matched points, based on the coordinates of each point in the left and right camera image plane, using the calibrated position of the optic centers as coordinate reference.
4. Calculation of spatial depth to the chosen scene points, using the calculated coordinate disparities from step (3), the calibrated camera parameters from step (1), and knowledge of the sensor chip dimensions for scaling the image coordinates to scene coordinates.

In this paper a new binocular technique is proposed, based on *composite image views* from two identical cameras with freely oriented, but parallel optic axes.

Only the matching step (2) above is fully shared with this new technique, while step (1) is reduced to measurement of image plane separation in *side views*. Step (3) is entirely absent and step (4) is replaced by depth from image ratio calculations, using distances to a vanishing point and image plane separation, but neither coordinate disparities, coordinate scale, camera focal length, camera base line nor knowledge of sensor chip dimensions. In section 2 the notion of composite binocular views is introduced, and it is explained, how these views may be used to determine the vanishing point of the spatial line connecting any two scene points with different and finite spatial depths. In section 3 a simple relation between relative spatial depth of two such scene points and their relative distances in the image plane to the identified vanishing point is derived. For both frontal views and side views, this relation allows for the determination of relative depth to the scene points in question without any knowledge of camera parameters, coordinate scale on the sensor chip nor coordinate disparities. In section 4 a relation between lens plane separation, absolute depths of two scene points and relative image plane distances to the vanishing point of the line connecting the points is derived. This relation neither involve coordinate disparities nor any other camera parameters but the image plane separation. The notion of side views is explained, and it is shown that such views allow unambiguous and unscaled absolute depths of two scene points. It is also shown that fronto parallel views do not have this property. One major inspiration for studying such side views has its origin in psychophysics, see for example Gibson (1974). Although no firm conclusion has been reached, concerning the human vision, this paper points out that within computational vision, non fronto parallel binocular views seem to contain useful information of a spatial scene.

2 Composite Image Views

In this section the notion of *composite images* from binocular views is introduced. Consider figure 1, where two scene points P_1 and P_2 with different and finite scene depths are viewed by two identical perspective cameras producing a left and a right image. Neither camera focal length, optic centers nor sensor chip sizes are assumed to be known for the cameras. The *composite image* is produced by placing the left and right image on top on each other using the image borders for registration. Since the cameras are assumed to be identical, such a registration will place the unknown optic centers on top of each other.

All image points from either image are part of the new *composite image*. The resulting composite image for the setup shown in figure 1 then consists of four image points: p_{1l} and p_{2l} from the left camera view and p_{1r} and p_{2r} from the right camera view. These four image points and the epipolar lines e_1 and e_2 are shown in figure 2. In the composite image is also indicated the left camera image ℓ_l and the right camera image ℓ_r of the spatial line $\mathcal{L}$ through the scene points P_1 and P_2. Each image of $\mathcal{L}$ has a vanishing point v. Since the vanishing point v lies on both image lines ℓ_l and ℓ_r it is found as the intersection of of these image lines in the composite image. For elementary estimation of

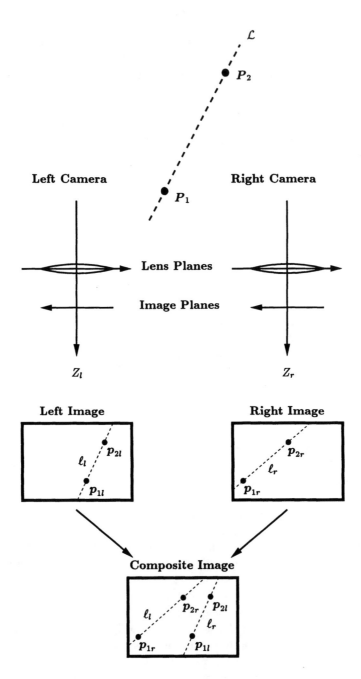

Fig. 1. The *composite image* is produced by placing the left and right images on top of each other.

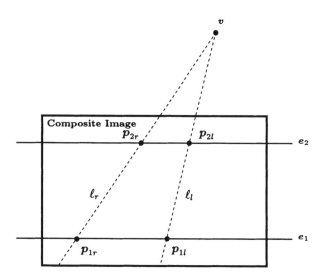

Fig. 2. The composite image produced from a binocular frontal view of two scene points P_1 and P_2. The points p_{1l}, p_{2l} are the image points from the left camera, and p_{1r}, p_{2r} are the image points from the right camera. The point v is the common vanishing point of the lines ℓ_1 and ℓ_2. The lines e_1, e_2 are epipolar lines.

vanishing points in various contexts see for example Ballard and Brown (1982), Naeve (1987), Arnspang (1989) and Henriksen (1993). For a contemporary, coherent and deep insight into mathematical description of visual perception see Koenderink (1990), and for several calculation examples see Haralick (1992). Figures 1 and 2 show the situation of *binocular frontal views*, where the lens planes of the left and right camera are identical and likewise the image planes. One consequence of this is parallel epipolar lines, as indicated in figure 2.

Figure 3 shows a *binocular side view*. In this situation the optic axes of the cameras are parallel as for the binocular frontal view shown in figure 1 but they are rotated an angle β about the vertical Y−axis. As a result of this rotation the lens planes of the cameras are separated by an amount δ. If the base line between the optic centers of the cameras has length Δ then the lens planes are separated by the amount $\delta = \Delta \sin \beta$.

The resulting composite image for a binocular side view of two scene points P_1 and P_2 with different and finite depths is shown in figure 4 If the line $\mathcal{L}$ is not parallel to the lens planes the images of the line ℓ_l and ℓ_r have a common vanishing point v in the composite image. Unlike the binocular frontal view of figure 1 and 2 the epipolar lines in figure 4 are not parallel, but meet at a point F (the epipole) in the composite image. If the two cameras were instances of a translating camera, F would correspond to the focus of expansion, see Ballard and Brown (1982), Arnspang (1989), Gibson (1974) and Haralick (1992) for a definition and calculation of the focus of expansion. In this paper the relations between the camera parameters, base line Δ, side view

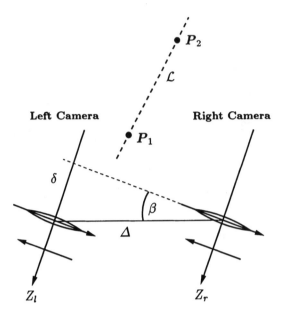

Fig. 3. In a *binocular side view* the optic axes are parallel but the cameras are rotated an angle β about the vertical Y−axis. The result is that the lens planes are separated by the amount δ. If the base line between the optic centers has length Δ then the lens plane separation is given by $\delta = \Delta \sin \beta$. The angle β is positive for clockwise rotation.

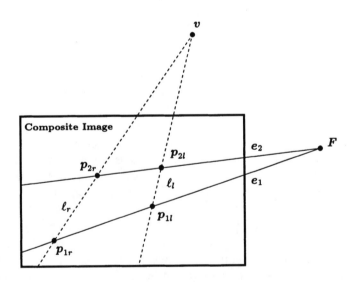

Fig. 4. The composite image produced from a binocular side view of two scene points P_1 and P_2 with different and finite depths. The points p_{1l}, p_{2l} are the image points from the left camera, and p_{1r}, p_{2r} are the image points from the right camera. The point v is the common vanishing point of the lines ℓ_1 and ℓ_2. The lines e_1, e_2 are epipolar lines, and the point F is the epipole.

angle β, optic center, focal length, and the epipole F are not discussed further. For the purpose of determining the absolute spatial depths to P_1 and P_2 only knowledge of the lens plane separation δ is necessary, as it will be shown. The quantity δ is assumed known throughout this paper, or alternatively the length of the base line Δ and side view angle β is assumed to be known, from which δ is readily calculated as $\delta = \Delta \sin \beta$.

In actual application oriented camera set ups it could be the case, that Δ is known a priori and β and thus δ may be measured from motor mechanical control information. Such possibilities or alternative visual calibration schemes for δ are left for further application oriented research. In psychophysics research, as for example Helmholtz (1990), it has been questioned whether a binocular side view significantly changes the visual perception in classic experiments, concerning for example perception of motion or shape. In a computational vision sense it is one major point of this paper that composite side views allow calculation of scene depths from data, which are not directly available in the frontal view case.

3 Relating Relative Scene Depth to Image Ratios

In this section it is shown that the relative depth of two scene points P_1 and P_2 with different depths is directly related to the relative distances of their image points (p_{1l}, p_{2l}) or (p_{1r}, p_{2r}) to the vanishing point v of the spatial line $\mathcal{L}$ through the points P_1 and P_2.

In figure 5 (left) is shown a camera and four colinear points, P_0, P_1, P_2, P_3, whose connecting line $\mathcal{L}$ is not parallel to the lens plane, and in figure 5 (right) is shown the image of the four scene points, p_0, p_1, p_2, p_3 and the vanishing point v of their connecting line ℓ.

From projective geometry it is known that the cross ratio of four collinear points is projectively invariant, see Haralick (1992), chapter 13. That is, the cross ratio of the four scene points P_0, P_1, P_2, P_3 equals the cross ratio of the their projections p_0, p_1, p_2, p_3 in the image

$$cr(P_0, P_1, P_2, P_3) = cr(p_0, p_1, p_2, p_3) \tag{1}$$

where the cross ratios are defined by

$$cr(P_0, P_1, P_2, P_3) = \frac{\mid P_1 P_2 \mid\mid P_0 P_3 \mid}{\mid P_0 P_1 \mid\mid P_2 P_3 \mid} = \left(\frac{B}{A}\right)\frac{A+B+C}{C} \tag{2}$$

$$cr(p_0, p_1, p_2, p_3) = \frac{\mid p_1 p_2 \mid\mid p_0 p_3 \mid}{\mid p_2 p_3 \mid\mid p_0 p_1 \mid} = \left(\frac{b}{c}\right)\frac{a+b+c}{a} \tag{3}$$

The cross ratio involves four points, but in order to only have two free points the scene points P_0 and P_3 are chosen to be at known locations. That is, point P_3 is chosen to be at infinity which means that its projection in the image becomes the vanishing point $p_3 \to v$. When P_3 approaches infinity the distance

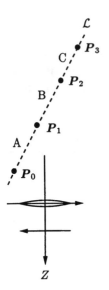

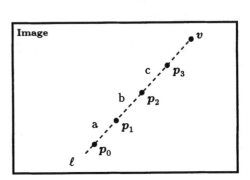

Fig. 5. Left: A pin-hole camera and four scene points P_0, P_1, P_2, P_3. The labels A, B, C denote the distances between the points along the connecting line $\mathcal{L}$. **Right:** The image of the scene points p_0, p_1, p_2, p_3 and the vanishing point v of their connecting line ℓ.

$C = \mid P_2 P_3 \mid \to \infty$ and equation (2) becomes

$$\lim_{C \to \infty} cr(P_0, P_1, P_2, P_3) = \lim_{C \to \infty} \left(\frac{B}{A} \right) \frac{A + B + C}{C} = \frac{B}{A} \qquad (4)$$

and the distance c in the image becomes $c = \mid p_2 p_3 \mid \to \mid p_2 v \mid$.

Further, the scene point P_0 is moved to a known location, namely to the image plane. When P_0 is moved to the image plane its Z−component goes to zero, and consequently its projection p_0 goes to infinity. That is, the distance $a = \mid p_0 p_1 \mid \to \infty$, and equation (3) becomes

$$\lim_{a \to \infty} cr(p_0, p_1, p_2, p_3) = \lim_{a \to \infty} \left(\frac{b}{c} \right) \frac{a + b + c}{a} = \frac{b}{c} \qquad (5)$$

When P_0 is moved to the image plane and P_3 is moved to infinity, as shown in figure 6, equations (1,4,5) yield

$$\frac{B}{A} = \frac{b}{c} \qquad (6)$$

If the scene depths of P_1 and P_2 are Z_1 and Z_2, and if the distances $\mid P_1 P_2 \mid$ and $\mid P_0 P_1 \mid$ are projected onto the Z−axis, equation (6) may be written

$$\frac{B}{A} = \frac{\mid P_1 P_2 \mid}{\mid P_0 P_1 \mid} = \frac{Z_2 - Z_1}{Z_1 - 0} = \frac{\mid p_1 p_2 \mid}{\mid p_2 v \mid} = \frac{b}{c} \qquad (7)$$

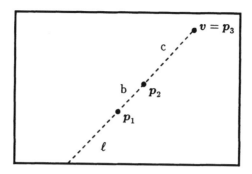

Fig. 6. The scene point P_0 has been moved to the image plane, and scene point P_3 has been moved to infinity, so only points P_1, P_2 have finite depths.

which is equivalent to

$$\frac{Z_2 - Z_1}{Z_1} = \frac{b}{c} \quad \Longleftrightarrow \quad \frac{Z_2}{Z_1} = \frac{b+c}{c} \qquad (8)$$

where b and c are the distances shown in figure 6.

If Z_2/Z_1 is denoted the *relative scene depth* of scene points P_1 and P_2, and $(b+c)$ and c are denoted the *image distances from p_1 and p_2 to the vanishing point v of their connecting line*, the following theorem holds:

Theorem 1. *The relative scene depth of two points equals the reciprocal ratio of the image plane distances to the vanishing point of their connecting line.*

Note, that theorem 1 is a two point theorem, which is valid for any perspective view of two scene points with different and finite scene depth and an identified vanishing point of their connecting line.

4   Relating Absolute Scene Depth to Image Ratios

In this section it is shown that a binocular side view allows determination of absolute depth without knowledge of neither camera focal length, image plane coordinate scale nor image coordinate disparities.

The relative depth theorem (theorem 1) allows determination of the relative scene depth of two spatial points P_1 and P_2 from composite image data (b_l, c_l, v) or from (b_r, c_r, v) measured from the left and right camera lens planes respectively. Below, the question is addressed, whether the absolute depths of P_1 and P_2 (assumed to have different and finite scene depths) can be obtained from the full composite image data set (b_l, c_l, b_r, c_r, v).

First, consider the composite image, figure 2, of a binocular frontal view, figure 1. Using equation (8) the following linear system is obtained, relating (b_l, c_l, b_r, c_r) to spatial depth Z_1 and Z_2 of the scene points P_1 and P_2

$$\begin{pmatrix} (b_l + c_l) & -c_l \\ (b_r + c_r) & -c_r \end{pmatrix} \begin{pmatrix} Z_1 \\ Z_2 \end{pmatrix} = \begin{pmatrix} 0 \\ 0 \end{pmatrix} \qquad (9)$$

For a binocular frontal view the epipolar lines e_1 and e_2 are parallel, which yields

$$\frac{|p_{2r}v|}{|p_{1r}v|} = \frac{|p_{2l}v|}{|p_{1l}v|} \quad \Longleftrightarrow \quad \frac{c_r}{b_r + c_r} = \frac{c_l}{b_l + c_l} \quad \Longleftrightarrow \quad b_r c_l - b_l c_r = 0 \quad (10)$$

which shows that the determinant of the linear system, equation (9) is always zero, and the system is always singular. The composite image data (b_l, c_l, b_r, c_r) cannot be used for determination of the absolute depths for a binocular frontal.

Next, consider composite image, figure 4, of a binocular side view, figure 3. In this case the relation between the Z–coordinates measured in the left and right images is given by

$$Z_l = Z_r - \delta \tag{11}$$

and using equation (8) yields the linear system of equations

$$\begin{pmatrix} (b_l + c_l) & -c_l \\ (b_r + c_r) & -c_r \end{pmatrix} \begin{pmatrix} Z_{1l} \\ Z_{2l} \end{pmatrix} = \begin{pmatrix} 0 \\ -\delta b_r \end{pmatrix} \tag{12}$$

For a binocular side view the epipolar lines e_1 and e_2 are not parallel. Therefore, equation (10) does not hold which means that

$$\begin{vmatrix} (b_l + c_l) & -c_l \\ (b_r + c_r) & -c_r \end{vmatrix} = b_r c_l - b_l c_r \neq 0 \tag{13}$$

and the linear system in equation (12) is regular. The composite image data (b_l, c_l, b_r, c_r) may then be used for determination of the absolute depth in a binocular side view, when the lens plane separation δ is known. A closed form solution for scene depths Z_{1l} and Z_{2l} to the scene points P_1 and P_2 is given by

$$Z_{1l} = \frac{-\delta b_r c_l}{b_r c_l - b_l c_r} \tag{14}$$

$$Z_{2l} = \frac{-\delta b_r (b_l + c_l)}{b_r c_l - b_l c_r} \tag{15}$$

Note, that equations (14,15) neither depends on camera focal length, coordinate disparities nor coordinate scale, as opposed to the classic depth from stereo paradigms found in for example Ballard and Brown (1982) and Grimson (1981). This result may be seen as a computational counterpart to the fact within psychophysics, that humans hardly use coordinate disparities for judging absolute depth, as discussed in Helmholtz (1990). The result that equations (14,15) may be used for binocular side views is also consistent with questions raised at Helmholtz (1990), concerning the possibility of adding capabilities to visual competences, when looking to the side. The question, whether such side view capabilities are qualitatively different from frontal view capabilities or just mutually transformable mathematical formulations, have not been concluded upon, neither in Helmholtz (1990) nor in this paper. Within computational vision, equations (14,15) obviously offers an alternative to the classic depth from stereo paradigms.

5 Conclusion

The classic depth from stereo problem has been revisited for the case of freely oriented, but parallel optic axes. Determination of relative and absolute depth has been addressed and reformulated by introducing the notion of composite images and binocular side views. By considering the composite image of a binocular view of two scene points with different but finite depths, a simple identification procedure has been devised for the vanishing point of the spatial line connecting the two scene points. It has then been shown that the relative scene depth of the two points equals the reciprocal of the ratio of the image distances to the vanishing point of their connecting line. Furthermore, it has been shown that in a composite image of a binocular side view, image distances to the vanishing point of the connecting line and knowledge of lens plane separation is sufficient for a closed form solution of the absolute spatial depths to any two scene points of different but finite scene depths. Laboratory experiment bench mark tests and industrial application oriented research, exploiting these alternatives to classic depth from stereo paradigms, seems natural choices for further research.

References

Ballard, D., Brown, C.M.: *Computer Vision.* Prentice-Hall, Inc., Eaglewood Cliffs, New Jersey, 1982.

Grimson, W.E.L.: *From Images to Surfaces: A Computational Study of the Human Early Visual System.* The MIT Press, Cambridge, Massachusetts, 1981.

Helmholtz: Helmholtz Seminars at Utrecht Biophysics Research Institute. 1990–1991.

Naeve, A., Eklundh, J.O.: On Projective Geometry and the Recovery of 3d Structure. In *Proceedings of the 1st Conference on Computer Vision*, page 128–135, London, June 1987.

Arnspang, J.: On the use of the Horizon of a Translating Planar Curve. *Pattern Recognition Letters*, 10(1):61–69, July 1989.

Henriksen, K.: Direct Determination of the Orientation and its Time Derivatives of a Moving 3-D Straight Line: Theory. Department of Computer Science, University of Copenhagen Universitetsparken 1, DK-2100 Copenhagen Ø, Denmark. DIKU-89-21, 1989.

Henriksen, K.: Orientation of a Moving Textured Straight Line. In *Proceedings of "Den Anden Danske Konference om Mønstergenkendelse og Billedanalyse"*, page 92–100, Copenhagen, August 1993.

Gibson, J.J.: *Perception of the Visual World.* Greenwood Press (reprint), Westport, Connecticut, 1974.

Koenderink, J.J.: *Solid Shape.* The MIT Press, Cambridge, Massachusetts, 1990.

Haralick, R.M., Shapiro, L.G.: *Computer and Robot Vision*, volume II. Addison-Wesley Publishing Company, Reading, Massachusetts, 1992.

A Modular Vision Guided System for Tracking 3D Objects in Real-Time

Markus Vincze, Minu Ayromlou, Wilfried Kubinger

Institute of Flexible Automation, Vienna University of Technology,
Gusshausstr. 27-29/361, 1040 Vienna, Austria,
{vm, ma, wk}@flexaut.tuwien.ac.at,
http://www.infa.tuwien.ac.at

Abstract. In industrial and commercial applications vision-based control of motion can only be feasible, if vision provides reliable control signals in real-time and when full system integration is achieved. A modular system architecture is built up around the geometrical features of the object. The initialisation is partly automated by using search functions to describe the task. The features found are then the basis for pose determination. Of particular need is a method of Edge-Projected Integration of Cues (EPIC) for robust feature tracking. A demonstration shows the robot following an object in 6 DOF with no special restrictions concerning room lighting, background are made.

1 Introduction

An essential quality of a vision, robotic, or integrated system to operate successfully in industrial or service applications is *real-time* capability of the system. That means, the overall system performance should not be limited by, e.g., exhaustive computations of the vision module. A considerable number of works present systems to control a mechanism via visual input. Analysing the present state-of-the-art, two major roadblocks are commonly identified.

1. *Integration of system components* like mechanism, control, and visual sensing and processing, is often neglected. However, this is essential to achieve a good overall performance and is also the basis for industrial applications.
2. Of all components assembled in a vision-based control system, the versatility of the vision system seems to be a limiting factor. In particular, the *robustness of the visual input* is weak. Applications are commonly restricted either to special environments and/or tasks (e.g. autonomous car driving [1]) or to prepared objects (markers, high contrast to background etc.).

The authors take the view that there is a need to deal with these two aspects to achieve 3D tracking of objects in real-time. This paper focuses on the advances made into this direction.

The remainder of the paper is organised as follows: The next section describes the modules integrated in our approach. Initialisation (section 3), tracking a wire-frame (section 4), and using EPIC for robust feature detection (section 5) are introduced. The results of an experiment, following the motion of a cube in 6 degrees of freedom, are provided in section 6. The conclusion contains a critique of the approach and an outlook to future work.

2 System for 3D Object Tracking

The goal of this work is to track an object in three-dimensional space. Using the results of tracking a control signal is derived to steer a mechanism, for example, to navigate a mobile platform or to handle parts with a robot arm. Commonly the object can be separated into a series of features, such as lines corner and circles, which are re-found at each tracking cycle. As a consequence, the *features* of the object constitute the central component of our system architecture for 3D object tracking (see Fig. 1).

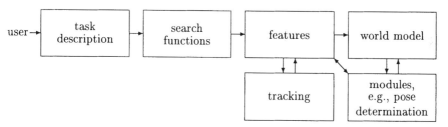

Fig. 1. Block diagram of principle strategy.

One goal of the tracking system is to continuously track the pose (position and orientation) of the object. Therefore the features tracked are selected accordingly.

Tracking is the vital component of a system for vision-based control of motion. Fundamental investigations of the dynamic performance of such a system lead to the requirement for tracking to operate at the highest possible cycle rate of the control system [5]. Using common CCD-cameras cycle rate is restricted by image acquisition to *field rate* ($50Hz$). To obtain field rate of the visual process, a windowing technique has been adopted.

A critical issue of a continuously operating system is the *initialisation*. Based on the feature-centred approach, *search functions* have been implemented that allow the user to specify a task (e.g. find object pose) as a simple *task description*. The task description uses the search functions to incorporate task and object knowledge (section 3). That means, given a target, the objective is to search for the features needed to fulfill the task.

The *world model* keeps track of the state of the features. It is image-based and stores therefore the topological model of the object found in the image. Using line, corner, and circle features the model is a wire-frame model. The advantage is the direct relation of the world model to a database of object models, which is commonly available from CAD-systems in wire-frame format.

The system architecture of Fig. 1 is modular. The goal is to render adding new *modules* as easy as possible. For example, pose determination is linked to the world model and the features. Pose determination can be further exploited to set preferences for feature search to avoid mistracking. Therefore the search can be biased to features that are best suited. As both feature tracking and feature search (driven by the visual search functions and the need of pose determination) operate in parallel, feature dependencies and new search results are constantly incorporated into the model while targets are in motion.

3 Search Functions as Primitives of Visual Descriptions

The heart of the initialisation procedure is an object description based on a set of search functions. These functions can be thought of as a high-level description language for search. The description also contains a goal, e.g., pose determination or object identification, which can direct and terminate the search. Executing the search program generates features which are entered into the world model.

The search language primitives are a small set of functions operating on images. Table 1 gives a list of basic functions.

generic search strategies
find_line(line_type or attributes)
find_region(region_type)

search strategies relative to lines
find_end_of_line(line_itself)
find_lines_at_end_of_line(line_end, line_types)
find_along_line(line_itself, direction)
find_in_direction_from_line(line_itself, direction)

search strategies relative to regions
find_within_region(line_type or region_type)
find_at_border_of_region(line_type or region_type)
find_around_region(line_type or region_type)
find_in_direction_from_region(line_type, direction)

Table 1. A set of basic search functions for image exploration.

For example, a task specification may be to determine the pose of a stapler with a red bar on top. One possible description of the task is as follows. The fact that a red item is searched for is used as a clue to find a candidate feature.

```
find_region(red)
find_at_border_of_region(line(mode right), direction: towards left)
repeat
    find_end_of_line(line_itself)
    if (end_of_line found)
        find_lines_at_end_of_line(line_itself, new_neighbor_lines)
until (structure_for_pose_found)
determine_pose
```

Fig. 2 illustrates the search pattern. At the end of one line five possible lines are found. Using mode (see 5) to validate the lines, only one line remains, the front top line. The line is added to the wire-frame model. As tracking proceeds the four lines of the top surface are found and finally closed.

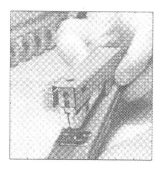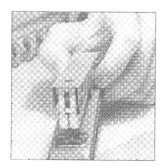

Fig. 2. Left, cycle 4: the edges found at the front end of the original line at the left of the red region. Right, cycle 19: top surface and the first line of the side surface found. The front four lines are sufficient to uniquely determine pose.

4 Tracking a Wire-Frame Structure

During the execution of the task description the wire-frame model containing the features of the object found in the image, and their relations/connections to each other, is built up. The search functions use this graph-based model as database and this structure for adding or deleting lines and corners.

Tracking the wire-frame model found in the image means the updating of all features and feature relations at each cycle. In detail, tracking proceeds as follows: The wire-frame contains a set of N lines $\mathcal{L}$ and a set of M corners $\mathcal{C}$. A line l_i is denoted by its states in the image, location x, y, orientation o and length len: $l_i(x, y, o, len) \in \mathcal{L}$ for $i = \{0, \cdots, N\}$. Similarly, a corner is given by $c_j(x, y) \in \mathcal{C}$ with $j = \{0, \cdots, M\}$. The relations between lines and corners in a wire-frame model are then given by

$$f_1 : \mathcal{L} \mapsto \mathcal{C}^m \qquad m = \{0, \cdots, 2\} \tag{1}$$
$$f_2 : \mathcal{C} \mapsto \mathcal{L}^n \qquad n = \{2, \cdots, \infty\} \tag{2}$$

where m indicates that a line can exist without any or up to two corners and n indicates the number of lines intersecting at a corner. A wire-frame $\mathcal{WF}$ is the union of lines and corners, that is $\mathcal{WF} = \mathcal{L} \cup \mathcal{C}$.

Updating the wire-frame at each tracking cycle using the data from the image $\mathbf{I}$ is then described by the following procedure:

update Wire-Frame $\mathcal{WF}$:
    update Line $\mathcal{L}$: $\forall l_i \in \mathcal{L}$ $f_3 : \mathcal{L} \times \mathbf{I} \mapsto \mathcal{L}$

        if $f_3(\mathcal{L}, \mathbf{I}) = l_i$ // line found
        else $f_3(\mathcal{L}, \mathbf{I}) = \emptyset$ // mistrack

    update Corner $\mathcal{C}$: $\forall c_j \in \mathcal{C}$ $f_4 : f_1(\mathcal{C}) \mapsto \mathcal{C}$

where f_3 is the result (line found or not found) of the tracking function outlined in section 5 and f_4 is the function of finding the intersection between all lines intersecting at one corner [3]. Tracking the features using the wire-frame update procedure is executed at each cycle before further search functions can be called.

5 EPIC: Edge-Projected Integration of Cues

The basic idea of EPIC to obtain robustness is to project cue values, such as intensity, colour, texture, or optical flow, shortly called *modes*, to the nearest edgel. Using the likelihood for these edgels renders subsequent line (or ellipse) detection more robust and effective. Lines are identified using a probabilistic [2] verification step, which adds the geometric constraint, as demonstrated in [6]. The resulting line tracker shows stable behaviour when lighting conditions vary (as edgels will still form a line although mode values change) and when background comes close to foreground intensity or colour (as the discontinuity still yields edgels though of low significance).

Tracking a line proceeds in three steps: (1) warping an image window along a line and edge detection, (2) integration of mode values to find a list of salient edgels, and (3) a probabilistic (RANSAC [2]) scheme to vote for the most likely line. These steps are outlined in the next three sections and constitute f_3 from the update wire-frame procedure.

5.1 Warping Image and Edge Detection

Every line l_i is tracked by warping a part of the image parallel to the line and is defined as a vector. Warping cuts out of the image a window oriented parallel to the edge (using the same principle as [4]). As the object moves, the edge must be re-found in the new window. Each line tracker holds an associated *state vector* comprising the basic line parameters (x, y, o, len) and two mode values m_{left} and m_{right}, here representing intensity (where left and right are defined through the vector and the values are extracted from the image $\mathbf{I}(x)$.

In a first step edgels are found by computing the first derivative of the intensity, $grad\ \mathbf{I}(x)$, using a Prewitt filter of size 8×1 [4] (other common filters give similar results). The positions of the k local maxima x_{Mk} define intervals along the line. The two ends of the tracker line are the leftmost and the rightmost limits of the intervals.

5.2 Integrating Edge and Mode Information

The second step is to project the modes values onto the edgel to find a likelihood value for each edgel. The mode value m for each interval is calculated using a histogram technique. The likelihood l_k that an edgel k is the correct edgel along the line to be tracked is evaluated to

$$l_k = \frac{1}{W} \sum_{i=1}^{n} w_i C_i \quad \text{with } i = 1, 2, 3, 4, \text{ and } W = \sum_{i=1}^{n} w_i, \tag{3}$$

where the w_i are weights for the cues C_i. In the present implementation the following four cues have been found to be most significant.

$$C_1 = \begin{cases} 1 : x_{Mk} > 8 \text{ [pixel values]} \\ 0 : \text{ otherwise} \end{cases} \tag{4}$$

$$C_2 = 1 - x_{Mk}^t / \max_k(x_{Mk}^t) \tag{5}$$

$$C_3 = 1 - (m_{left}^t - m_{left}^{t-1})/255 \qquad (6)$$

$$C_4 = 1 - (m_{right}^t - m_{right}^{t-1})/255 \qquad (7)$$

where the superscripts t and $t-1$ refer to the mode values at the present and previous tracking cycle, respectively.

The functionality of the cues is as follows: C_1 (weight $w_1 = 1$) selects practically all edges above a small threshold to eliminate noise. C_3 and C_4 select edgels with similar mode values as the edgels in the previous step. It can be easily seen that identical and similar mode values at cycles t and $t-1$ result in a high contribution to the likelihood. Now with the weights w_i for $i = 3, 4$ introduced in equation 3 it can be defined if either of the mode value is relevant for the calculation of likelihood. That means these weights are set according to where the visible surface of the object is situated. The mode value of the background have to be neglected. This makes the system robust to background changes.

The advantage of EPIC is that additional local cues operators can be easily integrated. First trials to incorporating colour added robustness to shadows and highlights.

5.3 Voting for Most Likely Line

In the final step, the edgels are used as input to the RANSAC voting scheme [2] for edge finding. The RANSAC idea is to randomly select two points to define a line and to count the votes given by the other edgels to this line. The line obtaining the highest number of votes is most likely the line searched for. The new states of the edge location and the mode values are calculated for this line and stored in the state vectors.

The investigation of the likelihood to find the correct line is the best justification for the combination of eq. (3) with the RANSAC scheme. Limiting the number of "good" edgels with liklihood g influences directly the likelihood l of a "good" line. One line is found with likelihood g^n for $n = 2$ (2 edgels selected) [2]. As edgels are searched locally in each line of the warped window, the two events are independent. Random selection is reapeated k times and therefore l is given by

$$l = 1 - (1 - g^n)^k. \qquad (8)$$

As a result just a few ms are needed for re-finding a line in a 40×40 window on our hardware setup (section 6).

6 Experiments

The feasibility and robustness of the approach is demonstrated with a robot grasping a cube. As seen in Fig. 3 there is no specific finishing or black cloth, the plate is particularly rough and has stains of rust. Lighting comes from two overhead neon lights and through the window behind the viewer.

A Pentium PC 133MHz executes image acquisition, image processing, and the short processing to determine relative pose to move the robot. A 486-PC runs RTLinux, a Linux version that provides real-time capabilities. An interface to the robot has been written in C++ such that the robot can be operated directly from the PC at a rate of 32 milliseconds. RTLinux provides the utility to supply signals to the robot when requested. The two PCs communicate the control signals over a link using TCP/IP protocol.

The cube is found using the task description [6]. The procedure is automatic. First cube coloured regions are searched for in the image at low resolution. By searching for the outline of this region, the top surface of the cube is found in a few steps. With the top surface four corners are available, which is sufficient information to determine the 6D pose. Using an extended Kalman filter the signal is smoothed and the control signal to steer the motion of the robot calculated. Fig. 3 shows snapshots from a descent motion.

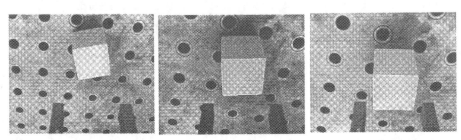

Fig. 3. Snapshots during the descent to grasp the cube. Note: Changing light intensity, due to shadows from the approaching robot, causes no problem for the line trackers.

In the last months the system dynamic performance of several controllers has been examined. Fig. 4 plots the relative pose values (object relative to gripper pose) in millimetres for position and degrees for orientation during descent. The left graph is the approach to a fixed pose by linearly moving towards the target pose. The right graph depicts the optimised fuzzy-controlled motion. As it can be seen the pose is reached rapidly, with little overshoot and with high accuracy (see table 2).

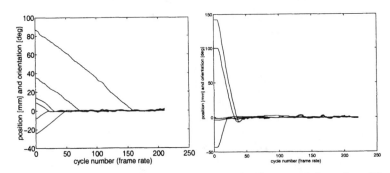

Fig. 4. Plots of the 6 components of pose during the descent of the robot. The entire motion is about 5 seconds. Left: Linear approach. Right: Controlled motion using a fuzzy-controller.

	position *mm*			orientation *deg*		
	x	y	z	α	β	γ
mean	-0.188	-0.2218	-0.1336	0.2831	-0.2317	0.08
std	0.5473	0.5626	0.2421	0.6040	0.072	0.1913

Table 2. Mean and standard deviation of pose accuracy obtained with one camera.

7   Conclusion

An integrated approach to vision-based control of motion has been presented.

One advantage of the approach is its *real-time* capability. The demonstration runs on a Pentium PC 133 MHz at field or frame rate for up to ten line features. The system architecture is based on the finding that there exists an *optimal system architecture* for a visual servoing (as well as a visual tracking) system [5]. It states that the system needs to operate at the highest possible rate and should use a certain number of pipelining steps depending on the overall latencies in vision and the mechanism. This is a guideline for system design.

The approach uses features as tracking primitives. It is the basis for a *modular architecture* that allows simple extensions and independent tracking of different features.Thus parallel processing can be fully exploited. The *image processing system* integrates tracking and detection in real-time and uses a wire-frame model as searching structure as well as for the object model. Moreover a simple and efficient *task description* technique is applied, which uses *search functions* at image level to find the features and thus the target.

Finally it can be stated that the cue integration approach EPIC (Section 5) is a good means to tackle the issue of *robustness* of the visual system. Although the demonstration object is simple, the environmental conditions handled reach beyond previous work. The following conditions can be handled: (1) gradually (6% each step) changing lighting conditions, (2) foreground/background separation at a minimum of 8 pixel values (= 3.2%), (3) partly occlusion without loosing the object and still calculating pose as long as any 4 lines are seen, and (4) automatic recovery after occlusion using the task description.

References

1. D.E. Dickmanns: *The 4D-Approach to Dynamic Machine Vision;* Conf. on Decision and Control, pp. 3770-3775, 1994.
2. M.A. Fischler, R.C. Bolles: *Random Sample Consensus: A Paradigm for Model Fitting;* Communications of the ACM Vol.24(6), pp.381-395, 1981.
3. W. Förstner, E. Gülch: *A Fast Operator for Detection and Precise Location of Distinct Points, Corners and Centers of Circular Features;* ISPRS Intercommission Workshop, Interlaken, 1987.
4. G. Hager, K. Toyama, *The XVision-System: A Portable Substrate for Real-Time Vision Applications,* Comp. Vision and Image Understanding 69(1), pp.23-37, 1998.
5. P. Krautgartner, M. Vincze: *Performance Evaluation of Vision-Based Control Tasks;* accepted for publication at IEEE ICRA, Leuven, 1998.
6. M. Vincze, M. Ayromlou: *Robust Vision for Pose Control of a Robot;* Proc. 22nd Workshop of the Austrian Association for Pattern Recognition ÖAGM '98, pp. 135-144, 1998.

Real-Time Optical Edge and Corner Tracking at Subpixel Accuracy

Stefan Brantner, Thomas Auer, and Axel Pinz

Computer Graphics and Vision, TU Graz,
Münzgrabenstr. 11, A-8010 Graz, Austria
tom@icg.tu-graz.ac.at,
WWW home page: http://www.icg.tu-graz.ac.at/~tom/stube/stube.html

Abstract. High position and orientation accuracy is required in augmented reality systems. This paper presents an extension to the well known XVision library for real-time tracking. We implement a real-time sub-pixel accuracy edge tracking algorithm estimating edge position and orientation based on "virtual edge" localization in a 5×5 window. In addition, we give a simple yet efficient heuristic for identifying corners. Experiments demonstrate results for sub-pixel edge and corner tracking superior to the original XVision primitives.

Keywords: real-time tracking, augmented reality, subpixel accuracy

1 Introduction

Vision-based augmented reality systems are demanding with regard to optical tracking. The precision required is rather high (in order to guarantee satisfying augmentation for the users), and tracking has to work in realtime.

In our multi-user augmented reality environment, the "Studierstube" [ABP99] we have incorporated an optical tracking system that fulfills both the needs for precision and time consumption.

In this paper we present the 2D optical tracking component of this system which is based on XVision [HT98]. We present a fast, reliable edge tracking primitive which improves tracking as provided by XVision with only minimal additional time consumption. Experimental results show the improvement for both jitter as well as accuracy.

2 The Tracking System

The tracking system for our multi-user augmented reality environment combines both optical and magnetic tracking, thus adding the robustness of the magnetic

---

* This work has been supported by the Austrian Science Fund (FWF) under contract number P-12074-MAT. The "Studierstube" is a joint project with the Vienna University of Technology. We want to thank Joachim Steinwendner for his input in designing the sub-pixel tracking algorithm.

tracker to the precision of the optical tracking, cf. [ABP99]. The magnetic tracking system consists of a standard Ascension *Flock of Birds* magnetic tracker, while the optical tracking system consists of the following two components:

- 2D tracking: features are tracked in the camera image.
- 3D computation: based on feature locations in 2D and corresponding 3D locations, position and orientation of the head-mounted camera are computed.

Our current 2D tracking system utilizes "signaled" landmarks, i.e., artificial features the locations of which have been determined by some external calibration procedure (we used a theodolite). Optical tracking is explained in detail in Section 3.1.

Once the positions of the features have been determined, the pose of the camera is computed.[1] In addition, the computed position (and orientation) are verified with the data from the magnetic tracker.

3 2D Optical Tracking

For optical tracking we use corners of black rectangles on a white background. The locations of these corners in 3D have been determined by external calibration. Based on the knowledge of these 3D positions and the approximate camera position and orientation (obtained from the magnetic tracker) the positions of the corners in the camera image can be predicted, thus allowing for rather small feature search areas. Tracking is done with XVision, a tracking library developed at the University of Yale, cf. [HT98], where each corner is composed of two edges. Subsequently we will first explain the edge tracking primitive used and second how to detect corners reliably.

3.1 Sub-pixel Tracking

XVision offers a very fast primitive for tracking edges. Unfortunately, the accuracy provided does not suffice for augmented reality applications, especially the jitter inherent in the system is too large. Thus, a new tracking primitive was added to XVision providing better accuracy and less jitter while still satisfying the need for fast computation: whereas the fastest edge tracking primitive in XVision takes approximately 1.5 msecs, our optimized primitive requires 2.5 msecs.

The new tracking primitive works very similar to a sub-pixel edge detection algorithm described in [SS98]: the pixel values in a small subwindow centered around the edge are compared with the grey values obtained by rendering a virtual edge in this subwindow. By slightly adjusting the position of the virtual edge the difference between real and computed gray values is minimized. In more detail, our algorithm works as follows:

---

[1] The algorithms used for computation of the pose have been described in [APG98], and a more detailed description can be found in [Bra98].

- determine position and angle of the edge with the XVision tracking primitive.
- compute a 5×5 pixel subwindow centered around the center of the edge (experiments showed that a 5×5 window is sufficient to cover the displacement of the edge position as determined by the XVision tracking primitive.).
- for each pixel within this subwindow, compute the gray value according to the current position and angle of the virtual edge. This is done by determining the distance between the center of the pixel and the edge, cf. Fig. 1a). Distance and angle yield the fraction of the pixel left (respectively right) of the edge. With the additional knowledge of the gray values left (respectively right) of the edge the gray value for the pixel is computed (note that the gray values are taken from the image). In order to speed this up, a lookup table for fractions left (respectively right) of the edge as a function of distance and angle is computed in advance.

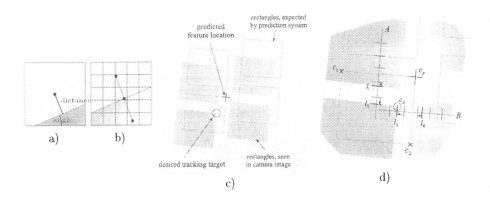

Fig. 1. Optimization of edge and corner location.

The differences between pixel values of the virtual edge image and real image data are used as an error function for edge position and angle. Thus, in order to minimize the tracking error, this error function has to be minimized. A line segment perpendicular to the edge to be tracked is used as initial search interval. Its center coincides with the center of the edge, and its initial length is four pixels, cf. Fig. 1b). By using Fibonacci optimization the interval is shrunk until it is smaller than a minimum size. Further, there is an upper bound for the number of iterations in order to limit time consumption.

We further implemented a tracking primitive optimizing both position and angle: As stated above, the gray value of each pixel depends on both position and orientation of the edge. Therefore, we also minimize the error function with respect to the angle of the edge. This is done by searching in an interval centered around the angle as computed by the XVision tracking primitive. Due to the fact that both optimizations are independent of each other, the algorithm should loop over both of these optimizations. Note however, that simple experiments showed that this does not improve the quality of tracking. Moreover, as will be shown in Section 4, angle optimization does not improve accuracy or reduce jitter. This is due to the fact that angular changes yield much smaller changes in gray values than small position changes.

3.2 Corner Recognition

As stated above, our system tracks corners of black rectangles. As we face a situation where similar objects are located close to each other in the camera image, the danger of ambiguous corner matches increases. Therefore we implemented measures to increase corner identification robustness by increasing the amount of knowledge available about a corner feature.

Due to errors in the magnetic tracker, predicted corner locations can be considerably wrong as depicted in Fig. 1c). Here several corners are located close to each other and the danger of a false match is relatively high.

For both edges making up a corner feature we determine an additional characteristic called the transition type. Since XVision edge-trackers provide direction information, we are in a position to unambiguously determine fore- and background brightness of an edge. Thus the transition type of an edge can be light-dark or dark-light.

Since we have high-level knowledge of the tracking targets, the expected transition types of the two edges making up an expected corner are also known in advance. Starting at the predicted corner position we employ several short and fast edges[2] that search for edges with respective directions in the neighborhood of the predicted position. Only edge segments, of which the transition type matches that of the model are accepted and combined into a single corner.

The simple process of recognizing a single corner with similar neighboring corners is depicted in Fig. 1d). Our algorithm predicts the corner location c_p and also knows expected locations for the two related corners c_1 and c_2. The search for the two edge-segments making up the desired corner is carried out along the lines A and B. During the search along A two edge-segments l_1 and l_2 are found. l_1 (light-dark) is rejected and l_2 (dark-light) accepted. The same applies to line B. l_4 is rejected and l_3 accepted. Finally we determine the intersection point and start the corner-tracker at the correct position c_d.

4 Experimental Results

In order to determine their usability for our augmented reality system we evaluated the characteristics of three tracking primitives:

- the original XVision tracking primitive (denoted by *simple edge*),
- edge tracking with position optimization (denoted by *position optimized edge*),

---

[2] XVision gains tracking speed by subsampling edges in either horizontal and vertical direction

– edge tracking with both position and angle optimization (denoted by *position and angle optimized edge*).

During edge tracking experiments we measured jitter for both real and simulated image data. When tracking corners we measured jitter and accuracy for simulated image data only.

4.1 Simulated Image Data

In order to determine the accuracy of the corner tracking algorithms we are using a "virtual camera" similar to [SHC⁺96] for obtaining ground truth: We rendered the features with the model parameters of the camera at chosen virtual locations. Thus, the results obtained from 2D optical tracking could be compared with corner locations as determined by means of the camera parameters.

On one hand, simulated camera image data can be quickly obtained and the simulation can be easily adapted to new feature or environment configurations. On the other hand, the camera is not accurately simulated (e.g., distortions can not be adequately modeled and anti-aliasing does not reproduce edges in real images correctly).

4.2 Edge Tracking

In order to determine the quality of the edge tracking primitives we measured standard deviation of both position and angle of tracked edges. Fig. 2a) depicts the standard deviation in pixels (i.e., the sum of the deviations for both x and y directions) for edges tracked on real image data: Both primitives using optimization show better results than *simple edge*. Among the optimized primitives, the deviation of *position optimized edge* is in general smaller than for *position and angle optimized edge*. The situation is similar for simulated image data. As depicted in Fig. 2b), *position optimized edge* has the smallest deviation in general, *simple edge* the largest and the data from *position and angle optimized edge* lies slightly above *position optimized edge*.

As depicted in Fig. 3a), results for angle deviations are different. Whereas angle deviations for both *simple edge* and *position optimized edge* are almost identical, angle deviations are in general much larger for *position and angle optimized edge*. As stated above, this is due to the fact that small position changes have more influence on the gray value of the pixels than relatively large changes in the edge angle. The optimization process does not eliminate position errors completely, but significantly reduces them. Thus, the second optimization actually tries to correct these remaining small position errors and therefore yields an angle result worse than without optimization. Note however, that the difference is not so significant when using simulated image data, as depicted in Fig. 3b). Whereas *position optimized edge* and *simple edge* also show nearly identical values, the difference to *position and angle optimized edge* is not as significant. In our opinion this is due to the fact that edges in simulated images are sharper than in real image data.

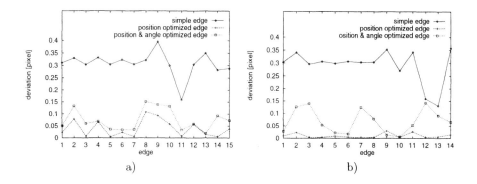

Fig. 2. Edge tracking: standard deviation of position data in pixel for all three tracking methods using real and simulated image data.

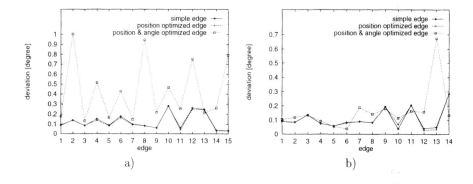

Fig. 3. Edge tracking: standard deviation of edge angles in degree for all three tracking methods using real and simulated image data.

4.3 Corner Tracking

Due to the fact that we are using corners as tracking primitives for our augmented reality system, we also evaluated the three primitives when used for tracking corners (recall that each corner is composed of two edges). In difference to the experiments for edge tracking, ground truth was available for simulated image data. Thus, we computed the difference between real corner locations and positions obtained by the tracking primitives. Fig. 4a) depicts average errors for all three methods. *position optimized edge* has much lower average errors in general than *simple edge*. With respect to maximum errors (as depicted in Fig. 4b)), the difference between *position optimized edge* (approximately 0.69 pixel) and *simple edge* (approximately 1.95 pixel) further increases.

Standard deviations of corner positions are depicted in Fig. 5. Similar to position accuracy, the difference between *position optimized edge* on one hand and *simple edge* on the other hand is significant. Whereas the deviation for

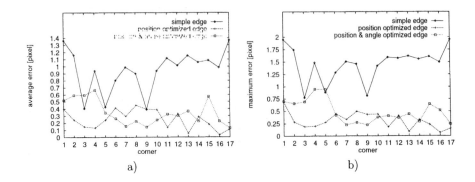

Fig. 4. Corner tracking: average and maximum error in pixel for all three tracking methods.

simple edge is larger than 0.5 pixel except for one data point, the deviation of *position optimized edge* is less than 0.06 pixel except for one data point.

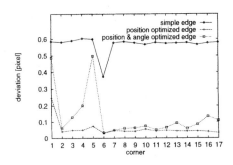

Fig. 5. Corner tracking: standard deviation in pixel for all three tracking methods.

4.4   3D Position Computation

In order to determine the influence of 2D corner tracking on the 3D pose computation, we tracked 3D position and orientation at nine different locations of the "virtual camera".[3] On one hand, tracking with the standard XVision primitive yields average position errors in the range from 6.5 mm to 33.5 mm, and maximum errors within 11.6 to 61.0 mm. For *position optimized edge* on the other hand, the average position error is less than 10 mm for all but one data point (19.45 mm), and maximum errors are within the range from 4.5 mm to 22.7 mm. Standard deviation of position data is at least 4 times smaller for *position*

---

[3] The computation of position and orientation is described in [ABP99].

optimized edge than for *simple edge*. For orientation errors, the situation is even better: Whereas average errors are only slightly smaller for *position optimized edge* than for *simple edge*, maximum errors are smaller by a factor of at least 1.6, and standard deviations are at least 5 times smaller. For *position optimized edge*, average angle errors lie within 0.003 and 0.014 radians, and maximum errors are within 0.003 and 0.017 radians. Standard deviations are less than 0.002 radians.

5   Conclusion

We presented a sub-pixel edge tracking primitive that is able to run in realtime. Compared to the standard tracking primitive in XVision

- the new edge primitive has less jitter in position and the same amount of jitter for angles, both for real and simulated image data;
- corners composed of sub-pixel edges have significantly lower corner position maximum and average errors than conventional XVision corner features;
- the amount of jitter (which is also crucial for our application) is significantly reduced by using position optimization.

We also showed that additional optimization of edge angles does not yield superior results, due to the fact that the edge model used for optimization is much more susceptible to position changes than to angle changes.

In addition we presented a simple yet efficient method to detect corners reliably.

References

[ABP99]   T. Auer, S. Brantner, and A. Pinz. The integration of optical and magnetic tracking for multi-user augmented reality. to appear in Proccedings 5th EUROGRAPHICS Workshop on Virtual Environments (EGVE'99), 1999.

[APG98]   T. Auer, A. Pinz, and M. Gervautz. Tracking in a multi-user augmented reality system. In *Proceedings of the IASTED International Conference on Computer Graphics and Imaging, Halifax, Canada*, pages 249–252, 1998.

[Bra98]   S. Brantner. A vision-based tracking technique for augmented reality. Master's thesis, Computer Graphics and Vision, Graz University of Technology, Münzgrabenstr. 11, 8010 Graz, Austria, 1998. ftp://ftp.icg.tu-graz.ac.at/pub/publications/tom/papers/brantner.ps.gz.

[HT98]   G.D. Hager and K. Toyama. Xvision: A portable substrate for real-time vision applications. *Computer Vision and Image Understanding*, 69(1):23–37, January 1998.

[SHC+96]   A. State, G. Hirota, D.T. Chen, W.F. Garrett, and M.A. Livingston. Superior augmented reality registration by integrating landmark tracking and magnetic tracking. In *Proceedings of SIGGRAPH'96, New Orleans, Louisiana*, pages 429–438. ACM SIGGRAPH, August 4-9 1996.

[SS98]   J. Steinwendner and W. Schneider. Algorithmic improvements in spatial sub-pixel analysis of remote sensing images. In *Pattern Recognition and Medical Computer Vision, 22nd Workshop of the Austrian Association for Pattern Recognition*, pages 205–214, May 1998.

Three-Dimensional Scene Navigation Through Anaglyphic Panorama Visualization

Shou Kang Wei, Yu Fei Huang, and Reinhard Klette

CITR, University of Auckland,
Tamaki Campus, Building 731, Auckland, New Zealand
{swei002, yhua002, r.klette}@cs.auckland.ac.nz
http://www.tcs.auckland.ac.nz

Abstract. Traditional panorama visualization techniques give only single eye's depth cue to the audience for 3D-world navigation. We introduce a new panorama visualization using color anaglyphs stereoscopic technique so that depth cues of both eyes are provided. Because the camera setups of panoramic and anaglyphic image acquisition are intrinsically incompatible, we measure the errors and support an error-controlled positioning of the camera. Furthermore, a real-time interactive navigator has been designed and implemented for navigation in the stereo panoramic virtual world. A new reprojection method is introduced and significantly improves the quality of anaglyphic panoramas. Additionally, we discuss a novel approach which enhances the navigator by allowing scene objects to be manipulated within the panoramic videos.

Keywords. 3D scene navigation, panorama, anaglyph.

1 Introduction

All animals with two eyes have binocular vision. Those animals with laterally placed eyes, such as humans, are able to integrate the information from the large area of binocular overlap and use it to analyse depth. In the 3D world the depth cues can be recognized by one or two eyes. A single eye can sense depth cues including perspective, image overlap, shading and others as discussed in [1], whereas both eyes perceive binocular disparity exclusively. The term *stereopsis* is used for describing the impression of depth introduced by binocular disparity.

In order to convey such impression to the audience with available media, many stereoscopic vision techniques have been developed for exploring such characteristics of the human vision system. The key idea among those techniques is to emulate human binocular visual effects by allowing only the left eye to see the left image, and the right eye to see the right image. The left and right images refer to the same viewing direction with slightly different sensor poses.

Many techniques, such as the *mirror stereoscope*, *prism stereoscope* and *polaroid stereoscope* require not only extra hardware such as mirrors, prisms, partition cards, polarizing filters, projectors or many others, but also accurate installations to ensure perception at the proper quality level. Details of those techniques can be found in [1–3]. They are neither applicable to visualize stereo images on a computer monitor nor suitable for panoramic scene exploration. So far there are at least two techniques applicable with modern computer devices. One of these is the *shutter stereoscope*, and the other is the *anaglyphic technique*. In the shutter system, the left and right images are presented in rapid alternation on a computer monitor. People view the display through electro-optic shutters that alternately occlude the two eyes in phase with the alternation of the display. Flicker may appear due to the slow frame rate used on moderate computers, so more expensive systems may be required. Contrarily, the anaglyphic technique suggests an affordable approach to stereo visualization. The only extra hardware it requires is a pair of 3D eyeglasses, which is composed of two color filters, conventionally red for the left eye, and green or blue for the right eye. The left and right images are superimposed into a single image, called *anaglyphic image*. The underlying principle is that only the red component in the anaglyphic image is seen through the red filter in the left eye and the green/blue component through the green/blue filter in the right eye. The human mind merges these two illusions into one and infers an understanding of scene structures according to routine experience.

This paper is structured as follows: Section 2 explains geometric difficulties raised in registration of stereo panoramic images. Section 3 provides an error controlling scheme such that unacceptably large errors due to those difficulties are minimized/avoided. In Section 4 a new reprojection approach and a new technique for embedding manipulable objects are illustrated. Conclusions are addressed in Section 5.

2   Problem of Registering Stereo Panoramic Images

A very important constraint for producing a full view panorama is that the *nodal point* (the *optical center*) of the camera must coincide with the rotation center. However, stereo image acquisition requires that both images are taken from slightly different positions. Therefore, combining both into a single stereo panorama image is impossible.

If the camera's optical center is aligned with the rotation axis, every 3D point P which is projected into two successive images may share the same y value in both image coordinate systems assuming proper image alignment, see eg. [6]. Otherwise, the point P may be projected into different rows in both images due to different distances between camera and scene objects. This is illustrated in Fig.1. A matching problem may severely occur if $|Q_1 - P_1| > |Q_2 - P_2|$.

This problem may be harmless to the resulting panoramic image if the distance between a closest scene object and the camera is large enough. In practice, the registeration process may deal with small sets of unmatched pixels such that the result may still be acceptable. However, to avoid an unacceptably large error we measure the maximum possible error and suggest an error controlling scheme such that an optimum camera position might be chosen in advance.

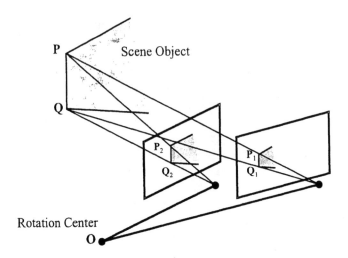

Fig. 1. The corresponding points, P_1-P_2 and Q_1-Q_2, lie on different rows in both images.

3 The Error Measurement and Control

We define the error e to be the vertical pixel difference of a pair of corresponding points in two consecutive images. To be more precise in the calculation we use physical distances (centimeters) in real numbers as measurements rather than pixel distances (discrete numbers). Assume that two consecutive images are taken and their cooresponding nodal points are equally away from the rotation center. Also assume that the camera's up-vector is always parallel to the rotation axis and it's optical axis is always perpendicular to the line defined by nodal point and rotation center.

These setting assumptions are illustrated in Fig.2(a) showing the XY-plane only. Let the rotation center **O** be the origin of the world coordinate system and assume that both optical centers **A** and **B** are r centimeters away from **O**. We have point $\mathbf{A} = (r, 0)$, and point $\mathbf{B} = (r \cos \theta, r \sin \theta)$ with respect to **O**, where θ is the rotation angle. Symbol α denotes the camera's horizontal or vertical field of view depending on how it is placed. Since $\alpha \geq \theta$ is required to ensure some overlap between images, point **B** must always lie inside the area bounded by line $\mathbf{AA}_l$ and line $\mathbf{AA}_r$. The darker gray area shown in Fig.2(a) is the overlapping section of two central projections from camera's **A** and **B** respectively. Within this area, the greatest error must occur on the plane that vertically cuts through the XY-plane along line **AC** passing through point **B**. Let **D** be an arbitrary point on line **AC**. Point **D** stands for a point in 3D space which can be seen by both camera **A** and **B** and which is d centimeter away from the rotation center. What we are interested in is the error that happens along the vertical-up line **DF** – a line perpendicular to the XY-plane, passing through point **D**.

Due to the distances of **DA** and **DB** being different, a point on line **DF** might project to different rows of images A and B. Therefore there will be a

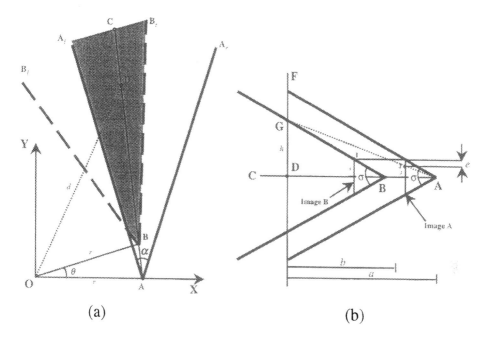

(a) (b)

Fig. 2. Geometry of two consecutive off-rotation-center cameras. (a)Top view. (b) Side view.

problem when matching two overlapping area in the image A and image B. The maximum error e should occur on the column corresponding to line **DF** in both images. Let us first discover the largest possible error e in terms of the known parameter d.

Let a and b be the distance of **AD** and **BD** respectively in Fig.2(b). Suppose **G** is an arbitrary point along vertical line **DF** h centimeters away from **D**. Moreover, point **I** and **J** are the projections of point **G** on image B and A respectively, where **I** is vertically i centimeter away from the center of image B and similarly **J** is j centimeters away. The error e is then defined as $(i - j)$ centimeters. In order to find the maximum e, value i must also be maximum. So point **I** happens to be on top/bottom boundary of image B.

Hence, the maximum error equals

$$e = (1 - \frac{b}{a})f \tan(\frac{\sigma}{2}). \tag{1}$$

Assuming d, a and b can be determined it follows that e can be calculated. Conversely, if we specify the maximum error e allowed in the stitching process for certain quality, the shortest distance d can also be calculated. Since e can be determined in advance and the camera can also be placed at d centimeter away from the closest object in the scene, we refer to this scheme as the error controlling.

As the positions of **A** and **B** are both known, the distance between them is also known, so we have the following relation between values a and b,

$$a = b + r\sqrt{2 \times (1 - \cos\theta)}. \tag{2}$$

Combine Eq.1 and Eq.2 we get,

$$a = \frac{rf}{e}\tan(\frac{\sigma}{2})\sqrt{2 \times (1 - \cos\theta)}, \tag{3}$$

where $e > 0$ as long as **A** $\neq$ **B**. Finally the distance between the rotation center **O** and point **D** should be at least equals to

$$d = \sqrt{r^2 + a^2 \pm \frac{2ra\sqrt{1 + t^2}}{1 + t^2}},$$

where $t = \tan(\frac{\theta}{2} + 90)$ and refer to a in Eq.3.

The value d is driven in terms of the parameters e (the error), r (the distance between camera and the rotation center), f (the camera focal length), θ (the rotation degree), and σ (the camera vertical field of view). Since they are known in advance and may be kept constant for making one complete full-view stereo panoramic image, by using the above formula for d we know that the closest object in the 360 degrees panorama view should be at least d centimeters away from the camera rotation center.

4   Stereo Panorama Navigator

A stereo panorama navigator is a navigation tool for exploring a photo-realistic virtual world. It allows the user to interact with the 3D scene dynamically according to viewer's own preference in real-time. In fact, this navigator is similar to most panoramic navigators except of the different considerations involved for its design. Besides, there is a new proposal for the navigator enhancement, which allows user manipulation of objects directly within the scene.

4.1   Lift-Up Reprojection

Instead of common cylinder reprojection which assumes the projection center at the center of a cylinder and heading up/down for the top/bottom view of a panorama, our reprojection allows the projection center to lift up and down. The reason for doing this is to preserve a relatively correct disparity proportion in the reprojecting area.

Figure 3 shows where the problem occurs for the head-up approach, and the improvement from the lift-up approach. In the head-up reprojection, the top of bamboo artwork circled has smaller disparity than the one of the bottom. Therefore, the top of bamboo artwork looks closer to the viewer while the bottom is relatively further away from the viewer. That is in fact inconsistent with

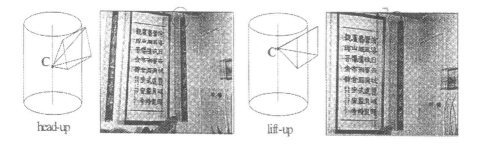

Fig. 3. Two reprojection models, lift-up and head-up, and their output results.

the actual scene. To avoid this, we use the lift-up reprojection, which gives a relatively correct disparity proportion and less distortion with respect to the quality of visualization.

In Fig. 4 we illustrate two rendered anaglyphic panoramic images, i.e. ours and the one provided by a commercial panorama navigator[1]. The main difference between both, besides of the disparity proportion correctness problem mentioned previously, is the vertical component of object borders in the scene. The commercial navigator provides jagged border lines around objects while we have crisp vertical borders, which is attributed to the different reprojection methods.

4.2 Embedding Manipulable Objects

When we walk through the virtual world, there should be many objects around us, e.g. a book on the table, a plant beside the door or a globe on the cabinet. So far with a panoramic navigator, we can view those objects in a different direction (for multi-nodes only), and perhaps with a close-up capability. But we cannot manipulate them directly in the scene, for instance, open the book, examine the plants or look up New Zealand from the globe on the cabinet. In this paper, we present one of possible ways of manipulating objects, which allows the user directly examining the object with predefined behaviors.

Object images should be taken within the same scene for the panorama. All the complexities such as illumination, shading, reflection, refraction and geometric constraints are then possibly consistent in the reference scene. Each object rotation, the simplest object motion, needs two shots for generating a stereo pair. After the digitization, we process them into anaglyphic images and place them into a single array if a single rotation is required; or into a multiple-dimensional array if more than one rotation is allowed. The frame size should correspond to the object size. The basic model is shown in Fig.5.

Since the manipulable objects are embedded in the panorama, the user interaction area for manipulating the objects and navigating the panorama overlaps. We define a rule such that they can behave independently, if the starting position of the mouse toggling is within the object area, the mouse event is passed

---

[1] http://www.livepicture.com

Fig. 4. The top (from our navigator) has crisp vertical borders of objects, while the bottom (from the commercial product) jagged. (See http://www.tcs.auckland.ac.nz/research/projects/demos)

to the object handler, else to the panorama handler. Since the object position is varied in the view port of the navigator, the navigator needs to dispatch the object position dynamically. So, even the panorama is auto-spinning, an additional function our navigator provides, we should still be able to manipulate the object desirably. The state of object is kept separately from the panorama so the object appears consistently and seamlessly with its surrounding objects/backgrounds. This approach vivifies the virtual world and increases the entertainment of navigation.

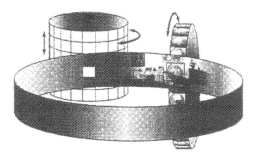

Fig. 5. The basic model for manipulable objects embedded in the panorama (1D and 2D cases).

5   Conclusions

This paper introduced a new panorama visualization which utilizes the human binocular visual characteristics and the capabilities of color comprehension through the color anaglyphs stereoscopic technique. The panoramic and the anaglyphic image acquisition models are known to be incompatible in their geometric configurations. We measure the errors and show that the errors can be controlled, i.e. to avoid unacceptable large errors.

We have also developed a real-time navigation tool allowing exploring the stereo panorama in multiple directions. Regarding the properties of anaglyphic images, the lift-up reprojection method is chosen supporting a better quality of rendered images.

The analyphic panorama approach produces a manifest stereo impression to human vision with no extra overhead and degradation in the performance compared with the traditional panorama visualization method. The applications of this approach also inherit the benefits of common panorama visualizations, which are affordable and applicable for a wide range of workstations and PCs.

References

1. I. P. Howard and B. J. Rogers. Binocular Vision and Stereopsis. *Oxford Psychology Series No.29.* Oxford University Press, New York, 1995.
2. D. Drascic. Stereoscopic Video and Augmented Reality. *Scientific Computing and Automation,* June 1993, pp 31-34.
3. H. von Helmboltz. Popular lectures on scientific subjects. *Scientific Computing and Automation,* 1893.
4. H.-C. Huang and Y.-P. Hung. Panoramic Stereo Imaging System with Automatic Disparity Warping and Seaming. *Graphical Models and Image Processing,* May 1998, 60(3):196-208.
5. L. McMillan and G. Bishop. Head-tracked stereo display using image warping. In *SPIE Proceeding 2409,* San Jose, CA, 1995, pp. 21-30.
6. S. E. Chen. QuickTimeVR - an Image-based Approach to Virtual Environment Navigation. In *Proceeding SIGGRAPH'95,* 1995, pp. 29-38.

Zero Phase Representation of Panoramic Images for Image Based Localization*

Tomáš Pajdla and Václav Hlaváč

Center for Machine Perception, Faculty of Electrical Engineering
Czech Technical University, Prague
Karlovo nám. 13, 121 35, Praha 2, Czech Republic
tel.:+420-2-24357348, fax.:+420-2-24357385
pajdla@cmp.felk.cvut.cz, http://cmp.felk.cvut.cz

Abstract. The paradigm for image based localization using panoramic images is elaborated. Panoramic images provide complete views of an environment and their information content does not change if a panoramic camera is rotated. The "zero phase representation" of cylindrical panoramic images, an example of a rotation invariant representation, is constructed for the class of images which have non-zero first harmonic in column direction. It is an invariant and fully discriminative representation. The zero phase representation is demonstrated by an experiment with real data and it is shown that the alternative autocorrelation representation is outperformed.

1 Introduction

Location of an observer is given by its position and orientation. Image based localization of an observer is a process in which the observer determines its location in an image map of the environment as the location in the image from the map which is the most similar to a momentary view. It is possible to do image based localization because images of an environment typically vary as a function of location.

Recently, it has been found that some species like bees or ants use vision extensively for localization and navigation [13, 14]. In contrary to techniques based on correspondence tracking, *they use whole images* to remember the way to food or nests [5, 3]. Attempts appeared to verify models explaining the behavior of animals by implementing mechanisms of ego-motion, localization, and navigation on mobile robots [12, 17, 16, 1].

Some animals like birds have developed eyes with a large field of view which allow them to see most of a surrounding environment and to use many cues around them for localization and surveillance [6, 8]. Panoramic cameras, combining a curved convex mirror with an ordinary perspective camera [17, 2, 15], provide large field of view. The advantage of real time acquisition of images

* This work was supported by the MSMT VS96049, GACR 102/97/0480, GACR 102/97/0855, and CVUT IG 309809603 grant.

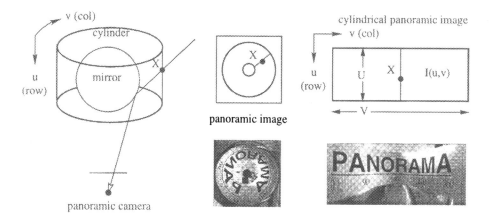

Fig. 1. A cylindrical panoramic image is obtained by projecting a panoramic image from the plane onto a cylinder and then by mapping the cylinder back to the plane by an isometry.

of whole environment suggests to use the panoramic camera for image based localization.

In this paper, we study image based localization using panoramic images. We concentrate on finding a proper representation of panoramic images in order to make the localization efficient. In contrary to approaches taken in [17, 16, 1], where rather straightforward methods were used to demonstrate image based localization, we develop the representation of panoramic images allowing an efficient storage of an image based map as well as fast localization while attaining the full discrimination between the positions.

2 Image based localization from cylindrical panoramic images

Cylindrical panoramic images are obtained by first projecting the panoramic image [15] from the plane onto a cylinder, Figure 1, and then mapped back into a plane by an isometry (cylinders are developable surfaces).

Cylindrical panoramic images can be modeled by a part of a cylinder $I(u, v)$ as shown in Figure 1. Function $I(u, v) : R^2 \rightarrow R$ is defined on an interval $[0, U] \times R$. It is periodic with a period V, $I(u, v) = I(u, v + V)$, where U and V are called vertical and horizontal image size respectively.

The rotation of the panoramic camera around its vertical axis by an angle φ induces the rotation of the cylinder $I(u, v)$ to a cylinder $J(u, v)$ around the axis z so that $J(u, v) = I(u, v - \varphi)$. The set of all such transformations of $I(u, v)$ forms the Abelian group of 1D rotations $SO(1)$. We will call the above action "shift" to be compatible with the notation used in Fourier analysis.

How do the cylindrical panoramic images change when the observer moves?

Firstly, a cylindrical panoramic image shifts if the observer rotates (changes its orientation) at the same position. The images obtained at the same position with different orientations are same in the sense that there exists a shift (independent of the scene geometry) which maps images exactly onto each other. Secondly, the image changes when observer's motion involves a translation (observer changes its position). In this case, there is no image transform which transforms two images at different positions onto each other because images depend on depth variations and occlusions in the scene.

It is the advantage of using a panoramic camera to have a complete view so that image information content does not change when changing the orientation. Ordinary cameras lack this property. Image based maps are better to construct from panoramic images as it is sufficient to store only one panoramic image at each position to describe the environment seen from that position under any orientation.

Images taken at close locations are likely to be very similar. In fact, it has been found that they are quite linearly dependent [7]. In order to obtain an efficient representation of the image based map of an environment, close images have to be as similar as possible. If so, they can be compressed by using principal component analysis [9] or a similar technique. The linear dependence of panoramic images, however, is disturbed by the rotation of the observer. An efficient map cannot therefore be directly constructed from images acquired under different non-related orientations of the observer. Each cylindrical panoramic image has to be shifted before creating the map so that close images are similar.

Finding the most similar image in the map to the momentary image naturally calls for a correlation based technique that compares a momentary image with each image in the map and selects the one with the highest correlation. Each comparison is done again by a correlation along the column direction in order to find the best matching orientation of the momentary view.

The search for the orientation can be avoided by finding a *shift invariant representation* of each class of images taken at the same position. It means that there is a procedure which gives the same representation for all images taken at one position irrespectively of their orientations. The procedure of finding an invariant representation has to be able to transform each image into the invariant representation just by using the information contained in the image alone. It is required in order to get an invariant representation of individual momentary views. If images in the map, as well as momentary views, are transformed into an invariant representation, a simple squared image difference evaluates their similarity.

In Section 3, the *shift invariant representation* of panoramic images is introduced and limitations of its construction are discussed. In Section 4, the *zero phase representation* (ZPR) is constructed for the class of images with non-zero first harmonic in the column direction. Section 5 demonstrates the ZPR by an experiment. The paper concludes with comparison of the ZPR to an alternative representation given in [1].

3 Shift invariant representations of panoramic images

The set of all cylindrical panoramic images $I(u,v)$, as defined in Section 2, can be partitioned into equivalence classes $I\backslash\varphi$ so that each class contains all images which can be obtained by shifts from one image. The equivalence classes correspond to the orbits of $SO(1)$ in the image space.

If we single out one image I^* for each class, we may use it to represent the whole class $I\backslash\varphi$. The ability to find a fixed representative image $I^*(u,v)$ is equivalent to the ability of finding the shift φ for each image $I(u,v)$ so that $I^*(u,v) = I(u,v-\varphi)$.

There is no "natural" representative image of a shift equivalence class of cylindrical panoramic images in the sense that it would be an intrinsic property of all images in the class. If it was so, it would mean that we could find φ from any image alone to transform it into the intrinsic representative image. In other words, there had to be some preferred direction which one could detect in the image. However, there cannot be any direction anyhow fixed if the images can take any values.

On the other hand, a natural representative image is something what we would really need as it is important to be able to obtain a representative image from each image alone. Therefore, it is necessary to choose a "reference direction" so that it is "seen" in each image. It can be done in each shift equivalence class independently but we prefer doing it so that the images taken at close positions have similar representative images of their shift equivalence classes. Only then we can gain from approximating many close images by a few of their principal components.

We do not want to prefer any direction by introducing something which could be seen in the scene like a reference object because it needs to modify the scene.

In some cases, we may find (or force) images $I(u,v)$ to come from a class which allows us to define a reference direction without augmenting the scene, e.g. the images have non-zero first harmonic in v direction when expanded into the Fourier series. The way, the reference direction is chosen, depends mainly on the class of the images we can get. For practical reasons, we might want to broaden the equivalence class of images from the shift equivalence class to the class which allows also for all linear brightness transformations, some noise, certain occlusions, etc. It may, in some cases, be possible but need not in others.

4 Zero phase representation

Let us assume that all images $I(m,n)$ have non-zero first harmonic. Image $I(m,n)$ is acquired by a digital camera and therefore m and n are integers, not real numbers. However, there is some function $I(u,v)$ behind, for which u and v are real. We just sample it at discrete points. If the camera rotates by a real angle α, we obtain $J(m,n) = \text{sampled}[I(u,v-\varphi)]$, where $\varphi = \frac{N}{2\pi}\alpha$, and M and N are the row and the column sizes of the image respectively.

As the first harmonic of $I(m,n)$ is assumed to be nonzero, its phase is affected by shifting $I(u,v)$. We can eliminate any unknown shift by shifting the interpolation of $I(m, n)$ and re-sampling it so that the first harmonic of the the resulting representative image of each shift equivalence class $I^*(m, n)$ equals zero. Let us show how to compute I^* in 1D.

Definition 1 (Shift). *Let function* $f(n): Z \to R$ *be periodic on the interval* $[0, \ldots, N]$*, i.e.*

$$f(n) = f(n + N) . \tag{1}$$

Mapping $\Phi_\varphi: R^Z \to R^Z$

$$\Phi_\varphi[f(n)] = \mathcal{F}^{-1}\{\mathcal{F}\{f(n)\} \, e^{-j\frac{2\pi}{N}\varphi k}\} , \tag{2}$$

where $F(k) = \mathcal{F}\{f(n)\}$ *denotes the Discrete Fourier Transform of* $f(n)$*, and is called the* shift *of* $f(n)$ *by phase* φ*.*

Definition 2 (Shift equivalence class). *Let function* $f(n): Z \to R$ *be periodic on interval* $[0, \ldots, N]$*. Then, the set*

$$S[f(n)] = \{g(n) \mid g(n) = \Phi_\varphi[f(n)], \, \forall \, \varphi \in R\} \tag{3}$$

is called the shift equivalence class *generated by* $f(n)$*.*

Definition 3 (Representative function of a shift equivalence class). *Let function* $f(n): Z \to R$ *be periodic on interval* $[0, \ldots, N]$*. Function* $r(n)$ *is a representative function of* $S[f(n)]$ *iff it generates* $S[f(n)]$*, i.e.*

$$S[f(n)] = \{g(n) \mid g(n) = \Phi_\varphi[r(n)], \, \forall \, \varphi \in R\} . \tag{4}$$

In other words, $r(n)$ is a representative function of an equivalence class iff it is its member, i.e. $r(n) \in S[f(n)]$.

Definition 4 (Shift invariant representation). *Let function* $f(n): Z \to R$ *be periodic on interval* $[0, \ldots, N]$*. Let* $\rho: R^Z \to R^Z$ *assign to each* $f(n)$ *a function* $s(n)$*. The* $s(n)$ *is called the* shift invariant representation *of* $f(n)$ *iff*

$$s(n) = \rho[\Phi_\varphi[f(n)]], \, \forall \, \varphi \in R . \tag{5}$$

Observation 1 *There are shift invariant representations of* $f(n)$ *which are not representative functions of* $S[f(n)]$*.*

Autocorrelation of a function $f(n)$ is an example of a shift invariant representation which is not a representative function of a shift equivalence class $S[f(n)]$. By shifting the autocorrelation function we get a shifted autocorrelation function but certainly not the original $f(n)$. Moreover, all different functions which have the absolute value of their Fourier transform equal to the absolute value of the Fourier transform of $f(n)$ have the same autocorrelation representations. So, the autocorrelation representation is quite ambiguous.

Definition 5 (Zero phase representation (ZPR)). *Let function* $f(n): Z \to R$ *be periodic on interval* $[0, \ldots, N]$. *Function*

$$f^*(n) = \mathcal{F}^{-1}\{\mathcal{F}\{f(n)\}\ e^{-j\phi[F(1)]k}\}. \tag{6}$$

where $F(k) = \mathcal{F}\{f(n)\}$ *is the Discrete Fourier Transform of* $f(n)$ *and* $\mathcal{F}^{-1}$ *is its inverse, is called the* zero phase representation *of* $f(n)$.

Lemma 1. *Let functions* $f(n)$, $g(n): Z \to R$ *be periodic on interval* $[0, \ldots, N]$ *with non-zero first harmonic, i.e.* $|F(1)| \neq 0$, $|G(1)| \neq 0$, *and let* $f^*(n)$, $g^*(n)$ *be defined by (6). Then,*

$$\exists \varphi \in R,\ g(n) = \Phi_\varphi[f(n)] \iff g^*(n) = f^*(n). \tag{7}$$

Proof. Lemma is verified by straightforward application of (2) and (6) on (7). See [10] for details. **Q.E.D.**

Observation 2 $f^*(n)$ *defined by (6) is a representative function of* $S[f(n)]$.

Observation 3 *If* $f(n)$ *has non-zero first harmonic then* $f^*(n)$ *defined by (6) is an invariant representation* $S[f(n)]$.

Observations 2 and 3 show that the ZPR is a good representation of the class of images which have non-zero first harmonic in column direction. The ZPR assures that the images taken at different positions will be represented differently and the images taken at the same place will have the same representative image.

Observation 4 *The first harmonic of* $f^*(n)$ *defined by (6) equals zero, i.e.* $\phi[F^*(1)] = 0$.

Observation (4) explains why the name "zero phase representation" has been chosen.

A straightforward generalization of 1D ZPR for 2D cylindrical panoramic images can be achieved by replacing the 2D FFT by a 2D DFT so that (6) is replaced by

$$I^*(m, n) = \mathcal{F}^{-1}\{\mathcal{F}\{I(m, n)\}\ e^{-j\phi[F(0,1)]l}\}, \tag{8}$$

where $F(k, l) = \mathcal{F}\{I(m, n)\}$ is a Discrete Fourier Transform of $I(m, n)$.

5 Experiment

Figure 2 shows real images taken with different orientations (a, c) and position (a, e) of a panoramic camera. Ideally, the images (a, c) should be related by a shift but this is violated by the holder of the camera which stays at the same place in the image because it rotates with the camera and by occlusions and changes in the scene. Figures 2 (b), (d), (f) show the ZPR of the images. The images (b) and (d) are quite correctly shifted so that their relative shift is almost zero as expected even though there were changes in the scene. The relative shift of the ZPR of the images (d) and (f) which were taken at different positions differs quite a lot.

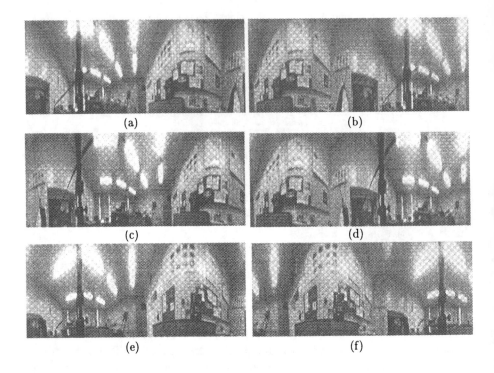

Fig. 2. (a) A panoramic image and its ZPR (b), (c) a panoramic image taken at the same position and different orientation and its ZPR (d), (e) a panoramic image taken at the different position and same orientation and its ZPR (f).

6 Conclusion

The fundamentals of image based robot localization have been given by providing the ZPR, an invariant and discriminative representation of cylindrical panoramic images with non-zero first harmonic in column direction. The idea of rotation invariant representation of panoramic images has independently been used by Aihara et. al [1] where independent autocorrelation of each row of a cylindrical panoramic image was used. The invariance for all images is achieved but it is not discriminative as images of many different scenes map into the same autocorrelation. In this respect, the ZPR is better suited for global localization as it is fully discriminative for the images with non-zero first harmonic in column direction.

In the future, we like to study a suitable representation of the images with zero first harmonic. The first experiments by Jogan and Leonardis [4] show that the ZPR provides images which can be compressed by principal component analysis. The question of further investigation is to find how smoothly the ZPR behaves when changing the position. More attention has to be paid to illumination changes and occlusions. Preliminary experiments [11] show that ZPR is not much affected by additive noise but it can strongly be affected by occlusions.

We thank to Tomáš Svoboda who acquired real panoramic images for the experiments, to Aleš Leonardis, Horst Bischof, and Radim Šára for discussions.

References

1. N. Aihara, H. Iwasa, N. Yokoya, and H. Takemura. Memory–based self–localization using omnidirectional images. In *14th Int. Conf. on Pattern Recognition, Brisbane, Australia*, volume II, pages 1799–1803. IEEE Computer Society Press, August 1998.
2. S. Baker and S.K. Nayar. A theory of catadioptric image formation. In *6th Int. Conf. on Computer Vision*, pages 35–42, India, January 1998. IEEE Computer Society, Narosa Publishing House.
3. T.S. Collet. Insect navigation en route to the goal: Multiple strategies for the use of landmarks. *J. Exp. Biol.*, 199.01:227–235, 1996.
4. M. Jogan and A. Leonardis. Mobile robot localization using panoramic eigen-images. In *4th Computer Vision Winter Workshop, Rastenfeld, Austria*. Pattern Recognition and Image Processing Group of the Vienna University of Technology, pages 13–23, February 1999.
5. S.P.D. Judd and T.S. Collet. Multiple stored views and landmark guidance in ants. *Nature*, 392:710–714, April 1998.
6. H.G. Krapp and R. Hengstenberg. Estimation of self-motion by optic flow processing in single visual interneurons. *Nature*, 384:463–466, 1996.
7. S.K. Nayar, S.A. Nene, and H. Murase. Subspace methods for robot vision. *RA*, 12(5):750–758, October 1996.
8. R.C. Nelson and J. Aloimonos. Finding motion parameters from spherical motion fields (or advantages of having eyes in the back of your head). *Biol. Cybern.*, 58:261–273, 1996.
9. E. Oja. *Subspace Methods of Patter Recognition*. Research Studies Press, Hertfordshire, 1983.
10. T. Pajdla. Robot localization using shift invariant representation of panoramic images. Tech. Rep. K335-CMP-1998-170, Czech Technical University, Dpt. of Cybernetics, Faculty of Electrical Engineering, Karlovo nám. 13, 12135 Praha, Czech Republic, November 1998. ftp://cmp.felk.cvut.cz/pub/cmp/articles/pajdla/Pajdla-TR-170.ps.gz.
11. T. Pajdla. Robot localization using panoramic images. In *4th Computer Vision Winter Workshop, Rastenfeld, Austria*. Pattern Recognition and Image Processing Group of the Vienna University of Technology, pages 1–12, February 1999.
12. M.V. Srinivasan. An image-interpolation technique for the computation of optical flow and egomotion. *Biol Cybern*, 71:401–415, 1994.
13. M.V. Srinivasan. Ants match as they march. *Nature*, 392:660–661, 1998.
14. M.V. Srinivasan and S.W. Zhang. Flies go with the flow. *Nature*, 384:411, 1996.
15. T. Svoboda, T. Pajdla, and V. Hlaváč. Epipolar geometry for panoramic cameras. In *5th European Conference on Computer Vision, Freiburg, Germany*, number 1406 in Lecture Notes in Computer Science, pages 218–232. Springer, June 1998.
16. K. Weber, S. Venkatesh, and M.V. Srinivasan. An insect–based approach to robotic homing. In *14th Int. Conf. on Pattern Recognition, Brisbane, Australia*, volume I, pages 297–299. IEEE Computer Society Press, August 1998.
17. Y. Yagi, K. Yamazawa, and M. Yachida. Rolling motion estimation for mobile robot by using omnidirectional image sensor hyperomnivision. In *13th Int. Conf. on Pattern Recognition, Vienna, Austria*, pages 946–950. IEEE Computer Society Press, September 1996.

Panoramic Eigenimages for Spatial Localisation [*]

Matjaž Jogan and Aleš Leonardis

Computer Vision Laboratory
Faculty of Computer and Information Science
University of Ljubljana
Tržaška 25, 1001 Ljubljana, Slovenia
e-mail: matjaz.jogan@mailcity.com

Abstract. Recent biological evidence suggests that position and orientation can be estimated from an adequately compressed set of environment snapshots and their relationships. In this paper we present a pure appearance-based localisation method using an eigenspace representation of panoramic images. We first review several types of rotational invariant representation of panoramic images in terms of their efficiency for an eigenspace-based localisation problem. Then, for each set of images an eigenspace from 25 location snapshots is built and analyzed. We evaluated simple localisation of images not included in the training set. The results show good prospects for the panoramic eigenspace approach.

1 Introduction

It is well known that a large number of animal species uses predominantly visual information to navigate in space. Most animals use visual information in combination with odometry, but in special cases, such as moving in the air or underwater, this is not possible. Several methods have therefore been implemented that use only vision to navigate. Nelson and Aloimonos [13] proposed that omnidirectional views could ease the task of estimating motion parameters and facilitate orientation. It seems reasonable to use omnidirectional views also for the localisation problem. Examples of using such an input can be found in the works of Yagi et al. [18] or Aihara et al. [1]. Simplified line-scan panoramic representations were proposed by Franz et al. [7], Francheschini et al. [6] and Chahl, Weber, Venkatesh and Shrinivasan [19, 3].

In pure appearance-based methods for navigation, which use simplified representations of views such as line-scan intensity rings [7, 19, 3], appearance cues are stored in memory to represent the explored space. Localisation is then performed by matching current views with those stored in the memory.

Recently, more evidence has been gathered which shows that, at some level, biological systems also perform appearance matching. Judd and Colett [9] reported that ants use multiple stored snapshots to learn a path from start to

---

[*] This work was supported by the Ministry of Science and Technology of Republic of Slovenia (Project J2-0414).

goal. Dill, Wolf, and Heisenberg [4] also reported that bees try to match retino-topically the incoming visual pattern with previously stored images. Appearance cues are found to serve either for matching or to be organized in higher cognitive schemes such as maps, as it is presumed for the hippo-campus brain area of mammals (e.g. Epstain [5]).

If we take instead of an intensity ring an iconic representation of the world, as in the case of snapshots taken with a wide visual field, it would be cumbersome to densely scan the environment and then perform matching operations to search for the most similar snapshot in the memory. In fact, because of the computational complexity and the amount of memory needed, this is not feasible even for a biological vision system. It is therefore obvious that the data must be compressed. An efficient method is the *Singular Value Decomposition (SVD)*, which can be used to find a low-dimensional representation of a set of images in terms of linear combinations (points) in eigenspace, spanned by orthogonal eigenvectors (eigenimages). If the data set distribution can be encompassed with a small number of eigenvectors, we can achieve significant dimensionality reduction. We can assume that two equally oriented images taken at close positions appear very similar, therefore we expect that (a) we can achieve a significant compression and (b) points in the eigenspace, representing neighboring positions, will also be close to each other. The SVD and similar methods have been widely used in appearance-based recognition problems [14, 17, 12, 2, 11].

The most straightforward application of eigenspaces therefore requires a representation of the visual input that would be easily aligned for matching. One way is to estimate the orientation from other sources such as light polarization, gyrocompass etc. We can also find a representation that would be rotation-invariant, such as row-correlation [1], but we loose the orientation specific information. An approach called *Zero Phase Representation (ZPR)* was recently proposed by Pajdla [15] which can be used to find a transformation that projects differently oriented but otherwise identical panoramic images into one representative image.

This paper is organized as follows. In section 2 we first review various rotationally invariant representations and evaluate them on our image set. In section 3 we describe the procedure of building eigenspaces with the SVD method and, finally, we present the results of localisation experiments in section 4. We conclude with a summary and outline future work.

2   Shift-invariant representations of panoramic images

2.1   Evaluation of panoramic images

Fig. 1 shows the map of the CMP Laboratory[2] with some examples of panoramic images taken at positions 1, 5, 12, and 15. From the original images, taken at random orientations, four sets of rotational independent images were generated.

---

[2] All panoramic images used in this paper were kindly provided by T.Pajdla and J.Černík from the Czech Technical University at Prague.

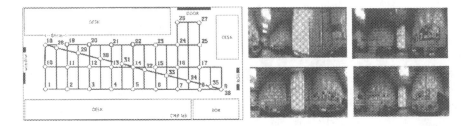

Fig. 1. Left: Map of the CMP lab with positions where the snapshots were taken. Right: Snapshots taken at positions 1, 5, 12, and 15. The center of each snapshot points towards West.

As our intention is to compress the image set by estimating the most significant eigenvectors and then present the training set as points in the eigenspace, we need a criterion to evaluate the image sets. A suitable criterion is the distance in the eigenspace. It has been shown [12, 14] that by estimating the *correlation* $i_p^T i_q$ of two images the distance between their projections in the eigenspace can be evaluated.

2.2 Manually aligned images

Manually aligned images represent the case in which the orientation of the sensor at the time of taking the snapshot is known and therefore the images can be shifted so that they are all oriented in the same direction. Since the orientation is the same, it is expected that images taken at small distances apart are strongly correlated, i.e., their correlation coefficient being higher than for images taken far apart. In Fig. 2 we can see how the correlation with the image obtained at position 30 varies with position for aligned images.

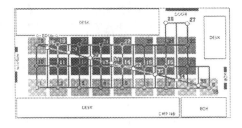

Fig. 2. Correlation with the snapshot taken at position 30 for manually aligned images. Darker areas indicate a higher level of correlation.

2.3 Autocorrelated images

Another way of achieving rotational invariance is by autocorrelating the images. We have two options. The first option is to perform autocorrelation by row and column directions, which results in not only rotational but also vertical invariance. It can be seen in Fig. 4(left) that the proximity of images based on the correlation of autocorrelated images seems weaker, compared with the performance of aligned sets. The second option is to correlate images just by row direction, which appeared in a similar context in [1]. It can be seen from Fig. 4(right) that the correlation values of row autocorrelated images can be compared to that of the fully autocorrelated ones.

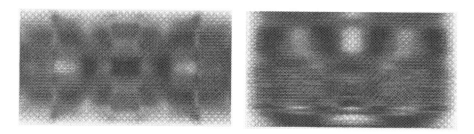

Fig. 3. Left: fully autocorrelated transform of the snapshot taken at position 1; Right: row autocorrelated transform of the same snapshot.

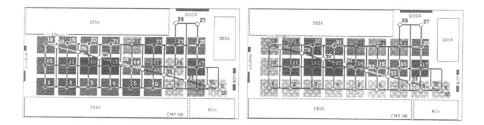

Fig. 4. Left: Correlation with snapshot taken at position 30 for fully autocorrelated images. Right: correlation with same snapshot for row autocorrelated images.

2.4 ZPR images

Zero Phase Representation (ZPR) images are obtained from original randomly oriented images by shifting the image so that its first harmonic in the *Discrete Fourier Transform* has a phase equal to zero [15]. This transformation produces one representative image for an equivalence class which, in our case, is the set of panoramic images taken under stable conditions at the same place but with possibly different orientations. If a panoramic image is represented as a 2-dimensional discrete function $I(row, col)$, then its ZPR representation $I_{ZPR}(row, col) = I(row, col - \phi)$ can be determined as

$$I_{ZPR}(row, col) = \mathcal{F}^{-1}\{\mathcal{F}\{I(row, col)\}e^{-j\Phi[F(0,1)]l}\} , \qquad (1)$$

where $F(k, l) = \mathcal{F}\{I(row, col)\}$ is a Fourier transform of I and $\Phi[F(0, 1)]$ denotes the shift ϕ of the image necessary to obtain an image whose first harmonic has a phase equal to zero [15].

Problems may arise due to the uneven resolution of the cylindrical panoramic image caused by the transformation process from the original panoramic image and because of possible self-occlusion of image content on the top and bottom of the images. Thus one needs to robustify the method by weighting the images [16], as it can be seen in Fig. 6(left). In Fig. 5 we can see how the ZPR performs on panoramic images taken at positions 6 and 7. It is expected that representative images of equivalence classes taken not far apart would be strongly correlated and thus close in the parametric eigenspace. Our tests show that this expectation is correct since our set of 35 images, when transformed by ZPR, results in a set of very similarly oriented images. In Fig. 6(right) one can see that the correlation distribution can be compared with that of the manually oriented set.

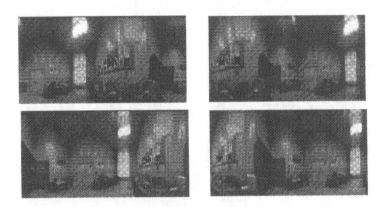

Fig. 5. Left: original snapshots taken at positions 6 and 7 in random orientation. Right: ZPR transforms of the images on the left.

3 Panoramic eigenspaces

3.1 Construction

Eigenspaces were constructed from 25 panoramic snapshots taken from the positions on the rectangular grid, labeled 1-25, resprectively. For testing, we used the remaining images.

 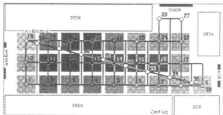

Fig. 6. Left: Weighted image taken at position 30. Right: Correlation with the snapshot taken at position 30 for weighted ZPR images. Darker areas indicate a higher level of correlation.

To reduce the computational complexity, eigenspaces were built with an algorithm that estimates the SVD of the smaller covariance matrix, that is, if A_{mn} is the matrix of n image vectors (normalized so that mean is in the origin), then the SVD of Q_{nn} is calculated:

$$Q_{nn} = A_{nm}^T A_{mn} = U_{nn} V_{nn} U_{nn}^T \ . \tag{2}$$

The eigenvectors obtained with further processing of the matrix U_{nn} are then sorted according to their corresponding eigenvalues from the diagonal of V_{nn}.

3.2 Comparison of eigenspaces

We can define the *energy* of an eigenvector as proportional to the magnitude of its eigenvalue. Then we can estimate the *energy* of an eigenspace as the cumulative sum of the energies of the eigenvectors included. If we then take an eigenspace of dimension p, $p < n$, we can relate the compression rate to the energy of the eigenspace.

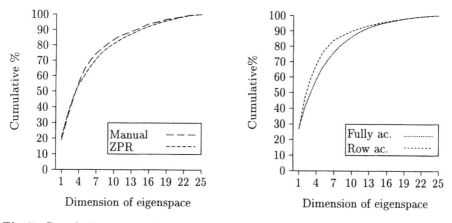

Fig. 7. Cumulative energy distribution for: manually aligned and ZPR of weighted image set (left); autocorrelated sets (right).

From the cumulative energy plots in Fig. 7, we can see the difference of energy distribution for eigenspaces based on manual and ZPR sets. ZPR performs slightly worse in terms of compactness of representation, which is mostly due to a greater dissimilarity between images taken from neighbor positions since orientation is not exactly uniform through the set.

Next, we compare the plots Fig. 7(left) with Fig. 7(right) and observe that we can for the correlated sets more energy can be captured in the first few eigenvectors. This would be favorable if our goal was to achieve a high compression rate, however when discrimination between the images is the primary objective, this is not necessarily the case. As it was previously stated, in the case of the autocorrelated image set similarity between images is not in strong relationship with proximity of the locations where the snapshots have been taken.

4 Experimental results of localisation

After the eigenspaces were built and analyzed we tested their performance in the localisation task. The test images were projected onto the eigenspace. As the correlation in Hilbert space equals the Euclidean distance in eigenspace [14, 12], the nearest neighbor can be used to estimate the most similar snapshot. We defined four criteria to measure the success rate:

I The **first criterion** tells whether the image from the training set representing the position nearest to that of the test image was successfully recognized as the nearest one.

II The **second criterion** tells whether the first *and* the second nearest positions were successfully recognized.

III The **third criterion** tells whether the *three* nearest positions were successfully recognized.

IV The **fourth criterion** is the weakest of all and tells whether *at least one* of the *four* nearest positions was recognized as the nearest one.

The first three conditions happen to be very strict since we use for the testing the images that were not included in the training set and are therefore taken at different (intermediate) positions. As we can see from Fig. 1(left), these positions, except for those numbered 26 and 27, lie on the diagonal line and comprise positions numbered from 28 to 35. In our case, there are some positions that lie near the middle of the quadrant and the difference of lengths to the nearest neighbors is small. It is obvious that if the first criterion fails, so do the second and the third. The fourth criterion however, checks only if one of the four positions surrounding the test position appears as the nearest.

In Fig. 8 we compare the performance on the manually aligned and the ZPR set of weighted images. The results for the manually aligned set seem good enough for approximate localisation. If we operate with a six-dimensional eigenspace, we correctly determine the three nearest positions (ordered) in 50% of the cases and the nearest position in 100% of the cases. An exact position estimation would of course require further processing and will not be discussed

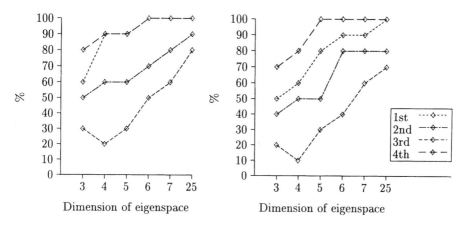

Fig. 8. Success rate for manually aligned (left) and ZPR of weighted image set (right).

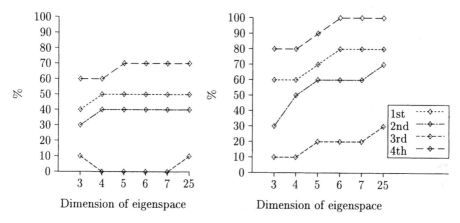

Fig. 9. Success rate for fully autocorrelated (left) and row autocorrelated image set (right).

in this paper. In Fig. 8(right) we can see that basically the same performance can be achieved by using ZPR transformed weighted images. As we can observe, eigenspaces of dimension less than 10 perform well enough and any further increase in the dimension leads to higher cost/performance rate.

In Fig. 9 we can see the same graphs for the fully autocorrelated and row autocorrelated images. As we can observe, the performance is not nearly as good as in the previous cases and it does not improve even with full dimensionality of the eigenspace.

In Fig. 10(left) we can see the result of position estimation for the snapshot number 30 from the manually aligned set of images. In Fig. 10(right) we can see how the method behaves while localising a snapshot taken at position 30 for the ZPR transformed weighted images. In this case, the result is as good as in the case of the manually aligned set.

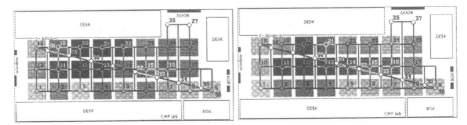

Fig. 10. Left: Localisation of snapshot 30 for the manually aligned set of images. Right: Localisation of snapshot 30 for the ZPR transformed weighted set of images. Darker shades denote smaller distance in the eigenspace.

5 Summary

In this paper we discussed the problem of appearance-based localisation. The eigenspace approach proves itself as a very viable one. Our tests show that the nearest positions can be estimated with significant certainty if we rely on similarity of equally oriented neighboring images. To achieve similar orientation of images taken with random orientation of the sensor, we tested the Zero Phase Representation which performed well enough for our limited image set without occlusions. It is however a matter of future research to analyze how it performs on larger sets and under unpredictable circumstances. We also tested the autocorrelated rotational invariant representations which do not preserve overall appearance. Our results show that the performance of such representations is inferior, mostly due to the fact that the relationship between the image correlation and distance of positions where the images are taken is not so explicit, as in the case oriented images. Note that, as opposed to [1], tests were performed only on the snapshots that were not in the training set, i.e., they were taken at slightly different positions.

In future work we plan to test additional sets of images, some with occlusions and under different illumination. On a wider set of images, multiple eigenspaces will be built and analyzed. On more densely sampled images, points in eigenspace will be interpolated to form a hypersurface and its characteristics will be analyzed. Strategies for calculating the exact position from localisation scores will be developed.

References

1. Aihara H., Iwasa N., Yokoya N., Takemura H.: Memory-based self-localisation using omnidirectional images. Anil K. Jain, Svetha Venkatesh, Brian C. Lovell, editors, 14th International Conference on Pattern Recognition, Brisbane, Australia, volume I. IEEE Computer Society Press (1998) 297-299
2. Black M. J., Jepson A. D.: EigenTracking: Robust matching and tracking of articulated objects using a view-based representation. International Journal of Computer Vision 26(1) (1998) 63-84

3. Chahl, J.S., Srinivasan M.V.: Visual computation of egomotion using an image interpolation technique. Biol.Cybern. Vol.74, No.5 (1996) 405-411

4. Dill M., Wolf R., Heisenberg M.: Visual pattern recognition in *Drosophila* involves retinotopic matching. Nature Vol 365 (1993) 751-753

5. Epstain, R., Kanwisher N.: A cortical representation of the local visual environment. Nature Vol.392 (1998) 598-601

6. Francheschini N., Pichon J.M. and Blanes C.: From insect to robot vision. Phil. Trans. R. Soc. Lond. B 337 (1992) 283-294

7. Franz M.O., Schölkopf B., Mallot H.A., Bülthoff H.H.: Where did I take that snapshot? Scene-based homing by image matching. Biol. Cybern. Vol.79, No.3 (1998) 191-202

8. Guerrero, J.J., Sagues C.: Tracking features with camera maneuvering for vision-based navigation. Journal of robotic systems Vol.15, No.4 (1998) 191-206

9. Judd S.P.D., Collett T.S.: Multiple stored views and landmark guidance in ants. Nature Vol.392 (1998) 710-714

10. Kurata J., Grattan K.T.V., Uchiyama H.: Navigation system for a mobile robot with a visual sensor using a fish-eyelens. Review of Scientific Instruments Vol.69 No.2, 1 (1998) 585-590

11. Leonardis A. and Bischof H.: Dealing with occlusions in the eigenspace approach. Proc. of CVPR'96, IEEE Computer Society Press (1996) 453-458

12. Nayar S. K., Nene S. A., Murase H.: Subspace methods for robot vision. IEEE Trans. on Robotics and Automation vol.12, No.5 (1996) 750-758

13. Nelson R.C. and Aloimonos J.: Finding motion parameters from spherical motion fields (or advantages of having eyes in the back of your head). Biol. cybernetics Vol.58 (1988) 261-273

14. Oja E.: Subspace methods of pattern recognition. Research Studies Press, Hertfordshire (1983)

15. Pajdla T.: Robot localization using panoramic images. Norbert Bräendle, editor, Proc. of the Computer Vision Winter Workshop, Rastenfeld, Austria. (1999) 1-12

16. Pajdla T.: Personal correspondence (1999)

17. Turk M. and Pentland A.: Face recognition using eigenfaces. Proc. Computer Vision and Pattern Recognition, CVPR'91 (1991) 586-591

18. Yagi Y., Nishizawa Y. and Yachida M.: Map-based navigation for a mobile robot with omnidirectional image sensor COPIS. IEEE Trans. on Robotics and Automation Vol.11, No.5 (1995) 634-648

19. Weber K., Srinivasan M.V.: An insect-based approach to robotic homing. ICPR'98 (1998) 297-299

Feature Grouping Based on
Graphs and Neural Networks

Herbert Jahn

Deutsches Zentrum für Luft- und Raumfahrt e.V. (DLR)
Institut für Weltraumsensorik und Planetenerkundung
Rudower Chaussee 5
D-12484 Berlin, Germany
Herbert.Jahn@dlr.de

Abstract. An attempt is made to develop a parallel-sequential method for feature grouping with special applications to gray value segmentation, grouping of contour points, and dot pattern clustering. These problems are closely connected with other preprocessing tasks such as edge preserving smoothing and edge detection which are also dealt with. The method is based on a special graph structure called Feature Similarity Graph which is defined via an adaptive feature similarity criterion and (Voronoi) neighborhood relations between the features. A recursive edge preserving method of feature averaging which is based on the similarity criterion is presented. Via the averaging procedure non-local processing is implemented which is necessary for efficient noise reduction. Network structures which are similar to the Cellular Neural Networks (CNN) can be used efficiently for implementing the nonlinear algorithm. One processing element or neuron is assigned to each feature and (Voronoi) neighbored features are connected with adaptive weights depending on feature similarity. It is demonstrated with few examples that the same grouping principles can be used for different tasks of segmentation and clustering.

1 Introduction

Feature grouping and segmentation techniques are an essential part of image preprocessing, and there is a vast amount of methods and algorithms in that field. In the human visual system these tasks are solved pre-attentively [1] using massively parallel neural processing structures. Therefore, it seems to be justified to look for artificial neural algorithms and structures for solving image preprocessing problems.

Here, an attempt is made to develop a parallel-sequential method for feature grouping with special applications to gray value segmentation, grouping of contour points, and dot pattern clustering. These problems are closely connected with other preprocessing tasks such as edge preserving smoothing and edge detection which are also dealt with.

In older attempts of the author [2–5, 17] problems of feature grouping have been dealt with using networks with special graph structures. To use graphs for the description of segments or clusters seems to be adequate because segments with a complicated (fuzzy) structure can be described efficiently. Especially, this is shown by the comprehensive work on irregular, stochastic and adaptive pyramids [6-9]. Because of its usefulness the graph structure is maintained in the approach presented here.

It turned out [5] that network structures which are similar to the Cellular Neural Networks (CNN) [10 – 11] can be used efficiently for dot pattern clustering. Because the problems of gray value segmentation and dot pattern clustering as special feature grouping problems are closely related an attempt was made [12] to apply CNN structures also for smoothing and segmentation of (gray value) images.

Here, the problem will be generalized. Gray values and dot coordinates (and other features, e.g. edge element directions, texel orientations and shape descriptors) are special cases of general features which have to be grouped to useful clusters or segments. Useful means here that the clustering results should be in accordance with that what we see because of the lack of general criteria human visual interpretation is used in many cases for assessing image processing results. Moreover, principles of human vision and perception can also give orientations for the development of image processing algorithms and therefore some results of perception research will be used here.

In section 2 the general approach is described. The special cases of gray value segmentation, border point grouping, and dot pattern clustering are dealt with and results are given in sections 3, 4 and 5, respectively. In the conclusions an outlook to necessary future developments is given.

2 Grouping of features

Let's consider M points $P_k = (x_k, y_k)$ (k=1,...,M). These points can be the pixel positions (i,j) (i,j=1,...,N) in case of gray value segmentation (then $M = N^2$), the positions of dots in case of dot pattern clustering or the positions of other already extracted image elements (e.g. directions of segment border points). The points P_k define the nodes k (or vertices) of a graph. The nodes can be connected by graph edges (or branches). In order to reduce the connectivity of the graph (and the neural network) a neighborhood $N(k)$ of each point P_k is introduced. P_k now can only be connected with its neighbors . To achieve a minimal but sufficient connectivity the Voronoi neighborhood [13] $N_V(k)$ can be used. This guarantees also the planarity of the graph. In case of pixels (i,j) in a square grid $N_V(i,j)$ is the 4-neighborhood. For randomly distributed points the Voronoi neighborhood generalizes the 4-neighborhood and can be considered as a "natural" neighborhood. In case of segment border points which have at most two neighbors the 8-neighborhood is used.

Now a feature vector f_k is assigned to each graph node k. For gray value segmentation the 1D feature is the gray value $g_{i,j}$ whereas for dot pattern clustering the dot position itself defines the 2D feature vector: $f_k^T = (x_k, y_k)$. In case of border point direction the unit vector $(\cos\varphi_k, \sin\varphi_k)$ is used as feature vector. Other examples (not considered here) are the orientation, size, and certain shape features of texture elements (texels).

The connection of two graph nodes k and k' is governed by a similarity criterion

$$\|\mathbf{f_k} - \mathbf{f_{k'}}\| \leq t_{k,k'} \tag{1}$$

where $\|\mathbf{f}\|$ is the norm of the vector $\mathbf{f}$ ($\|\mathbf{f}\| = |f|$ in case of an 1D feature f) and $t_{k,k'}$ is an adaptive threshold. (1) is justified by observations from perception research [1] which tell us that not the features themselves are responsible for segmentation but the feature differences (feature contrast). Furthermore, the discriminability of two neighbored features depends on the feature variability within some neighborhood of both the features. In [2 – 5] some feature variability measures have been considered. It turned out that (up to now) the following threshold $t_{k,k'}$ gives the best results:

$$t_{k,k'} = \mu \operatorname{MIN}\{\langle\Delta\mathbf{f_k}\rangle,\langle\Delta\mathbf{f_{k'}}\rangle\} + \varepsilon \tag{2}$$

In (2) μ and ε are two parameters which depend on the kind of the feature (see below). $\langle\Delta\mathbf{f_k}\rangle$ is defined as follows:

$$\Delta\mathbf{f_k} = \operatorname{MIN}_{k' \in N(k)} \|\mathbf{f_k} - \mathbf{f_{k'}}\| \tag{3}$$

is the minimal feature difference (nearest neighbor feature distance) within the neighborhood $N(k)$ of P_k. $\langle\Delta\mathbf{f_k}\rangle$ is a mean value of $\Delta\mathbf{f_k}$ taken over some neighborhood of P_k which depends on the kind of the feature. The measure of feature variation

$$\operatorname{MIN}\{\langle\Delta\mathbf{f_k}\rangle,\langle\Delta\mathbf{f_{k'}}\rangle\} \tag{4}$$

used in (2) has the advantage that it never crosses feature discontinuities and therefore measures the feature variation in segments but not across segments. In ideal images with stepwise constant features one has $\operatorname{MIN}\{\langle\Delta\mathbf{f_k}\rangle,\langle\Delta\mathbf{f_{k'}}\rangle\} = 0$. Then the threshold (2) reduces to $t_{k,k'} = \varepsilon$. Therefore, ε can be interpreted as the minimal resolvable feature difference. For instance, if $\mathbf{f_k}$ is a texel orientation (angle) then there exists a minimal angle difference which can be resolved by the human visual system. In busy (noisy) image regions $\mu\operatorname{MIN}\{\langle\Delta\mathbf{f_k}\rangle,\langle\Delta\mathbf{f_{k'}}\rangle\}$ determines the similarity threshold whereas in smooth ones ε is prevalent. Generally, $t_{k,k'}$ varies within the image (or dot pattern), and it adapts to changing image statistics and structure.

If for two graph nodes k, k' the feature similarity criterion (1) is fulfilled then the nodes become connected by a graph edge. For graphical representation one can connect the points P_k and $P_{k'}$ by a straight line (if there is a limited number M of points as in certain dot patterns or segment borders). For algorithmic purposes the neighbor adjacency list NAL(k) [14] can be used for graph definition. NAL(k) contains all nodes k' which are connected with k. Finally, the clusters or segments can be defined as the connected components of the graph which can be called the Feature Similarity Graph (FSG).

If this method is applied to images or patterns one obtains good results only for nearly ideal images or patterns (with stepwise constant features and low noise). In noisy images the method fails because only local information is used for graph construction. One needs regional and even global information. To exploit such information an adequate smoothing procedure is used which smoothes the features within regions but not across regions (edge preserving smoothing). In order to proceed from local to regional and (sometimes) global processing a recursive smoothing procedure is introduced which uses the FSG connectivity.

To develop such a procedure feature averaging over the (Voronoi) neighborhood is considered. Let n_k be the number of (Voronoi) neighbors of point P_k. Then

$$\langle \mathbf{f_k} \rangle = \frac{1}{n_k + 1} (\mathbf{f_k} + \sum_{k' \in N(k)} \mathbf{f_{k'}}) \tag{5}$$

is the mean value of the feature $\mathbf{f_k}$. An equivalent (recursively written) notation of (5) is

$$\mathbf{f_k}(t+1) = \mathbf{f_k}(t) + \frac{1}{n_k + 1} \sum_{k' \in N(k)} [\mathbf{f_{k'}}(t) - \mathbf{f_k}(t)] \tag{6}$$

$$(t=0,1,2,....).$$

The initial condition is $\mathbf{f_k}(0) = \mathbf{f_k}$.

Because of its linearity, the recursive algorithm (6) with increasing recursion level (or discrete time) t diminishes the resolution of the image/pattern and blurs the edges more and more. It is a special case of linear diffusion and can be used in the multi-scale analysis [15] of images. But here we do not want to blur edges and to smooth out image/pattern details. Therefore, the feature differences in (6) must be weighted properly to prevent that. Introducing weights $w_{k,k'}$ the following scheme is obtained:

$$\mathbf{f_k}(t+1) = \mathbf{f_k}(t) + \alpha_k(t) \cdot \sum_{k' \in N(k)} w_{k,k'}(t) [\mathbf{f_{k'}}(t) - \mathbf{f_k}(t)] \tag{7}$$

In (7) the $\alpha_k(t)$ are parameters which must be chosen properly to guarantee stability. Constant and small enough $\alpha(t) = \alpha_0$ (e.g. $\alpha_k = 1/(n_k+1)$ give stability. But, as usual for learning algorithms stability can be enforced and convergence accelerated using a decreasing sequence of values $\alpha(t)$.

For choosing the weights $w_{k,k'}$ the similarity criterion (1) is used:

$$W_{k,k'} = \begin{cases} 1 & if~(1)~is~fulfilled \\ 0 & elsewhere \end{cases}$$

(8)

With these weights encouraging results can be obtained. But, in images with strong noise the averaging power of (7), (8) is not sufficient. To enhance the smoothing capability of the method a degree of similarity is introduced. With edge strength $x_{k,k'} = \|f_k - f_{k'}\|/t_{k,k'}$ the similarity criterion (1) is equivalent to $x_{k,k'} \leq 1$. Then the function

$$s_0(x) = \begin{cases} 1 & if~0 \leq x \leq 1 \\ 0 & if~x > 1 \end{cases} \qquad (9)$$

can be interpreted as a degree of similarity. More generally, each non-increasing function $s(x)$ with $s(0)=1$ and can be defined as a degree of similarity between neighbored features f_k and $f_{k'}$. The function

$$s_1(x) = \frac{1}{1 + x^2} \qquad (10)$$

turned out to be useful but many other functions are possible too.
With the function $s(x)$ the weights $w_{k,k'}$ used in (7) can be written as

$$w_{k,k'} = s(x_{k,k'}). \qquad (11)$$

Via these weights a more general definition of the Feature Similarity Graph can be given. Each FSG node k is connected with all (Voronoi) neighbor nodes by a FSG edge. To each graph edge between k and k' the weight $w_{k,k'}$ is assigned. If the weights (11) are given by the function s_0 (9) then the FSG is a "hard" graph with connections (weight 1) or no connections (weight 0) between the nodes. But, if another (soft or fuzzy) similarity degree $s(x)$ (e.g. (10)) is used then the FSG is a "soft" or fuzzy FSG. Now the following processing scheme can be introduced:
- For each point P_k its (Voronoi) neighbors are determined.
- The smoothing scheme (7) is applied recursively up to the recursion level t_{max} where stability is reached. Especially, the feature variation measures (4) must be calculated for neighbored points P_k, $P_{k'}$. The smoothed features $\langle f_k \rangle = f_k(t_{max})$ are the result of the procedure.
- Now the fuzzy or soft FSG can be generated by assigning the degree of similarity $s(x_{k,k'})$ to connected FSG nodes k, k'.
- The hard FSG can be obtained from the soft FSG if the degree of similarity $s(x_{k,k'})$ is binarized using the threshold $s = 1/2$. The connected components of the hard FSG are the (hard) segments of the image or dot pattern. These segments can be labeled with integers m, the numbers of the segments, using a graph traversal algorithm [14].
The algorithm can also be used for edge extraction. A "hard" edge is situated between two neighbored points P_k, $P_{k'}$ if the corresponding FSG nodes are not connected by a graph edge, i.e. if (1) is not fulfilled. It follows that there is always an edge if P_k and $P_{k'}$ belong to different segments. But there can also be (non-closed)

edges inside a segment. Furthermore, via the degree of similarity an edge strength $x_{k,k'}$ can be assigned to every connection between (Voronoi) neighbors.

The nonlinear difference equations (7) can be implemented with a discrete time neural network if a neuron is assigned to each node k. The smoothed feature value is the state of neuron k at recursion level t, and the original features f_k define the initial state of the network. Each neuron k is connected to its (Voronoi) neighbors. Therefore, the network is closely related to the discrete time version of a Cellular Neural Network originally introduced by Chua and Yang [10]. The essential difference is the used nonlinearity. In most neural networks including the CNN the nonlinearity acts on the state variables whereas in the network represented by (7) the nonlinearity acts on the differences of the state variables of neighbored neurons. Furthermore, in (7) the nonlinearity is not a constant function, as e.g. the sigmoid function, but it depends on the statistics of the input signal (via the $\langle \Delta f_k \rangle$).

The method now is applied to gray value segmentation, border point grouping and dot pattern clustering.

3 Grey value segmentation

Now the points P_k are given by the pixels (i,j), and the features f_k are the gray values $g_{i,j}$. The threshold (2) is given by

$$t_{i,j;i',j'} = \mu \cdot \mathrm{MIN}\{\langle \Delta g_{i,j} \rangle, \langle \Delta g_{i',j'} \rangle\} + \varepsilon \tag{12}$$

Experiments with various images have shown that $\mu = 1$ and $\varepsilon = 5$ are good parameter values for 8 bit images. For the window used for averaging the $\Delta g_{i,j}$ the size 11 × 11 is adequate. Further minor specifications can be found in [12]. With the similarity degree s_1 (10) the smoothing algorithm (7) is closely related to the discrete version of Perona - Malik anisotropic diffusion [16]. Some results are shown in figures 1- 4 (see also [17]).

It should be said that the same algorithm which was applied here to the original image gray values $g_{i,j}$ can also applied to other feature images, e.g. to the gradient image $|\nabla g_{i,j}|$ ([18]). Then e.g. the border of the noise in figure 1 can be found.

Fig. 1. Background with square and additive noise (S/N = 1)

a) original   b) smoothed image (35 iterations)   c) 2 biggest segments

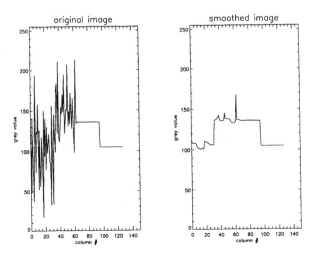

Fig. 2. Original and smoothed gray value profiles of figure 1

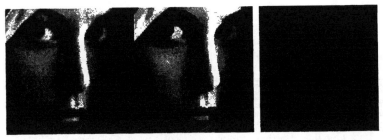

Fig. 3. Lena a) Original image b) smoothed image (76 iterations) c) Segment borders

Fig. 4. Berlin Television Tower a) Original b) Smoothed (62 iterations) c) Segment borders

4 Border point grouping

An image segment with 4-connectivity as defined above has a closed contour of border points. Here, for simplicity, only outer borders of area segments are considered, i.e. line segments are excluded. Then, each border point Q has two border point neighbors Q' and Q'' which are located in the 8-neighborhood of Q. Figure 5 shows a segment with gray marked 8-connected border points and assigned directions. The directions φ can have the values 0^0, 45^0, 90^0, 135^0, 180^0, 225^0, 270^0, and 315^0 as indicated by the arrows.

Fig. 5. Segment with border points (gray) and assigned directions

As feature vector the unit vector $\mathbf{n_k}^T = (\cos\varphi_k, \sin\varphi_k)$ is used. The distance between two 8-neighbored vectors $\mathbf{n_k}$ and $\mathbf{n_{k'}}$ is given by the square root of

$$\| \mathbf{n_k} - \mathbf{n_{k'}} \|^2 = 2(1 - \mathbf{n_k} \bullet \mathbf{n_{k'}}) = 2(1 - \cos\phi) \qquad (13)$$

where ϕ is the angle between vectors $\mathbf{n_k}$, $\mathbf{n_{k'}}$. After smoothing the smoothed normal vector $\langle \mathbf{n_k} \rangle$ is normalized to unit length. The following border examples have been processed with $\mu = 2$, $\varepsilon = 0.2$.

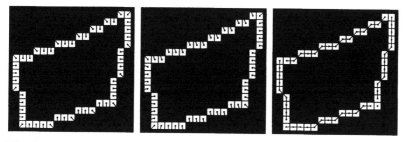

Fig. 6. Border point segmentation

a) Border points with directions b) Smoothed directions c) Feature similarity graph

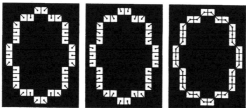

Fig. 7. Border point segmentation

These few examples show that because of smoothing corners can be detected properly (figure 6) and that "round" borders are not split into pieces (figure 7).

4 Dot pattern clustering

Here the features are the dot coordinates (x_k, y_k). The similarity of points P_k, $P_{k'}$ now is better called proximity (as in Gestalt theory) and indeed the clustering method can explain the Gestalt law of proximity [3 – 5]. The proximity criterion (1) can be written as

$$d(k,k') \leq \mu \cdot MIN\{d_k, d_{k'}\} + \varepsilon \qquad (14)$$

Here d_k is the Euklidean distance between dot P_k and its nearest neighbor, and $d(k,k')$ is the Euklidean distance between the points P_k and $P_{k'}$. Good values are $\mu = 2$ (see [3 – 5]) and $\varepsilon = 0$. In the results shown in figures 8, 9 the d_k are not averaged. See [5] for more details.

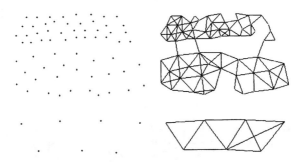

Fig. 8. Dot clusters and Feature Similarity Graph without smoothing (2 clusters)

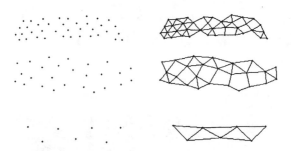

Fig. 9. Dot clusters and Feature Similarity Graph with smoothing (125 iterations, 3 clusters)

Figure 9 shows that because of the smoothing procedure the separation of the two upper clusters becomes possible. For other successful clustering results see [5].

5 Conclusions

It was shown that feature grouping can be carried out with CNN like processing structures and a properly defined graph, the Feature Similarity Graph. Of course, the presented solution is not a final one. It shows only a direction of possible future research. There is hope that other features can also dealt with the presented method (see [3] for a first approach related to texel orientation and size). To apply the method to texture segmentation adequate texel shape descriptors must be chosen and their similarity relations formulated.

References

1. H.-C. Nothdurft, "The Role of Features in Preattentive Vision: comparison of Orientation, Motion and Color Cues", Vision Res., Vol. **33**, pp. 1937 - 1958 (1993)
2. H. Jahn, "Image segmentation with a layered graph network", SPIE Proceedings, Vol. **2662**, pp. 217-228 (1996)
3. H. Jahn, "A graph structure for image segmentation", SPIE Proceedings, Vol. **3026**, pp. 198-208 (1997)
4. H. Jahn, "A graph structure for grey value and texture segmentation", Computing Suppl. **12**, pp. 73-82, Springer, Wien, 1998
5. H. Jahn, "Dot pattern clustering using a cellular neural network", SPIE Proceedings **3346**, pp. 298-307, 1998
6. J.M. Jolion, and A. Montanvert, "The Adaptive Pyramid: A Framework for 2D Image Analysis", CVGIP: Image Understanding **55**, pp. 339 - 348, 1992
7. W.G. Kropatsch, and S.B. Yacoub, "A revision of pyramid segmentation", Proc. of ICPR'96, pp. 477-481, 1996
8. P. Meer, "Stochastic Image Pyramids", CVGIP **45**, pp. 269 - 294, 1989
9. P. Nacken, "Image Segmentation by Connectivity Preserving Relinking in Hierarchical Graph Structures", Pattern Recognition **28**, pp. 907 - 920, 1995
10. L.O. Chua and L. Yang, "Cellular neural networks: Theory", IEEE Trans. on Circuits and Systems, Vol. **35**, pp. 1257 – 1272, 1988
11. T. Roska and J. Vandevalle, Eds., Cellular Neural Networks, J. Wiley&Sons, West Sussex, 1993
12. H. Jahn, "Image preprocessing and segmentation with a cellular neural network", SPIE Proceedings, Vol. **3304**, pp. 120-131, 1998
13. K. Voss, Discrete Images, Objects, and Functions in Zn, Springer, Berlin/Heidelberg, 1993
14. T. Pavlidis, Structural Pattern Recognition, Springer, Berlin, 1977
15. T. Lindeberg and B. M. ter Haar Romeny; "Linear Scale Space I: Basic Theory", in: B. ter Haar Romeny (Ed.), Geometry-Driven Diffusion in Computer Vision, pp. 1 - 38, Kluwer Academic Publishers, Dordrecht 1994
16. P. Perona, T. Shiota, and J. Malik, "Anisotropic diffusion", in: B. ter Haar Romeny (Ed.), Geometry-Driven Diffusion in Computer Vision, pp. 73 - 92, Kluwer Academic Publishers, Dordrecht, 1994
17. H. Jahn, A Neural Network for Image Smoothing and Segmentation, Lecture Notes in Computer Science **1451**, pp. 329-338, 1998
18. H. Jahn, W. Halle, "Texture segmentation with a neural network", SPIE Proceedings, Vol. **3646**, "Nonlinear Image Processing X", pp. 92 – 99, 1999

Frame-Relative Critical Point Sets in Image Analysis

Stiliyan N. Kalitzin, Joes Staal, Bart M. ter Haar Romeny, Max A. Viergever

Image Sciences Institute, Utrecht University

Abstract. We propose a new computational method for segmenting topological sub-dimensional point-sets in scalar images of arbitrary spatial dimensions. The technique is based on computing the homotopy class defined by the gradient vector in a sub-dimensional neighborhood around every image point. The neighborhood is defined as the linear envelope spawned over a given sub-dimensional vector frame. In the paper we consider in particular the frame formed by an arbitrary number of the first largest principal directions of the Hessian. In general, the method segments ridges, valleys and other critical surfaces of different dimensionalities.

Because of its explicit computational nature, the method gives a fast way to segment height ridges in different applications. The so defined topological point sets are connected manifolds and therefore our method provides a tool for *feature grouping*. We have demonstrated the grouping properties of our construction by introducing in two different cases an extra image coordinate. In one of the examples we considered the scale as an additional coordinate and in the second example, local orientation parameter was used for grouping and segmenting elongated structures.

Keywords: Point Sets, Topology, Image Analysis, Grouping

1 Introduction and related works

In this paper we introduce a topological quantity that characterizes the neighborhood of every pixel in scalar images of arbitrary spatial dimensions. This quantity is an integer number that can single out special points such as extrema and saddle points (all being critical points). It can also be generalized to define the membership of the point to extended sub-dimensional structures such as ridges or edges. We introduced the topological homotopy class number in [2, 3] in relation to its importance for the deep-structure image analysis and application to multi-scale segmentation. In its essence this number reflects the behavior of the gradient image vector in a close neighborhood around the given point. In the simplest one-dimensional case, the topological class of a point is defined as the difference of the sign of the signal derivative taken from both sides of the point. Clearly this number is zero everywhere except in the local extrema.

The proper generalization to higher dimensions is provided by the homotopy class $\pi_{D-1}(S^{D-1})$ that parameterizes the space of non-equivalent (non-deformable smoothly into each other) mappings between two $D-1$ dimensional spheres (D is the number of dimensions of the image). The mapping is defined by the normalized gradient vector taken on a closed surface (homotopic to a sphere) surrounding the test point. One can show that the set of non-equivalent mappings for $D > 1$ can be labeled with an integer number. This is particularly evident in the $D = 2$ example where the topological number reduces to the well-known winding number [4] indicating the number of times the gradient vector rotates around its origin when a point is circumventing the test point. Extrema points have winding number $+1$, saddle points are with winding number -1 and for regular points the number vanishes. A detailed discussion on the 2D case is presented in [2] and in the next section we give a concise summary of the construction as well as of its generalization to arbitrary image dimensions.

In addition to segmentation of critical points, the method is extended in this paper for localization of points lying on *relative critical sets* [1, 7]. To this end we introduce a *relative homotopy class* of a given test image point defined as the homotopy class calculated on a linear sub-space in the neighborhood of the test point. This sub-space can be defined as a linear envelope spawned over the vectors of a given vector frame field. A particularly important case is that of the frame formed by a subset of the eigenvectors of the Hessian (the matrices of the second image derivatives). In this case we obtain a constructive definition for *topological ridges*. We can classify such a point set by the number of the Hessian eigenvectors defining the subspace and by the signs of the corresponding eigenvalues. Alternatively we can define the relative critical sets by using for instance any globally defined (image independent) frame field.

Another interesting case is the hyper-dimensional critical set defined relative to the gradient vector itself and segmented by the zero-crossings of the second derivative in the direction of the gradient. This point set can be interpreted as a *topological edge*.

Critical points and various topological point-sets play an essential role in uncommitted image analysis as revealed in [9, 10]. These topological structures were studied in the context of multi-scale image analysis [8, 11]. They form a sort of a topological back-bone on which the image structures are mounted.

2 Homotopy numbers

In [2] we give self-contained definitions of our quantities as well as a detailed proof of their essential properties. Here we present the highlights of the construction and introduce some brief notations that provide the natural basis for the introduction of a topological number associated to any singular image point.

Suppose P is a point in the image and V_P is a region around P which does not contain any singularities (points where the gradient vector vanishes) except possibly P. We define a quantity characterizing the image in the surrounding of the point P. Let S_P be a closed hyper-surface, topologically equivalent to a $D - 1$ dimensional sphere, such that it is entirely in V_P and our test point P is inside the region W_P bounded by S_P. In other words, $P \in W_P : S_P = \partial W_P$.

At any **non-singular** point $A = (x_1, \ldots, x_D)$ we can define a $(D - 1)$ form:

$$\Phi(A) = \xi_{i_1} d\xi_{i_2} \wedge \cdots \wedge d\xi_{i_D} \epsilon^{i_1 i_2 \ldots i_D}, \xi_i = \frac{\partial_i L}{(\partial_j L \partial_j L)^{1/2}} \tag{1}$$

where $\epsilon^{i_1 i_2 \dots i_D}$ is the totaly antisymmetric tensor.

An important property of the $(D-1)$ form Φ is that it is a closed form, or $d\Phi(A) = 0$. This property ensures the correctness of the definition of the topological quantity

$$\nu(P) = \oint_{A \in S_P} \Phi(A). \tag{2}$$

The integral above is the natural integral of a $(D-1)$ form over a $(D-1)$ dimensional manifold without border.

If W is a region where the image has no singularities, then the form Φ is defined for the entire region W. After applying the generalized Stokes theorem and using the fact that Φ is closed, we obtain that the topological quantity (2) is identically zero for non-singular points. Therefore the topological number (2) can be used to localize the set of singular points in the image.

For one-dimensional signals $L(x)$, the topological number (2) of a point $P = x$ is ultimately simple: $\nu_P \equiv \text{sign}(L_x)_B - \text{sign}(L_x)_A$ for any $A, B : A < P < B$ in the close vicinity of P. In other words, the topological number of a point is reduced to the difference between the signs of the image derivative taken from the left and from the right. Obviously, $\nu_P = 2$ for local minima, $\nu_P = -2$ for local maxima and $\nu_P = 0$ for regular points or in-flex singularities. For two-dimensional images the topological number 2 labels the equivalent class of mappings between two unit circles (see the Introdiction). This label also is known as the winding number. Clearly, the winding number of any closed contour must be an integer multiple of 2π. It is possible to show, see [2], that in extremal points (minima or maxima) the winding number is $+1$, in non-degenerate saddle points it is -1 and in regular points it is 0. In degenerate saddle points, named also as "monkey saddles" the topological number is $-n+1$ where n is the number of ridges or valleys converging to the point.

3 Critical sets relative to vector sub-frames

3.1 General construction

The general idea is to select at every image point P a *local linear sub-space* K_P and to project the gradient vector L_i onto it:

$$L_\alpha(x) = h_\alpha^i(x) L_i(x). \tag{3}$$

Here $h_\alpha^i, \alpha = 1, \dots, D_K$ are the local frame vectors defining the linear sub-space K with dimension $D_K < D$.

The next step is to compute the *relative homotopy number* $\nu(K)$ in point $P \in R^D$ in analogy with (2). where now the closed surface S_P around P is of dimension $D_K - 1$ and lies in the sub-space $K_P \equiv LE(h_\alpha(P))$ where $LE()$ stands for linear envelope. The $(D_K - 1)$ form Φ is computed as in (1) but from the vector field 3. As we already mentioned, the local linear space K_P is defined as the linear envelope over the vectors $h_\alpha(P), \alpha = 1, \ldots, D_K$ taken in the test pont P. We assume that the set of vectors h_α is of maximal rank D_K.

Now we define our central construct, the topological critical set, relative to the sub-frame h.

Definition 1. Let $h(x)$ be a local non-degenerate sub-frame of dimension (rank) D_K and let the local linear envelope of this frame be K_x. The point set

$$PS(h) : \{x; x \in R^D, \nu_x(K_x) \neq 0\} \tag{4}$$

is the relative critical point set (RCPS) associated to the local frame h.

Clearly definition 1 implies $L_\alpha(P) = 0$ when $P \in PS(h)$. Therefore our RCPS is equivalent to those defined in [1,7].

It is clear that if $D_K = D$, or in other words if the local sub-frame is complete, the relative homotopy class is just the full homotopy class. The manifold defined in definition (1) will then be of dimension 0. In fact this set is the set of all critical points in the image.

In addition to the feature that the topological number (2) is non-zero, the RCPS can be characterized by the value of this number and eventually by some characteristic properties of the sub-frame h.

Now we address the question of the *local topological structure* of the RCPS associated with an arbitrary sub-frame h_α^i. We show in what follows that:

Proposition 2. *If point P belongs to the relative critical point set from definition (1) and if the Hessian in P is non-singular, then the point set is locally isomorphic to a linear space of dimension $D - D_K$.*

The following proof uses some techniques from differential geometry that provide the adequate covariant formalism.

Proof. Let the sub-frame $h_a^i, a = D - D_K + 1, \ldots, D$ be chosen in such a way that the system of vectors h_α, h_a represents a complete frame in the D-dimensional space of the image and $h_a^i h_\alpha^i = 0$ (summing over all repeated indices is assumed) for all a, α from the corresponding ranges. If the point P belongs to the RCPS defined by the sub-frame h_α then obviously $L_\alpha \equiv h_\alpha^i \partial_i L(P) = 0$. If P has coordinates x^i then the infinitesimally close point with coordinates $x^i + \delta x^i$ will belong to the same RCPS if $\delta x^A \nabla_A L_\alpha = 0$ where the index A takes all values of $1, \ldots, D$; $\delta x^A \equiv (h^{-1})_i^A \delta x^i$ and ∇_A is the *covariant derivative* induced by the frame h_A (such that $\nabla_A h_B = 0$). Applying these notations we obtain the following equation for the variation δx^i:

$$\delta x^i h_\alpha^j \partial_i \partial_j L(P) = 0. \tag{5}$$

Using the completeness of the system h_A and the orthogonality between h_a and h_α we obtain:

$$\delta x^i \partial_i \partial_j L(P) = C_a h_a^j, \tag{6}$$

where C_a are $D - D_K$ arbitrary constants. If the Hessian $H_{ij} \equiv \partial_i \partial_j L(P)$ is non-singular it can be inverted giving the following form of the RCPS coordinate variations:

$$\delta x^i = (H^{-1})^{ij} h_a^j C_a. \tag{7}$$

Clearly the system $g_a^i \equiv (H^{-1})^{ij} h_a^j$ has rank $D - D_K$ as it is obtained by a non-singular linear transformation from the sub-frame h_a. Therefore, the variations in eq. (7) belong to the $D - D_K$ dimensional linear envelope over the sub-frame g_a.

The above proposition motivates the introduction of RCPS as topological manifolds. It is clear from the proof that these manifolds are continuous for all points where both the image Hessian and the sub-frame are defined and non-singular.

   In the following sub-section we give some important examples of RCPS associated with a subset of the principal eigenvectors of the local Hessian.

3.2   Definition and detection of topological ridges

Let the local Hessian field $H_{ij} \equiv \partial_i \partial_j L(x)$ have eigenvalues $\lambda_1, \ldots, \lambda_D$ with corresponding eigenvectors $h_1, \ldots, h_D$. We can assume that the eigenvalues are labeled in decreasing order of their absolute values: $|\lambda_1| > \cdots > |\lambda_D|$. There are different ways to select a sub-set of eigenvectors to form the sub-frame h in the definition 1. In what follows the following definition is most suitable for the interpretation of the RCPS as height ridges and their generalizations.

Definition 3. A topological ridge set $R^{(m_+, m_-)}(L)$ of co-dimension $m_+ + m_- \equiv D_K$ is defined as a RCPS associated with the Hessian eigenvectors $h_1, \ldots, h_{D_K}$ corresponding to the D_K largest by absolute value of the eigenvalues $\lambda_\alpha, \alpha = 1, \ldots, D_K$. Excluded are points where the Hessian is degenerate so that there is no unique set of λ_α. In the so defined point set only those points are included where there are exactly m_+ positive and m_- negative Hessian eigenvalues.

This definition puts a natural label on the topological ridge. For example if $m_+ = 0, D_K = m_-$ we obtain a classical height ridge, if $m_- = 0, D_K = m_+$ we have a valley. In the general cases where $m_+ \neq 0; m_- \neq 0$ we can talk of "saddle ridges". The definition extends the $D_K = D$ case where the ridge is of dimension zero. Then we have the possibilities of maxima, minima and saddle points of different signature. Note that the classification implied by definition 3 is richer than the one induced from the value of the topological number alone. For non-degenerate critical points the homotopy class is equal to the sign of the determinant of the Hessian (see detailed proof in [2]). Obviously a lot of different signatures discriminated by definition (3) will have the same homotopy number.

   As a consequence from proposition (2), the topological ridges are manifolds of dimension $D - D_K$. In this case the vectors h_α are eigenvectors of the Hessian and we can choose the frame h_a, h_α from the proof (2) to be the entire system of eigenvectors of H_{ij} (and therefore also of H^{-1}).

In this case we have $H_{ij}h_\alpha^j \equiv \lambda_\alpha h_\alpha^i$ for all $\alpha = 1, \ldots, D_K$ and therefore equation 5 takes the form

$$\lambda_\alpha \delta x^i h_\alpha^i = 0, \text{ no summation over } \alpha. \tag{8}$$

It is easy to see that if $\lambda_\alpha \neq 0$ for all α then eq. (7) now takes the simpler form:

$$\delta x^i = h_a^i C_a. \tag{9}$$

Therefore, as an addition to the general RCPS properties, the topological ridges are locally *orthogonal* to the vector system h_α.

A different definition may include the eigenvectors corresponding to the first D_K eigenvalues of the Hessian ordered by their *signed value*. Such scheme is extensively studied in [7]. The corresponding classification in that case will partially overlap with ours and namely for the cases of $R^{m,0}$ and $R^{0,m}$ RCPS (strict valleys or ridges). For the mixed signatures the two schemes will segment out different topological manifolds. We note that our general constructive approach from definition (1) can be used for both of the definitions. A comparison of the different vector frame field choices and their relations will appear in a forthcoming publication.

A direct consequence of the definition of topological ridges as relative critical sets is the relation:

$$m^+ \geq m'^+; m^- \geq m'^- \rightarrow R^{(m_+, m_-)}(L) \in R^{(m'_+, m'_-)}(L). \tag{10}$$

The importance of this inclusion is discussed in the next section where it provides a way for detection of both critical lines and their annihilation points in one-dimensional signals. Another practical application includes detection of the optimal scale for elongated image structures and simultaneously establishing a link to the finest scale.

For completeness we finally include the system of topological edges defined as the point set where the quantity $\eta(x) = \text{sign}(\xi^i \xi^j H_{ij}(x + \rho\xi)) - \text{sign}(\xi^i \xi^j H_{ij}(x - \rho\xi))$ takes value -2 for some small ρ. This point sets are nothing else but the (negative) zero-crossings of the second image derivative in the direction of the gradient vector ξ.

4   Examples and applications to image processing

We first give a summary of the possible RCPS configurations in the different signal dimensions. In the simples 1D case we have only two possible sub-sets, the local minima and the local maxima.

In two dimensions, the possible cases are given in table 1.

Finally the 3D case is classified in the table 2: Scipping the trivial 1D case, our first example uses a 2D MRI image. The complete relative critical set system is given in figure 1. Note the inclusion relations between the different RCPS.

m_+	0	1	2
m_-			
0	regular	negative ridge	minimum
1	positive ridge	saddle	x
2	maximum	x	x

Table 1. The possible point sets for two dimensional signals. An 'x' stands for an impossible configuration.

m_+	0	1	2	3
m_-				
0	regular	negative surface	negative string	minimum
1	positive surface	saddle string	saddle point	x
2	positive string	saddle point	x	x
3	maximun	x	x	x

Table 2. The possible point sets for three dimensional signals. An 'x' stands for an impossible configuration.

The second illustration of our method uses also the 2D technique but instead of an original 2D signal, we consider a scale-space built over a one-dimensional signal $L(x)$.

$$L(x,t) = \frac{1}{N_t} \int dx' e^{\frac{-(x-x')^2}{4t}} L(x'), \qquad (11)$$

where t is the scale and N_t is a normalization factor. Next we have taken the second natural derivative $t\partial_x{}^2 L(x,t)$ for the reason that we will comment shortly.

We can now localize the ridges $R^{1,0}$, $R^{0,1}$ as well as the 2D critical point sets $R^{2,0} \cup R^{0,2}$ (this is the set of all extrema) and $R^{1,1}$ (the set of all saddle points). We see that the inclusion property 10 assigns the set of saddle points as the set of annihilation points for the 1D critical trajectories (ridges and valleys from a 2D point of view). Another appearance of the inclusion relation we find the extremal points lying over the ridges. We can interpret these points as scale selectors in the sence that they indicate local extrema of the second natural derivative of the signal both in space, x, and scale, t, directions. If we have instead chosen just the second signal derivative, no extrema would have been found as a consequence of the smoothening property of the transformation (11).

As a 3D example we consider two different superstructures built over a 2D image given in the first frame of figure 3. In one of the computations we take the 2D analog of the scale space (11) of the natural Laplacian, just as in the 1D

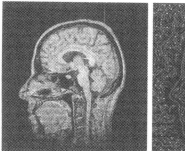

Fig. 1. *Left frame:* An MRI 256x256 sagittal image of a human head. *Right frame:* Complete set of 2D ridges (red) and valleys (blue), edges (green) and singular points (extrema: yellow, saddle points: purple). The calculations are done for a spatial scale of 2 pixels.

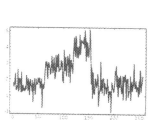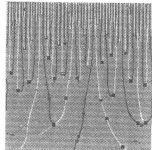

Fig. 2. *Left frame:* One dimensional signal of 256 data points. *Rigth frame:* RCPS of the scale-space of the second natural derivative of the signal. White corresponds to ridges, black to valleys, blue to saddle points, green to maxima and purple to minima.

case, and as a second 3D signal we consider the orientation bundle

$$F(x, y, \theta) = \int dx' dy' \Phi(x - x', y - y', \theta) L(x', y') \qquad (12)$$

where $\Phi(x, y, \theta)$ is any orientation filter. For the example in figure 3 we have chosen for an anisotropic Gaussian kernel with scale ratio of 10 along the selected orientation θ.

In figure 3 we have depicted only the most interesting for this application case of positive strings relative to a suitably chosen two-dimensional vector frame. We used two different frame-fields in detecting the RCPS.

In the scale-space case (bottom left frame) one of the frame vectors was selected to point always in the direction of the scale (z-axis) and the second vector was selected as the eigenvector corresponding to the highest eigenvalue of the 2D Hessian. In the orientation space x, y, θ the vector frame was chosen as explained in figure 3. We clearly see that strings in the scale space construction represent elongated objects "lifted" at their optimal scale. In the case of orientation bundle, the elongated objects (blood vessels in the examples) are segmented at their optimal orientation (the vertical dimension). Crossing lines thus become separated as their local orientations at the crosspoint are different.

In the last example we used fully the flexibility of our definition (1). While for genuine 3D signals a vector frame field is provided naturally by the system of the Hessian eigenvectors, for 3D signals obtained by some pre-processing constructions a specific frame field can be more suitable for the interpretation of the results.

5 Conclusions

In the present paper we proposed a constructive definition of relative critical sets in images of any spatial dimensions. The definition is very flexible because it associates critical sets to an arbitrary chosen local vector frame field. Depending on the visual task, different model structures can be identified with the relative critical sets. As a consequence our construction can be purely intrinsic (defined only by the image structures), or it can involve externally specified frames. The last situation may be useful for involving additional model information. In this paper we demonstrated the examples of two of the most-popular intrinsic cases: the sets of ridges and edges.

Acknowledgments

This work is carried out in the framework of the NWO research project STW/4496.

References

1. D. Eberly, R. Gardner, B. Morse, S. Pizer, and C. Scharlach. Ridges for image analysis. Submitted to Journal of Mathematical Imaging and Vision, July 1993, 1993.
2. S. N. Kalitzin, B. M. ter Haar Romeny, A. H. Salden, P. F. M. Nacken, and M. A. Viergever. Topological numbers and singularities in scalar images. scale-space evolution properties. *Journal of Mathematical Imaging and Vision*, 9, 1998, pages 253-269
3. S. N. Kalitzin, B. M. ter Haar Romeny, and M. A. Viergever. On topological deep-structure segmentation. In *Proc. Intern. Conf. on Image Processing*, pages 863–866, Santa Barbara, California, October 26-29 1997.
4. M. Kass, A. Witkin, and D. Terzopoulos. Snakes: Active contour models. In *Proc. IEEE First Int. Comp. Vision Conf.*, 1987.
5. T. Lindeberg. *Scale-Space Theory in Computer Vision*. The Kluwer International Series in Engineering and Computer Science. Kluwer Academic Publishers, Dordrecht, the Netherlands, 1994.

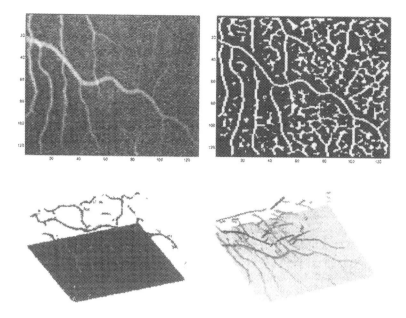

Fig. 3. *Top Left frame:* Vessel structure of the optical image of human retina. *Top Right frame:* Two-dimensional ridge structure defined by the RCPS $R^{(0,1)}$. *Bottol Left frame:* Scale space strings (green) of the natural Laplacian of the image. We fixed one vector of the local frame field in the direction of increasing scale. *Bottom Right frame:* Orientation space strings (blue lines) obtained as RCPS in the orientation bundle (12). The 2D frame field is taken to be orthogonal to the vector $(\cos(\theta),\sin(\theta),0)$ where θ (measuring the vertical axis) is the angle argument of the orientation filter in (12).

6. M. Morse and S. S Cairns. *Critical Point Theory in Global Analysis and Differential Topology.* Academic Press, New York and London, 1969.

7. J. Damon. Generic Structure of Two Dimensional Images Under Gaussian Blurring. To appear in SIAM Journal of Applied Mathematics 1998.

8. Peter Johansen. On the classification of toppoints in scale-space. *Journal of Mathematical Imaging and Vision,* 4(1):57–68, 1994.

9. J. J. Koenderink. The structure of images. *Biological Cybernetics,* 50:363–370, 1984.

10. T. Lindeberg. Scale-space behaviour of local extrema and blobs. *Journal of Mathematical Imaging and Vision,* 1(1):65–99, March 1992.

11. O. F. Olsen. Multi-scale watershed segmentation. In L. M. J. Florack John Sporring and Peter Johansen, editors, *Classical Scale-Space Theory.* Kluwer Academic Publishers, 1997. In press.

A Fast Traffic Sign Recognition Algorithm for Gray Value Images

C. Schiekel

Forschungsinstitut für anwendungsorientierte Wissensverarbeitung (FAW)
Helmholtzstr. 16, D-89081 Ulm, Germany
email: schiekel@faw.uni-ulm.de

Abstract. A fast shape recognition method is presented for detecting traffic signs in monochrome images. Based on Marr's *primal sketch* we used a hierarchical grouping method to link low-level features of pixels like gradient orientations to high-level features like triangles and ellipses. The method is fast, because it is based on local pixel operations instead of analyzing the whole shape. Examining the local area around a pixel for parts of geometrical shapes, results are symbolic representation which are linked back to the pixel, i.e. the pixel is part of the geometrical shape. A unique description of the geometrical shape is calculated from pixels lying on the shape.
Key words: traffic sign recognition, grouping, shape detection

1 Introduction

In this paper we provide a contribution to fast shape recognition illustrated in the field of traffic sign recognition. We present an algorithm which detects shapes of German traffic signs in monochrome images.

1.1 Motivation

In recent years a great variety of research has been done in the field of traffic sign recognition. One problem of the existing approaches is that the robustness of traffic sign recognition algorithms in twilight is not sufficient. This problem is one of camera sensitivity. While in daylight images being recorded with a CCD camera provide good information about traffic signs, twilight makes great demands on the camera. Using a color camera there is too few information available, because the color information is not sufficiently present in dark images. Using a standard gray value CCD camera is no solution, because the contour of the traffic signs is less visible. The new *high dynamic range CMOS* (HDRC) gray value camera significantly improves the image quality in traffic situations with high brightness changes, i.e. hard shadow, entrance of tunnels and back lighting. In our approach we detect a traffic sign in an image by a hierarchical edge pixel grouping method. The algorithm is fast, because methods are based on grouping pixels instead of grouping lines or parts of a shape (see also [7]).

1.2 Related work

Most of the published work uses color information to segment the traffic signs in the image. Few methods use monochrome information.

A classical approach is pursued by [8]. Based on edge images various methods are used to detect triangular and round shapes. With the generalized Hough Transformation [1][8] the processing time is 15 sec for a 256x256 sized image on a SparcStation ELC.

A traffic sign recognition system using both monochrome and color information has been published ([3][5][9][11]). Regions including shape information are extracted from the hierarchical structure code [4] and color information from the color structure code [9] which are classified as round or polygonal shape. The system was implemented on an expensive multiprocessor system with four transputers T805 and four PowerPC 601 allowing 200ms cycle time ([5]).

Seitz et al. describe [10] a hierarchical feature matching algorithm, which is based on a representation of the scene by local orientations. A hierarchical template matching algorithm compares predefined templates with the gradient orientation image to locate the searched sign. The main difference between our approach and the method from Seitz is that Seitz uses templates to find edge pixels of a sign while our approach groups edge pixels. The advantage is more flexibility with respect to different sized shapes.

Previously published approaches using monochrome information don't meet real-time demands when run on standard hardware.

The paper is organized as follows. In section 2 the algorithm for detecting traffic signs is described. The results are presented in section 3. A conclusion and an outlook are given in the last section.

2 Method

The presented approach for detecting traffic signs in gray value images exploits shape characteristics. Traffic signs in Germany have different shapes. For example speed limit signs are circular while warning signs are triangular (Fig. 1 a)). To detect traffic signs, the different shapes of the traffic signs have to be detected in the gradient magnitude image.

The algorithm is able to detect ellipses, triangles and rectangles depending on the grouping constraints for the shape provided to the algorithm. In this paper we present the grouping constraints for the triangle shape.

For detecting a shape in the image, the center point of the shape and a unique description of the shape are needed. The grouping constraints for equilateral triangular shapes (Fig. 3 a)) are summarized as follows. We identify pixels which lie on the boundary of the triangle and which are as close as possible to corners. Later these pixels serve as corner approximations (Fig. 1 b)).

The remainder of this section is organized as follows. In section 2.1 a preprocessing task is described, where edge pixels are selected. Only those edge pixels are segmented which lie on a triangular shape. This procedure is described in

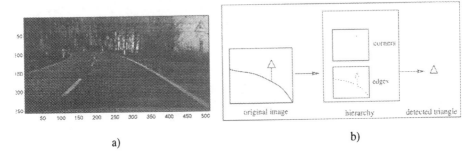

a)   b)

Fig. 1. a) Example of a triangle traffic sign pictured with the HDRC camera. b) Schematic overview of the triangle detection process

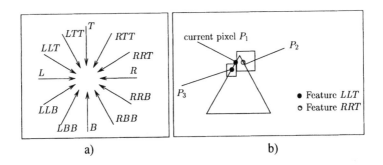

a)   b)

Fig. 2. a) This sketch shows the relationship between gradient orientations and labels. The gradient orientations are shown by arrows. b) The sketch shows an example how the edge pixels are grouped around a current pixel to decide if the current pixel P_1 is close to the upper corner of a triangle. The edge pixels have to have special features, namely *LLT* and *RRT* respectively

section 2.2. In section 2.3 the coordinate of the center of the shape and the shape parameters are computed, which uniquely describe the shape.

2.1 Preprocessing

In the preprocessing stage edge pixels are determined and an initial symbolic label is linked to edge pixels depending on their gradient orientation.

The original image $I(x,y)$ is therefore convoluted with two Sobel filters S_X, S_Y resulting in the gradient magnitude $G(x,y)$ and the gradient orientation $G_o(x,y)$ [6]. Edge pixels are identified by threshold the magnitude image. Due to noise, isolated edge pixels are thinned with the *non-maxima suppression* method [2]. The directions of the gradient are quantized into $\{-5/6\pi, -2/3\pi, \ldots, \pi\}$ and a value for the background. Each orientation is labeled with a symbolic representation being shown in Fig. 2 a). The output image is called $G_{out}(x,y)$.

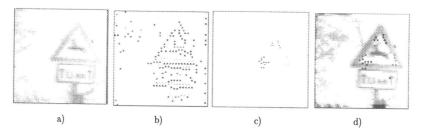

a) b) c) d)

Fig. 3. In a) a clipping of the warning sign is shown. In b) and c) the hierarchy of the warning sign, the first level and the last level respectively are shown. The original image a) overlayed with the segmented edge pixels c) is shown in d). Increasing gray value of the segmented pixels indicate different features. In this example three different features upper corner, left corner and right corner of a triangle have been found

2.2 Segmentation

In this subsection an algorithm is described being segmented edge pixels, which lie on an upward equilateral triangle corresponding to the perimeter of the warning signs in Germany.

The pixels are segmented by two steps that are organized hierarchically. At each level of the hierarchy an image of edge pixels is generated, with higher levels corresponding to larger edge segments. For the two step hierarchy of triangle identification, the lower level identifies the edge feature, i.e. the preprocessed image, while the upper level then identifies the corner feature for each pixel. Therefore the feature of one pixel is modified from level to level.

The quintessence of the method is to decide on which segment of a shape the pixels lie. Therefore pixels are grouped locally. Starting by a current pixel the relative position of the other pixels to the current pixel is the basis of the decision on which shape the current pixel lies on (see Fig. 2 b)). Depending on the label of the current pixel a local area is determined by a predefined set of rectangles and vectors (see table 1). The upper left corner of each rectangle is positioned relatively to the current pixel by a vector. The origin of the coordinate system in the image is the upper left corner of the image with increasing x-axis to the right and increasing y-axis to the bottom. In each rectangle pixels are searched with a special label (see table 1). If this search is successful then the current pixel and the founded pixels are grouped together and are represented by a pixel being linked to a new label and is set at the same position as the current pixel in the next level. The new label is a representation of the segment of the shape, on which the current pixel lies. The new labels for an upward equilateral triangle are UC, LIC and RIC which means upper corner, lower left corner and lower right corner, respectively.

An example of the grouping constraints of an upward equilateral triangle is given in table 1. The constraints for one segment of a shape are given in one row of the table.

The algorithm is described below. The name of the main function is *Find_Pixel_-On_Shape*. The input parameters of the function are the preprocessed image, grouping constraints of the geometric shape and the number of levels of the hierarchy. The output parameter is an image containing the segmented edge pixels. An example of the hierarchy and of the segmented pixels are shown in Fig. 3.

Find_Pixel_On_Shape
```
INPUT:

  - the image G_out from the preprocessing step
  - the grouping constraints of the shape
  - number of levels of the hierarchy

BEGIN
set the first level to G_out
FOR every hierarchy level DO
  initialize the next hierarchy level
  FOR ALL pixels in the current level DO
    determine the feature of the current pixel
    FOR ALL segments of the shape with the feature of the...
        ...current pixel in the grouping constraints DO
      determine the grouping constraints for the current segment...
        ...of the shape, i.e. the set of rectangles ...
        ...and features to be searched for
      set counter i to 0
      FOR ALL rectangles DO
        determine a set F containing all feature of pixels...
          ...being inside the current rectangle
        IF the searched feature is in F THEN
          increase counter i
        END
      END
      IF the counter i is equal to the number of rectangles...
          ...for the current segment of shape THEN
        link the new feature to the current pixel
        set the current pixel in the next level
      END
    END
  END
  set the current level to the next level
END
END
OUTPUT:

  - the last hierarchy level
```

Table 1. The grouping constraints of an upward equilateral triangle

| current label | Local area/rectangles | | | new label |
	relative position	size	searched label	
{LLT}	(−4; 0)	(4; 4)	{LLT}	UC
	(0; −4)	(6; 6)	{RRT}	
{LLT}	(0; −4)	(4; 4)	{LLT}	LIC
	(−4; 0)	(6; 6)	{B}	
{RRT}	(0; 0)	(4; 4)	{RRT}	UC
	(−6; −4)	(6; 6)	{LLT}	
{RRT}	(−4; −4)	(4; 4)	{RRT}	RIC
	(−4; 0)	(6; 6)	{B}	
{B}	(−6; −4)	(6; 6)	{B}	RIC
	(−4; −6)	(6; 6)	{RRT}	
{B}	(−4; −6)	(6; 6)	{LLT}	LIC
	(0; −4)	(6; 6)	{B}	

2.3 Shape Parameter

In this subsection, the center point of the sign and the parameter of the shape are calculated from the position of the segmented pixels. The parameters are needed to describe the shape uniquely.

For example the equilateral triangle has one parameter, that is the length of one of the three edges. To calculate the length of an edge and the position of the center point of the triangle, two of the three corners of the triangle are needed, for instance the upper and the left corner. Starting with a pixel labeled with feature UC, the left corner is determined as follows. A local area is defined by a rectangle being positioned relatively to the current pixel (Fig. 4, left image). The upper right corner of the rectangle has the same position as the current pixel. Within the rectangular area, pixels labeled as LIC meaning *lower left corner* are searched for. The position of the left corner is determined by the mean value. It is calculated from the coordinates of all pixels inside the rectangle having the feature LIC. The parameter of the triangle and the coordinates of the center point are easy to calculate geometrically. According to this strategy, the right corner and the upper corner are determined (Fig. 4).

To distinguish between the segmented pixels of two different triangles which are not superimposed, the parameters are clustered by the position of the center points of the triangles. Each cluster contains the parameters of one triangle.

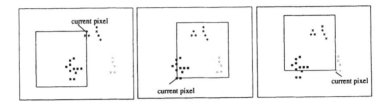

Fig. 4. The three images show how the rectangle is positioned relatively to the current pixel, in order to find the other corner. Increasing gray values mean different features, i.e. bright gray means *RIC*, middle gray means *UC* and dark gray means *LIC*

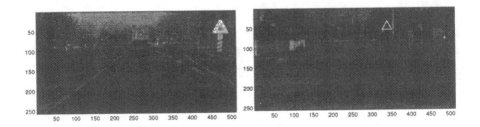

Fig. 5. Two examples of triangular traffic sign detection

Two mean values are computed for each cluster, these are the center point and the shape parameter.

3   Experimental Results

The algorithm has been evaluated on sequences recorded by the Robert BOSCH GmbH with the HDRC-camera being developed in the project BASTA. The camera was fixed behind the windscreen of a car. The sequences were taken from country roads. The images are sized up to 512 by 256 pixels and the gray value resolution is 8 bit.

We analyzed 66 sequences containing different traffic signs. Each sequence has between 200 and 1400 images. 22 sequences contain warning signs, 40 sequences contain speed limit signs and 6 sequences contain city limit signs.

We evaluated the algorithm with the triangular, elliptical and rectangular model. The recognition rate for triangular and elliptical traffic signs is 95% and for rectangular traffic signs is 93%. The detection rate of city limit signs is less, because the sign background discrimination is less distinct than for instance for speed limit signs. In Fig. 5 two examples of triangluar shape detection are shown.

Determining the false detection rate is difficult. In gray valued real images there are many objects of triangular, elliptical and rectangular shape. For instance branches of trees or advertisement signs can look like triangles or circles. Therefore without a classification of the detected objects determining the false detection rate is not useful.

The hardware we use is a standard hardware like the pentium II 333 MHz. This allows for one shape to run one cycle within 200 ms per image. For every further shape additionally 80 ms are needed.

4 Conclusion

The presented algorithm is able to detect different geometrical shapes in gray value images with real-time capabilities. The presented results show, that the algorithm is usable for online traffic sign recognition in gray value images.

5 Acknowledgment

The work is done in the project BASTA, sponsored by the *Bundesministerium für Bildung und Wissenschaft, Forschung und Technologie* under the co-ordination of the Robert Bosch GmbH. The Robert BOSCH GmbH friendly provided the image sequences.

References

1. D. H. Ballard. Generalizing the hough transform to detect arbitrary shapes. *Pattern Recognition*, 13(2):111–122, 1981.
2. H. Bässmann and Ph. W. Besslich. Konturorientierte Verfahren in der Bildverarbeitung. *Springer Verlag*, 1989.
3. B. Besserer, S. Estable, and B. Ulmer. Multiple knowledge sources and evidental reasoning for shape recognition. *Proceedings of 4th ICCV, Los Alamitos, CA*, pages 624–631, 1993.
4. U. Bücker, H. Austermeier, G. Hartmann, and B. Mertsching. Verkehrssszenenanalyse in hierarchisch codierten Bildern. *Mustererkennung 15. DAGM-Symposium, Lübeck, September*, 15:694 – 701, 1993.
5. S. Estable, J. Schick, F. Stein, R. Janssen, R. Ott, W. Ritter, and Y. J. Zheng. A real-time traffic sign recognition system. *Proceedings of intelligent vehicle, Paris October*, pages 213–218, 1994.
6. Bernd Jähne. Digitale Bildverarbeitung. *Springer Verlag*, 1997.
7. T. Kämpke, C. Schiekel, and M. Schuster. Nonparametric Multilevel Thresholding. *International Conference Signal and Image Processing, Las Vegas, Nevada Oct*, 1998.
8. G. Piccioli, E. De Micheli, and M. Campani. A robust method for road sign detection. *Computer Vision ECCV*, 1994.
9. L. Priese, J. Klieber, R. Lakemann, and V. Rehrmann. New results on traffic sign recognition. *Proceedings of the Intelligent Vehicles Symposium Paris October*, pages 249 – 254, 1994.
10. P. Seitz, G. K. Lang, B. Gilliard, and J. C. Pandazis. The robust recognition of traffic signs from a moving car. *Mustererkennung 13 - DAGM Symposium*, 13:287 – 294, 1991.
11. Y. J. Zheng, Werner Ritter, and Reinhard Janssen. An adaptive system for trafic sign recognition. *Proceedings of intelligent vehicle, Paris October*, pages 165–170, 1994.

A Geometric Approach to Lightfield Calibration

R. Koch[1], B. Heigl[2], M. Pollefeys[1], L. Van Gool[1], and H. Niemann[2]

[1] Center for Processing of Speech and Images (PSI), K.U.Leuven, Belgium
[2] Lehrstuhl für Mustererkennung, Universität Erlangen-Nürnberg, Germany
email: Reinhard.Koch@esat.kuleuven.ac.be, heigl@informatik.uni-erlangen.de

Abstract. Lightfield rendering allows fast visualization of complex scenes by view interpolation from images of densely spaced camera viewpoints. The lightfield data structure requires calibrated viewpoints, and rendering quality can be improved substantially when local scene depth is known for each viewpoint. In this contribution we propose to combine lightfield rendering with a geometry-based structure-from-motion approach that computes camera calibration and local depth estimates. The advantage of the combined approach w.r.t. a pure geometric structure recovery is that the estimated geometry need not be globally consistent but is updated locally depending on the rendering viewpoint. We concentrate on the viewpoint calibration that is computed directly from the image data by tracking image feature points. Ground-truth experiments on real lightfield sequences confirm the quality of calibration.

1 Introduction

There is an ongoing debate in the computer vision and graphics community between geometry-based and image-based scene reconstruction and visualization methods. Both methods aim at realistic and fast rendering of 3D scenes from image sequences.

Geometric reconstruction approaches generate explicit 3D scene descriptions with polygonal (triangular) surface meshes. A limited set of camera views of the scene is sufficient to reconstruct the 3D scene. Texture mapping adds the necessary fidelity for photo-realistic rendering to the object surface.

Image-based rendering approaches like lightfield rendering [14] and the lumigraph [6] have lately received a lot of attention, since they can capture the appearance of a 3D scene from images only, without the explicit use of 3D geometry. Thus one may be able handle scenes with complex geometry and surface reflections that can not be modeled otherwise. Basically one caches all possible views of the scene and retrieves them during view rendering.

Both approaches have their distinct advantages and weak points. In this contribution we discuss the combination of image-based rendering with a geometric structure-from-motion approach to obtain lightfields from image sequences of a freely moving camera. The necessary camera calibration and local depth estimates are obtained with the structure-from-motion approach. We will first give a brief overview of image-based rendering and geometric reconstruction techniques. We will then focus on the calibration problem for lightfield acquisition from hand-held camera sequences. Experiments on lightfield calibration and geometric approximation conclude this contribution.

2 Image-based rendering

Image-based rendering techniques allow to capture a scene with a principally unlimited geometric complexity, with complex lighting and specular surface reflections. The view rendering depends only on the efficiency of data access and not on the scene complexity, hence rendering in constant time is possible. The price to pay for this advantage is a very high amount of data and a tedious image acquisition. In fact, one has to obtain the plenoptic function of the scene space with viewing rays in all possible positions, which is a 5-dimensional function. Perfect rendering is possible only if all viewing rays of a newly rendered view intersect the focal centers of originally acquired views. Interpolation between viewpoints will cause a distortion that is dependent on the scene geometry as well. The amount of views to be acquired is limited by the storage requirements, since a dense view sampling of a scene might easily generate Gigabytes of image data. Therefore one must try to compress the data efficiently by removing the inherent redundancy. Since the approach is strictly image-based, no viewpoint extrapolation is possible. Furthermore the geometry is encoded implicitly in the data and there is no way to change geometric scene properties e.g. for animations.

Recently two equivalent realizations of the plenoptic function were proposed in form of the lightfield [14], and the lumigraph [6]. They handle the case when we observe an object surface in free space, hence the plenoptic function is reduced to four dimensions (light rays are emitted from the 2-dimensional surface in all possible directions). The 4-D lightfield data structure employs a two-plane parameterization (see fig. 1). Each light ray passes through two parallel planes with plane coordinates (s, t) and (u, v). Thus the ray is uniquely described by the 4-tuple (u, v, s, t). The (s, t)-plane is the *viewpoint plane* in which all camera focal points are placed on regular grid points. The (u, v)-plane is the *focal plane* where all camera image planes are placed with regular pixel spacing. The optical axes of all cameras are perpendicular to the planes. This data structure covers one side of an object. For a full lightfield we would need to construct six such data structures on a cube around the object.

New views can be rendered from this data structure by placing a virtual camera on an arbitrary viewpoint and intersecting the viewing ray r with the two planes at (s, t, u, v). The resulting radiance is a simple radiance lookup for r. This, however, applies only if the viewing ray passes through original camera viewpoints and pixel positions. For rays passing in between the (s, t) and (u, v) grid coordinates an interpolation is applied that will degrade the rendering quality depending on the scene geometry. In fact, the lightfield contains an implicit geometrical assumption: The scene geometry is planar and coincides with the focal plane. Deviation of the scene geometry from the focal plane causes image warping. If depth information for each view is available, a specific geometrical warping can compensate the image distortion. Heidrich et al. [8] introduce a

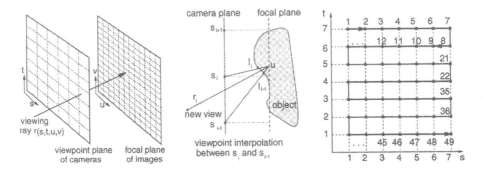

Fig. 1. Left: 4-D lightfield data structure with (s, t) viewpoint plane and (u, v) focal plane. Center: Rendering of novel views by interpolation of viewing ray r between the grid coordinates in the (s, u) slice. The radiance is interpolated from the object radiance at positions $l(s_i, u)$ and $l(s_{i-1}, u)$. Image distortion occurs if the object surface deviates from the focal plane. Right: Tracking path for camera calibration along the (s, t)-viewing grid. Measurements are performed sequentially from image 1 to 49.

warping-based refinement from a depth-compensated lightfield to synthesize intermediate views. They construct a dense lightfield from a sparse set of ray-traced synthetic images. This approach allows interactive visualization of complex ray-traced scenes that is split into the initial off-line ray-tracing of few images and the online refinement for lightfield rendering. The problem is facilitated by the fact that calibration and depth estimation is obsolete since we deal with synthetic scenes. The ray-tracer delivers all necessary depth information as side product of the rendering. The discussion above reveals two major problems when acquiring lightfields from real image sequences:

– the need to directly obtain camera calibration from the image data, and
– the need to estimate local depth for view interpolation.

The original lumigraph approach [6] already tackles both problems. A calibration of the camera is obtained by incorporating a background with a known calibration pattern into the scene. The known specific markers on the background are used to obtain camera parameters and pose estimation [19]. It provides no means to calibrate the images from image data only. For depth integration the object geometry is approximated by constructing a visual hull from the object silhouettes. The hull approximates the global surface geometry but can not deal with local concavities. Furthermore, the silhouette approach is not feasible for general scenes and viewing conditions since a specific background is needed. This approach is therefore confined to laboratory conditions and does not provide a general solution for arbitrary scenes.

3 Camera calibration and geometric reconstruction

The problem of simultaneous camera calibration and depth estimation from image sequences (structure-from-motion, SFM) has been addressed for quite some time in the computer vision community. In the case of known intrinsic camera parameters, the camera pose as well as the scene structure can be estimated from correspondences in the 2D image sequence up to an unknown scale factor. Longuet–Higgins [15] first demonstrated how to obtain structure and camera pose from eight point correspondences in one image pair. The uniqueness of this external calibration was proven in [18]. It exploits the basic relationship between image correspondences of a rigid scene, the Essential matrix E. The approach has been extended in several works, e.g. [10, 4] to an arbitrary number of point correspondences and views using non-linear optimization methods. Faugeras [5] and Hartley [7] later demonstrated that a projective reconstruction is possible from image matches alone even if the camera is totally uncalibrated.

A 3D scene reconstruction system using structure-from-motion was proposed by Beardsley et al. [1] who obtained projective calibration and sparse 3D structure by robustly tracking salient feature points throughout an image sequence. We have extended their method to obtain metric reconstructions (Euclidean reconstruction up to global scale) for fully uncalibrated sequences with methods of self-calibration [16]. For dense structure recovery a stereo matching technique was applied between image pairs of the sequence to obtain a dense depth map for each viewpoint. From this depth map a triangular surface wire-frame is constructed and texture mapping from the image is applied to obtain realistic surface models [11]. To summarize, we obtain a metric scene reconstruction in a 3-step approach:

1. Camera pose calibration is obtained by robust tracking of salient feature points over the image sequence,
2. local dense depth maps for all viewpoints are computed from correspondences between adjacent image pairs of the sequence,
3. a global 3-D surface mesh approximates the geometry, and surface texture is mapped onto it to enhance the visual appearance.

3.1 Combining Lightfield rendering and SFM

If we compare lightfield rendering and SFM, we see a considerable overlap. Both approaches require a good camera calibration and the estimation of local depth maps from the image data. For a geometric reconstruction we then need to combine all local depth estimates into a globally consistent surface model with a unique surface texture. This may be difficult to obtain for complex geometries and reflectivities. It would be better to compute depth maps only and to switch the geometry and surface texture depending on the current rendering viewpoint. And this is precisely what the lightfield approach can do once calibration and depth maps are given [8]. We therefore propose to combine the first two steps of our structure-from-motion approach with lightfield rendering.

The calibration is facilitated for the lightfield approach since we use densely spaced viewpoints where the adjacent images are rather similar. The camera viewpoints are tracked sequentially (along the 1-D viewing path that the camera takes). However, for a lightfield we are obtaining a 2-D viewing surface with the camera viewpoints as nodes of this grid in the s, t-plane. With a moving camera we can scan this viewing surface row by row in a sequential fashion (see also Fig. 1, right). The camera poses are estimated by tracking salient image features throughout the sequence. Salient image feature points are matched using robust (RANSAC) techniques for that purpose. At first feature correspondences are found by extracting intensity corners in different images and matching them using a robust corner matcher [17]. In conjunction with the corner matching a restricted calibration of the setup is calculated. This allows to eliminate matches which are inconsistent with the calibration. The matching is started on the first two images of the sequence. The calibration of these views defines a metric coordinate system in which the projection matrices of the other views are retrieved one by one. A depth triangulation of the corresponding image matches will give a 3D estimate of salient scene points. In subsequent views we utilize this 3D estimate to predict correspondences and to verify them throughout the sequence. The estimated 3D feature points also define a coarse estimate of 3D scene structure. The intrinsic camera parameters were calibrated offline [19] and the approach of [16] has been modified to estimate the camera poses only. This allows robust metric reconstruction from any camera motion.

Once we have retrieved the metric calibration of the cameras we can use image correspondence techniques to estimate scene depth. For dense correspondence matching a disparity estimator based on the dynamic programming scheme of Cox *et al.* [2] is employed. It operates on rectified image pairs where the epipolar lines coincide with image scan lines. The rectification is easily obtained for each pair of adjacent viewpoints by projective mapping of the image planes onto standard parallel stereo geometry. The matcher searches at each pixel in one image for maximum normalized cross correlation in the other image by shifting a small measurement window (kernel size 5x5 or 7x7) along the corresponding scan line. The algorithm employs an extended neighborhood relationship and a pyramidal estimation scheme to reliably deal with very large disparity ranges [3]. It was further extended to multi-viewpoint depth analysis [11]. This allows to obtain locally consistent dense depth estimates for each viewpoint.

4 Experimental Results

In order to test the approach we used a calibrated robot arm for image acquisition. This allows us to obtain ground truth information for the camera pose. The intrinsic parameters were estimated off-line before the experiments.

The camera is mounted on the arm of a robot of type SCORBOT-ER VII. The position of its picker arm is known from the angles of the 5 axes and the dimensions of the arm. Optical calibration methods have to be applied to determine the relative position of the camera to the picker arm. This is done by the hand/eye-calibration method of [20]. The main problem is to determine the position along the optical axis of the camera. The repetition error of the robot is 0.2 mm and 0.03 degrees, respectively. Because of the limited size of the robot, we are restricted to scenes with maximum size of about 100 mm in diameter. For testing we used a scene with planar motion (compliant to the viewpoint plane) and a scene with spherical motion.

Fig. 2. Planar motion on a 7×7 viewpoint grid. Left: One of the input images. Middle: Corresponding dense depth map (color coded: dark=near,light=far, black=undefined). Right: 3D surface model and calibrated camera grid. The little pyramids symbolize the estimated camera poses.

Planar motion: To show the applicability for natural scenes, we chose a small cactus together with two mirrors. This scene has a non-trivial geometry with occlusions, spikes, reflections, etc., see fig. 2. In the planar case, we controlled the robot so that the center of projection moved within a planar 7×7 grid, the optical axis always intersecting one central point in the middle of the scene. The grid had spacing between the views of 13.3×16.6 mm. The orthogonal distance of the grid to the central point was 200 mm. This setup allows ground truth comparison of the camera calibration method. During calibration we tracked the camera positions sequentially row by row, moving like a snake from image 1 to 49 over all views (see fig. 1,right). We did not consider the connectivity between the rows which would additionally stabilize the tracking. The calibrated sequence allows to reconstruct dense depth maps for each viewpoint. From the depth map we obtain local geometry for image warping. Results of the 3D surface modeling are shown in fig. 2.

A quantitative evaluation of the camera calibration can be computed if we compare the length of the estimated displacements between adjacent camera positions (the baseline) with the baseline as given by the robot. Since the SFM algorithm generates only metric estimates (arbitrary scale for the baseline between the first two cameras) we scaled the estimates to the known robot baseline of 13.3 mm between the first two cameras. We could then measure all baselines between adjacent cameras.

The result is summarized in table 4. We have to distinguish between the statistics of row and column displacements. Since the camera moved sequentially row by row (see fig. 1,right), only the camera positions along the rows were estimated directly. The adjacency between columns was not exploited which causes an increased column baseline error. Still, the column statistics show a very good agreement with the expected value and no significant error accumulation was noticed. These figures document the stability and accuracy of the proposed calibration method. The overall performance of the calibration is within the range of the robot arm accuracy.

Spherical motion The proposed system can work with any camera motion and is not confined to planar viewing planes. To test this we performed a spherical robot motion. In the spherical case the robot sampled a 8×8 spherical grid

Table 1. Statistical distribution of the baseline length between camera positions.

Displacement Statistics	ground truth robot baseline[mm]	measured baselines mean [mm]	standard deviation [mm]	repeatability of robot [mm]
rows	13.30	12.65	0.478	0.2
columns	16.60	16.99	0.931	0.2

with a radius of 230 mm. The viewing positions enclosed a maximum angle of 45 degrees. The results of the camera calibration and geometric estimation are shown in fig. 3. The estimated camera positions are equally spaced on the viewing sphere and the geometry shows quite some detail.

Fig. 3. Spherical motion. Left: One of the input images. Center: Dense depth map of scene. Right: 3D surface model and calibrated camera grid of spherical scene.

5 Conclusions and Further Work

We have employed a geometric structure-from-motion approach for lightfield calibration and local depth estimation which can be used to improve lightfield rendering. Image acquisition as well as rendering quality will profit from this integration. Most notably, we were able to generate lightfields from image sequences of freely moving cameras.

We are currently working to further integrate both approaches. The ongoing research has delivered additional results which we could not include here but would like to refer to. The inherent two-dimensional relationship between lightfield images has been exploited, resulting in a robust calibration of a 2D viewpoint mesh over all views [12] and improved depth reconstruction from the viewpoint mesh [13]. The image-based rendering approach has been adapted to incorporate local depth estimates and to render images from irregular viewpoint meshes of hand-held camera sequences [9].

Acknowledgments

We gratefully acknowledge financial support from the Belgian project IUAP 04/24 'IMechS', and the German Science Foundation DFG SFB-603,C2.

References

1. P. Beardsley, P. Torr and A. Zisserman: 3D Model Acquisition from Extended Image Sequences. In: B. Buxton, R. Cipolla(Eds.) Computer Vision - ECCV 96, Cambridge, UK., vol.2, pp.683-695. LNCS, Vol. 1064. Springer, 1996.
2. I. J. Cox, S. L. Hingorani, and S. B. Rao: A Maximum Likelihood Stereo Algorithm. Computer Vision and Image Understanding, Vol. 63, No. 3, May 1996.
3. L.Falkenhagen: Hierarchical Block-Based Disparity Estimation Considering Neighborhood Constraints. Intern. Workshop on SNHC and 3D Imaging, Rhodes, Greece, Sept. 1997.
4. O. Faugeras, S.Maybank: Motion from Point Matches: Multiplicity of solutions. IJCV 4(3), June 1990.
5. O. Faugeras: What can be seen in three dimensions with an uncalibrated stereo rig. Proc. ECCV'92, pp.563-578.
6. S. Gortler, R. Grzeszczuk, R. Szeliski, M. F. Cohen: The Lumigraph. Proceedings SIGGRAPH '96, pp. 43–54, ACM Press, New York, 1996.
7. R. Hartley: Estimation of relative camera positions for uncalibrated cameras. Proc. ECCV'92, pp.579-587.
8. W. Heidrich, H. Schirmacher, and H.-P. Seidel: A Warping-Based Refinement of Lumigraphs. Proc. WSCG '99, N.M. Thalman and V. Skala(eds.), Pilsen, 1999.
9. B. Heigl, R. Koch, M. Pollefeys, J. Denzler, L. Van Gool: Plenoptic Modeling and Rendering from Image Sequences taken by a Hand–Held Camera. Submitted to DAGM symposium, Bonn, Germany.
10. T.S. Huang, A.N. Netravali: 3D Motion estimation. in: H. Freeman (ed.), Machine vision for three-dimensional scenes, Academic Press, 1990.
11. R. Koch, M. Pollefeys, and L. Van Gool: Multi Viewpoint Stereo from Uncalibrated Video Sequences. Proc. ECCV'98, Freiburg, June 1998.
12. R. Koch, B. Heigl, M. Pollefeys, L. Van Gool, H. Niemann: Calibration of Handheld Camera Sequences for Plenoptic Modeling. Submitted to ICCV'99, Corfu.
13. R. Koch, M. Pollefeys, L. Van Gool: Robust Calibration and 3D Geometric Modeling from Large Collections of Uncalibrated Images. Submitted to DAGM Symposium 99, Bonn, Germany.
14. M. Levoy, P. Hanrahan: Lightfield Rendering. Proceedings SIGGRAPH '96, pp. 31–42, ACM Press, New York, 1996.
15. H.C. Longuet–Higgins: A computer algorithm for reconstructing a scene from two projections. Nature vol. 293 (81), pp. 133–135, 1981.
16. M. Pollefeys, R. Koch and L. Van Gool: Self-Calibration and Metric Reconstruction in spite of Varying and Unknown Internal Camera Parameters. Proc. ICCV'98, Bombay, India, Jan. 1998. Also accepted for publication in: International Journal on Computer Vision, Marr Price Special Issue, 1998.
17. P.H.S. Torr: Motion Segmentation and Outlier Detection. PhD thesis, University of Oxford, UK, 1995.
18. R. Y. Tsai, T. S. Huang: Uniqueness and Estimation of Three–Dimensional Motion Parameters of Rigid Objects with Curved Surfaces. PAMI 6(1), pp. 13–27, 1984.
19. R.Y.Tsai: A Versatile Camera Calibration Technique for High-Accuracy 3D Machine Vision Metrology using off-the-shelf Cameras and Lenses. IEEE Journal Robotics and Automation RA-3,4 (Aug. 1987), 323-344.
20. R.Y. Tsai and R.K. Lenz: Real Time Versatile Robotics Hand/Eye Calibration Using 3D machine vision. Proceedings of ICRA 88, 554–561, IEEE Press, Philadelphia, April 1998.

Coding of Dynamic Texture on 3D Scenes

Bert DeKnuydt, Stef Desmet, Luc Van Eycken and Luc Van Gool*

PSI/VISICS, Kard. Mercierlaan 94, B-3001 Heverlee, Belgium
Tel. +32-16-321880, Fax. +32-16-321986
{deknuydt,desmet,veycken,vangool}@esat.kuleuven.ac.be
http://www.esat.kuleuven.ac.be/psi/visics

Abstract. As cheap and powerful 3D render engines become commonplace, demand for nearly realistic 3D scenes is increasing. Besides more detailed geometric and texture information, this presupposes the ability to map dynamic textures. This is obviously needed to model movies, computer and TV screens, but also for example the landscape as seen from inside a moving vehicle, or shadow and lighting effects that are not modeled separately. Downloading the complete scene to the user, before letting him interact with the scene, becomes very unpractical and inefficient with huge scenes. If the texture is not a canned sequence, but a stream, it is altogether impossible. Often a back channel is available, which allows on demand downloading so the user can start interacting with the scene immediately. This can save considerable amounts of bandwidth.

Specifically for dynamic texture, if we know the viewpoint of the user (or several users), we can code the texture taking into account the viewing conditions, i.e. coding and transmitting each part of the texture with the required resolution only.

Applications that would benefit from view-dependent coding of dynamic textures, include (but are not limited to) multiplayer 3D games, walkthroughs of dynamic constructions or scenes and 3D simulations of dynamic systems. In this paper, a scheme based on an adapted OLA (Optimal Level Allocation) video codec is shown. Important data rate reductions can be achieved with it.

Keywords: Texture coding, dynamic texture.

1 Introduction

The coding process of texture to be mapped on 3D-scenes can benefit from knowledge of rendering parameters, such as spatial configuration, lighting, haze, motion blur and resolution of the output device. If a back channel is present, the most obvious parameter to take account with is the viewpoint: clearly, if a huge texture map will be rendered on an object far away, or at a very sharp angle, there is no need to transmit it to full detail: many texture pixels will be rendered to one image pixel. See [5] and [4] for the static case. Especially for dynamic texture, huge bitrate reductions can be obtained by coding the texture locally to only the resolution and quality needed.

---

* This research has been sponsored by the "Fund for Scientific Research"

As is shown in figure 1, exploiting this information means that for each part of the texture, we have to calculate the distance, slant and tilt angle with which it will be mapped. The calculation is based in the 3D scene information available at the transmitter and the viewpoint returned to the transmitter via the back channel. If there are multiple viewers, one could either send each one a different texture stream optimal for his own viewpoint, or one could generate a single texture stream that satisfies all viewers, at the expense of a higher bitrate. This, together with any other relevant information, allows us to calculate the lowest horizontal and vertical resolution needed in the texture map, such that the rendered scene will be of high enough quality. Finally, we can code the texture with a video coding engine that is able to code each part of the map just good enough for the required resolution.

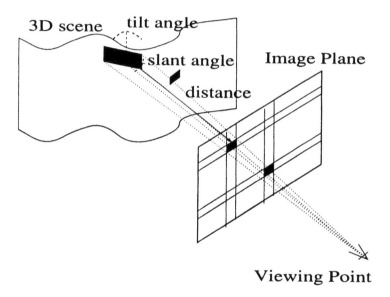

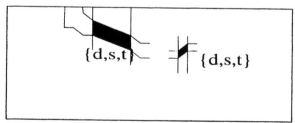

Fig. 1. Scene overview

2 Coding Engine

In general, any video coder could be used for the coding of dynamic texture. However, DCT based coders, like MPEG1, MPEG2 and even MPEG4, do not have enough flexibility to easily change the resolution in specific image areas and directions. More specifically, one can only drop higher order DCT coefficients to reduce the resolutions, which is a rather crude mechanism, often resulting in severe ringing.

Ideally, we would like a coder, which allows to specify an arbitrary quality criterion for each image area. In this case, the quality criterion would include a directional variable resolution filter. The OLA coder is capable of dealing with any quality criterion.

2.1 Standard OLA Coder

The standard OLA video coder starts by analyzing the input images and assigning attributes of visual importance (like "texture" or "visually important edges") to each area. In this application, 16 × 16 blocks were used, but the algorithm is not limited to this. The selected attributes correspond to a classification, which in its turn selects an appropriate visually relevant quality criterion.

Then, for each image block, the best prediction is chosen out of two motion compensated predictions (one from the previous and one from the second previous image) from the prior images and an intra prediction. The motion vector selection algorithm has half pixel resolution and can handle illumination changes. The prediction selection and the motion vectors are predicted and the error is sent to an adaptive arithmetic coder (AAC). See [2].

The appropriate quality criterion is now evaluated and the result compared with the required quality. If quality is not sufficient, the average value of the prediction error is quantized and transmitted. The criterion is now re-evaluated and an OLA coding step is applied. This is done iteratively until the criterion is fulfilled (or the maximum number of iterations, normally 5, is reached). The number of iterations is predicted and AAC coded.

The OLA coding step itself consists of the optimal selection of two levels, which are quantized and transmitted. Each pixel in the prediction error is now updated with one of these two levels, or with zero. The selection map is adaptively scanned, predicted and AAC coded. See 2.2

Optionally, a motion vector deblocking and an image deblocking scheme are applied, inside the coding loop. See [1].

Although the encoding scheme is rather complex, mainly because of the analysis steps, the decoding complexity is very low, which would make it suitable for downloading.

2.2 OLA Coding Engine

One of the most simple ways of coding one can imagine, consists of allocating a few levels, you could call them quantization levels, and assigning each pixel to the most appropriate one. This concept, shown in algorithm 1, is quite powerful, since it allows easy introduction of arbitrary quality criteria.

Algorithm 1: OLA base algorithm

```
regions = DIVIDE_INTO_REGIONS(image)
foreach region in regions
    if OBTAINED_Q(region) < REQUIRED_Q(region)
        nr_levels = DETERMINE_NR_LEVELS(region)
        foreach lvl in [1 : nr_levels]
            value_level[lvl] = DETERMINE_VALUE(region,lvl)
        foreach pixel in region
            value_to_code = original[pixel] - predicted[pixel]
            level_to_code = DETERMINE_LEVEL(value_to_code)
            coded[pixel] + = value_level[level_to_code]
```

Initially, one would set the number of levels to 2, as is done in the traditional block truncation coding. This has some drawbacks however. The first is that we cannot be sure that one additional iteration will improve the quality for a single pixel. Suppose the previous iteration coded some pixel values perfectly. If the total quality was not sufficient, we make another iteration and the pixel values are now "updated" to an incorrect value, as the levels will consist of a negative and a positive value. This means that the local quality does not necessarily improve monotonously while iterating. Visually this gives a rather unpleasing granular type of noise. So we decided to set the number of levels to 3, with the middle level fixed to zero. Only now we are sure to improve local quality with each iteration, but still retain the advantage of only having to optimize two (independent) levels. This works best if we start off with a error histogram more or less centered around zero. We can easily obtain this by coding the DC level of the block prior to starting up OLA iterations.

Deduction of the optimal level choice for 3 levels, one of which fixed to zero Suppose we have a general distribution $F(i)$ of values to assign to one of three levels, called L_l, L_0 and L_u, the lower (negative) level, the zero level and the upper (positive) level. Our goal function or penalty function is a sum of weighted errors over all assignments. We take a square function as weighting. The square error function corresponds to the PSNR criterion. Note that in practice we will use different quality criteria, but it is in general impossible to take into account the exact ones for optimizing the levels, without needlessly complicating the codec.

So our penalty for an assignment is

$$P(L_l, L_0, L_u) = \int_{-\infty}^{\frac{L_l+L_0}{2}} (i - L_l)^2 F(i)di + \int_{\frac{L_l+L_0}{2}}^{\frac{L_u+L_0}{2}} (i - L_0)^2 F(i)di \qquad (1)$$

$$+ \int_{\frac{L_u+L_0}{2}}^{\infty} (i - L_u)^2 F(i)di \qquad (2)$$

The level L_0 was fixed to zero, for reasons described above. Therefore, we can split up the penalty in two independent parts.

$$P_l(L_l) = \int_{-\infty}^{\frac{L_l}{2}} (i - L_l)^2 F(i)di + \int_{\frac{L_l}{2}}^{0} i^2 F(i)di$$

$$P_u(L_u) = \int_{0}^{\frac{L_u}{2}} i^2 F(i)di + \int_{\frac{L_u}{2}}^{\infty} (i - L_u)^2 F(i)di$$

Let's continue with the second of these similar expressions. It can be rewritten as:

$$P_u(L_u) = \int_{0}^{\infty} i^2 F(i)di - 2L_u \int_{\frac{L_u}{2}}^{\infty} iF(i)di + L_u^2 \int_{\frac{L_u}{2}}^{\infty} F(i)di$$

We now take the derivative to L_u and force it to zero.

$$0 = \frac{dP_u(L_u)}{dL_u}$$

After some simplification, we obtain

$$0 = -\int_{\frac{L_u}{2}}^{\infty} iF(i)di + L_u \int_{\frac{L_u}{2}}^{\infty} F(i)di$$

This can be rewritten as

$$L_u = \frac{\int_{\frac{L_u}{2}}^{\infty} iF(i)di}{\int_{\frac{L_u}{2}}^{\infty} F(i)di}$$

In words, this means the L_u is the average of the population above $\frac{L_u}{2}$. An analogous expression holds for L_l.

The value for L_u has to be found iteratively. It can be shown that after a few iterations the following expression almost always converges.

Bitmap Coding This is a critical point of the compression algorithm, as it generates the largest part of the total bitstream.

The data we have to code consist of $n \times m$ rectangular blocks (usually 8×8, 16×8 or 16×16) with l (usually 3) levels. Coding has to be reversible.

We start by scanning the two dimensional block into a one dimensional array. We could of course do a simple left-to-right, top-to-bottom scanning, but, as neighboring pixels in the block have some correlation, it is better to try to increase the locality of the scanning. We can take some advantage of our knowledge of the visual attributes; we would ideally have to maximize the vertical locality for image blocks with the vertical edge attribute, horizontal locality for those with horizontal edges and the global locality for all others. This results in the three scanning patterns illustrated in figure 2. The leftmost and center one are called horizontal and vertical oxen scan[1], the rightmost one is the two dimensional Hilbert curve of order 3. In [3], Hilbert curves are proven to be superior in distance-preserving mapping than Peano or Reflected Binary Gray codes. The disadvantage is that they are defined only for all dimensions equal to 2^n.

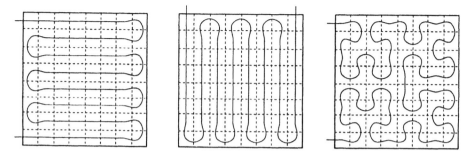

Fig. 2. Horizontal and vertical oxen and Hilbert curve scanning patterns for 8 × 8 blocks

Simulations have shown that we can safely replace the Hilbert one by the horizontal oxen for luminance and by vertical oxen for chrominance without performance degradation. For blocks with both the horizontal and vertical attribute switched on, we take the horizontal oxen scan.

We simply apply standard adaptive arithmetic coding to the input data, or we insert a differential encoding step, using the previous value for prediction, starting with a zero (the most frequent value).

2.3 Practical Implementation of OLA

Startup and DC level handling The "block-to-code" is constructed and its quality measured. If it is sufficient, the block is skipped. Otherwise, its DC level is coded.

The average DC level of the error can be determined simply by adding the errors over all pixels and dividing the result by the number of pixels in the block. There several minor improvements that can be made to this approach when dealing with changes lighting conditions.

The DC level of the error is quantized (rounded) to 8 bits precision. If the block was an INTRA block, the DC level could be quantized further to a smaller number of bits. This is not done for the moment, as there is no worthwhile performance benefit. The average DC level in this case behaves statistically like the pixels of an image, as it is nothing else than a crude subsampling. Therefore, we cannot expect to gain much improvement by simple coding tricks.

If, on the contrary, the block was coded INTER and we use standard non-DC-corrected motion prediction, then we can assume that the DC value of the block will in general have a small absolute value. Here we are better off reserving only a small number of bits to code the DC level perfectly. We spend 3 bits for the Iclasses "nothing" and "masking", and only 2 for the others, this both in luminance and chrominance. If the DC level is out of range, which happens only very sporadically when motion prediction fails completely, we have to saturate it at the maximum

---

[1] The correct term would be boustrophedonic: (Gr. turning like oxen in plowing; to turn.) Relating to an ancient mode of writing, in alternate directions, one line from left to right, and the next from right to left (as fields are plowed), as in early Greek and Hittite. (Webster Dictionary)

available level. The first OLA iteration will then be suboptimal, because it has to code the remainder of the DC bias.

If on the other hand we use DC-corrected motion estimation, DC levels can be quite large, even in INTER mode. We can use the same system as before, but the number of saturated values becomes large, and the gain of the DC-correction in the motion compensation is lost. It is therefore better to retain one level as an escape code. After the escape code, we can send the quantized DC level. It turns out that 4 to 5 bit is enough for this. At the switch-over between plain and escape coding, the determination level is slightly biased from the center towards the level with the shorter code length.

After coding and transmitting the level, we can adjust the remaining error values to code by adding the DC level, and clipping the result to ITU-R601 levels. The iteration count for the block is updated to 1.

Level determination and coding Now, the quality of the block is evaluated with the appropriate error criterion. If the block is good enough we skip to the next one.

We iterate the level optimization formulae, until convergence. In all but a few exotic cases, 3 iterations are enough.

When the levels are determined as described, they each span a range of 7 bits. In general, it is not needed to reserve 7 bits for them, as high absolute values are highly improbable and don't need to be transmitted perfectly, because they only occur in the beginning of the iteration process, and slight inaccuracies will be corrected later anyhow. So we can quantize these levels before transmission with only a few bits.

After trial and error, the following scheme turned out to be best. For absolute values lower than 32, we code the value directly and perfectly with 5 bits. For values outside this range, we emit the escape code of 5 zeros, followed by 5 other bits representing the value quantized with the appropriate stepsize.

The Borstelmann wrap-around addition Now, we determine which level we actually assign to a pixel error value to code, by wrap adding the level with the pixel value already coded, normalize the result to IRU-R range, and check which one is closest to the original. If more than one level reaches the same quality, the zero level is preferred. This improves later bitmap compression. The wrap adding can improve coding performance considerably in the initial iterations.

Iteration Count Coding The number of iterations done per image block is transmitted separately. First the maximum number it determined and transmitted with a fixed number of bits. The remainder is then coded with AAC.

"Just good enough" coding As was described, the main idea behind OLA is to code an image part better step by step, until a quality criterion if fulfilled. Sometimes, the increase in quality an additional iteration results in, can be rather large (typically 6 dB), especially for the higher iteration counts. This does not come without a price: large quality improvements cost large amounts of bits. Because of this mechanism, we often overshoot the required quality. This is a waste of bits and results in an uneven quality. The overshoot problem becomes increasingly worse, if the required quality is low or is dynamic, as is the case in dynamic texture coding.

The evident solutions consists of coding "just good enough" for the quality criterion. The simplicity and flexibility of the OLA coding allows for this.

Several algorithms are envisageable to reduce both bitrate and quality in small steps, depending on the quality criterion and the bitmap coding that are used. One should take care though, not to force ZERO pixels to LOW or HIGH, as this can result in flickering.

The most obvious one is to increase the number of ZERO pixels, by applying a morphological closing with variable mask size on the bitmap. In case of arithmetic coding, it is better to apply it after scanning, because this better reduces bitrates and indirectly takes into account the visual attributes.

3   Extended Quality Criterion

Because of the information returned via the back channel, we know the viewpoint of the observer. We can easily calculate the distance, slant and tilt angle with respect to the image plane, of each part of the texture to be mapped. We express them in a visibility number, 0 meaning not visible (object at infinite distance, object occluded, or mapping plane perpendicular to the image plane) and 1 meaning fully visible (mapping plane coincides with the image plane).

Then, we can than determine a mapping filter that corresponds to the projection of the texture back to the image plain. The filter is a directional subsampling filter. As we are only interested in the quality after mapping, we can include it in the quality criterion for the OLA coding step. Because of linearity of the system, we can apply the filter to obtain the coding errors after mapping from those in the texture domain. The errors after mapping are added up to determine the final quality of that image part.

In the codec, the filter is implemented as a variable size, separable Gaussian filter. Theoretically, the size of this filter can become infinite. In practice we have to limit the size to about the coding block size. Mirroring is applied at the block borders to reduce bias.

From the filtered errors, the PSNR could be calculated, but a slightly more complex measure is taken. The errors are sent though a lookup table consisting of three parts. First there is a dead zone around zero, to reduce the impact of (camera) noise, followed by a quadratic part, in accordance with the PSNR criterion, and finally a linear part, to limit the effect of large single pixel errors. The accumulated sum of these values returns a quality measure.

3.1   Configuration

In the setup of the coder, a number of possibilities remain open. For instance, what to do with texture mapped on a object that becomes occluded. We decided to continue sending only motion vectors, in order to minimize the data we have to send, when the object would reappear.

4 Simulation

4.1 Room Walkthrough

We constructed a virtual room, where a movie is playing in a flat television screen on the back wall. Meanwhile, a person enters the room from a door at the backside, walks around the table and finally takes a seat in the couch, close to the screen. The action takes 60 seconds.

If we code the movie as plain video material, we need on average 3.17 Mbit/sec for MPEG2 like quality. If we code it as dynamic texture, only 0.79 Mbit/sec is needed. Rendered, both approaches result in the same quality.

In figure 3, you see a frame from the rendered scene, in figure 5 the texture for the virtual television screen, once coded as video (top), once as dynamic texture (bottom). Finally, in figure 4 you see the momentary bitrates for the respective ways of coding.

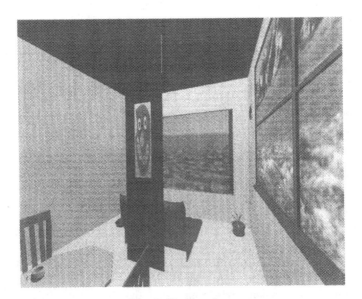

Fig. 3. Rendered scene

5 Conclusion

In complex 3D scenes, the exploitation of scene information, and especially the viewing point, can greatly reduce the bitrate needed to transmit dynamic (and static) textures. This implies a coder able to deal with other quality criteria than plain PSNR. Existing DCT based (video) coders are not well suited for this application, mainly because the coding does not take place in space domain. More flexible, or ideally downloadable, (de)coding schemes are urgently needed.

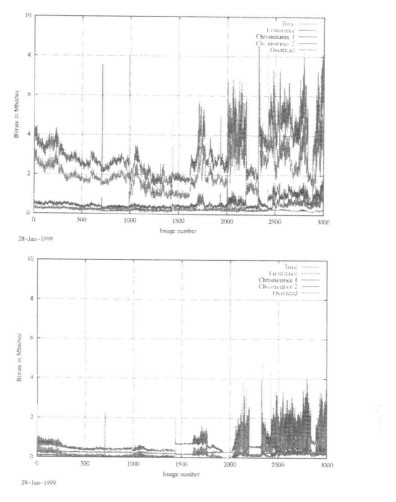

Fig. 4. Bitrates for coding as video (top) and as dynamic texture (bottom)

References

1. Bert DeKnuydt, Luc Van Eycken, and André Oosterlinck. A simple deblocking scheme suitable for use inside the codec loop. In *IEEE Proceedings, Melecon '94, Antalya, Turkey*, volume 1, pages 9-12, April 1994.
2. Stef Desmet, Bert DeKnuydt, Nan Li, Luc Van Eycken, and André Oosterlinck. A simple algorithm to extract realistic motion fields out of video sequences. In *Proceedings, EOS-SPIE International Symposium on Fiber Optic Networks and Video Communications, Berlin, Germany*, volume 1977, pages 248-254, April 1993.
3. Christos Faloutsos and Shari Roseman. Fractals for secondary key retrieval. *CS-TR-2242*, May 1989.
4. Stefan Horbelt, Fred Jourdan, and Touradj Ebrahimi. View dependent texture coding for transmission of virtual environment. In *Proceedings, Conference on Image Processing and its Applications, Dublin, Ireland*, volume 1, pages 433-437, June 1997.
5. Fred Jordan and Touradj Ebrahimi. View-dependent texture for transmission of virtual environment. MPEG4-meeting, doc. ISO/IECJTC/SC29/WG11/M1501, Maceió,

Fig. 5. Screen coded as video (top) and as dynamic texture (bottom)

Image Stitching - Comparisons and New Techniques

Chia-Yen Chen and Reinhard Klette

CITR, University of Auckland,
Tamaki Campus, Building 731, Auckland, New Zealand
{yen, reinhard}@citr.auckland.ac.nz
http://citr.auckland.ac.nz

Abstract. This work describes the steps involved in the generation of a panoramic image, i.e., image acquisition, image registration and image merging. Different approaches for each of these steps are discussed and compared. The resultant images from image registration and image merging are quantitatively evaluated to provide an indication to the performance of the different methods. The methods provided in this work can be used to generate panoramic images for the use of interactive panoramic viewing of images, architectural walk-through, and other applications associated with the modelling of 3D environments.

1 Introduction

Recently a specialised form of image mosaicing known as image stitching, has become increasingly common [10], especially in the making of panoramic images. Stitched images are used in applications such as interactive panoramic movies, architectural walk-through, multi-node movies and other applications associated with modelling the 3D environment using images acquired from the real world.

The first step in the generation of a panoramic image is *image acquisition*, where a series of overlapping images is acquired. The second step, *image registration*, is then performed to find the translations which align adjacent images. Finally, in *image merging*, the images are merged to form a panoramic image using the obtained translations, and the intensity values of the images are adjusted if necessary.

The series of images, I_1, I_2, ..., I_k, for the generation of panoramic images can be acquired by either camera rotations, or camera translations. The images are taken with each movement of the camera until the desired range has been covered. It has been suggested to have at least 50% of the image overlap with the previous image and the other 50% of the image overlap with the following image [4] to improve the quality of the resultant panoramic image.

Acquiring images by camera rotation requires the camera to be fixed at one spot and rotated with respect to the vertical axis in one chosen direction. Ideally, the Y-axis should pass the optical centre of the camera and the camera should not be rotated in directions other than the specified direction. Due to the simplicity of this method, it is usually the preferred image acquisition method for the generation of panoramic images.

In acquiring images by camera translations, the camera is shifted in a direction parallel to the imaging plane. With this method, it is important to keep the image planes parallel to the direction of camera translation. However, as the distance between the camera and the object increases, so does required translation distance. Therefore, image acquisition by camera translation may become difficult for objects that are distant from the camera.

A common problem in image acquisition is the *intensity shift* between adjacent images due to the variation in the lighting conditions. The appearance of high-lights also degrade the quality of the acquired images.

Distortions caused by lens aberrations also need to be corrected. Perspective distortions caused by wide-angle lens, such as pincushion and barrel distortions, may also be corrected by the use of a grid [2]. Such corrections are required in order to register the acquired images later on.

This paper discusses the processes of image registration and image merging. It is our objective to evaluate several image stitching methods such that a quantitative indication of the performances of the various methods can be provided for reference.

2 Image Registration

The process of image registration aims to find the translations to align two or more overlapping images such that the projection from the view point through any position in the aligned images into the 3D world is unique.

An image registration method usually consists of four main components [3]. The *feature space* refers to the set of features to be used in the comparison of the images, for example, the intensity values, contours and textures. The *similarity measure* is a function which returns a scalar value to provide an indication of the similarities between features. The *search space* is a set of possible transformations for aligning the images, such as translations or rotations. The *search strategy* is the algorithm that decides how to select the next transformations from the search set. By varying the contents of these four components, different registration methods with different behaviours can be constructed.

2.1 Used Components

Over the years, a number of image registration methods have been proposed. These methods usually involve pattern matching to find the transformations required to align the images [1, 3, 5, 6, 11]. In this section, selected components from previously proposed methods are used to construct image registration methods. A vector will be used to represent the different components used in each registration method. Table 1 shows the different components used in image registrations.

For the calculation of the similarity measure, let $W_k(i, j)$ be a single, fixed window defined on image I_k and $W_{k+1,(u,v)}$ be a moving window defined on image I_{k+1}, centered at position (u, v). Let $S_k(u, v)$ be the similarity measure for position (u, v). The first similarity measure calculates $S_k(u, v)$ by

$$S_k(u, v) = \sum_{i=1}^{m} \sum_{j=1}^{n} \left| W_k(i, j) - W_{k+1,(u,v)}(i, j) \right| \tag{1}$$

The second measure $S_k(u, v)$ are squared differences

$$S_k(u, v) = \sum_{i=1}^{m} \sum_{j=1}^{n} \{ W_k(i, j) - W_{k+1,(u,v)}(i, j) \}^2 \tag{2}$$

The third measure $S_k(u, v)$ is the correlation

$$S_k(u, v) = \sum_{i=1}^{m} \sum_{j=1}^{n} W_k(i, j) W_{k+1,(u,v)}(i, j) \tag{3}$$

and the fourth measure is the standard deviation of all intensity differences

$$W_k(i, j) - W_{k+1,(u,v)}(i, j), \text{ for } i = 1, ..., m \text{ and } j = 1, ..., n. \tag{4}$$

Once the similarity measures have been calculated for all possible positions (u, v) on I_{k+1}, the optimal matching position, denoted by (u^*, v^*), is selected such that the value of $S_k(u^*, v^*)$, is optimal. The transformation required to align I_k and I_{k+1} can then be obtained from (u^*, v^*).

One of the search spaces used include all possible translations for registering adjacent images. The other search space uses an exhaustive search for the first pair of images and restricts following translations according to the translations required to align previous images, while allowing some variations.

The search strategies used include an exhaustive search strategy which calculates the similarity measure for each transformation in the search set. Another search strategy is a 2D binary step search [12] which determines the next position to calculate the similarity measure according to the similarity measure at the current position and a specified set of near-by pixels.

Table 1. Components used in image registration methods.

No.	Feature space	Similarity measure	Search space	Search strategy
1	averaged intensities	absolute differencs	all translations	exhaustive search
2	binary edges	squared differences	restrict set of translations	2D binary step search
3		correlation product		
4		standard deviation of intensity differences		

Table 2. Vector encodings for image registration methods.

No.	Method	No.	Method	No.	Method	No.	Method	No.	Method
1	[1,1,1,1]	4	[1,3,1,1]	7	[1,1,1,2]	10	[1,3,2,1]	13	[1,4,2,2]
2	[2,1,1,1]	5	[1,4,1,1]	8	[2,1,2,1]	11	[1,4,2,1]		
3	[1,2,1,1]	6	[1,1,2,1]	9	[1,2,2,1]	12	[2,1,2,2]		

2.2 Used Registration Methods

Table 2 gives a list of image registration methods which may be obtained by varying the four different components from the *basic image registration method*, [1,1,1,1]. Each image registration method is given by a vector that contains four elements, respectively representing the feature space, similarity measure, search space and the search strategy. The numberings of the components are given in Table 1. For example, the method indicated by [1,1,1,1] means that the method has a feature space containing the averaged intensities, a similarity measure calculated by the absolute difference of intensity values, a search space containing all of the possible translations, and an exhaustive search strategy which calculates the similarity measures for all of the translations within the search space.

2.3 Comparison of Methods

The images registered by the image registration method mentioned in Section 3.2 are used to evaluate the performance of each method quantitatively. Equation 5 calculates the sum of differences between the translations obtained by the image registration method and the manually obtained translations

$$d_{ver} = \sum_{i=1}^{k} \{t_{ver}(i) - t_{m,ver}(i)\}^2 \text{ and } d_{hor} = \sum_{i=1}^{k} \{t_{hor}(i) - t_{m,hor}(i)\}^2 \quad (5)$$

where $t_{ver}(i)$ and $t_{hor}(i)$ are the obtained vertical and horizontal translations and $t_{m,ver}(i)$ and $t_{m,hor}(i)$ are the manually obtained translations for the ith image. Results for one example of a panoramic image (see Fig. 1) are shown in Table 3. The methods are ranked according to the magnitudes of the differences. The ranking of 1 is given to the method which has the smallest overall difference with the manually obtained translations. Fig. 1 shows the images registered by method number 8.

3 Image Merging

Image merging is the process of adjusting the values of pixels in two registered images, such that when the images are joined, the transition from one image to the next is invisible. At the same time, the merged images should preserve the quality of the input images as much as possible.

Table 3. Summary of differences and ranking of methods for the image in Fig. 1

Method	d_{ver}	d_{hor}	$d_{ver} + d_{hor}$	Rank
1	3664	6696	10360	12
2	44	240	284	2
3	1552	5440	6992	11
4	13892	12196	26088	13
5	26	282	308	4
6	23	700	723	5
7	2870	2779	5649	10
8	10	240	250	1
9	151	708	859	6
10	499	1768	2267	8
11	17	282	299	3
12	77	1925	2002	7
13	166	2142	2308	9

3.1 Image Merging Methods

Table 4 shows four methods used to merge adjacent images. The first image merging method was proposed in [8]. It has been applied to the red, green and blue channels separately to merge colour images.

The *seam* is defined as a line between two columns in the final panoramic image where images I_k and I_{k+1} contribute on either side of the line.

Method 1: Let n be the number of pixels on either side of the seam that are to be adjusted and let e_k be the intensity difference across the seam. $H_{k,k+1}$ is the image obtained by merging images I_k and I_{k+1}. The intensity values of $H_{k,k+1}$ are given by

$$H_{k,k+1}(i, r_k - j) = I_k(i, r_k - j) - (n - j)e_k(i) \text{ and} \tag{6}$$
$$H_{k,k+1}(i, l_k + j) = I_{k+1}(i, l_{k+1} + j) + (n - j)e_k(i),$$

where r_k and l_{k+1} are the indices of columns immediately on the right and left hand side of the seam.

Although the first merging method can remove the seam between joined images, the intensities are adjusted only with respect to the intensity difference

Fig. 1. Example of a panoramic image (Tamaki Campus) where the images are registered by method 8. See `citr.auckland.ac.nz/research/cvu/projects/demos`.

Table 4. Image merging methods.

No.	Intensity adjustment
1	with respect to intensity differences across the seam
2	with respect to median of intensity differences across the seam
3	with respect to corresponding pixels in overlapping regions
4	with respect to corresponding pixels in filtered overlapping regions

across the seam. A large fluctuation in the magnitude of the difference causes the whole row to be much brighter or darker than its neighbouring rows. Therefore, intensity discontinuities in the vertical direction occur in the merged images.

Method 2: In the second merging method, the median of intensity differences is used to avoid selecting large fluctuations in the intensity differences. Let $m_k(i)$ represent the median of intensity differences for row i, it is given by

$$m_k(i) = median\{e_k(i-c), e_k(i-c+1), ..., e_k(i+c)\} \tag{7}$$

The values of m_k then substitute e_k in Equation 6 for the calculation of the intensities in the merged image.

Method 3: Another way of adjusting the intensity values in the merged image is to consider the intensity values in the overlapping regions of the images. In the third image merging method, the contributions of the intensities are linearly varied across a neighbourhood of the seam. The intensities in the merged image are calculated by

$$H_{k,k+1}(i, r_k + j) = \frac{n-j}{2n} \cdot I_k(i, r_k + j) + \frac{n+j}{2n} \cdot I_{k+1}(i, l_{k+1} + j - 1) \tag{8}$$

This method successfully removes the seam in cases where the overlapping regions of image I_k and I_{k+1} differ by an intensity shift. However, the relative positions and shapes of the features within the overlapping regions may also vary due to object movement, parallax errors in image acquisition, or failure to find the ideal translations between images during image registration. In such cases, the image merging method may produce an effect know as the *double exposure effect*.

Method 4: In method number four, a low-pass filter is applied on the overlapping region to extract the low frequency components that contribute to the overall intensities of the image, the intensity values in the merged image are then calculated according to the intensity values in the filtered images to avoid the double exposure effect. In our work, the contributions to the merged image from the filtered images are set to be less than 50%. Equation 9 calculates the intensities in the merged image using the median filtered images M_k and M_{k+1},

$$H_{k,k+1}(i, r_k + j) = \begin{cases} \frac{(n-j)M_k(i,r_k+j)}{2n} + \frac{(n+j)I_{k+1}(i,l_{k+1}+j-1)}{2n}, & \text{if } \frac{n-j}{2n} \leq 0.5 \\ \frac{(n-j)I_k(i,r_k+j)}{2n} + \frac{(n+j)M_{k+1}(i,l_{k+1}+j-1)}{2n}, & \text{if } \frac{n+j}{2n} < 0.5 \end{cases} \tag{9}$$

This method is able to reduce the appearance of the double exposure effect and remove the seam between the joined images.

3.2 Comparison of Image Merging Methods

In this section the comparison of the image merging methods has been done with respect to two criteria, the elimination of the seam and the preservation of image quality.

It has been found that all of the above image merging methods can eliminate visible seams. The qualities of the merged images are evaluated by the differences between the contrast of the merged and the original images. Contrast is used to mean the intensity differences between neighbouring pixels and the contrast of images H and I are respectively represented by H' and I'. The image merged with original images I_k and I_{k+1} is represented by $H_{k,k+1}$. The performance measure d_k denotes the sum of differences in contrast between $H_{k,k+1}$ and I_k, and d_{k+1} denotes the sum of differences in contrast between $H_{k,k+1}$ and I_{k+1}. d_k and d_{k+1} are calculated by

$$d_k = \sum_{i=1}^{H} \sum_{j=1}^{n} \left| H'_{k,k+1}(i, s - n + j) - I'_k(i, s - n + j) \right| \qquad (10)$$

$$d_{k+1} = \sum_{i=1}^{H} \sum_{j=1}^{n} \left| H'_{k,k+1}(i, s + j) - I'_{k+1}(i, +j) \right|$$

Table 5 contains the mean and variance of the performance measure obtained for several pairs of images.

By using the contrast differences to measure the image quality of the merged images, the second image merging method has been found to have the best performance for our example of panoramic images.

4 Resultant Panoramic Images

From the evaluations conducted, the best performing image registration and image merging methods are used to generate several panoramic images from series of images. One of the panoramic images stitched by the implemented image stitcher is shown in Fig. 2. The image has been divided into two parts to fit onto one page.

Table 5. Mean and variance of contrast differences.

	Method 1		Method 2		Method 3		Method 4	
	d_k	d_{k+1}	d_k	d_{k+1}	d_k	d_{k+1}	d_k	d_{k+1}
mean $(\times 10^3)$	118.2	60.9	8.0	7.5	30.2	29.1	24.4	23.6
variance $(\times 10^6)$	21217.0	161.5	6.1	3.7	58.3	23.2	34.0	7.2

Fig. 2. Example of a resultant panoramic image.

5 Conclusions and Future Directions

In this work, we have discussed different methods for achieving image registration and merging. The methods have also been quantitatively evaluated.

Image stitching provides a cost effective and flexible alternative to acquire panoramic images using a panoramic camera. The generated panoramic images can be used for 360 degree interactive panoramic movies, such as QuickTime VR, or multi-node panoramic movies.

In conclusion, we hope that our work is able to provide some fundamental ideas to those wishing to further investigate the process of image stitching and the image stitcher in the future.

References

1. D. I. Barnea and H.F. Silverman, "A class of algorithms for fast digital registration", *IEEE Trans. Comput.*, **C-21**, 1972, pp. 179–186.
2. R. N. Bracewell, *Two dimensional imaging*, Prentice–Hall, New Jersey, 1995.
3. L. G. Brown, "A survey of image registration techniques", *Computing Surveys*, **vol. 24**, no. 4, 1992, pp.325–376.
4. S. E. Chen, "QuickTime - An image-based approach to virtual environment navigation", *Proc. SIGGRAPH*, Aug. 1995, pp. 29-37.
5. C. Chen and R. Klette,"An image stitcher and its application in panoramic movie making", *Proc. DICTA'97*, Albany, Auckland, Dec. 1997, pp.101–106.
6. R. Kasturi and M. M. Trivedi, *Image analysis applications*, Marcel Dekker, New York, 1990.
7. R. Klette and P. Zamperoni, *Handbook of image processing operators*, John Wiley and Sons, Chichester, 1996.
8. D. L. Milgram,"Computer methods for creating photomosaics", *IEEE Trans. Comput.*, **C-24**, Nov. 1975, pp.1113–1119.
9. D. L. Milgram, "Adaptive techniques for photomosaicking", *IEEE Trans. Comput.*, **C-26**, Nov. 1977, pp.1175–1180.
10. H. Shum and R. Szeliski, www.research.microsoft.com, 1997.
11. M. Sonka, V. Hlavac and R. Boyle, *Image processing, analysis and machine vision*, Chapman and Hall, London, 1993.
12. G. Tziritas and C. Labit, *Advances in image communication*, vol. 4 , Elsevier, 1994.

Procrustes Alignment with the EM Algorithm

Bin Luo [1,2]   Edwin. R. Hancock [1]

[1]Department of Computer Science,
University of York,York YO1 5DD, UK.
[2]Anhui University, P.R. China
luo,erh@minster.cs.york.ac.uk

Abstract. This paper casts the problem of point-set alignment via Procrustes analysis into a maximum likelihood framework using the EM algorithm. The aim is to improve the robustness of the Procrustes alignment to noise and clutter. By constructing a Gaussian mixture model over the missing correspondences between individual points, we show how alignment can be realised by applying singular value decomposition to a weighted point correlation matrix. Moreover, by gauging the relational consistency of the assigned correspondence matches, we can edit the point sets to remove clutter. We illustrate the effectiveness of the method matching stereogram. We also provide a sensitivity analysis to demonstrate the operational advantages of the method.

1 Introduction

The problem of point pattern matching has attracted sustained interest in both the vision and statistics communities for several decades. For instance, Kendall [8] has generalised the process to projective manifolds using the concept of Procrustes distance. Ullman [15] was one of the first to recognise the importance of exploiting rigidity constraints in the correspondence matching of point-sets. Recently, several authors have drawn inspiration from Ullman's ideas in developing general purpose correspondence matching algorithms using the Gaussian weighted proximity matrix. For instance Scott and Longuet-Higgins [11] locate correspondences by finding a singular value decomposition of the inter-image proximity matrix. Shapiro and Brady [12, 13], on the other hand, match by comparing the modal eigenstructure of the intra-image proximity matrix. In fact these two ideas provide some of the basic groundwork deformable shape models of Cootes *et al* [5] and Sclaroff and Pentland [10] build. This work on the coordinate proximity matrix is closely akin to that of Umeyama [16] who shows how point-sets abstracted in a structural manner using weighted adjacency graphs can be matched using an eigen-decomposition method. These ideas have been extended to accommodate paratererised transformations [17] which can be applied to the matching of articulated objects [18]. More recently, there have been several attempts at modelling the structural deformation of point-sets. For instance, Amit and Kong [4] have used a graph-based representation (graphical templates) to model deforming two-dimensional shapes in medical images.

Lades *et al* [9] have used a dynamic mesh to model intensity-based appearance in images.

Broadly speaking the aim of point pattern matching is to recover the transformation between image and model co-ordinate systems. In order to estimate the transformation parameters a set of correspondence matches between features in the two co-ordinate systems is required. In other words, the feature points must be labelled. Posed in this way there is a basic chicken-and-egg problem. Before good correspondences can be estimated, there need to be reasonable bounds on the transformational geometry. Yet this geometry is, after all, the ultimate goal of computation. This problem is usually overcome by invoking constraints to bootstrap the estimation of feasible correspondence matches. If reliable correspondences are not available, then a robust fitting method must be employed [13, 12]. This involves removing rogue correspondences through outlier rejection.

The idea underpinning this paper is to provide a new framework for the maximum likelihood alignment of point-sets which allows linkage between alignment and correspondence. Specifically, we aim to realise iterative Procrustes alignment using the EM algorithm. The possibility of hidden or missing data provides a natural way of representing the unknown correspondences between individual points. In the maximisation step of the algorithm, we align the points so that they minimise the weighted Procrustes distance. In the expectation step the positional residuals are used to estimate correspondence matching probabilities used in the weighting process. Editing to correct structural errors in the point-sets can be performed on the basis of the consistency of the pattern of matches. These three processes are interleaved and iterated to convergence.

2   Point-Sets

Our goal is to recover the parameters of a geometric transformation $\Phi^{(n)}$ that best maps a set of image feature points $\mathbf{w}$ onto their counterparts in a model $\mathbf{z}$. In order to do this, we represent each point in the image data set by a position vector $\mathbf{w}_i = (x_i, y_i)^T$ where i is the point index. This vector represents the two-dimensional point position in a homogeneous coordinate system. We will assume that all these points lie on a single plane in the image. In the interests of brevity we will denote the entire set of image points by $\mathbf{w} = \{\mathbf{w}_i, \forall i \in \mathcal{D}\}$ where $\mathcal{D}$ is the point index-set. The corresponding fiducial points constituting the model are similarly represented by $\mathbf{z} = \{z_j, \forall j \in \mathcal{M}\}$ where $\mathcal{M}$ denotes the index-set for the model feature-points z_j.

Later on we will show how the two point-sets can be aligned using singular value decomposition. In order to establish the required matrix representation of the alignment process, we construct two co-ordinate matrices from the point position vectors. The data-points are represented by the following matrix whose columns are the co-ordinate position vectors,

$$D = (\, \mathbf{w}_1 \quad \mathbf{w}_2 \quad .... \quad \mathbf{w}_{|D|} \,) \tag{1}$$

The corresponding point-position matrix for the model is

$$M = (\, z_1 \quad z_2 \quad .... \quad z_{|M|} \,) \tag{2}$$

One of our goals in this paper is to exploit structural constraints to improve the recovery of transformation parameters from sets of feature points. We abstract the representation of correspondences using a bi-partite graph. Because of its well documented robustness to noise and change of viewpoint, we adopt the Delaunay triangulation as our basic representation of image structure [14, 7]. We establish Delaunay triangulations on the data and the model, by seeding Voronoi tessellations from the feature-points [1–3].

The process of Delaunay triangulation generates relational graphs from the two sets of point-features. An example is shown in Figure 1. More formally, the point-sets are the nodes of a data graph $G_D = \{\mathcal{D}, E_D\}$ and a model graph $G_M = \{\mathcal{M}, E_M\}$. Here $E_D \subseteq \mathcal{D} \times \mathcal{D}$ and $E_M \subseteq \mathcal{M} \times \mathcal{M}$ are the edge-sets of the data and model graphs.

3 Dual Step EM Algorithm

The aim in this paper is to show how the Procrustes alignment of the two point-sets can be realised using the EM algorithm. The ultimate goal of the alignment process is to identify point-to-point correspondences between the data and the model. Moreover, we are interested in the case where there are significant structural differences between the two-point sets due to the addition of noise or the occlusion and drop-out of certain feature-points.

The EM algorithm provides a natural framework for recovering the required correspondences. The method is concerned with finding maximum likelihood solutions to problems posed in terms of missing or hidden data. In the alignment problem it is the correspondences which are missing and the transformation parameters that need to be recovered. The utility measure underpinning the method is the expected log-likelihood function. Under the assumption that the positional errors between the aligned point-sets are Gaussian, then the maximum likelihood problem becomes one of minimising a weighted squared error measure or Procrustes distance. The weights used to control the different positional errors are in fact the *a posteriori* probabilities of the point correspondences. The EM algorithm iterates between two interleaved computational steps. In the expectation step the *a posteriori* correspondence probabilities are estimated from the current position errors by applying the Bayes formula to the Gaussian distribution functions. In the maximisation step the alignment parameters are estimated so as to maximise the expected log-likelihood function. This is equivalent to minimisation of the weighted error measure. Here we realise the maximisation step by adopting a matrix representation of the point-sets together with their putative correspondence probabilities and applying singular value decomposition to recover the alignment parameters. In practice we iterate the Procrustes alignment on a weighted correspondence matrix.

3.1 Mixture Model

The idea underpinning the EM algorithm is to construct a mixture model over the hidden data to explain the distribution of the observed data. The ultimate goal is the set of maximum likelihood parameters Φ which explain the observed distribution of data. In our alignment problem, the observed data are the position vectors belonging to the set **w**. The parameters are the translation, rotation and scaling required by the Procrustes alignment of the point-sets.

The method commences by assuming that the different observations are independent of one-another. As a result we can factorise the joint conditional likelihood of the data over the individual point position vectors, i.e.

$$p(\mathbf{w}|\Phi) = \prod_{i \in D} p(w_i|\Phi) \tag{3}$$

The next step is to focus on the probability distribution $p(w_i|\Phi)$. Here we assume that the observed data-point positions have arisen from the model-points via a measurement process. However, the original model point is hidden from us. We must therefore entertain the possibility that each data point may have originated via measurement from any of the model-points. This situation is expressed probabilistically by constructing a mixture model over the set of hidden model-data associations or correspondences. As a result, we write

$$p(w_i|\Phi) = \sum_{j \in M} p(w_i|z_j, \Phi)\pi_{i,j} \tag{4}$$

where $p(w_i|z_j, \Phi)$ is the probability distribution for the data-point position measurement or observation w_i to have originated from the model-point z_j under the set of alignment parameters Φ. The quantity $\pi_{i,j}$ is the mixing proportion required for the model-point z_j in explaining the observation w_i.

With these ingredients, the complete likelihood function that has to be maximised is

$$\mathcal{L} = \prod_{i \in D} \sum_{j \in M} p(w_i|z_j, \Phi)\pi_{i,j} \tag{5}$$

The idea underpinning the EM algorithm is to accommodate the hidden data, be re-couching the maximisation of the likelihood function in terms of the expected log-likelihood function. It was Dempster, Laird and Rubin [6] who originally showed that maximising the expected value of the log-likelihood function under hidden or missing data, was equivalent to maximising the following quantity

$$Q(\Phi^{(n+1)}|\Phi^{(n)}) = \sum_{i \in D} \sum_{j \in M} P(z_j|w_i, \Phi^{(n)}) \ln p(w_i|z_j, \Phi^{(n+1)}) \tag{6}$$

According to this viewpoint the *a posteriori* probabilities available at iteration n of the algorithm are used to compute the expectation value of the log-likelihoods of the missing data at iteration $n + 1$.

3.2 Maximisation

To develop a useful alignment algorithm we require a model for the measurement process. Here we assume that the observed position vectors, i.e. w_i are derived from the model points through a Gaussian error process. Suppose that the revised estimate of the position of the point w_i under the set of alignment parameters $\Phi^{(n)}$ is $w_i^{(n)}$. According to our Gaussian model of the alignment errors,

$$p(w_i|z_j, \Phi^{(n)}) = \frac{1}{2\pi\sqrt{|\Sigma|}} \exp\left[-\frac{1}{2}(z_j - w_i^{(n)})^T \Sigma^{-1}(z_j - w_i^{(n)})\right] \quad (7)$$

where Σ is the variance-covariance matrix for the point measurement errors. Here we assume that the position errors are isotropic, in other words the errors in the x and y directions are identical and uncorrelated. As a result we write $\Sigma = \sigma^2 I_2$ where I_2 is the 2x2 identity matrix. With this model, maximisation of the expected log-likelihood function $Q(\Phi^{(n+1)}|\Phi^{(n)})$ reduces to minimising the weighted square error measure

$$\mathcal{E} = \sum_{i \in \mathcal{D}} \sum_{j \in \mathcal{M}} w_{i,j}^{(n+1)}(z_j - w_i^{(n)})^T(z_j - w_i^{(n)}) \quad (8)$$

where we have used the shorthand notation $w_{i,j}^{(n+1)}$ to denote the *a posteriori* correspondence probability $P(z_j|w_i, \Phi^{(n)})$.

We would like to recover the maximum likelihood alignment parameters by applying Procrustes normalisation to the two point-sets. This involves performing singular value decomposition of a point-correspondence matrix. In order to develop the necessary formalism, we rewrite the weighted squared error criterion using a matrix representation. Suppose that $W^{(n)}$ is the data-responsibility matrix whose elements are the *a posteriori* correspondence probabilities $w_{i,j}^{(n)}$. With this notation the quantity $\mathcal{E}$ can be expressed in the following matrix form

$$\mathcal{E} = Tr[M^T M] - 2Tr[D^{(n+1)} W^{(n)} M^T] + Tr[D^{(n+1)T} D^{(n+1)}] \quad (9)$$

Since the first and third terms of this expression do not depend on the alignment of the point-sets we can turn our attention to maximising the quantity $\mathcal{F} = Tr[D^{(n+1)} W^{(n)} M^T]$, where $D^{(n+1)}$ is the revised matrix of point-positions which we aim to estimate via maximisation of $\mathcal{E}$. This quantity can be thought of as a weighted measure of overlap or correlation between the point-sets under the current alignment estimate. It is worth pausing to consider its relationship with measures exploited elsewhere in the literature on point pattern matching. The quantity $M^T D$ is simply the standard measure of overlap that is minimised in the work on least-squares alignment [17]. The matrix W, on the other hand, is just the inter-image proximity matrix used by Scott and Longuet-Higgins [11]. So, the utility measure delivered by the EM algorithm play a synergistic role. The inter-image point proximity matrix weights the least-squares criterion.

The quantity $\mathcal{F}$ can be maximised by performing a singular value decomposition. The procedure is as follows. The matrix $D^{(n+1)} W^{(n)} M^T$ is factorised into a product of three new matrices U, V and Δ, where Δ is a diagonal matrix whose elements are either zero or positive, and U and V are orthogonal matrices. The factorisation is $D^{(n+1)} W^{(n)} M^T = U\Delta V^T$.

The matrices U and V define a rotation matrix Θ which aligns the principal component directions of the point-sets M and D. The rotation matrix is equal to $\Theta = VU^T$.

With the rotation matrix to hand we can find the Procrustes alignment which maximises the correlation of the two point sets. The procedure is to first bring the centroids of the two point-sets into correspondence. Next the data points are scaled to that they have the same variance as those of the model. Finally, the scaled and translated point-sets are rotated so that their correlation is maximised.

To be more formal the centroids of the two point-sets are $\mu_D^{(n)} = E(w_i^{(n)})$ and $\mu_M = E(z_j)$. The corresponding covariance matrices are $\Sigma_D^{(n)} = E[(w_i^{(n)} - \mu_D^{(n)})(w_i^{(n)} - \mu_D^{(n)})^T]$ and $\Sigma_M = E[(z_j - \mu_M)(z_j - \mu_M)^T]$.

With ingredients the update equation for re-aligning the data-points is

$$w_i^{(n+1)} = \mu_M + \frac{Tr\Sigma_M}{Tr\Sigma_D}VU^T(w_i^{(n)} - \mu_D^{(n)}) \tag{10}$$

3.3 Expectation

In the expectation step of the EM algorithm the *a posteriori* probabilities of the missing data (i.e. the model-graph measurement vectors, z_j) are updated by substituting the revised parameter vector into the conditional measurement distribution. Using the Bayes rule, we can re-write the *a posteriori* measurement probabilities in terms of the components of the corresponding conditional measurement densities

$$P(z_j|w_i, \Phi^{(n+1)}) = \frac{\alpha_{i,j}^{(n)} p(w_i, z_j|\Phi^{(n)})}{\sum_{j' \in \mathcal{M}} \alpha_{j'}^{(n)} p(w_i, z_{j'}|\Phi^{(n)})} \tag{11}$$

The mixing proportions are computed by averaging the *a posteriori* probabilities over the set of data-points, i.e.

$$\alpha_{i,j}^{(n+1)} = \frac{1}{|\mathcal{D}|} \sum_{i \in \mathcal{D}} P(z_j|w_i, \Phi^{(n)}) \tag{12}$$

3.4 Structural Editing

The final step in the matching process is to edit the data point-set to remove unmatchable points which are noise or clutter. The aim here is to measure the consistency of the arrangement of correspondence matches on each neighbourhood of the Delaunay graph. We meet this goal by comparing the matched edges of the data-graph with those in the model graph.

In order to gauge the consistency of match, we represent the state of correspondence match between the nodes of the data-graph and those of the model graph using a function $f^{(n)} : \mathcal{D} \to \mathcal{M}$. The statement $f^{(n)}(i) = j$ means that the data-graph node i is matched to the model-graph node j at iteration n of the algorithm. The matches are assigned on the basis of the maximum *a posteriori* correspondence probabilities. In other words, $f^{(n)}(i) = \arg\max_{j' \in \mathcal{M}} P(z_j|w_i, \Phi^{(n)})$.

The pattern of the assigned matches is used to compute a probability of compatible correspondence arrangement. To compute this probability, we appeal to the model of structural pattern error recently reported by Wilson and Hancock [19] for graph-matching by discrete relaxation. Accordingly, we assume that the probability of erroneous edge insertion is P_e. By counting the number of consistently matched edges in the neighbourhood of each node we can the measure of consistency of correspondence match. The number of consistently matched edges for the correspondence match $f^{(n)}(i) = j$ is $H_{i,j} = \sum_{(i,i') \in E_D} (1 - \epsilon_{i,i'})$, where the consistency of each edge is measured by the quantity

$$\epsilon_{i,i'} = \begin{cases} 1 & \text{if } (j, f(i')) \in E_M \\ 0 & \text{otherwise} \end{cases} \tag{13}$$

With this consistency measure to hand, the probability of compatible correspondence match is

$$\zeta_{i,j}^{(n)} = \frac{\exp\left[-\mu H_{i,j}\right]}{\sum_{j' \in \mathcal{M}} \exp\left[-\mu H_{i,j'}\right]} \tag{14}$$

where $\mu = \ln \frac{1-P_e}{P_e}$. This probability is used to make decisions concerning the deletion of nodes from the data graph that fail to find a consistent correspondence match. The node i is deleted if $\zeta_{i,j}^{(n)} < P_e$. Once a point has been deleted the remaining points are retriangulated.

4 Experiments

In this section we provide some experimental evaluation of the new alignment method. This is divided into two sections. First, we provide a simulation study to provide some sensitivity characteristics. second, we provide some real world experiments on stereo images.

To evaluate the robustness of the novel approach, we furnish a sensitivity study which compares the new iterative alignment method with the standard non-iterative Procrustes alignment. We have used randomly generated point-sets. We have added controlled fractions of clutter to the random point-sets. Figure 1a shows the comparison. When the clutter level is less than 10%, the EM alignment method always outperforms non-iterative Procrustes alignment.

Our next experiment aims to determine the accuracy of the new alignment method under increasing levels of clutter. From figure 1b we can see that if the structural corruption is lower than 10%, the average registration error is reasonably low(under 2%).

Finally, we provide some examples on real-world data. Here we use pairs of stereo images of an office scene to test the proposed algorithm. Figure 2 shows the correspondences using conventional Procrustes alignment. There exist significant correspondence errors. When the EM method is used, most of the false correspondences are removed (Figure 3).

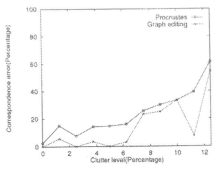

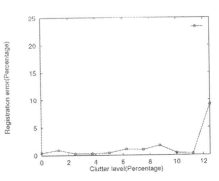

(a) a. Comparison of the correspondence error

(b) b. Registration error subject to
the clutter level

Fig. 1. Sensitivity study

Fig. 2. Correspondence after Procrustes alignment

5 Conclusions

In conclusion, we have shown how the process of Procrustes alignment can be
formulated as a maximum likelihood estimation problem using the EM algorithm. This interpretation leads to a new point-set similarity measure in which
point correspondence probabilities weight the standard least-squares point overlap distance. In other words, our new measure of point-set similarity combines
the ideas already developed by Scott and Lonquet-Higgins, and, Umeyama in a
single statistical utility measure. Moreover, our new method both allows structural constraints to be imposed on Procrustes alignment and provides a frame-

Fig. 3. Correspondence after graph editing

work for point-set editing to remove noise and clutter. The method leads to more accurate point-set alignment.

References

1. N. Ahuja, Dot Pattern Processing using Voronoi Neighbourhoods, *IEEE PAMI* , **4**(1982), 336-343.
2. N. Ahuja and B. An and B. Schachter, Image Representation using Voronoi Tessellation, *CVGIP*, **29**(1985), 286-295.
3. N. Ahuja and M. Tuceryan, Extraction of Early Perceptual Structure in Dot Patterns: Integrating Region, Boundary and Component Gestalt, *CVGIP*, **48**(1989), 304-356.
4. Amit, Y. and Kong, A., Graphical Templates For Model Registration, *PAMI*, **18**(1996), 225-236.
5. T.F Cootes and C.J. Taylor and D.H. Cooper and J. Graham, Active Shape Models - Their Training and Application, *CVIU*, **61**(1995), 38–59.
6. A.P. Dempster and N.M. Laird and D.B. Rubin, Maximum-likelihood from incomplete data via the EM algorithm, *J. Royal Statistical Soc. Ser. B (methodological)*, **39**(1977), 1-38.
7. O.D. Faugeras and E. Le Bras-Mehlman and J-D. Boissonnat, Representing Stereo Data with the Delaunay Triangulation, *Artificial Intelligence*, **44**(1990), 41-87.
8. D. G. Kendall, Shape manifolds,Procrustean metrics, and complex projective spaces, *Bulletin of the London Mathematical Society*, **16**(1984), 81-121.
9. M. Lades and J.C. Vorbruggen and J. Buhmann and J. Lange and C. von der Maalsburg and R.P. Wurtz and W.Konen, Distortion-invariant object-recognition in a dynamic link architecture, *IEEE Transactions on Computers*, **42**(1993), 300–311.
10. S. Sclaroff and A.P. Pentland, Modal Matching for Correspondence and Recognition, *IEEE PAMI*, **17**(1995), 545-661.
11. G.L. Scott and H.C. Longuet-Higgins, An Algorithm for Associating the Features of 2 Images, *Proceedings of the Royal Society of London Series B-Biological*, **244**(1991), 21-26.
12. L.S. Shapiro and J. M. Brady, Feature-based Correspondence - An Eigenvector Approach, *IVC*, **10**(1992), 283-288.
13. L.S. Shapiro and J. M. Brady, Rejecting Outliers and Estimating Errors in an Orthogonal-regression Framework, *Phil. Trans. Roy. Soc. A*, **350**(1995), 403-439.
14. M. Tuceryan and T Chorzempa, Relative Sensitivity of a Family of Closest Point Graphs in Computer Vision Applications, *Pattern Recognition*, **25**(1991), 361-373.
15. S. Ullman, The Interpretation of Visual Motion, *MIT Press*,(1979).
16. S. Umeyama, An Eigen Decomposition Approach to Weighted Graph Matching Problems, *IEEE PAMI*, **10**(1988), 695-703.
17. S. Umeyama, Least Squares Estimation of Transformation Parameters between Point sets, *IEEE PAMI*, **13**(1991), 376-380.
18. S. Umeyama, Parameterised Point Pattern Matching and its Application to Recognition of Object Families, *IEEE PAMI*, **15**(1993), 136-144.
19. R.C. Wilson and E.R. Hancock, Structural Matching by Discrete Relaxation, *IEEE PAMI*, **19**(1997), 634-648.

Structural Constraints for Pose Clustering

Simon Moss and Edwin R. Hancock

Department of Computer Science,
University of York,
York, Y01 5DD, UK.

Abstract. This paper describes a structural method for object alignment by pose clustering. The idea underlying pose clustering is to decompose the objects under consideration into k-tuples of primitive parts. By bringing pairs of k-tuples into correspondence, sets of alignment parameters are estimated. The global alignment corresponds to the set of parameters with maximum votes. The work reported here offers two novel contributions. Firstly, we impose structural constraints on the arrangement of the k-tuples of primitives used for pose clustering. This limits problems of combinatorial background and eases the search for consistent pose clusters. Secondly, we use the EM algorithm to estimate maximum likelihood alignment parameters. Here we fit a mixture model to the set of transformation parameter votes. We control the order of the underlying mixture model using a minimum description length criterion.

1 Introduction

Pose clustering [10, 11, 5, 7] is a voting method that allows recognition to be achieved via object alignment. Although it shares the feature of voting with techniques such as geometric hashing [15] and the generalised Hough transform [1], it has the unique feature of explicitly recovering the transformation parameters required to align the component parts of shapes. The idea underpinning the method is to construct a distributed object representation. Both model and data are decomposed into subparts or k-tuples of object-primitives. By bringing each data k-tuple into correspondence with each k-tuple of the model, a set of alignment parameters is generated. The different parameter values resulting from the putative k-tuple correspondences are used as votes from the global alignment parameters. Recognition is effected by searching for local maxima in the vote accumulator. The main advantage over direct alignment or template matching methods [12, 14, 6] is robustness to the addition of clutter or partial occlusion and the use of a distributed object representation.

One of the problems which limits pose clustering is that of combinatorial background. The main problem stems from the fact that all model and data k-tuples must be permuted and compared in order to explore the set of potential primitive correspondences before the set of alignment votes can be accumulated. This not only implies a considerable computational overhead, it also means that considerable ingenuity must be expended in searching for the set of consistent alignment parameters which may become swamped by background. The aim in this paper is to consider how relational constraints on the arrangement of the k-tuples can be used to control the problems of background.

The idea underpinning our method is as follows. We commence by imposing a graph-structure on the raw image primitives used to construct the k-tuples. If the primitives are represented by points (e.g. corners, line-centres or region centroids) then the graph could be a neighbourhood structure such as the N-nearest neighbour graph, the Delaunay triangulation, the Gabriel graph or the relative neighbourhood graph. If the primitives are line-segments then the constrained Delaunay graph or a Gestalt arrangement graph would be more appropriate. Once the relational graph has been computed, then it may be used to impose constraints on the arrangement of primitives into k-tuples. We investigate two alternatives here. The simplest is to restrict our search to pairs of primitives connected by edges in the relational graph. A more sophisticated alternative is to use the cyclic ordering of the edges in the neighbourhood graph as a constraint. By imposing structural constraints on the set of votes we can both limit the amount of computation required and reduce the degree of ambiguity. Once the restricted set of votes is to hand, then the search for consistent sets of model-votes can be posed as a statistical density estimation process. Here we use the EM algorithm [2] to fit a Gaussian mixture model to the local maxima of the vote accumulator.

2 Representation

The task confronting us is to find all the ways in which a set of model primitives can be aligned with their counterparts in a set of data. The model-primitives have index-set $\mathcal{M}$ while the data-primitives have index-set $\mathcal{D}$. Each primitive is characterised by a vector of attributes; in the case of the data-primitives the attributes are denoted by the set $A_D = \{\underline{x}_i^D; i \in \mathcal{D}\}$ while in the case of the model the attribute-set is $A_M = \{\underline{x}_j^M; j \in \mathcal{M}\}$. As a concrete example, in the case where the primitives are points, then the attribute-vector represent coordinates in 2D or 3D. In the case of line-primitives, the attribute vector could be conveniently represented using the line centre-points and the line-orientation [6].

We would like to transform the attribute information into a space of transformation parameters which can be used for the purposes of recognition. These parameters represent possible transformations that bring the model and data primitives into alignment with one-another. To meet this goal we proceed as follows. We first select an appropriate image transformation. The transformation $T(\phi)$ is specified by a vector of parameters ϕ. Depending on the chosen transformation and the nature of the image attributes, the transformation can be estimated using appropriately sized subsets of the model and data primitives. For instance, in the case of affine transformation between 2D point-sets, then three primitives are sufficient. In order to facilitate the transformation to the parameter representation, we extract all tuples of the appropriate size from the data and model primitive sets.

Suppose that k primitives are necessary to estimate the parameter-vector ϕ for the transformation T. Furthermore, let $\underline{t}_i^D = \{u_1, ..., u_k\} \subset \mathcal{D}$ denote the index-set of the primitives forming one of the data k-tuples. The set of available data k-tuples is denoted by $\Theta_D = \{\underline{t}_i^D\}$. The corresponding model k-tuple is denoted by the set of primitives $\underline{t}_j^M = \{v_1, ..., v_k\} \subset \mathcal{M}$ and the complete set of model k-tuples is denoted by $\Theta_M = \{\underline{t}_j^M\}$.

The idea underpinning pose clustering is to pair each of the data and model k-tuples with one-another in turn. However, ad $initio$ we do not know the correspondences between individual primitives constituting the different k-tuples. In order to estimate the relevant transformation parameters we must therefore explore the permutation structure between the different data and model k-tuples under study. The procedure is as follows: For each pair of k-tuples we generate a set of correspondence assignments between the individual primitives. For the k-tuples $\underline{t}_i^D$ and $\underline{t}_j^M$ the set is denoted by $\chi_{i,j}$. Each member $\kappa = \{(u_{l_1}, v_{m_1}), ..., (u_{l_k}, v_{m_k})\} \subseteq \underline{t}_i^D \times \underline{t}_j^M$ of $\chi_{i,j}$ is itself a set of Cartesian pairs which associate the individual primitives in the two k-tuples. According to this notation, the Cartesian pair (u_l, v_m) means that data-primitive $u_l \in \mathcal{D}$ is paired with the model primitive $v_m \in \mathcal{M}$ by the permutation κ. In other words κ is a set of correspondence assignments between data and model primitives. The individual correspondences, i.e. (u_l, v_m), are found by permuting the order of the model k-tuple primitives with respect to those in the data k-tuple. The complete set of k-tuple correspondences available for estimating the set of transformation parameter votes is therefore, $\mathcal{P} = \bigcup_{(i,j) \in \Theta_D \times \Theta_M} \chi_{i,j}$. For every k-tuple permutation belonging to $\mathcal{P}$, we use the set of correspondences between the data and model primitives to estimate a vector of transformation parameters. In other words, the putative set of correspondences κ between the data k-tuple $\underline{t}_i^D$ and the model k-tuple $\underline{t}_j^M$ is used to estimate the parameter vector ϕ_κ that bring the individual primitives forming the two k-tuples into alignment. The measurements used to compute the parameter-vector are $a_i^D = \{\underline{x}_l^D; l \in \underline{t}_i^D\}$ and $a_j^M = \{\underline{x}_m^M; m \in \underline{t}_j^M\}$. Hence, the individual attributes vectors for the paired primitives satisfy the equation $\underline{x}_v^M = T(\phi_\kappa, \underline{x}_u^D)$. The entire set of available pairwise transformation parameter estimates is denoted by $\Phi = \{\phi_\kappa; \kappa \in \mathcal{P}\}$.

It is the process of exploring the permutation structure of the set of k-tuple correspondences that poses the main computational bottleneck to pose clustering. The alternative ways of constructing the correspondence-set $\mathcal{P}$ is the subject of the next section of this paper.

3 Tuple Pairing

The idea underpinning this paper is to investigate whether relational constraints can be used to restrict and refine the set of transformation votes belonging to the set Φ. To commence we must establish a relational structure on the set of primitives in the model and the data. There are a number of alternatives reported in the literature. For point-features, these include the N-nearest neighbour graph, the Delaunay graph, together with the Gabriel graph and the relative neighbour graph (which are both obtained by pruning the Delaunay graph). However, as recently demonstrated by Tuceryan and Chorzempa [13], it is the Delaunay graph that is the most robust of these structures. That is to say the edge-set of the Delaunay graph is the least sensitive to the addition and deletion of random nodes. Moreover, since the Delaunay graph is planar, we can exploit the cyclic ordering of edges as an additional structural constraint.

We have investigated three different strategies for generating the set of measurement tuple pairings χ. These are listed below

- **Permutations:** Here the set tuple primitive-pairings is constructed by taking the complete set of possible permutations of the individual k-tuples. This is the conventional strategy adopted in the literature. There is no attempt to either impose structure on the feature-sets nor to impose constraints on the consistency of the pairings.
- **Structural Gating:** The aim here is to impose a neighbourhood structure on the feature-sets. For instance, if we are dealing with point-sets, then we commence by either constructing an N-nearest neighbour graph or a Delaunay triangulation. The edges of the graph are adjacency relations. Suppose that $G_D = (\mathcal{D}, E_D)$ is the graph representing the model feature points, where E_D is the set of edges of the neighbourhood graph. Similarly, let $G_M = (\mathcal{M}, E_M)$ represent the neighbourhood graph for the model-points. We exploit the graph structure in two ways. Firstly, the sets of primitives used to construct the different k-tuples are interconnected by edges, i.e. $(u_1, u_2) \in \underline{t}_j^D \Rightarrow (u_1, u_2) \in E_D$. Secondly, the set of putative correspondences $\mathcal{P}$ is constructed so that the edge structure of the graphs is used as a constraint. In other words, edges in the data-graph are mapped onto edges in the model graph. As a result $(u_1, v_1) \in \kappa \wedge (u_2, v_2) \in \kappa \Rightarrow (u_1, u_2) \in E_D \wedge (v_1, v_2) \in E_M$.
- **Cyclic Ordering:** The aim here is to preserve the cyclic ordering of the edges in the neighbourhood graphs when pairing the tuples to estimate pose parameters. Here we appeal to the idea recently developed by Wilson and Hancock [16] of using a dictionary of structure preserving mappings as a constraint representation for matching relational graphs. Each k-tuple now consists of a centre-node together with the set of nodes connected to it by edges of the Delaunay graph. In other words, $\underline{t}_i^D = i \cup \{i'; (i, i') \in E_D\}$. The set of correspondences is generated so that both the edge-structure and the cyclic ordering of the non-centre nodes is preserved. This process is realised by placing the non-centre nodes in a list. The ordering of this list is the cyclic order of the primitives on the image plane around the centre-node. Each set of k-tuple correspondences is such that the centre nodes are always paired with one another. The neighbourhood nodes are paired so that the cyclic ordering of the two lists is preserved.

In order to illustrate the effects of the different constraints, we now show some examples of the pose-votes accumulated using each of the schemes described above. We choose as our example the problem of estimating the parameters of Euclidean transformation from point-sets..In Figure 1 we show the set of accumulated votes in the scale and rotation dimensions of the transformation-space. Specifically, Figure 1a shows the set of votes obtained with the permutation set, Figure 1b is the result obtained using structural gating, and Figure 1c is the result obtained with cyclic ordering constraints. In each plot there is a clear spike due to a single dominant pose cluster. The main effect of each of the refinements is to increase the strength of the cluster with respect to the background. This means that the transformation parameters are more easily estimated. In the simple example shown here a straightforward thresholding procedure would suffice. However, even with the use of structural constraints in some pose parameter estimation tasks the structure of the votes may be more complicated due to multiple model instances. For this reason, in the next section of this paper we present a mixture model which can be used to fit Gaussian kernels to the accumulator peaks and hence locate consistent sets of pose parameters.

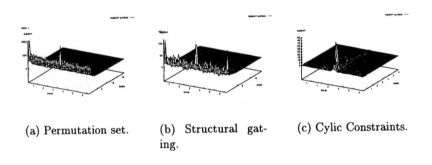

(a) Permutation set.   (b) Structural gating.   (c) Cylic Constraints.

Fig. 1. Pruning the pose votes.

4   Estimating Pose Parameters

Our aim is to model the distribution of estimated transformation parameters that result from the pose clustering process. Each element of the set Φ represents a vote for a particular alignment hypothesis. The modes of the distribution represent instances where several paired k-tuples vote for the same hypothesis. If there are several instances of the model contained within the data, then multiple modes can be anticipated. Our aim in this paper is to model this multi-modal structure using a mixture model. Suppose that we have K mixing components modelling the different modes of our transformation parameter-space. The mixture distribution for the observed parameter-vector ϕ is

$$P(\phi) = \sum_{\alpha=1}^{K} p(\phi)|\omega_\alpha, \theta_\alpha) P(\theta_\alpha) = \sum_{\alpha=1}^{K} g(\phi, \omega_\alpha) \pi_\alpha \qquad (1)$$

In the above formula $p(\phi|\omega_\alpha, \theta_\alpha)$ is the conditional density function and $P(\theta_\alpha)$ is the class-prior for the component of the mixture model indexed α. The conditional density function is modelled by the kernel-function $g(\phi, \omega_\alpha)$. The choice of kernel-function depends on the problem in-hand and is deferred until later. Suffice to say that the shape kernel is controlled by a hyper-parameter vector ω_α which regulates position and width. Finally, the class prior is modelled by the mixing proportion π_α.

The process of model alignment is couched as one of estimating the set of kernel hyper-parameters $\Omega = \{\omega_\alpha; \alpha = 1, K\}$. We pose this problem as one of maximising the likelihood function

$$\mathcal{L}(\Phi|\Omega) = \prod_{\phi \in \Phi} \sum_{\alpha=1}^{K} g(\phi, \omega_\alpha) \pi_\alpha \qquad (2)$$

We solve this density estimation problem [9] using the EM algorithm [2]. The steps are as follows

- **Expectation:** In this step we compute the *a posteriori* measurement probabilities using the most recently available estimates of the kernel-functions

$$P(\theta_\alpha|, \omega_\alpha, \phi) = \frac{g(\phi, \omega_\alpha) \pi_\alpha}{\sum_{\beta=1}^{K} g(\phi, \omega_\beta) \pi_\beta} \qquad (3)$$

The mixing proportions are found by averaging the it a posteriori probabilities over the set of observations i.e. $\pi_\alpha = \frac{1}{|\Phi|} \sum_{\phi \in \Phi} P(\theta_\alpha|\omega_\alpha, \phi)$.
- **Maximisation:** In the maximisation step we seek the set of kernel parameters $\Omega^{(n+1)} = \arg\max_\Omega Q(\Omega|\Omega^{(n)})$ that maximise the expected log-likelihood function

$$Q(\Omega^{(n+1)}|\Omega^{(n)}) = \sum_{\phi \in \Phi} \sum_{\alpha=1}^{K} P(\theta_\alpha|\omega_\alpha^{(n)}, \phi) \log g(\phi, \omega_\alpha^{(n+1)}) \qquad (4)$$

- **Mixture Kernels:** If we assume that measurement errors in the raw attribute information follow a multivariate Gaussian distribution, then provided that we are using a linear co-ordinate transformation then the distribution in the parameter-vectors will also be Gaussian. This is certainly the case for the Euclidean and affine transformation of point-sets. Moreover, certain non-linear transformation, such as that resulting under perspective geometry, can be approximated in a linear fashion. Notwithstanding these difficulties and in the interests of tractability, we pursue the Gaussian case using kernel functions of the following form

$$g(\phi, \omega_\alpha) = \frac{1}{(2\pi)^{\frac{d}{2}} \sqrt{|\Sigma|}} \exp[-\frac{1}{2}(\phi - \mu_\alpha)^T \Sigma^{-1}(\phi - \mu_\alpha)] \qquad (5)$$

where μ_α is the mean parameter-vector, Σ_α is the variance-covariance matrix for the mixing kernel indexed α and d is the length of the parameter vector ϕ..

– **Model-order Selection:** The final issue that remains to be addressed is that of model-order selection [4, 9]. As with all density estimation we must prove a means of selecting the number of parameters, or kernels, that we use to model the data. Here we use the minimum description length (MDL) criterion [8]. Suppose that σ^2 is the total sample variance and σ_α^2 is the variance for the mixing component indexed α, then the MDL criterion for K mixing components is given by

$$MDL(K) = -\frac{1}{2}K \ln\left[|\Phi|\right] + \left[\sigma^2 - \sum_{\alpha=1}^{K} \sigma_\alpha^2\right] \qquad (6)$$

The model order is selected by varying K until $MDL(K)$ is minimised.

5   Experiments

Our experimental evaluation of the new matching scheme is based on matching 2D line-segment patterns under Euclidean transformation. Details of the transformation process our outside the space limitations of this paper. However, a full description can be found in a recent account which involves the application of a conventional EM approach to image registration in low quality radar imagery [6]. Each line segment is characterised by its centre-point location (x_i, y_i) and its orientation θ_i. From the relative position and relative orientation of pairs of line-segments drawn from the model and data images, we estimate the four parameters of Euclidean transformation, i.e. the two components of translation together with the isotropic scale and the overall rotation. In other words our alignment problem involves applying the EM algorithm to a 4-dimensional parameter-space. In order to provide structural constraints for the alignment process, we have seeded Delaunay triangulations from the centre-points of the lines.

The evaluation of the alignment process involves aligning a digital map against linear structures segmented from radar images. The data used in this study is described fully in the recent paper by Wilson and Hancock [16]. Here the line-segments represent hedge-structures in a rural landscape. The problem is a challenging one since there is significant clutter and the segmentation process used to locate the line segments is far from perfect. To underline this point, there are 95 line segments in the radar data and only 30 segments in the digital map. The radar data is shown in Figure 2a, the extracted linear segments are shown in Figure 2b and the digital map is in Figure 2c. In Figure 2d we show the final alignment of the map and radar data. Despite the relatively poor radar-image segmentation and the high degree of clutter, the overall alignment is good.

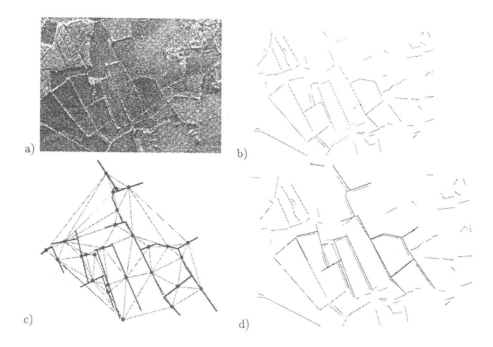

Fig. 2. Aerial image alignment: a) Original image, b) Extracted hedge features, c) Map lines, d) Final alignment.

6 Conclusions

We have presented a new pose clustering algorithm for object alignment. There are two novel ideas underpinning the work. Firstly, we use structural constraints provided by a Delaunay triangulation to limit the set votes that are used in the voting process. Two sets of constraints are used. The first of these are provided by graph-edges. The second set is provided by cyclic ordering constraints on the graph neighbourhoods. In practice, it is the second set of constraints that prove to be the most effective in limiting the combinatorial background in the voting process. Secondly, we adopt a density estimation framework and use the EM algorithm to locate pose clusters.

There are a number of ways in which which we intend to extend the ideas presented in this paper. he most pressing of these is to consider the issue of model-order selection more closely. Here we have used an MDL criterion [8]. We have two avenues in mind for future investigation. Firstly, we intend to explore the bias-variance structure of the mixture model and to set the model-order so as to provide the best tradeoff between model-bias and fit-variance. The second avenue is more ambitious. Richardson and Green [3] has recently shown how reversible-jump Markov chain Monte-Carlo can be used to stochastically sample variable order mixture models. This would provide a principled way off locating alignment parameters that are globally optimal.

References

1. Ballard D., " Generalising the Hough Transform to Detect Arbitrary Shapes", *Pattern Recognition*, **13**, pp. 111-122, 1981.
2. Dempster A.P., Laird N.M. and Rubin D.B., "Maximum-likelihood from Incomplete Data via the EM Algorithm", *J. Royal Statistical Soc. Ser. B (methodological)*,**39**, pp 1-38, 1977.
3. Richardson S. and P.J. Green, "On Bayesian analysis of mixtures with an unknown number of components", *Journal of the Royal Statistical Society Series B-Methodological*, .**59**, pp.731-758, 1997.
4. Geman D., E. Bienenstock and R. Doursat, "Neural networks and the bias variance dilemma", *Neural Computation,*, 4, pp.1-58, 1992.
5. Hel-Or Y. and Werman M, "Pose Estimation by Fusing Noisy Data of Different Dimensions", *IEEE PAMI*, **17**, pp. 195–201, 1995.
6. Moss S. and Hancock E.R., "Registering Incomplete Radar Images with the EM Algorithm", *Image and Vision Computing*, to appear, 1997.
7. Olson C.F., "Probabilistic Indexing for Object Recognition", *IEEE PAMI*, **17**, pp. 518–522, 1995.
8. Rissanen J., "Minimum Description Length Principle", *Encyclopedia of Statistic Sciences*, **5**, pp. 523–527, 1987.
9. Silverman B.W., "Density Estimation for Statistics and Data Analysis", *Chapman-Hall*, 1986.
10. Stockman G., "Object Recognition and Localisation via Pose Clustering", *CVGIP*, **40**, pp. 361–387, 1987.
11. Stockman G, Kopstein S and Benett S., "Matching Images to Models for Registration and Object Detection", *IEEE PAMI*, **4**, pp. 229–241, 1982.
12. Ullman S., " An Approach to Object Recognition: Aligning Pictorial Descriptions", *Cognition*, **32**, pp. 193-254, 1989.
13. Tuceryan M. and T. Chorzempa, "Relative Sensitivity of a Family of Closest-point Graphs in Computer Vision Applications", it Pattern Recognition, **25**, pp. 361–373, 1991.
14. Viola P. and Wells W., "Alignment by Maximisation of Mutual Information", *Proceedings of the Fifth International Conference on computer Vision*, pp. 16–23, 1995.
15. Woolfson H., " Model-Based Object Recognition by Geometric Hashing", ECCV90(526-536).
16. Wilson R.C. and Hancock E.R., "Relational Matching by Discrete Relaxation", *IEEE PAMI*, **19** pp. 634–648, 1997.

Subpixel Stereo Matching by Robust Estimation of Local Distortion Using Gabor Filters *

Peter Werth[1,2], Stefan Scherer[1], and Axel Pinz[1]

[1] Institute for Computer Graphics and Vision, Graz University of Technology,
Inffeldgasse 16/2, A-8010 Graz, Austria
werth@icg.tu-graz.ac.at
http://www.icg.tu-graz.ac.at
[2] Erich-Schmid-Institut of Materials Science, Austrian Academy of Sciences,
Jahnstr. 12, A-8700 Leoben, Austria

Abstract. Area-based stereo matching algorithms are based on the support of a surrounding area to establish a point-to-point correspondence in two images. Two important problems arise in this context: How to obtain subpixel information and how to choose the optimal surrounding area.

In this paper we present a non-iterative two-step algorithm for subpixel accurate stereo matching by using an adaptive window. In contrast to existing algorithms the window is not restricted to a rectangle but can be of any general shape. Starting from an initial sparse disparity estimate, the first step is to find the general shape of the window. This is performed by estimating the local disparity of each pixel in a box of maximum size using a bank of Gabor filters, and by applying a consistency constraint. In the second step the projective distortion is computed using the masked window. The performed experiments show the accurate and robust behavior of the proposed algorithm.

Keywords. Subpixel Stereo Matching, Adaptive Window, Multi-channel Gabor Filtering

1   Introduction

Shape-from-stereo is one of the important techniques in recovering the third dimension of a real world scene projected onto a 2D image plane. Prerequisite is the determination of the visual correspondence between two points referring to the same scene point known as homologue points.

One major group of algorithms to establish this correspondence consider a certain area around selected points. For the sake of simplicity and computational expense this area is mostly of rectangular shape. Common techniques as normalized cross-correlation or minimum SSD do only provide pixel accuracy and suffer from projective distortion within the supporting area, from now on referred to

* This work was supported by the Austrian Science Foundation (FWF) under grant P12278-MAT.

as the window. Therefore techniques were developed to improve the accuracy by least squares methods applied to the window [3], [4], [9]. Others propose to adjust the size of the window itself [8], [2] or provide both, a subpixel refinement of the correspondence and an estimation of the window [12]. A well known technique of the latter category is the iterative adaptive window algorithm proposed by Kanade and Okutomi [7]. A work closely related to ours was published by T. Lindeberg [10]. There he mentions that the local projective deformations contain essential information for binocular stereo matching algorithms.

In this paper we present a non-iterative two-step algorithm for accurate stereo correspondence. Starting from initial homologue points the optimal shape of the window is estimated. This shape is not restricted to a rectangle, but can be of any shape. This can be achieved by enforcing a new global consistency constraint on the local disparities. A bank of oriented, even-symmetric Gabor filters, originally designed for texture segmentation [6], is used to determine these disparities. In the second step the knowledge of the local disparities is used to establish a subpixel refinement by estimating a projective transformation between the stereoscopic windows. A strategy for how and where to find initial disparity estimates are given in the next section.

2   Strategies to Establish an Initial Correspondence

Correspondences can be established at image positions ignoring the content of the image itself, e.g. at every n'th position. On the other hand, these positions can be selected regarding information contained in the image, leading to the so called points of interest. These points have to represent image features which are discrete, and do not form a continuous structure like pixels belonging to edges or texture. The points of interest for the proposed algorithm are determined by first applying the well known corner detector introduced by Harris and Stephens [5] on a regular grid over the image plane. Each grid point defines an exclusive surrounding area, whose best corner-like points of interest will be selected. This guarantees a nearly uniform distribution of the selected locations, since they are lying on a so called structured grid.

In order to determine the approximate correspondences of the points of interest a minimization of the sum of squared differences (SSD) with a fixed window size of 15×15, is applied. Martin and Crowley show in [11] that SSD minimizes the probability of error for white Gaussian additive noise. Trapp shows in [13] that for direct use of the phase-differences the Fourier shift theorem must strictly apply. That is not true for our case, so that's why we proceed with a minimum SSD method.

3   Multi-channel Gabor Filtering

In the following we want to summarize the mathematical background of a bank of Gabor filters, mainly following the work of Jain and Farrokhnia [6]. Image features coding information about scale and orientation are derived from the responses of the filter bank. In contrast to the original work we do not use the image features for texture classification, but to represent a location (ξ, η) within the local window (see section 4). The impulse response of a real, even-symmetric Gabor filter is given by

$$h(\xi, \eta) = \frac{1}{2\pi\sigma_\xi\sigma_\eta} e^{-\frac{1}{2}\left(\frac{\xi^2}{\sigma_\xi^2} + \frac{\eta^2}{\sigma_\eta^2}\right)} \cos(2\pi u_0 \xi) \tag{1}$$

where u_0 is the frequency of the sinusoidal plane wave in the $0°$ direction. σ_ξ and σ_η are the shaping parameters of the Gaussian envelope function along the two axes of the window coordinate system, and are calculated as follows:

$$\sigma_\xi = \frac{\sqrt{\frac{\ln 2}{2}}}{\pi u_0} \cdot \frac{(2^{B_f} + 1)}{(2^{B_f} - 1)}, \qquad \sigma_\eta = \frac{\sqrt{\frac{\ln 2}{2}}}{\pi u_0 \tan\left(\frac{B_\phi}{2}\right)}. \tag{2}$$

The filters sensitive to the desired orientations are obtained by rotating the image coordinate system. Since all filter outputs are computed in the frequency domain, the stereo images have to be transformed by a 2D Fourier transform (FFT). It is noteworthy that the input for the multi-channel filtering process is a window which is appropriately sized around the given point of interest. This ensures that the passband of the highest frequency filter is located inside the window.

Concerning the bank, a filter is placed every $\phi = 15°$ with an orientation bandwidth of $B_\phi = 10°$. The number of radial frequency bands is determined by the lowest wavelength reasonable for disparity estimation of $2\sqrt{2}$ pixels, and by the highest of $(nw_{max}/4)\sqrt{2}$ pixels, where nw_{max} must be specified as the maximum pixel width of the window. The relative frequency bandwidth is $B_f = 1$ octave. These parameters provide an approximately uniform covering of the frequencies of interest, in a sufficient number of orientation discretization. A detailed description of the filter bank parameter is given in [6]. After the filtering with all modulation transfer functions (MTF) corresponding to the Fourier domain representation of each single filter, an inverse FFT is performed on all filter responses. Fig. 1 shows the half-peak supports of the modulation transfer functions of the used filter set, with overlapping regions set to black. The inverse FFT is done for each filter, since the essential normalization is done much faster in the spatial domain.

Fig. 1: The applied Gabor filter bank in the frequency domain.

4 The Two-Step Algorithm for Subpixel Matching

In section 2 we showed how the initial disparity map is obtained. In the last section we presented the mathematical background for building up a filter bank. In this section the two-step algorithm for subpixel matching is presented.

4.1 Step One - Finding the Window Shape

The first step is to find the general shape of the window. In order to determine this shape a dense disparity map is estimated within a maximum window. For every pixel (ξ, η) in the left window the corresponding pixel in the right window is determined. This local matching is performed by first assigning a feature vector **r** to every pixel in the left window, for which a correspondence has to be found. Each dimension d_k of this feature vector represents a single filter response in the spatial domain of the multi-channel Gabor filter bank. The window coordinates of the pixels are treated as features as well, in order to incorporate the probability of spatial adjacency. The feature vector is normalized by setting the energy of each local neighborhood as well as the energy of the filter coefficients to unity. Thus the features are weighted properly and the output can be treated as invariant against irregular intensity variations.

The local matches are now determined by minimal distance matching of the reference vectors in the left window and the vectors in the right window. In order to enhance robustness, in the following a new consistency constraint is introduced. This constraint can be seen as a simplified combination of the well known order constraint [1] and the epipolar constraint. The application of the constraint leads then either to a masked window with consistent, thus unambiguous, information, or to a complete rejection of the initial disparity estimate.

First, for each pixel in the left window the direction to its corresponding pixel in the right window is calculated, leading to an angle $\alpha(\xi, \eta)$ for each pixel position. In Fig. 2 a) some of the resulting vectors are shown. Using these vectors, the median angle α_m of all angles $\alpha(\xi, \eta)$ is calculated in the range $(-\pi, \pi)$ by

$$\alpha_m = \text{median}\left\{\arctan\left[\frac{d_h(\xi, \eta)}{d_v(\xi, \eta)}\right]\right\} , \tag{3}$$

where $d_h(\xi, \eta)$ and $d_v(\xi, \eta)$ are the horizontal and vertical disparities in pixel, respectively. ξ and η range from zero to $(x-1)$ and $(y-1)$, where x and y are the dimensions of the window. Once α_m is known, an angle $\alpha_r(\xi, \eta)$ can be defined

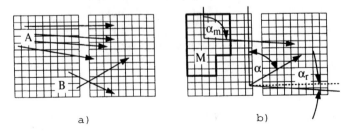

Fig. 2. Obtaining the shape of the window a) Local disparities within the window; region A represents a consistent region, region B is inconsistent. b) Final window shape M by applying the introduced constraint.

for each local disparity by

$$\alpha_r(\xi,\eta) = \left| \arcsin \left(\frac{1/2}{\sqrt{d_v^2(\xi,\eta) + d_h^2(\xi,\eta)}} \right) \right| , \qquad (4)$$

This angle defines a threshold for the final shape of the window, where the factor $(^1/_2)$ defines the maximum allowed deviation from the median direction. We now use α_r to define the mask M, which represents the final shape of the window as follows:

$$M(\xi,\eta) = \begin{cases} 1 & : & (\alpha_m - \alpha_r(\xi,\eta)) < \alpha(\xi,\eta) < (\alpha_m + \alpha_r(\xi,\eta)) \\ 0 & : & elsewhere \end{cases} . \qquad (5)$$

Fig. 2 b) illustrates the determination of the mask M where the region A, shown in Fig. 2 a) does not violate the introduced consistency constraint, while the region B does violate the constraint and thus does not contribute to M. Fig. 3

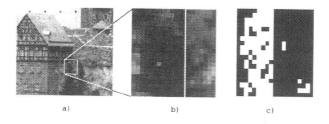

a)   b)   c)

Fig. 3. Adaptive window on a real image. a) Part of a real image. b) The template; the white line denotes the border between wall and background. c) Resulting window shape with correctly detected occlusion.

shows the mask obtained from a real image. Fig. 3 b) shows the window to be matched. The vertical white line denotes the border between the wall and the background shown in Fig. 3 a). In the window of the right image, the right region results in an occlusion, detected by our algorithm and masked out in the final shape of the window (Fig. 3 c)).

4.2   Step Two - Estimating the Projective Transformation

Once the mask is determined, an eight-parameter projective transformation $\mathbf{T}$ between the two windows is calculated. As outlined by Lindeberg in [10], the projective transformation by which textures are deformed when mapped onto the image plane of a camera, represents useful geometric information. For each corresponding point given in homogeneous 2D coordinates, a (3×3) transformation matrix $\mathbf{T}$ may be given. Note that by normalization of the ninth parameter t_{33} an arbitrary 3D transformation can be interpreted as a projective transformation of homogeneous 2D coordinates. The 2D position can thus be written as shown in equation 6.

$$x^r = \frac{t_{11}x^l + t_{12}y^l + t_{13}}{t_{31}x^l + t_{32}y^l + 1} \qquad y^r = \frac{t_{21}x^l + t_{22}y^l + t_{23}}{t_{31}x^l + t_{32}y^l + 1} \qquad (6)$$

In general, a projective transformation specifies the relation between a plane in the three-dimensional real world scene and its perspective mapping on a two-dimensional image plane. Supposing there are n correspondences masked as valid within the window, the system to solve is directly derived from Eq. 6 and can be written in closed form as

$$\mathbf{A} \cdot \mathbf{t} = \mathbf{b} , \qquad (7)$$

where $\mathbf{t}$ denotes the vector of the elements of the transform $\mathbf{T}$. The system is highly overdetermined and is thus solved by a least squares approach, by computing the pseudo-inverse of $\mathbf{A}$, leading to $\mathbf{t}_{LS}$.

$$\mathbf{t}_{LS} = (\mathbf{A}^T\mathbf{A})^{-1}\mathbf{A}^T\mathbf{b} . \qquad (8)$$

Provided that at least 4 of the n available observations are linearly independent in 2D, the parameters of the transformation can be determined. The refined stereo correspondence (x_s, y_s) is now calculated by applying the transformation to the homogeneous coordinates of the position estimate from section 2.

5 Experimental Results

In order to verify the accuracy and robustness of the proposed algorithm, images with known ground truth were used. The images shown in Fig. 4 are images

a)   b)

Fig. 4. Images with known ground truth. The images are provided by the Calibrated Imaging Laboratory at Carnegie Mellon University.

provided by the Calibrated Imaging Laboratory at Carnegie Mellon University. Fig. 4a) shows the "castle" scene and Fig. 4b) the "plantex" scene. In Tab. 1 the accuracy analysis is summarized. In the case of the castle image the disparities are known with subpixel accuracy for a total of 28 points. The proposed algorithm classified three points as outliers and therefore a disparity was obtained for 25 points. Further the root-mean-square errors in the horizontal, vertical and

Table 1. Results on images with known ground truth.

image	npt	rms_x	rms_y	rms_e	max_e	rms_e^*	rms_{KO}
castle	25	0.14	0.06	0.15	3.21	0.08	0.16
plantex	9	0.08	0.07	0.03	0.34	-	0.32

Euclidean direction are shown. The max_e-values indicate the maximum error in the Euclidean direction. The value of 3.21 means that the algorithm for subpixel matching failed for this single point. If this single mismatch is not considered, a substantially improved rms_e^* of 0.08 is reached. For comparison, experiments with the adaptive window matching algorithm developed by Kanade and Okutomi [7] were performed. As this algorithm requires horizontal epipolar lines, only the root-mean-square error in the horizontal direction is reported as rms_{KO}. It has to be mentioned that the adaptive window matching algorithm detected 4 points as outliers what means that a total of 24 points would have contributed to the root-mean-square error.

In the case of the plantex image the disparities for a total of 9 points is known. No outliers were detected and no mismatches occurred, what lead to the low root-mean-square error of 0.03. The adaptive window matching algorithm lead to a root-mean-square error of 0.42.

Table 2. Results on images with 5 % salt & pepper noise.

image	npt	rms_x	rms_y	rms_e	max_e	rms_{KO}
castle	18	0.14	0.13	0.19	2.67	0.29
plantex	5	0.24	0.22	0.32	2.43	0.37

In order to evaluate the robustness of the proposed algorithm, experiments on the same images as shown in Fig. 4 with added salt-and-pepper noise and white Gaussian noise were performed. It has to be mentioned that the SSD matching for obtaining the initial pixel to pixel correspondence failed in the cases of present noise. As the focus is on subpixel refinement, the pixel accurate positions were provided manually. The results for salt-and-pepper noise are summarized in Tab. 2. Finally, the results with the images corrupted by 25% additive Gaussian noise of $\sigma = 10$ are shown in Tab. 3. The ground truth data provides an accuracy

Table 3. Results on images with 25 % additive Gaussian noise with $\sigma = 10$.

image	npt	rms_x	rms_y	rms_e	max_e	rms_{KO}
castle	24	0.11	0.07	0.13	1.59	0.27
plantex	9	0.21	0.07	0.19	1.48	0.30

in the order of a tenth of a pixel. The proposed algorithm shows a robust and accurate behavior within the scope of the available ground truth.

6 Conclusion and Further Work

In this paper we present an initial approach to subpixel stereo matching. The proposed algorithm uses a sparse disparity map, obtained using point of interest detection along with minimum SSD matching. For each of those initial point correspondences a local dense disparity map is estimated by calculating a feature vector for each pixel from the responses of a multi-channel Gabor filter. On this set of local disparities a new consistency constraint is applied. This delivers a masked window, in which regions of non-consistent disparities are deleted. Using the remaining consistent information, the parameters of a projective transformation are estimated, leading to subpixel accuracy. Experiments were performed on images with known ground truth. The proposed algorithm shows a robust and accurate behavior. In order to reduce the computational complexity, an accurate filter selection scheme is subject to further work. Additionally an estimation of the projective transformation by a least median of squares (LMedS) method may enhance its robustness. The initial satisfying results motivate for the proposed extensions and invites the application of the algorithm to occlusion detection.

References

1. N. Ayache. *Artificial Vision for Mobile Robots.* MIT Press, 1991.
2. Y. Boykov, O. Veksler, and R. Zabih. A Variable Window Approach to Early Vision. *IEEE Transactions on Pattern Analysis and Machine Intelligence,* 20(12):1283–1294, 1998.
3. W. Förstner. On The Geometric Precision Of Digital Correlation. *Int. Arch. of Photogrammetry and Remote Sensing,* 24(3):176–189, 1982.
4. A. Grün. High Accuracy Object Reconstruction With Least Squares Matching. In *Bildverarbeitung'95,* pages 277–295. Technische Akademie Esslingen, 1995.
5. C. Harris and M. Stephens. A Combined Corner And Edge Detector. Technical report, Plessey Research Roke Manor, United Kingdom, 1988.
6. A. K. Jain and F. Farrokhnia. Unsupervised Texture Segmentation Using Gabor Filters. *Pattern Recognition,* 24(12):1167–1186, 1991.
7. T. Kanade and M. Okutomi. A Stereo Matching Algorithm with an Adaptive Window: Theory and Experiment. *IEEE Transactions on Pattern Analysis and Machine Intelligence,* 16(9):920–932, 1994.
8. Z.-D. Lan and R. Mohr. Robust matching by partial correlation. Technical Report N RR-2643, INRIA, August 1995.
9. Z.-D. Lan and R. Mohr. Direct linear sub-pixel correlation by incorporation of neighbor pixels' information and robust estimation of window transformation. *Machine Vision and Applications,* 10(5,6):256–268, 1998.
10. T. Lindeberg. A Scale Selection Principle for Estimating Image Deformations. *Image and Vision Computing,* 16(14):961–977, December 1998.
11. J. Martin and J.L. Crowley. Comparison of Correlation Techniques. In *Proceedings Conference on Intelligent Autonomous Systems,* 1995.
12. S. Scherer, P. Werth, and A. Pinz. The Discriminatory Power of Ordinal Measures - Towards a New Coefficient. In *IEEE Conference on Computer Vision and Pattern Recognition,* 1999.
13. R. Trapp, S. Drüe, and G. Hartmann. Stereo Matching with Implicit Detection of Occlusions. In *Proc. European Conference on Computer Vision,* 1998.

Author Index

Lecture Notes in Computer Science

For information about Vols. 1–1594
please contact your bookseller or Springer-Verlag